Lecture Notes in Physics

Springer-Verlag Berlin Heidelberg GmbH

The Editorial Policy for Proceedings

The series Lecture Notes in Physics reports new developments in physical research and teaching – quickly, informally, and at a high level. The proceedings to be considered for publication in this series should be limited to only a few areas of research, and these should be closely related to each other. The contributions should be of a high standard and should avoid lengthy redraftings of papers already published or about to be published elsewhere. As a whole, the proceedings should aim for a balanced presentation of the theme of the conference including a description of the techniques used and enough motivation for a broad readership. It should not be assumed that the published proceedings must reflect the conference in its entirety. (A listing or abstracts of papers presented at the meeting but not included in the proceedings could be added as an appendix.)
When applying for publication in the series Lecture Notes in Physics the volume's editor(s) should submit sufficient material to enable the series editors and their referees to make a fairly accurate evaluation (e.g. a complete list of speakers and titles of papers to be presented and abstracts). If, based on this information, the proceedings are (tentatively) accepted, the volume's editor(s), whose name(s) will appear on the title pages, should select the papers suitable for publication and have them refereed (as for a journal) when appropriate. As a rule discussions will not be accepted. The series editors and Springer-Verlag will normally not interfere with the detailed editing except in fairly obvious cases or on technical matters.
Final acceptance is expressed by the series editor in charge, in consultation with Springer-Verlag only after receiving the complete manuscript. It might help to send a copy of the authors' manuscripts in advance to the editor in charge to discuss possible revisions with him. As a general rule, the series editor will confirm his tentative acceptance if the final manuscript corresponds to the original concept discussed, if the quality of the contribution meets the requirements of the series, and if the final size of the manuscript does not greatly exceed the number of pages originally agreed upon. The manuscript should be forwarded to Springer-Verlag shortly after the meeting. In cases of extreme delay (more than six months after the conference) the series editors will check once more the timeliness of the papers. Therefore, the volume's editor(s) should establish strict deadlines, or collect the articles during the conference and have them revised on the spot. If a delay is unavoidable, one should encourage the authors to update their contributions if appropriate. The editors of proceedings are strongly advised to inform contributors about these points at an early stage.
The final manuscript should contain a table of contents and an informative introduction accessible also to readers not particularly familiar with the topic of the conference. The contributions should be in English. The volume's editor(s) should check the contributions for the correct use of language. At Springer-Verlag only the prefaces will be checked by a copy-editor for language and style. Grave linguistic or technical shortcomings may lead to the rejection of contributions by the series editors. A conference report should not exceed a total of 500 pages. Keeping the size within this bound should be achieved by a stricter selection of articles and not by imposing an upper limit to the length of the individual papers. Editors receive jointly 30 complimentary copies of their book. They are entitled to purchase further copies of their book at a reduced rate. As a rule no reprints of individual contributions can be supplied. No royalty is paid on Lecture Notes in Physics volumes. Commitment to publish is made by letter of interest rather than by signing a formal contract. Springer-Verlag secures the copyright for each volume.

The Production Process

The books are hardbound, and the publisher will select quality paper appropriate to the needs of the author(s). Publication time is about ten weeks. More than twenty years of experience guarantee authors the best possible service. To reach the goal of rapid publication at a low price the technique of photographic reproduction from a camera-ready manuscript was chosen. This process shifts the main responsibility for the technical quality considerably from the publisher to the authors. We therefore urge all authors and editors of proceedings to observe very carefully the essentials for the preparation of camera-ready manuscripts, which we will supply on request. This applies especially to the quality of figures and halftones submitted for publication. In addition, it might be useful to look at some of the volumes already published. As a special service, we offer free of charge LaTeX and TeX macro packages to format the text according to Springer-Verlag's quality requirements. We strongly recommend that you make use of this offer, since the result will be a book of considerably improved technical quality. To avoid mistakes and time-consuming correspondence during the production period the conference editors should request special instructions from the publisher well before the beginning of the conference. Manuscripts not meeting the technical standard of the series will have to be returned for improvement.

For further information please contact Springer-Verlag, Physics Editorial Department II, Tiergartenstrasse 17, D-69121 Heidelberg, Germany

Sadruddin Benkadda   George M. Zaslavsky (Eds.)

Chaos, Kinetics and Nonlinear Dynamics in Fluids and Plasmas

Proceedings of a Workshop
Held in Carry-Le Rouet, France, 16–21 June 1997

 Springer

Editors

Sadruddin Benkadda
LPIIM, Equipe Turbulence Plasma
CNRS-Université de Provence
Centre Universitaire de Saint-Jérôme
F-13397 Marseille Cedex 20, France

George M. Zaslavsky
Courant Institute of Mathematical Sciences
New York University
251 Mercer Street, New York 10012, USA
and
Physics Department
New York University
4 Washington Place, New York 10003, USA

Cataloging-in-Publication Data applied for.

Die Deutsche Bibliothek - CIP-Einheitsaufnahme

Chaos, kinetics and nonlinear dynamics in fluids and plasmas :
proceedings of a workshop held in Carry-Le Rouet, France, 16 - 21
June 1997 / Sadruddin Benkadda ; George M. Zaslavsky (ed.).

(Lecture notes in physics ; Vol. 511)
  ISBN 978-3-662-14202-8   ISBN 978-3-540-69180-8 (eBook)
  DOI 10.1007/978-3-540-69180-8

ISSN 0075-8450
ISBN 978-3-662-14202-8

Typesetting: Camera-ready by the authors/editors
Cover design: *design & production* GmbH, Heidelberg
SPIN: 10644199   55/3144-543210 - Printed on acid-free paper

Preface

As our understanding of chaotic dynamics becomes deeper, and our encounters with phenomena of chaos become more extensive, we realize that numerous natural processes represent a mixture of regular and erratic parts of the dynamics. We deal with incomplete chaos where the presence of dynamical, or coherent, structures plays a crucial role. There are different indications of the partial coherency of real chaotic processes which are named in different ways in the literature as: coherent structures, intermittency, flights, trappings, ballistic modes, etc. From one point of view, this means a serious complication for routine investigation of chaotic dynamics and its applications. From another viewpoint, this means the loss of universality that is attributed to a "normal" or "pure" chaos. More accurately, we can say that there may be a number of different classes with a specific universality within each. In particular, this is how one arrives at Lévy-type processes or fractional or even multifractional kinetics rather than habitual Gaussian or Poissonian processes. New ideas and new tools are bound to widen our possibilities in the understanding and description of chaotic processes to help us gain new insights in the origin of turbulence.

Most of the material of the book is based on the invited talks at the workshop held in Carry-Le Rouet in the summer of 1997. Some of the articles are written especially for this edition. The book includes a number of related subjects overlapping via common ideas concerned with more specific understanding of the nonuniversality of the chaotic dynamics and utilization of this information in the kinetics of particles, fluids, and plasmas. The workshop was sponsored by the following institutions and organizations: Commissariat à l'Energie Atomique (CEA), CNRS, Comission of the European Union (fusion programme), Conseil Général du Département des Bouches-du-Rhône, Direction de la Recherche et des Etudes Techniques (DRET), Ministère de la Recherche et des Technologies, Université de Provence, and the US Department of Navy. Its Programme Committee included S. Benkadda, M. Shlesinger, and G.M. Zaslavsky.

We would like to express our deep gratitude to all contributors of the volume, who have worked hard to make the issue readable and (we hope) useful.

Marseille, France
New York, NY, USA
May 1998

S. Benkadda
G.M. Zaslavsky

Contents

IV. Kinetics and Statistics

Part 1:

Dynamics and Chaos

Dynamics in a Neigborhood of Separatrices of an Area-Preserving Map

Dmitry Treschev

Department of Mechanics and Mathematics, Moscow State University, Leninski Gori, Moscow 119899, Russia

Abstract. We discuss the global structure of the separatrix branches in a two-dimensional area-preserving map and present some formulas estimating the width of stochastic layers, provided the map is near-integrable. The concept of the separatrix map is also discussed.

1 Introduction

Let T be an area-preserving diffeomorphism of a two-dimensional manifold M and $\hat{z} \in M$ a hyperbolic fixed point of the map T. The point $\hat{z}$ generates four asymptotic curves (below they are called branches or separatrix branches). We denote the stable branches by $\Gamma^s_{1,2}$ and the unstable ones by $\Gamma^u_{1,2}$. We assume that the point $\hat{z}$ does not belong to $\Gamma^{s,u}_{1,2}$.

Poincaré noticed [17] that in general the separatrices intersecting, form a very complicated network. Dynamics in the vicinity of this network is highly unstable and irregular. Because of this it is accepted to call this vicinity the stochastic layer and to characterize dynamics in the stochastic layer by the word "chaos".

The structure of such a chaos is weakly understood. It is known that in the stochastic layer there exists an invariant hyperbolic set on which T is isomorphic to the Smale horseshoe. However, the measure of this set vanishes and the question what behavior is typical for trajectories in the stochastic layer remains open.

In this paper we discuss global structure of the separatrix branches and present some formulas estimating the width of stochastic layers in near-integrable maps.

2 Closure of asymptotic curves

We mention here two questions well-known in mathematical folklore and related to the problems we deal with in this paper. Suppose that stable and unstable separatrices do not coincide: $\Gamma_1^s \cup \Gamma_2^s \neq \Gamma_1^u \cup \Gamma_2^u$. and belong to a compact invariant set.

• Does the closure of the separatrices $\overline{\Gamma}_1^s \cup \overline{\Gamma}_1^u \cup \overline{\Gamma}_2^s \cup \overline{\Gamma}_2^u$ have positive measure?

• If the answer to the first question is positive, is the map T restricted to this set ergodic?

Positive answers to this questions would give us a set of positive measure which carries a chaotic dynamics. Apparently, both questions are very difficult to answer. We present here a simpler result concerning closures of separatrix branches.

Theorem 1 *Suppose that T is C^1-smooth and the following conditions hold.*
(1) The set $\Gamma_1^s \cap \Gamma_1^u$ is not empty.
(2) The curves Γ_1^s and Γ_1^u lie in an invariant domain $D \subset M$. The closure $\overline{D}$ is compact.
Then the closure of the unstable branch $\overline{\Gamma}_1^u$ contains the stable one Γ_1^s.

Corollary 2.1 *Since conditions of the theorem are symmetric with respect to Γ_1^s and Γ_1^u, the inclusion $\Gamma_1^u \subset \overline{\Gamma}_1^s$ holds. Hence, the sets $\overline{\Gamma}_1^s$ and $\overline{\Gamma}_1^u$ coincide.*

Let p and q be hyperbolic periodic points of T and let i and j respectively be their periods. Since p and q are hyperbolic fixed points for the maps T^i and T^j, we can define the branches $\Gamma_{1,2}^{s,u}(p)$ и $\Gamma_{1,2}^{s,u}(q)$. We put

$$W_{1,2}^{s,u}(p) = \cup_{k=0}^{2i-1} T^k(\Gamma_{1,2}^{s,u}(p)), \quad W_{1,2}^{s,u}(q) = \cup_{k=0}^{2j-1} T^k(\Gamma_{1,2}^{s,u}(q)).$$

Theorem 2 *Suppose that the following conditions hold.*
(1′) The set $W_1^s(p) \cap W_1^u(q)$ is not empty.
(2′) The set $W_1^s(q) \cap W_1^u(p)$ is not empty.
(3′) The asymptotic manifolds $W_1^s(p)$, $W_1^u(p)$, $W_1^s(q)$, and $W_1^u(q)$ belong to an invariant domain $D \subset M$. The closure $\overline{D}$ is compact.
Then $\overline{W}_1^u(p) \cup \overline{W}_1^s(p) = \overline{W}_1^u(q) \cup \overline{W}_1^s(q)$.
Furthermore, in the case $p = q$ the equality $\overline{W}_1^s(p) = \overline{W}_1^u(p)$ holds.

Note that the manifold M in Theorems 1–2 can be not compact, not orientable. It can even have a boundary.

Remark 2.1 *In the case $p = q$ conditions (1′) and (2′) coincide.*

Recall that points of the sets

$$\left(W_1^s(p) \cup W_2^s(p)\right) \cap \left(W_1^u(q) \cup W_2^u(q)\right), \quad \left(W_1^s(q) \cup W_2^s(q)\right) \cap \left(W_1^u(p) \cup W_2^u(p)\right)$$

are called homoclinic if p and q lie on the same periodic trajectory, and heteroclinic otherwise.

Takens proved [18] that if M is compact, for any hyperbolic periodic point p of C^1-generic area-preserving self-map T of M the set of homoclinic points is dense on $W_{1,2}^{s,u}(p)$. The word "generic" is understood in the sense of Baire category. Note that methods of [18] are essentially restricted to the C^1-topology.

The following conjecture was formulated (in a weaker form) by Poincaré.

Conjecture 1 *If $p = q$ and conditions of Theorem 2 hold, the set of homoclinic points is dense on $W_1^s(p)$ and on $W_1^u(p)$.*

Mather [13] proved that if M is compact, for a C^r-generic ($r \geq 4$) area-preserving map any two branches of a hyperbolic periodic point have the same closure.

Oliveira [15] obtained the following results related to the problems in question. Let T be a C^1 area-preserving diffeomorphism of a compact orientable surface. Assume L and K are branches of a hyperbolic fixed point with $L = K$ or $L \cap K = \emptyset$. If $K \cap \omega(L) \neq \emptyset$ then $K \subset \omega(L)$. (Here as ususal, $\omega(L)$ is the ω-limit set of L.)

This result implies, [15] the following assertion. Let M be a compact orientable surface and $1 \leq r \leq \infty$. Then $L \subset \omega(L)$ for any branch L of a C^r-generic area-preserving map.

3   Width of a stochastic layer

Now assume that T is close to an integrable map. We assume also that separatrix branches of the hyperbolic fixed point $\hat{z}$ look as shown on Fig. 1. Three invariant curves $\gamma_\pm$ and γ_0 closest to the separatrices, form the boundary of the stochastic layer. The width w of the stochastic layer is one of important quantities characterizing chaotic properties of T in the vicinity of the separatrices.

It turns out that under some natural assumptions the following relation holds:

$$w/d \sim 1/\lambda. \tag{3.1}$$

Here d is the width of a lobe domain D bounded by segments of the separatrices and $\lambda > 0$ is logarithm of the larger multiplier at the hyperbolic fixed point $\hat{z}$. The symbol $\sim$ means that if we have a smooth family T_ε, of analytic symplectic maps, where T_0 is integrable then for sufficiently small ε

$$C_1/\lambda(\varepsilon) < w(\varepsilon)/d(\varepsilon) < C_2/\lambda(\varepsilon),$$

where C_1 and C_2 are positive constants. Here the family T_ε must satisfy certain regularity conditions [21]. We believe that these conditions are generic. In some cases the genericity can be proved.

6

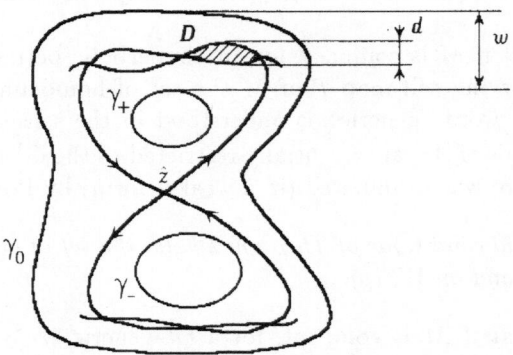

Figure 1: Separatrices of the hyperbolic fixed point $\hat{z}$ and the stochastic layer.

If $\lambda \sim 1$, we see that the quantities w and d are of the same order. However, if the stochastic layer appears when a resonant invariant curve of an integrable map disintegrates λ is close to zero. Hence, in this case w is much greater than d. This situation is typical for exponentially small separatrix splitting. If $\lim_{\varepsilon \to 0} \lambda(\varepsilon) = 0$ and T satisfies a certain symmetry condition, we expect that the estimate

$$\lim_{\varepsilon \to 0} \frac{w(\varepsilon)\lambda(\varepsilon)}{d(\varepsilon)} = \frac{4\pi}{k_0}, \tag{3.2}$$

holds, where

$$k_0 = \inf\{k' : \text{ for all } k > k' \text{ the standard map}$$
$$\begin{pmatrix} I \\ \varphi \end{pmatrix} \mapsto \begin{pmatrix} J \\ \psi \end{pmatrix} = \begin{pmatrix} I + \frac{k}{2\pi}\sin(2\pi\varphi) \\ \varphi + J \end{pmatrix}$$
$$\text{has no invariant curve homotopic to the circle } I = 0\}.$$

More precisely, let us put

$$k_1 = \inf\{k' : \text{ there exists } k < k' \text{ such that the standard}$$
$$\text{map has no invariant curve homotopic to the circle } I = 0\}.$$

A well-known conjecture states that $k_0 = k_1$. The formula (3.2) holds provided this conjecture is valid. The constant $k_0 = 0.971635\ldots$ was evaluated numerically in [8, 12, 16].

Note that to have a clear meaning for the relation (3.2) one should define precisely the quantities w and d. This can be done, for example, in terms of normal coordinates, see [21] for details.

It is possible to present an invariant version of the relation (3.1). Let $\mathcal{A}$ be the symplectic area of the stochastic layer and $\mathcal{A}_D$ the area of a lobe. Then we have:

$$\mathcal{A} \sim \frac{\mathcal{A}_D \log \mathcal{A}_D^{-1}}{\lambda^2}. \tag{3.3}$$

The corresponding analog of the formula (3.2) (under the same assumptions) is

$$\lim_{\varepsilon \to 0} \frac{\mathcal{A}\lambda^2}{\mathcal{A}_D \log \mathcal{A}_D^{-1}} = \frac{8\pi^2}{k_0}. \tag{3.4}$$

In the general (non-symmetric) case the fractions

$$\frac{w(\varepsilon)\lambda(\varepsilon)}{d(\varepsilon)}, \qquad \frac{\mathcal{A}\lambda^2}{\mathcal{A}_D \log \mathcal{A}_D^{-1}}$$

do not have limits as $\varepsilon \to 0$ but oscillate between two positive constants.

Formulas (3.1)–(3.4) can be regarded as relations between w, $\mathcal{A}$ and the quantities $\lambda, d, \mathcal{A}_D$. In the perturbative situation the last ones are standard to compute: to obtain the functions $d(\varepsilon)$ and $\mathcal{A}_D(\varepsilon)$ one can use the Poincaré-Melnikov theory or its generalizations to the case of exponentially small splitting [9, 10, 3, 7, 19, 20]; $\lambda(\varepsilon)$ in the main approximation is usually evaluated easily.

Particular cases of the relation (3.1) were discovered in [2, 6, 23], but no rigorous proofs were presented there. In these papers such useful object as the separatrix map was introduced. Analyzing this map, Dovbysh [4] proved the estimate $w/d \leq$ const provided $\lambda \sim 1$. In paper [1] the estimate $w/d \geq$ const is established in the same case. Lazutkin [11] has obtained the estimate (3.1) for separatrices of the standard map.

Consider as an example a pendulum with vertically periodically oscillating suspension point. Hamiltonian of the system is as follows:

$$H(\widehat{q}, \widehat{p}, t, \varepsilon) = \widehat{p}^2/2 + \Omega^2 \cos \widehat{q} + \varepsilon\theta(\omega t) \cos \widehat{q}. \tag{3.5}$$

Here $\widehat{q} = \widehat{q} \bmod 2\pi$ is the angle between the pendulum and the vertical, $\widehat{p}$ is the corresponding momentum, $\Omega > 0$ is the "interior frequency" of the system (Ω^2 equals gravity acceleration divided by the length of the pendulum), ω is the frequency of the suspension point oscillations and the parameter ε is proportional to the amplitude of the oscillations multiplied by ω^2. The form of the oscillations is determined by the 2π-periodic function θ. First, let us assume that ε is small and the other parameters in the system are of order 1.

The Poincaré map in this system has the hyperbolic fixed point $\widehat{q} = \widehat{p} = 0$. It is easy to calculate:

$$\lambda(\varepsilon) = 2\pi\Omega/\omega + O(\varepsilon).$$

Width of the stochastic layer around the "eight-like" separatrix structure of the point $\widehat{q} = \widehat{p} = 0$ is of order ε provided $\theta, \Omega, \omega \sim 1$ and θ is not constant. Under the same assumptions the area $\mathcal{A}$ of the stochastic layer is of order $\varepsilon \log \varepsilon^{-1}$.

Now consider the system with Hamiltonian (3.5) under the following assumptions:

$$\Omega = 1, \quad \omega = 1/\varepsilon, \quad \theta(s) = 2\varepsilon^{-1}B\cos s.$$

This means that the frequency of the suspension point oscillation is large ($\sim 1/\varepsilon$) and the amplitude is of order ε^2.

Area of the corresponding stochastic layer layer can be estimated as follows:

$$\frac{4\pi^2}{\varepsilon^4 k_0}e^{-\pi\varepsilon^{-1}/2}Bf(B^2)(1+o(1)) \le \mathcal{A} \le \frac{4\pi^2}{\varepsilon^4 k_1}e^{-\pi\varepsilon^{-1}/2}Bf(B^2)(1+o(1)),$$

where $f(z) = \sum_0^\infty f_n z^n$ is an entire real-analytic function [20, 19]. Here are the values of several coefficients f_n: $f_0 = 2$,

$$\begin{array}{llll}
f_1 &=& 0.65856738\ldots, & f_2 &=& 6.651741\ldots\cdot 10^{-2}, \\
f_3 &=& 3.21010\ldots\cdot 10^{-3}, & f_4 &=& 9.03367\ldots\cdot 10^{-5}, \\
f_5 &=& 1.6620\ldots\cdot 10^{-6}, & f_6 &=& 2.1534\ldots\cdot 10^{-8}, \\
f_7 &=& 2.070\ldots\cdot 10^{-10}, & f_8 &=& 1.53\ldots\cdot 10^{-12}.
\end{array}$$

Recall that apparently, $k_0 = k_1$.

4 Separatrices and the Poincaré recurrence theorem

In this section we prove Theorem 1 in the topologically simple situation: when M is a plane, or a cylinder, or a sphere. The case of other surfaces and Theorem 2 are analyzed in [22].

Below we can assume that the map T preserves each branch $\Gamma_{1,2}^{s,u}$. Indeed, if T does not satisfy this condition, we just change T by T^2.

Let $U \subset D$ be any open set such that $U \cap \Gamma_1^s \neq \emptyset$. Suppose that

$$U \cap \Gamma_1^u = \emptyset. \tag{4.1}$$

Then Theorem 1 is proved as soon as we obtain a contradiction with its conditions. Considering if necessary instead of U a smaller domain (which will be denoted also by U for brevity), we can assume that $U = U^+ \cup U^0 \cup U^-$, where $U^\pm$ are open, connected, and $U^0 \subset \Gamma_1^s$ is a connected interval (see Fig. 2). Let $\hat{I}$ be the minimal connected piece of Γ_1^s such that $\hat{z}$ is its endpoint and $U^0 \subset \hat{I}$. We can assume that

$$U \cap T(\hat{I}) = \emptyset, \tag{4.2}$$
$$U \cap T^{-1}(\hat{I}) = U^0. \tag{4.3}$$

These equalities mean that U^0 and $U^\pm$ are sufficiently small.

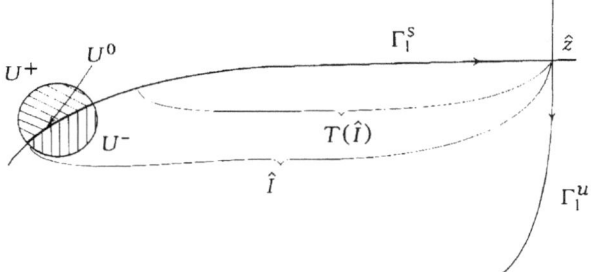

Figure 2 : Scheme of the minimal connected piece of the stable branch.

Lemma 4.1 *There exists a natural $l > 1$ such that*

$$U^+ \cap T^l(U^+) \neq \emptyset, \quad U^- \cap T^l(U^-) \neq \emptyset.$$

Proof of Lemma 4.1. Consider the map

$$T \times T : M \times M \to M \times M, \quad T \times T(z_1, z_2) = (T(z_1), T(z_2)).$$

This map preserves the measure $\sigma \times \sigma$, where σ is the area on M. The set $U^+ \times U^-$ lies inside the compact invariant set $\overline{D} \times \overline{D}$. Hence, according to the Poincaré Recurrence Theorem, for infinitely many naturals l

$$(T \times T)^l(U^+ \times U^-) \cap (U^+ \times U^-) \neq \emptyset.$$

The lemma is proved.

There exists a smooth closed curve γ satisfying the following properties.
(a) $\gamma \subset U'$, $\qquad U' = U \cup T^l(U)$.
(b) The set $\gamma \cap T^l(U^0)$ consists of a single point z' and the curves γ and $T^l(U^0)$ intersect at z' transversely.

The curve γ goes along the set $T^l(U^+)$ from the point z' to the set U^+. Then γ passes through the interval U^0 to the set U^-, goes to the set $T^l(U^-)$ and returns to the point z' (see Fig. 3).

According to property (a) and equality (4.1) we have:

$$\gamma \cap \Gamma_1^u = \emptyset. \tag{4.4}$$

Lemma 4.2 *There exists an interval $I \subset \Gamma_1^s$ satisfying the following two properties.*
(A) *Endpoints of I are $\hat{z}$ and z_0, where z_0 is homoclinic.*
(B) $I \cap U' = T^l(U^0)$.

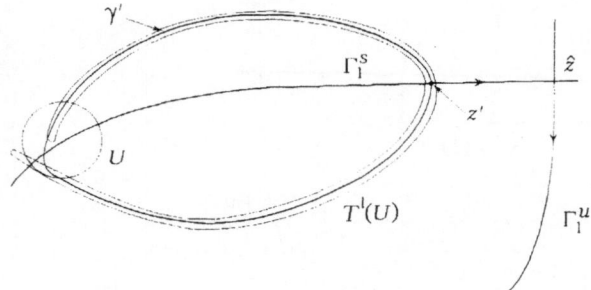

Figure 3 : Scheme of the γ curve as defined in lemme 4.1.

Proof of Lemma 4.2. The interval $I' = T^{l-1}(\widehat{I}) \setminus T^l(\widehat{I})$ contains a homoclinic point. Indeed, otherwise we have a contradiction with assumption (1) because of the relation $\Gamma_1^s = \cup_{k \in \mathbf{Z}} T^k(I')$. We denote this homoclinic point by z_0. Hence, I is defined by (A).

Now Lemma 4.2 follows from two inclusions:

$$T^l(U^0) \subset (I \cap U') \subset T^l(U^0).$$

The first one is obvious. Let us check the second one. We have: $I \subset T^{l-1}(\widehat{I})$. This relation together with the one $l > 1$ imply

$$I \cap U \subset T^{l-1}(\widehat{I}) \cap U \subset T(\widehat{I}) \cap U = \emptyset$$

due to (4.2). Analogously,

$$I \cap T^l(U) \subset T^{l-1}(\widehat{I}) \cap T^l(U) \subset T^l(T^{-1}(\widehat{I}) \cap U) = T^l(U^0).$$

Here we have used assumption (4.3). Lemma 4.2 is proved.

According to property (A) the points $\hat{z}$ and z_0 can be connected by an interval I^u of the unstable separatrix branch Γ_1^u. The curve $\sigma = I \cup I^u$ is closed. According to the definition of γ and property (B) the curves I and γ have exactly one common point (the point z') and intersect at z' transversely. The curves I^u and γ do not intersect because of (4.4). Hence, the curves σ and γ are transversal to each other and have exactly one common point. Here we arrive at a contradiction because in the topologically simple case (i.e., on a plane, on a cylinder, or on a sphere) any two curves transversal to each other have an even number of intersection points.

5   The separatrix map

Estimates (3.1)–(3.4) are proved in [21]. The main tool used in the proofs is the separatrix map. In this section we describe our construction of the separatrix map. This construction differs from the original one [23, 6], Seems, our methods are more convenient when one needs estimates of errors of the main approximation.

Below we always assume that the map T is real-analytic in z.

In the vicinity of the point $\hat{z}$ the map T is a hyperbolic rotation. More precisely, the following lemma holds.

Lemma 5.1 *It is possible to choose in a neighborhood of the point $\hat{z}$ on M real-analytic symplectic coordinates (x, y) such that the following assertions hold.*

(1) Coordinates of $\hat{z}$ vanish.

(2) The map T has the form

$$(x, y) \to L(x, y) = (x\mathcal{M}, y/\mathcal{M}), \qquad \mathcal{M} = \mathcal{M}(xy), \tag{5.1}$$

where the function $\mathcal{M}$ is real-analytic and $\mathcal{M}(0) = \mu > 1$.

On a formal level existence of such coordinates was established by Birkhoff. They are called normal. Convergence of the procedure introducing the normal coordinates was established in [14]. The map (5.1) has the first integral xy.

Remark 5.1 *The normal coordinates (x, y) are not defined uniquely. For any real-analytic at zero function $r(z)$ $(r(0) \neq 0)$ the coordinates $x' = x/r(xy)$, $y' = y\, r(xy)$ are also normal.*

Remark 5.2 *If the map T depends smoothly on a parameter (say, ε), the normal coordinates also can be chosen depending smoothly on ε while the point $\hat{z} = \hat{z}(\varepsilon)$ remains hyperbolic.*

Below we assume that inside the higher loop of the "eight" (see Fig. 1) we have $x > 0$, $y > 0$.

The normal coordinates (x, y) can be continued to a neighborhood of the separatrices according to the following inductive procedure. Suppose that the point $z \in N$ has the coordinates (x, y). Then we define coordinates of the points $T(z)$ and $T^{-1}(z)$ by $L(x, y)$ and $L^{-1}(x, y)$ respectively. Far from the point $\hat{z}$ we have at least two different continuations of the coordinates (by T^k and by T^{-k}, see Fig. 4). Hence, we have two gluing transformations:

$$U^+ : \{(x, y) : y \text{ is small}, x > 0, x \sim 1\} \to \{(x, y) : y > 0, y \sim 1, x \text{ is small}\},$$
$$U^- : \{(x, y) : y \text{ is small}, x < 0, x \sim 1\} \to \{(x, y) : y < 0, y \sim 1, x \text{ is small}\}. \tag{5.2}$$

The map U^+ corresponds to the higher loop of the "eight" and U^- to the lower one. The maps L and $U^\pm$ obviously commute:

$$U^\pm \circ L = L \circ U^\pm. \tag{5.3}$$

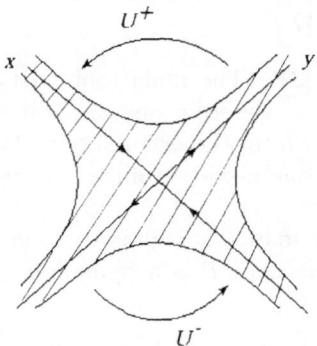

Figure 4: Continued normal coordinates and the gluing maps

Proposition 5.1 *Suppose that the map T has an analytic integral $F : N \to N$ (i.e. $F = F \circ T$). Then in normal coordinates $F = F(xy)$. Moreover, if the critical point $x = y = 0$ of the function F is non-degenerate, the gluing maps preserve the product xy and*

$$U^{\pm}(x, y) = \left(xy(\alpha_{\pm}^2/x + y\kappa_{\pm}(xy))^{-1}, \alpha_{\pm}^2/x + y\kappa_{\pm}(xy)\right)$$

Here $\alpha_{\pm}$ are positive constants and $\kappa_{\pm}(r)$ are functions analytic at the point $r = 0$.

Let us assume that the T_{ε} is a smooth family of analytic maps such that the fixed point $\hat{z}(0)$ of the map T_0 is hyperbolic, T_0 is integrable and separatrices of T_0 form a figure eight curve. The gluing maps have the form

$$U_{\varepsilon}^{\pm} = \left(xy(\alpha_{\pm}^2/x + y\kappa_{\pm}(xy))^{-1}, \alpha_{\pm}^2/x + y\kappa_{\pm}(xy)\right) + \left(\hat{f}(x, y, \varepsilon), \hat{g}(x, y, \varepsilon)\right). \quad (5.4)$$

According to the relation (5.3),

$$\hat{f}_{\pm}(x\mathcal{M}, y/\mathcal{M}, \varepsilon) = \mathcal{M}\hat{f}(x, y, \varepsilon), \quad \hat{g}_{\pm}(x\mathcal{M}, y/\mathcal{M}, \varepsilon) = \mathcal{M}^{-1}\hat{g}(x, y, \varepsilon). \quad (5.5)$$

Proposition 5.2 *If the maps $U^{\pm}$ are analytic in x and y in the domains*

$$\{x_0 < \pm x < X_0, \quad |y| < Y_0\}, \quad (5.6)$$

for some positive x_0, X_0 and Y_0, the following relations hold:

$$\hat{f}_{\pm}(x, y, \varepsilon) = x f_{\pm}(x, y, \varepsilon), \quad \hat{g}_{\pm}(x, y, \varepsilon) = y g_{\pm}(x, y, \varepsilon).$$

The functions $f_{\pm}, g_{\pm}$ are analytic in the domains (5.6) and

$$f_{\pm}(x\mathcal{M}, y/\mathcal{M}, \varepsilon) = f_{\pm}(x, y, \varepsilon), \quad g_{\pm}(x\mathcal{M}, y/\mathcal{M}, \varepsilon) = g_{\pm}(x, y, \varepsilon). \quad (5.7)$$

We put

$$f_{\pm}(x, y, \varepsilon) = \nu_p m(\log |x| / \log \mathcal{M}(0,0)) + O(y) + O(\varepsilon). \tag{5.8}$$

The functions $\nu_p m(\varphi)$ are obviously 1-periodic in φ.

In the vicinity of the separatrices the maps T_ε generate the so-called separatrix maps S_ε. [1]

To define the maps S_ε, we identify any two points z_1 and z_2 of N whose normal coordinates (x_1, y_1) and (x_2, y_2) satisfy the relation

$$x_2 = \mathcal{M}^n(x_1 y_1, \varepsilon) x_1, \quad y_2 = \mathcal{M}^{-n}(x_1 y_1, \varepsilon) y_1, \quad n \in \mathbf{Z}. \tag{5.9}$$

(The parameter ε is assumed fixed.) After this identification the maps S_ε are determined by the gluing maps. Below we assume that the argument y in the gluing maps is small and x is of order one. By using (5.4)–(5.8), we can write:

$$\begin{aligned}
U_\varepsilon^{\pm}(x, y) &= \left(\alpha_{\pm}^{-2} x (xy + \varepsilon \nu_{\pm}(\log |x| / \log \mathcal{M}(0,0))), \alpha_{\pm}^2 / x \right) \\
&\quad + (O(|y| + \varepsilon)^2, O(|y| + \varepsilon)).
\end{aligned} \tag{5.10}$$

Let us introduce the action-angle variables for the corresponding separatrix map. We put

$$I = \mathcal{M} \varepsilon^{-1} xy, \quad \varphi = \mathcal{M}^{-1} \log |x|. \tag{5.11}$$

It is easy to check that $dI \wedge d\varphi = \varepsilon^{-1}(1 + O(xy)) dy \wedge dx$. In the variables (I, φ) the identification (6.1) takes the form $\varphi = \varphi \bmod 1$.

We restore the information about sign of the variable x (lost in (5.11)) by adding the sign $+$ or $-$ to the coordinate system (I, φ) according to the value of $\text{sign}(x)$. In the variables (I, φ, σ), $\sigma \in \{+, -\}$ the maps S_ε have the form

$$\begin{aligned}
S_\varepsilon(I, \varphi, \sigma) &= (J, \psi, \rho), \\
J &= I + \lambda \cdot \nu_\sigma(\varphi) + \varepsilon \lambda \cdot O(1 + |I|/\lambda)^2, \\
\psi &= \varphi + \frac{1}{\lambda} \left[\log \frac{\varepsilon}{\alpha_\sigma^2 \lambda} + \log |J| + \varepsilon \cdot O(1 + |I|/\lambda)^2 \right], \\
\rho &= \sigma \, \text{sign}(J).
\end{aligned} \tag{5.12}$$

Here we have used the equalities (5.10)–(5.11) and the estimate $O(y) = O(\varepsilon I/\lambda)$. Note that similar formulas are contained in [5].

In the main approximation the dependence of the separatrix map on $\log \varepsilon$ is periodic. Indeed, for any integer n the change $\varepsilon \to e^{n\lambda} \varepsilon$ preserves the map

$$\begin{aligned}
J &= I + \lambda \cdot \nu_\sigma(\varphi), \\
\psi &= \varphi + \frac{1}{\lambda} \left[\log \frac{\varepsilon}{\alpha_\sigma^2 \lambda} + \log |J| \right], \\
\rho &= \sigma \, \text{sign}(J).
\end{aligned}$$

---

[1] Note that the separatrix map can be defined also in the case when $\hat{z}(\varepsilon)$ is hyperbolic only for $\varepsilon > 0$. This case is typical for exponentially small separatrix splitting. See [21] for the details.

Earlier this property was pointed out in [24].

Note that for large values of the action variable the separatrix map is close to integrable. Indeed, let us put $I = I_0(1 + \lambda u)$ and $J = I_0(1 + \lambda v)$, where $|I_0|$ is large. The separatrix map becomes close to the following one:

$$v = u, \quad \psi = \varphi + \frac{1}{\lambda}\left[\log\left|\frac{I_0\varepsilon}{\alpha_\sigma^2\lambda}\right| + \log(1 + \lambda v)\right], \quad \rho = \sigma\,\mathrm{sign}(I_0).$$

Due to this it is possible to use perturbation arguments when studying the case of large I.

The work was partially supported by Russian Foundation of Basic Research (Grant 96-01-00747) and by INTAS 93-339-ext.

References

[1] Ahn T., Kim G. and Kim S. (1996): Analysis of the separatrix map in Hamiltonian systems, Physica **89D**, p. 315–328

[2] Chirikov B. V. (1979): A universal instability of many-dimensional oscillation systems, Phys. Rep., v. 52, no. 5, p. 263–379

[3] Delshams A. and Seara T. M. (1992): An asymptotic Expressions for the Splitting of Separatrices of the Rapidly Forced Pendulum, Comm. Math. Phys., v. 150, p. 433–463

[4] Dovbysh S. A. (1989): Structure of the Kolmogorov set near separatrices of a plane map, Math. Zametki, v. 46, no. 4, p. 112–114 (in Russian)

[5] Dovbysh S. A. (1987): Intersection of asymptotic surfaces in the perturbed Euler-Poinsot problem, Prikl. Mat. Mekh., v. 51, no. 3, p. 363–370 (in Russian)

[6] Filonenko N. N., Sagdeev R. Z. and Zaslavsky G. M. (1967): Destruction of magnetic surfaces by magnetic field irregularities, Nuclear Fusion, v. 7, p. 253–266

[7] Gelfreich V. G. (1997): Reference systems for splitting of separatrices, Nonlinearity, V. 10, P. 175–193.

[8] Greene J. M. (1979): A method for determining a stochastic transition, J. Math. Phys., v. 20, p. 1183–1201

[9] Lazutkin V. F. (1984): Splitting of separatrices for standard Chirikov's mapping, VINITI no. 6372-84, 24 Sept. (in Russian)

[10] Lazutkin V. F., Schachmannski I. G. and Tabanov M. B. (1989): Splitting of separatrices for standard and semistandard mappings, Physica D, v. 40, p. 235–248

[11] V. F.Lazutkin V. F. (1990): On the width of a stability zone near separatrices of the standard map, Dokl. Akad. Nauk SSSR, v. 313, no. 2, p. 268–272 (in Russian)

[12] MacKay R. S. (1983): A renormalization approach to invariant circles in area-preserving maps Physica **7D**, p. 283–300

[13] Mather J. N., Invariant subsets for area-preserving homeomorphisms of surfaces, Mathematical Analysis and Applications, Part B. Advances in Mathematics. Supplementary Studies, Vol. 7B, Leopoldo Nachbin, ed.

[14] Moser J. (1956): The analytic invariants of an area preserving mapping near hyperbolic fixed point, Commun. Pure Appl. Math., v. 9, p. 673–692

[15] Oliveira F. (1987): On the generic existence of homoclinic points, Ergodic Th. Dynam. Sys., V. 7, P. 567–595.

[16] Olvera A. and Simó C. (1987): An obstruction method for the destruction of invariant curves Physica **26D**, p. 181–192

[17] Poincaré H., Les méthodes nouvelles de la mécanique celeste, V. 1–3: Gauthier-Villars, Paris, 1892, 1893, 1899.

[18] Takens F. (1972): Homoclinic points of conservative systems, Inv. Math. V. 18, P. 267–292.

[19] Treschev D. (1997): Separatrix Splitting for a Pendulum with Rapidly Oscillating Suspension Point, Russian J. of Math. Phys., V. 5, No. 1, P. 63–98.

[20] Treschev D. (1996): An averaging method for Hamiltonian systems, exponentially close to integrable ones, Chaos v. 6, no. 1, p. 6–14

[21] Treschev D. (1998) Width of stochastic layers in near-integrable two-dimensional symplectic maps. to appear in Phys. D.

[22] Treschev D., Closures of asymptotic curves in two-dimensional symplectic maps. Preprint.

[23] Zaslavsky G. M. and Filonenko N. N. (1968): Stochastic instability of trapped particles and the conditions of applicability of the quasilinear approximation, Zh. Eksp. Teor. Fiz., v. 54, p. 1590–1602, (in Russian)

[24] Zaslavsky G. M. and Abdullaev S. S. (1995): Scaling properties and anomalous transport of particles inside the stochastic layer. Phys. Review E, V. 51, No. 5, P. 3901–3910.

On Smooth Hamiltonian Flows
Limited to Ergodic Billiards

Dmitry Turaev[1,2] and Vered Rom-Kedar[1]

[1] The Department of Applied Mathematics and Computer Science, The Weizmann
Institute of Science, P.O.B. 26, Rehovot 76100, Israel. vered@wisdom.weizmann.ac.il
[2] Current address: Weierstrass Institute for applied analysis and stochastics, Mohren-
str 39, 10117 Berlin, Germany. turaev@wias-berlin.de

AMS No.: 58F15, 82C05, 34C37, 58F05, 58F13, 58F14.

Abstract. Sufficient conditions are found so that a family of smooth Hamiltonian
flows limits to a billiard flow as a parameter $\epsilon \to 0$. This limit is proved to be
C^1 near non-singular orbits and C^0 near orbits tangent to the billiard boundary.
These results are used to prove that scattering (thus ergodic) billiards with tangent
periodic orbits or tangent homoclinic orbits produce nearby Hamiltonian flows with
elliptic islands. This implies that ergodicity may be lost for smooth potentials which
are arbitrarily close to ergodic billiards. Thus, in some cases, anomoulous transport
associated with stickiness to stability islands is expected

1 Introduction

The billiard model is concerned with the motion of a point particle traveling
with a constant speed in a region and undergoing elastic collisions at the
region's boundary. This motion is very much like in that of a real billiard
table - the main difference is that there is no friction in the model (so the
ball never stops nor rolls). In the two-dimensional setting of our model, the
ball is actually a small disk (a two-dimensional ball). Different shapes of the
billiard table, and the number of balls that one considers influence the type
of motion a ball may execute. Ergodic billiards are billiard tables in which
the balls execute a uniformly disordered motion: all possible positions and
velocities are realized by the traveling billiard balls (for almost all initial
positions).

The billiard problem has been extensively studied both in its classical and
quantized formulation. Numerous applications lead to study such a model
problem; First, there exist direct mechanical realizations of this model (e.g.
the motion of N rigid d-dimensional spheres in a d-dimensional box may
be reduced to a billiard problem, possibly in higher dimensions [21, 22, 7].
See also [6] for the inelastic case.). Second, it serves as an idealized model
for the motion of charged particles in a potential, a model which enables
the examination of the relation between classical and quantized systems,

see [14] and references therein. Finally, and most important, this model has
been suggested [21] as a first step for substantiating the basic assumption of
statistical mechanics - the ergodic hypothesis of Boltzmann (see especially
the discussion and references in [22, 24]).

In all the applications of this model, of special interest are so-called *scattering* billiards, i.e., billiards in a complement to the union of a finite number
of closed convex regions. For example - the two-dimensional idealization of a
gas in the form of a lattice of rigid disks produces a scattering billiard ("the
Sinai billiard"). The motion in a scattering billiard is highly unstable thus
produces strong mixing in the phase space. More precisely, it has been shown
[21, 11, 1] that the corresponding dynamical system is (non-uniformly) hyperbolic, it is ergodic with respect to the natural invariant measure and it
possesses K-property. Based on this theory, statistical properties of various
scattering systems have been analyzed (see [5, 4]).

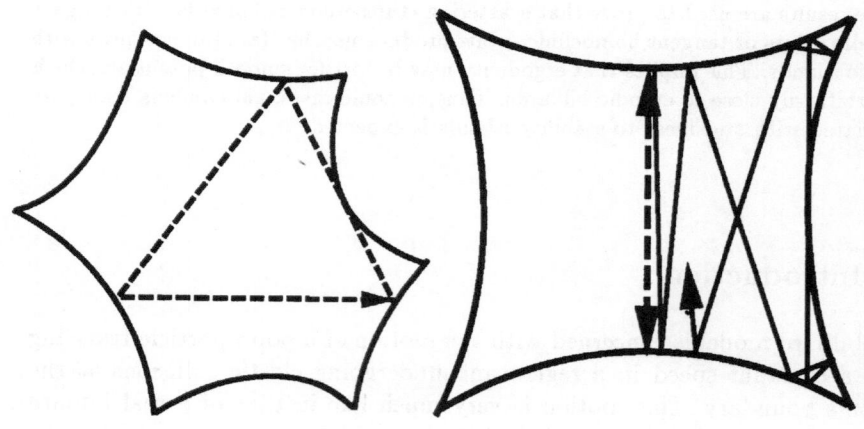

Fig. 1.1. Tangent trajectories

a) Singular (tangent) periodic trajectory
b) – – – – –non-singular periodic trajectory,
————Tangent homoclinic trajectory to the periodic orbit.

Do small perturbations ruin the ergodicity property of a scattering billiard? In this paper we consider the perturbation caused by the "natural"
smoothening of a billiard flow, by which the step-function potential at the
billiard boundary is replaced by a family of smooth potentials approaching
the step function, preserving the correct reflection law near the boundary.
We stress that the billiard reflection rule ("angle of reflection equals angle
of incidence") appears as a limit only, and the billiard itself is, of course,
an idealized model to the real motion. Therefore, the problem of relating
the statistics manifested by the billiard dynamical systems to actual physi-

cal applications must inevitably include the study of the smoothening of the billiard potential.

The influence of such smoothening is a non-trivial question, since the dynamical system associated with the billiard we consider (in the simplest setting, this is a two-dimensional area-preserving mapping [21]) is singular. In particular, as explained more precisely in section 2.1, singularities appear near trajectories which are tangent to the billiard's boundary - like the ones shown in figure 1.1. Thus, even though the scattering billiard is hyperbolic almost everywhere, theoretically, there exists a possibility that the singular set (e.g. singular periodic orbits) will produce stability islands under small perturbation. While such a phenomenon seems to be quite common, general theory does not exist. Indeed, it is clear that the results are not straightforward - namely it is not true that all smooth systems approaching a singular hyperbolic and mixing system have stable periodic orbits nor is the converse - that they have the same ergodic properties as the singular system. (As an example, consider an analogous problem for one dimensional maps; For a family of tent maps of an interval which are known to be ergodic and mixing, the ergodicity property may be easily destroyed in an arbitrarily close smooth family: if the maximum of the interval image produces a periodic orbit, it is clearly stable. However, the smooth one-dimensional map does not always possess stable periodic orbits: there may be a positive measure set of parameter values for which the smooth maps are ergodic and mixing [16]).

In this paper we prove that, indeed, *a perturbation of a scattering billiard to a smooth Hamiltonian flow may create stability islands near singular periodic and homoclinic orbits of the billiard.* An important ingredient of the proof is the established connection between the limiting smooth Hamiltonian flows and the singular billiard flow. This connection, which seems to be fundamental for understanding the applicability and limitations of the billiards to more realistic models of particle motion has not been previously formalized (to the best of our knowledge), and has received surprisingly little attention.

In the physics community it has been assumed to exist; For example, in [15] the qualitative behavior of orbits of the diamagnetic Kepler problem has been analyzed by studying the four-disk billiard system which has similar spatial structure. Furthermore, in that paper, the correspondence between elliptic periodic orbits of the smooth Hamiltonian system and singular periodic orbits of the modeling billiard was noticed. Nevertheless, our analysis reveals non-trivial requirements on smooth potentials approaching the billiard potential, which are essential for the dynamics of the corresponding Hamiltonian system to follow the dynamics of the billiard flow. Therefore, a rigorous proof of a correspondence between billiard and "smooth" orbits can not be immediate.

Mathematically, Marsden [19] has studied a more general question of the behavior of the symplectic structure when a family of smooth Hamiltonians approaches a singular limit, and related these problems to the general study of distributions on manifolds. In this setting, he showed that some properties of the smooth Hamiltonians are preserved by the singular one. For example, he proved that if the families of Hamiltonians are uniformly mixing then the mixing property carries to the singular system as well. Here we investigate the other direction of the above result - namely given a singular system which is mixing - what can be said on the natural family of smooth Hamiltonian which approaches this limiting system.

Recently, an example of another kind of smooth analogue of a scattering billiard with elliptic islands was constructed [9]; namely, for the motion of a point-wise particle in a finite-range smooth potential, where the potential's support consists of a finite number of non-overlapping disks on a plane torus. It was shown that in this geometry the smooth potential effect is to create a finite-length-travel along the scattering disks, and this produces focusing shifts near tangent trajectories even in the limit of high energies. Thus, it was proved that for any given energy level, there exists an arrangement of the disks for which elliptic islands exist. Here, a completely different approach is taken, which in particular, does not assume any specific geometry of the scatterers nore that the potential is of a finite-range.

Another type of natural perturbation of a billiard is achieved by a deformation of the billiard's boundary (in a non-smooth fashion for scattering billiards with a piece-wise smooth boundary). While such deformations have been extensively studied numerically, we are not aware of theoretical approaches for studying the near-ergodic regime. On the other end, perturbations of near-integrable billiards may be studies using Melnikov technique [8].

Traditionally, transport properties of the extended Sinai billiard were studied in terms of the decay of the correlation function [5]. More recently (see [27] and references therein), Poincaré recurrences and stickiness in phase space of both Sinai billiards and Casini billiards were numerically studied. It has been demonstrated that the appearance of sticky islands for some parameter values causes anomoulous transport - specifically power-law decay for the Poincaré recurrences distribution. To produce the anomoulos transport a parameter controlling the shape of the billiard was carefully tuned to produce self-similar sticky island structure. Moreover, it has been observed that such a tuning is possible near any parameter value for which islands exist. Here, we prove that islands may be produced by smoothening of the billiard boundary. Combining these results implies that by tuning the smoothening one can obtain sticky islands and thus *anomoulous transport for the Lorenz gas model with arbitrarily sharp smooth potentials.*

The general scheme of the paper is as follows: In 2.1 we introduce the billiard flow in a general domain, and describe its nature near regular and tangent collision points and its relation to the standard billiard map. Then, in 2.2, we introduce a class of one-parameter families of Hamiltonians and formulate sufficient conditions on this class so that as the parameter $\epsilon \to 0$

they approach the billiard flows. In section 2.3 some examples of families of smooth Hamiltonians satisfying our assumptions are presented. In section 2.4 we formulate the main theorems which establish in which sense the Hamiltonian flows approach the billiard flow. In section 3 we utilize these theorems to prove the existence of elliptic islands in Hamiltonian flows which approximate scattering (Sinai) billiards; First, we study the phase space structure of the billiard map near singular periodic orbits and near singular homoclinic orbits. We prove that existence of such orbits implies the appearance of a non-smooth analogue of the Smale horseshoe, similar to the horseshoe in the Hénon map. Then, using the closeness results of section 2.4 we establish that if a singular periodic orbit/homoclinic orbit exists for the billiard map, then necessarily there exist nearby Hamiltonians with elliptic periodic orbits. The appearance of persistent singular homoclinics and singular (tangent) periodic orbits for scattering billiards is conjectured and the former is numerically demonstrated. Section 4 is devoted to a discussion on the implication of these results. In appendix A examples showing the necessity of some of the conditions imposed on the family of Hamiltonians are presented.

2 Closeness of plane billiards and smooth Hamiltonian flows

2.1 Billiard flow

Consider an open bounded region D on a plane with a piecewise smooth (C^{r+1}, $r \geq 2$) boundary S. On S there is a finite set C of so-called *corner points* $c_1, c_2, \ldots$ such that the arc of the boundary that connects two neighboring corner points is C^{r+1}-smooth. Let us call these arcs *the boundary arcs* and denote them by $S_1, S_2, \ldots$. The set C includes all the points where the boundary loses smoothness and all the points where the curvature of the boundary vanishes. Thus, the curvature has a constant sign on each of the arcs S_i. Being equipped with the field of inward normals, the arc is called *convex* if its curvature is negative (with respect to the chosen equipment) and it is called *concave* if its curvature is positive (see figure 2.1).

Consider *the billiard flow* on $\bar{D}$ which describes the motion of a point mass moving with a constant velocity between consecutive elastic collisions with S. The phase space of the flow is co-ordinatized by (x, y, p_x, p_y) where (x, y) is the position of the particle in $\bar{D}$ and (p_x, p_y) is the velocity vector:

$$\dot{x} = p_x \quad \dot{y} = p_y. \tag{2.1}$$

Henceforth, to distinguish between the phase space and the configuration space $\bar{D}$ we reserve the term "orbit" for the orbits in the phase space and the term "trajectory" for the projection of an orbit to the (x, y)-plane.

The flow is defined by the condition that the velocity vector (p_x, p_y) is constant in the interior, and at the boundary it changes by the elastic reflection rule so $p_x^2 + p_y^2 = const$ and the angle of reflection equals the angle of incidence with the opposite sign. Taking the point of reflection as the origin of the coordinate frame and the boundary's normal at that point as the y-axis, the reflection rule is simply

$$p_x \rightarrow p_x, p_y \rightarrow -p_y; \tag{2.2}$$

namely, the angle of incidence ϕ is $arctan p_y/p_x$. This law is well defined only when the normal can be well defined: it is invalid at the corners where the boundary looses its smoothness.

Generally, the incidence angle ϕ belongs to $[-\frac{\pi}{2}, \frac{\pi}{2}]$, but if the boundary is convex, $|\phi| < \frac{\pi}{2}$. If the boundary arc is concave, it is possible to have $\phi = \pm\frac{\pi}{2}$ (figure 2.1) which corresponds to a trajectory tangent to S.

A special case is a tangent trajectory ($\phi = \pm\frac{\pi}{2}$) which reaches the boundary at an inflection point. One can easily see that any close trajectory undergoes an unboundedly large number of collisions before leaving a small neighborhood of the inflection point, and for the trajectory tangent to the boundary at the inflection point itself there is no reflection at all (figure 2.1). The trajectory is terminated at the moment of such tangency and the corresponding orbit of the flow is not defined for greater times. That is the reason for excluding the inflection points from consideration by putting them into the corner set.

Denote points in the phase space of the billiard flow as $q \equiv (x, y, p_x, p_y)$ and the time t map of the flow as $b_t : q_0(x_0, y_0, p_{x0}, p_{y0}) \mapsto q_t(x_t, y_t, p_{xt}, p_{yt})$. Recall that the reflection law is not defined at the corner points; thus, by writing $q_t = b_t q_0$, we mean, in particular, that the piece of trajectory that connects (x_0, y_0) and (x_t, y_t) is on a finite distance of the corner set C. At the same time we allow the trajectory to have one or more points of tangency with concave components of S.

A point $q(x, y, p_x, p_y)$ in the phase space is called *an inner point* if $(x, y) \notin S$, and *a collision point* if $(x, y) \in (S \backslash C)$. Obviously, if q_0 and $q_t = b_t q_0$ are inner points, then q_t depends continuously on q_0 and t. Otherwise, if q_t is a (non-tangent) collision point, the velocity vector undergoes a jump: denoting by $q_{t-0} = b_{t-0} q_0$ and $q_{t+0} = b_{t+0} q_0$ the points just before and just after the collision, it follows that (p_{xt+0}, p_{yt+0}) and (p_{xt-0}, p_{yt-0}) are related by the elastic reflection law. To avoid ambiguity we assume that at a collision point the velocity vector is oriented inside D; thus, we put $b_t \equiv b_{t+0}$.

Further, if q_t is an inner point and if the piece of trajectory that connects (x_0, y_0) and (x_t, y_t) does not have tangencies with the boundary, then q_t depends C^r-smoothly on q_0 and t. On the other hand, it is well known [21] that the map b_t loses smoothness at any point q_0 whose trajectory is tangent to the boundary at least once on the interval $[0, t]$. Indeed, choosing coordinates so that the origin is a point on a concave boundary arc S_i, the y-axis is the normal to S_i and the x-axis is tangent to S_i, the arc is locally given by the equation

$$y = -x^2 + \dots$$

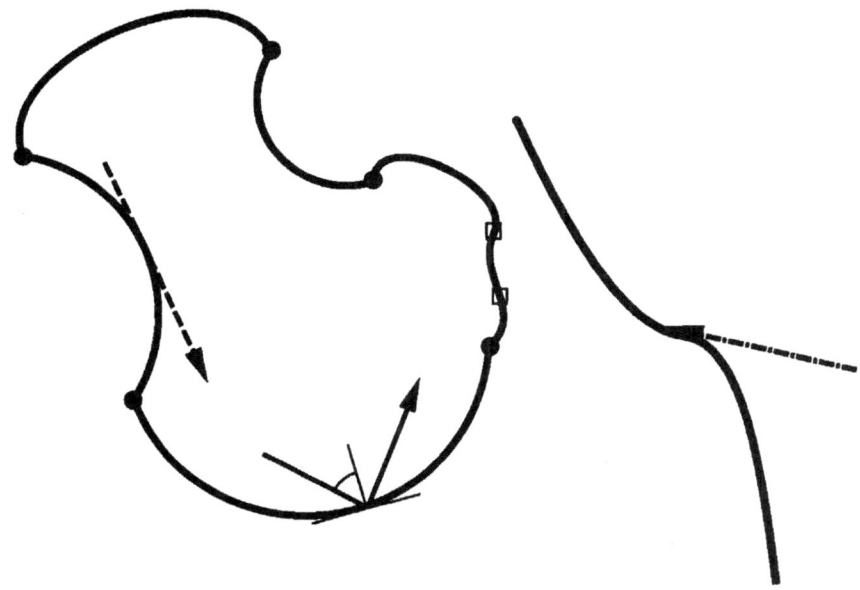

Fig. 2.1. Billiard flow

a) • - standard corner points, □ - inflection corner points
$S_{1,3,5}$ - concave boundary arcs, $S_{2,4,6,7}$ - convex arcs
—— Regular reflection, - - - Tangent trajectory
b) — · —· Tangent trajectory terminated at an inflection point

It follows that for small $\delta > 0$ the time $t = \delta$ map of the slanted line ($x_0 = -\delta/2 + ay_0, p_{x0} = 1, p_{y0} = 0$) has a square root singularity in the limit $y_0 \to -0$ which corresponds to the tangent trajectory (see figure 2.2; $a \neq 0$ for graphical purposes):

$$(x_\delta, y_\delta, p_{x\delta}, p_{y\delta}) = \begin{cases} (\frac{1}{2}\delta + ay_0, y_0, 1, 0) & \text{at } y_0 \geq 0 \\ (\frac{1}{2}\delta + ay_0 + O(\delta y_0), 2\sqrt{-y_0}\delta + O(\delta y_0), 1 + O(y_0) \\ \quad , 2\sqrt{-y_0} + O(y_0)) & \text{at } y_0 \leq 0 \end{cases}$$

If q_0 and $q_t = b_t q_0$ are inner points, then for arbitrary two small cross-sections in the phase space, one through q_0 and the other through q_t, *the local Poincaré map is defined by the orbits of the billiard flow. If no tangency to the boundary arcs is encountered between q_0 and q_t, then the Poincaré map is locally a C^r-diffeomorphism.*

One can easily prove that the same remains valid if q_0, or q_t, or both of them are collision points, provided the corresponding cross-sections are composed of the nearby collision points. In fact, the collision set (the surface

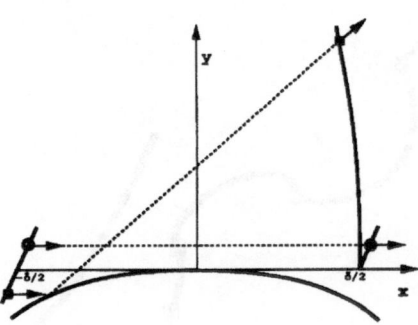

Fig. 2.2. Singularity near a tangent trajectory

$(x, y) \in S$ in the phase space) provides a *global* cross-section for the billiard flow. The corresponding Poincaré map relating consecutive collision points is called *the billiard map*. A point on the surface is determined by the position s on the boundary S and by the reflection angle ϕ which yields the direction of the outgoing velocity vector (the absolute value of the velocity does not matter because $p_x^2 + p_y^2$ is a conserved quantity - the energy - and it may be taken arbitrary by rescaling the time). The initial conditions, corresponding to a trajectory directed to a corner or tangent to a boundary arc at the moment of the next collision, form *the singular set* on the (s, ϕ)-surface. Generically, the singularity set is a collection of lines which may be glued at some points. The billiard map is a C^r-diffeomorphism outside the singular set; it may be discontinuous at the singular points. Near a singular point corresponding to the tangent trajectory the continuity of the map can be restored locally by taking two iterations of the map on a half of the neighborhood of the singular point (see figure 2.2). The obtained map will, nevertheless, be non-smooth at the singular point, having the square root singularity described above.

2.2 Class of smooth Hamiltonians

Formally, the billiard flow may be considered as a Hamiltonian system of the form

$$H_b = \frac{p_x^2}{2} + \frac{p_y^2}{2} + V_b(x, y) \tag{2.3}$$

where the potential vanishes inside the billiard region D and equals to infinity outside:

$$V_b(x, y) = \begin{cases} 0 & (x, y) \in D \\ +\infty & (x, y) \notin D \end{cases} \tag{2.4}$$

Clearly, this is an approximate model of the motion of a pointwise particle in a *smooth* potential which stays nearly constant in the interior region and grows very fast near the boundary. However, it is not obvious immediately when (and in which sense) this motion is indeed close to the billiard motion. We examine this question in this section and describe a class of potentials for which the billiard approximation (2.4) is correct in some reasonable sense.

Consider a Hamiltonian system associated with

$$H = \frac{p_x^2}{2} + \frac{p_y^2}{2} + V(x,y;\epsilon) \tag{2.5}$$

where the potential $V(x,y;\epsilon)$ tends to zero inside the region D as $\epsilon \to 0$ and it tends to infinity outside. Specifically, we require that

I. *For any compact region $K \subset D$ the potential $V(x,y;\epsilon)$ diminishes along with all its derivatives as $\epsilon \to 0$:*

$$\lim_{\epsilon \to +0} ||V(x,y;\epsilon)|_{\{(x,y)\in K\}}||_{C^{r+1}} = 0. \tag{2.6}$$

The growth of the potential to infinity across the boundary is a more delicate issue. The crucial construction here is that V is evaluated along the level sets of some *finite* function near the boundary. Namely, putting the set C of corner points c_i out of consideration, we suppose that in a neighborhood of the set $(\bar{D}\backslash C)$ there exists a function $Q(x,y;\epsilon)$ which is C^{r+1} with respect to (x,y) and it depends continuously on ϵ (in C^{r+1}-topology) at $\epsilon \geq 0$. Specifically, $Q(x,y;\epsilon)$ along with its derivatives have a proper limit as $\epsilon \to 0$. Assume that

IIa *On the boundary, the function $Q(x,y;0)$ is constant between any two neighboring corner points:*

$$Q(x,y;\epsilon = 0)|_{(x,y)\in S_i} \equiv Q_i \tag{2.7}$$

We call Q a *pattern function*. For each boundary component S_i, for Q close to Q_i, let us define *a barrier function* $W_i(Q;\epsilon)$ which does not depend explicitly on (x,y) and assume that:

IIb *There exists a small neighborhood N_i of the arc S_i on which the potential V is given by W_i evaluated along the level sets of the pattern function Q:*

$$V(x,y;\epsilon)|_{(x,y)\in N_i} \equiv W_i(Q(x,y;\epsilon);\epsilon) \tag{2.8}$$

IIc *The gradient of V does not vanish in a finite neighborhood of the boundary arcs:*

$$\nabla V|_{(x,y)\in N_i} \neq 0 \tag{2.9}$$

which is equivalent to the following conditions

$$\nabla Q|_{(x,y)\in N_i} \neq 0 \tag{2.10}$$

and

$$\frac{d}{dQ}W_i(Q;\epsilon) \neq 0. \tag{2.11}$$

Conditions **IIa,b,c** formalize the requirement that the direction of the gradient of the potential must be normal to the boundary as $\epsilon \to +0$. Obviously, this is necessary for having a proper reflection law in the limit: if the reflecting force has a component tangent to the wall, then the tangent component p_x of the momentum will not be preserved during the collision (see (2.2)).

Now we may describe the rapid growth of the potential across the boundary in terms of the barrier functions W_i only. Choose any of the arcs S_i and henceforth suppress the index i. Without loss of generality assume $Q = 0$ on S. By (2.10), the pattern function Q is monotonically increasing across S and we assume Q is positive inside D near S and negative outside (otherwise, change inequalities in (2.12) to the opposite ones). Assume

III *As $\epsilon \to +0$ the barrier function increases from zero to infinity across the boundary S_i:*

$$\lim_{\epsilon \to +0} W(Q;\epsilon) = \begin{cases} +\infty & Q < 0 \\ 0 & Q > 0 \end{cases} \tag{2.12}$$

Note that according to **I.** and **IIb.**, for any $Q_0 > 0$

$$\lim_{\epsilon \to +0} \|W(Q,\epsilon)|_{Q \geq Q_0}\|_{C^{r+1}} = 0. \tag{2.13}$$

Clearly, it will cause no troubles if one allows W to take infinite values: by (2.11), the function W is monotonic and if it is infinite at some Q, it is infinite for all smaller Q; on the other hand, trajectories always stay in the region where W is bounded: since the energy given by (2.5) is conserved, the value of the potential is bounded by the initial value of H. We will study limiting behavior (as $\epsilon \to +0$) of the smooth Hamiltonian system (2.5) in a given, *fixed* energy level, $H = H^*$. This implies that all trajectories stay in the region $W \leq H^*$ for any ϵ. It follows that the symbol $+\infty$ in (2.12) may be replaced by any value greater than H^*.

It is immediately evident that the particle in the potential V satisfying condition **I** moves in the interior of D with essentially constant velocity along a straight line until it reaches a thin layer near the boundary S where the potential runs from small to very large values (the smaller the value of ϵ, the thinner the boundary layer). By virtue of condition **III**, if the particle enters the layer near an interior point of some boundary arc (corner points are not considered in this paper), it can not penetrate the layer and go outside - because fixing the value of the energy bounds the potential from above. Thus, the particle is either reflected, exiting the boundary layer near the point where it entered, or it might, in principle, stick into the layer, traveling along the boundary far away from the entrance point. As simple arguments show (see the proof of theorem 1 below), condition **II** guarantees that when a reflection does occur it will be of the right character, approximately preserving the tangential component (p_x) of the momentum and changing sign of the normal component (p_y). However, as argued below, and shown by an example in Appendix A, conditions **I-III** are insufficient for preventing the existence of non-reflecting trajectories. Since such finite length travels along the boundary layer must be forbidden in the limit $\epsilon \to 0$, we impose an additional restriction on the shape of the potential near the boundary. Denote the normal force function by $F(Q;\epsilon) = \dfrac{d}{dQ}W(Q,\epsilon)$ and require the following:

IV *The normal force is a monotonic function of Q:*

$$W''(Q) \equiv F'(Q) \geq 0. \tag{2.14}$$

(According to condition **III**, since W decays rapidly across $Q = 0$, it follows that its derivative $F(Q)$ is close to $-\infty$ at small Q. Then, as Q grows, $F(Q) \to 0$ by (2.13). Thus, $F(Q)$ can not be strictly decreasing function and the monotonicity of $F(Q)$ is indeed equivalent to the positiveness of $F'(Q)$.)

To see how a violation of the monotonicity condition can lead to the appearance of non-reflecting trajectories suppose that for arbitrarily small ϵ there is an interval of values of Q, arbitrarily close to the boundary, on which the graph of absolute value of $F(Q)$ is as shown in figure 2.3: it grows from zero to very large values, then decays back to nearly zero at a value Q_ϵ which approaches zero as $\epsilon \to 0$, and only after that it grows to infinity. Since the force is the gradient of the potential and, according to condition **II**, it is proportional to $F(Q)$ whereas the distance to the wall is proportional to Q, it follows that the graph of the normal component of the reflecting force *vs* the distance to the wall has the same shape as in figure 2.3. Thus, the initial velocity of the particle can be taken such that the normal component of the velocity is completely damped when moving through the region of the first peak of $F(Q)$, leading to the trapping of the particle in the zone where the reflecting force is nearly zero with the normal component of velocity close to zero too. In this case the distance to the wall will change very slowly and the particle may stay at a small distance to the wall for a long time, travelling along the boundary instead of making reflection. An explicit example of such trapping in a circular billiard is presented in Appendix A. In fact, the geometry of the boundary plays a crucial role here: one can show that the finite length travels along a *concave* boundary arc are forbidden even for the non-monotonic $F(Q)$ (though the reflection time may still be unboundedly large in this case).

Conditions **I-IV** guarantee, as is precisely formulated in section 2.4, a correct reflection law only in the C^0-topology and not in the C^1-topology. As this issue is very important for the sequel, we explain its intuitive implication now. Let us take a point (x_0, y_0) and momentum (p_{x0}, p_{y0}) as initial conditions for an orbit of the Hamiltonian system (2.5) and let us take the same initial conditions for the billiard orbit. Consider a time interval t for which the billiard orbit collides with the boundary S only once, at some point (x_c, y_c) (see figure 2.4). Here, the incidence angle ϕ^{in} is the angle between the vector $(x_0 - x_c, y_0 - y_c)$ and the inward normal to S at the point (x_c, y_c); the reflection angle ϕ^{out} is the angle between the vector $(x_t - x_c, y_t - y_c)$ and the normal, where (x_t, y_t) is the point reached by the billiard trajectory at the

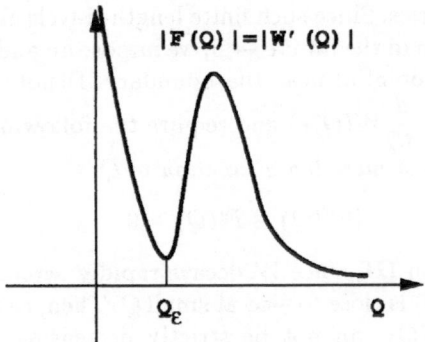

$|\mathbf{F}(\mathbf{Q})| = |\mathbf{W}'(\mathbf{Q})|$

Q_ε

Q

Fig. 2.3. Non-monotonic normal force

time t. In the same way one may define the incidence and reflection angles for the trajectory of the Hamiltonian system where (x_0, y_0) and (x_c, y_c) are taken the same as for the billiard trajectory and $(x_t(\epsilon), y_t(\epsilon))$ is now defined by the Hamiltonian flow (see figure 2.4). We expect the trajectory of the Hamiltonian system to be close to the billiard trajectory; in particular, it should demonstrate a correct reflection law

$$\phi^{in}(\epsilon) + \phi^{out}(\epsilon) \approx 0$$

for sufficiently small ϵ. Note, however, that $(\phi^{in} + \phi^{out})$ is a function of the initial conditions. Conditions **I-IV** give only C^0-closeness of these functions to zero and to ensure a C^1-correct reflection law we need the following additional condition on $W(Q)$:
V *There exists an* $\alpha \in (0,1)$ *such that the following holds for any interval* $[Q_1(\epsilon), Q_2(\epsilon)]$ *on which* $W(Q)$ *is bounded away from zero and infinity for all* ϵ:

$$\lim_{\epsilon \to 0} \frac{W''(Q)}{|W'(Q)|^{3+\alpha}} = 0, \tag{2.15}$$

uniformly on the interval $[Q_1, Q_2]$.
This condition is used directly in the proof of theorem 1 (see [25]). To give the reader a feeling of how the smoothness may be lost, consider a *one-dimensional* reflection described by the equation $\ddot{Q} + W'(Q; \epsilon) = 0$ where $Q \geq 0$, $W(0; \epsilon) = +\infty$, $\lim_{\epsilon \to 0} W(Q; \epsilon) = 0$ at $Q > 0$. Here, Q is the position of a particle moving inertially until a collision with the wall at $Q = 0$, after which the particle reflects elastically and moves back. The time of collision is given by $\tau = \int_{Q^*}^{1} \frac{\sqrt{2}dQ}{\sqrt{H - W(Q)}}$ where H is the value of energy and $Q^*(\epsilon)$ is such that $W(Q^*; \epsilon) = H$. Differentiation with respect to H gives

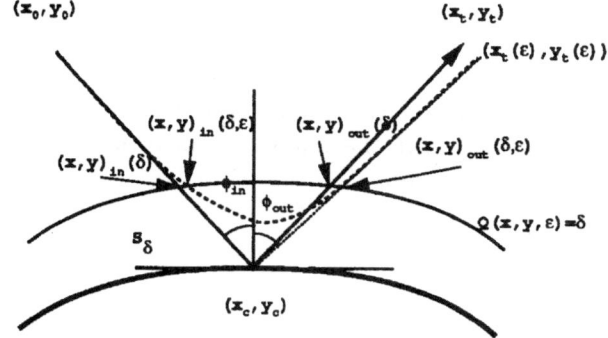

Fig. 2.4. Reflection by Hamiltonian flow

$$\frac{d\tau}{dH} = \frac{\sqrt{2}}{W'(Q^*)}(H - W(1))^{-1/2}$$

$$- \frac{1}{\sqrt{2}W'(Q^*)} \int_{Q^*}^{1} \frac{W'(Q^*) - W'(Q)}{W(Q^*) - W(Q)}(H - W(Q))^{-1/2}dQ. \quad (2.16)$$

Note that $(W'(Q^*) - W'(Q))/(W(Q^*) - W(Q)) \approx W''/W'$, therefore restrictions should be imposed on W'', like in condition **V**, to have $d\tau/dH$ bounded.

2.3 Examples for smooth Hamiltonians limiting to billiards

Conditions **I-V** are in fact quite general, and they are fulfilled by many reasonable choices of the pattern and barrier functions. For the pattern function, consider any smooth function Q depending on two variables (x, y). Corners are created at the singularities of the level sets and at the points of inflection.

For the barrier function conditions **I-V** need to be fulfilled. For example, the following barrier functions $W(Q, \epsilon)$ satisfy them (for $\beta > 0$):

$$\frac{\epsilon}{Q^\beta}, \quad (1 - Q^\beta)^{\frac{1}{\epsilon}}, \quad \epsilon e^{-\frac{1}{Q^\beta}}, \quad \epsilon |\ln Q|^\beta, \quad \epsilon \ln \ldots |\ln Q|.$$

One may easily produce more examples because there is no restriction on the growth rate: given any potential V satisfying conditions **I-V** the potential $\psi(V)$ also satisfies these conditions provided ψ is a smooth monotonic function of V such that $\psi(0) = 0$, $\psi(\infty) = \infty$.

In section 3 we consider the billiard corresponding to the following family of pattern functions:

$$Q(x, y; \gamma) = \gamma(\frac{1}{x^2 + (y - \frac{1}{\gamma})^2 - R^2} + \frac{1}{x^2 + (y + \frac{1}{\gamma})^2 - R^2}$$

$$+ \frac{1}{(x - \frac{1}{\gamma})^2 + y^2 - R^2} + \frac{1}{(x + \frac{1}{\gamma})^2 + y^2 - R^2})^{-1} \quad (2.17)$$

where $R^2 = 1 + (1 - \frac{1}{\gamma})^2$ and γ is a parameter (not necessary small). The billiard domain is bounded by the level set $Q(x, y) = 0$. For $\gamma \to 0$ this defines a square whereas for $\gamma > 0$ it defines a concave shape bounded by the four circles of radius R which intersect at the four corner points $(x, y) = (\pm 1, \pm 1)$.

Taking the barrier function in the simplest form $W(Q, \epsilon) = \frac{\epsilon}{Q}$ produces the following Hamiltonian system:

$$H_{\gamma,\epsilon}(x, y, p_x, p_y) = \frac{1}{2}p_x^2 + \frac{1}{2}p_y^2 +$$
$$\epsilon\Big(\frac{1}{\gamma(x^2 + y^2 - 2) + 2(1 - y)} + \frac{1}{\gamma(x^2 + y^2 - 2) + 2(1 + y)}$$
$$+ \frac{1}{\gamma(x^2 + y^2 - 2) + 2(1 - x)} + \frac{1}{\gamma(x^2 + y^2 - 2) + 2(1 + x)}\Big)$$

Notice that for $\gamma \to 0$, the square geometry produces separable - hence integrable - Hamiltonian flow. This is, of course, a very interesting limit, which is not studied in this paper. Notice also that here the limit $\epsilon \to 0$ is equivalent to the limit $H \to \infty$ with ϵ held fixed.

2.4 Closeness theorems

Denote the Hamiltonian flow of (2.5) by $h_t(\epsilon)$. Given t and ϵ, the flow maps a phase point $q_0 \equiv (x_0, y_0, p_{x0}, p_{y0})$ to $q_t(\epsilon) \equiv (x_t(\epsilon), y_t(\epsilon), p_{xt}(\epsilon), p_{yt}(\epsilon))$. We will call $q_t(\epsilon)$ *the smooth orbit* of q_0 and will examine how close is it to *the billiard orbit* $b_t q_0 \equiv q_t(0)$. The corresponding trajectories $(x_t(\epsilon), y_t(\epsilon))$ and $(x_t(0), y_t(0))$ on the (x, y)-plane will be called *the smooth* and, respectively, *the billiard* trajectories.

Let (x_c, y_c) be the first point of collision of the billiard trajectory with the boundary S; by definition, $(x_c, y_c) = (x_0, y_0) + (p_{x0}, p_{y0})t_c$, where $t = t_c$ is the moment of collision. Since the potential V is nearly zero in the interior of the billiard domain D, the smooth orbit of q_0 is arbitrarily close (as $\epsilon \to 0$) to the billiard orbit before the collision: namely, the point $(x_t(\epsilon), y_t(\epsilon))$ moves with essentially constant velocity until reaching a small neighborhood of (x_c, y_c). Take a small $\delta > 0$ and consider *the boundary layer* $S_\delta \equiv \{|Q(x, y; \epsilon) - Q(x_c, y_c; \epsilon)| \le \delta\}$, where Q is the pattern function. For any small δ, if ϵ is sufficiently small, the smooth trajectory enters the boundary layer at some time $t_{in}(\epsilon)$. Denote $q_{in}(\epsilon) = q_{t_{in}}(\epsilon)$; by definition, $|Q(x_{in}(\epsilon), y_{in}(\epsilon); \epsilon) - Q(x_c, y_c; \epsilon)| = \delta$. The closeness of the billiard and the smooth orbits (before the collision) implies the existence of the limits (see figure 2.4)

$$t_{in} \equiv \lim_{\epsilon \to 0} t_{in}(\epsilon), \quad q_{in} \equiv \lim_{\epsilon \to 0} q_{in}(\epsilon);$$

moreover,

$$\lim_{\delta \to 0} t_{in} = t_c, \quad \lim_{\delta \to 0}(x_{in}, y_{in}) = (x_c, y_c), \quad \lim_{\delta \to 0}(p_{x,in}, p_{y,in}) = (p_{x0}, p_{y0}).$$

Analogously, denote the moment when the smooth trajectory exits the boundary layer as $t_{out}(\epsilon)$ (we will prove that such a moment exists) and denote the corresponding value of $q_t(\epsilon)$ as $q_{out}(\epsilon)$. The time interval $(t_{out}(\epsilon) - t_{in}(\epsilon))$ will be called *the collision time*. For fixed δ, the limiting values of the introduced quantities as $\epsilon \to 0$ will be denoted as t_{out}, q_{out} (the existence of the limits is given by Theorem 1 below).

It is natural to call the relation between the limits q_{out} and q_{in} *the reflection law*. By definition, $q_{out}(\epsilon)$ and $t_{out}(\epsilon)$ are functions of q_{in}. If the convergence of $\lim_{\epsilon \to 0}(q_{out}, t_{out})(\epsilon)$ is uniform in some neighborhood of a given q_{in}, then the reflection law is C^0. If, moreover, there is a uniform convergence for the derivatives with respect to q_{in}, then these limit to $\dfrac{\partial(q_{out}, t_{out})}{\partial q_{in}}$, so the reflection law is C^1.

Note that the relation between the reflection laws corresponding to different values of δ is found trivially for the billiard flow, and it is absolutely the same for the Hamiltonian flow because it limits to the billiard flow out of any fixed boundary layer. Therefore, no information is lost if one considers the limit of the reflection law as $\delta \to 0$, as it is done in the following theorem.

Theorem 1. *For the Hamiltonian system (2.5) where the potential $V(x, y; \epsilon)$ satisfies conditions **I-IV**, if initial conditions q_0 are such that for the billiard orbit $b_t q_0$ the point of reflection is not a corner: $(x_c, y_c) \in S\backslash C$, then for any sufficiently small δ the limits (as $\epsilon \to 0$) q_{out} and t_{out} are well defined. As $\delta \to 0$, the collision time tends to zero:*

$$\lim_{\delta \to 0}(t_{out} - t_{in}) = 0 \tag{2.18}$$

and the limiting C^0 reflection law is:

$$(x_{out}, y_{out}) = (x_{in}, y_{in})$$
$$(p_{x,out}, p_{y,out}) + (p_{x,in}, p_{y,in}) = 2(p_{x,in}e_x + p_{y,in}e_y)(e_x, e_y) \tag{2.19}$$

where $\bar{e} = (e_x, e_y)$ is the unit vector tangent to the boundary at the point (x_c, y_c).

*If, additionally, condition **V** is fulfilled and the ingoing velocity vector $(p_{x,in}, p_{y,in})$ is not tangent to the boundary at the point (x_c, y_c), then the reflection law is C^1.*

One may check that the above reflection law is exactly the reflection law associated with the billiard flow. In other words, theorem 1 says that

$$\lim_{\delta \to 0}\lim_{\epsilon \to 0} ||(q_{out}(\epsilon), t_{out}(\epsilon)) - (q_{out}(0), t_{out}(0))|| = 0 \tag{2.20}$$

where the norm is C^0- or C^1-norm in a small neighborhood of q_{in}. Since out of the boundary layer the Hamiltonian flow limits to the billiard flow as $\epsilon \to 0$, this local result implies immediately the following global version.

Theorem 2: *If q_0 and $q_t = b_t q_0$ are inner phase points, then, as $\epsilon \to 0$, the time t map $h_t(\epsilon)$ of the flow defined by Hamiltonian (2.5) where $V(x, y; \epsilon)$ satisfies assumptions* **I-IV** *limits to the map b_t in the C^0-topology in a small neighborhood of q_0. If, additionally, condition* **V** *is fulfilled and if the billiard trajectory of q_0 has no tangencies to the boundary for the time interval $[0, t]$, then $h_t(\epsilon) \to b_t$ in the C^1 sense.*

Theorem 2 follows from theorem 1, and vice versa. The proof of the theorems (in fact, a C^r-convergence proof) is given in [25]. Namely, the following is proved there

$$\lim_{\delta \to 0} \limsup_{\epsilon \to 0} \|(q_{out}(\epsilon), t_{out}(\epsilon)) - (q_{out}(0), t_{out}(0))\| = 0 \qquad (2.21)$$

which is formally weaker than (2.20), but it is, obviously, also sufficient for the validity of theorem 2.

The general idea of the proof is as follows (see details in [25]). By condition **II**, the gradient of the potential is close to normal to the boundary near the point of reflection. This implies, almost immediately, that the tangential component p_x of the momentum is approximately preserved during the collision. Essentially, this means that the motion described by the Hamiltonian system (2.5) can be thought as a sum of two almost independent motions: inertial motion parallel to the boundary and reflection in the normal direction. In the limit $\epsilon \to 0$, the parallel motion prevails in some sense for the nearly tangent trajectories, whereas for the non-tangent trajectories its contribution can be neglected. Thus, in both cases the consideration is essentially one-dimensional and this makes the proof of the C^0 part of theorem 1 pretty simple. The proof of the C^1 version is more involved and it requires estimates of some integrals along the orbit of the Hamiltonian system, necessary for the evaluation of the solution of the linearized equations.

A more specified way to formulate closeness of the Hamiltonian system under consideration to the billiard approximation is to use the Poincaré sections. Let q_0 and $q_t = h_t(\epsilon)q_0$ ($\epsilon \geq 0$) be inner phase points and ω_0 and ω_1 be small surfaces transverse to the flow near q_0 and q_t. Then the flow defines the local Poincaré map $h_{t_f}(\epsilon) : \omega_0 \to \omega_1$ where $t_f(\epsilon)$ is *the flight time* from ω_0 to ω_1. The Poincaré map preserves the foliation of the cross-sections by the levels of equal energy. Therefore, reduced Poincaré maps are defined taking fixed energy levels on ω_0 onto the levels of the same energy on S_1. For $\epsilon > 0$ (respectively $\epsilon = 0$) the reduced Poincaré map is a two-dimensional area-preserving C^r-diffeomorphism (respectively - almost everywhere C^r-diffeomprphism). Obviously, the flow is recovered by the set of reduced Poincaré maps along with the corresponding flight times, and vice versa. Thus, theorem 2 admits the following reformulation.

Theorem 3. *If q_0 and $q_t = b_t q_0$ are inner phase points and ω_0 and ω_1 are small cross-sections through q_0 and q_t respectively, then at all small ϵ the Hamiltonian flow (2.5) satisfying conditions* **I-IV** *defines the reduced Poincaré map of the the energy level of q_0 in ω_0 into ω_1. As $\epsilon \to 0$ this map* *limits (in C^0) to the reduced Poincaré map of the billiard flow as does the flight time. In addition, if condition* **V** *is satisfied and the segment of billiard trajectory between q_0 and q_t does not have tangencies to the boundary of the billiard domain, then the convergence is C^1.*

The last theorem allows one to utilize persistence theorems regarding two-dimensional area preserving diffeomorphisms in order to establish relations between periodic orbits of the billiard flow and of the Hamiltonian flows under consideration.

Recall that an orbit (e.g., a periodic orbit) of the billiard flow is called *non-singular* if its trajectory in the (x, y)-plane does not have tangencies with the boundary of the billiard domain (and by definition the trajectory cannot hit a corner either). For a non-singular periodic orbit, for a cross-section through an inner point on it, the reduced Poincaré map of the billiard flow is locally a diffeomorphism and the intersection of the periodic orbit with the cross-section in the phase space is a fixed point of the diffeomorphism. Generally, the fixed point is either hyperbolic or elliptic. Fixed points of both types are preserved under small smooth perturbations in the class of area preserving diffeomorphisms. Thus, theorem 3 implies the following statement.

Corollary 1 - persistence of periodic orbits: *If a non-singular periodic orbit L_0 of the billiard flow is hyperbolic or elliptic, then at ϵ sufficiently small the Hamiltonian flow $h_t(\epsilon)$ has a unique continuous family of hyperbolic or, respectively, elliptic periodic orbits L_ϵ in the fixed energy level of L_0 which limit to L_0 as $\epsilon \to 0$.*

If L_0 is hyperbolic, the local stable $(W^s_{\mathrm{loc}}(L_\epsilon))$ and unstable $(W^u_{\mathrm{loc}}(L_\epsilon))$ manifolds of L_ϵ depend continuously on ϵ (as smooth manifolds) and limit to $W^s_{\mathrm{loc}}(L_0)$ and $W^u_{\mathrm{loc}}(L_0)$ respectively. The global stable and unstable manifolds - $W^u(L_\epsilon)$ and $W^s(L_\epsilon)$ - are obtained as the continuation of $W^s_{\mathrm{loc}}(L_\epsilon)$ and $W^u_{\mathrm{loc}}(L_\epsilon)$ by the orbits of the flow. Note that for the billiard flow, by applying the continuation process tangencies to the boundary and corner points are bound to be encountered by some points belonging to the manifolds. Using local cross-sections as above, it is easy to see that the following result holds.

Corollary 2 - extensions of stable and unstable manifolds: *Any piece K_0 of $W^u(L_0)$ or $W^s(L_0)$ obtained as a time $t > 0$ shift of some region in $W^u_{\mathrm{loc}}(L_0)$ (respectively, a time $t < 0$ shift of some region in $W^s_{\mathrm{loc}}(L_0)$) is a C^0- or, if no tangencies to the boundary are encountered in the continuation process, C^1-limit of a family of surfaces $K_\epsilon \subset W^u(L_\epsilon)$ (resp. $K_\epsilon \subset W^s(L_\epsilon)$).*

The above persistence results apply only to non-singular periodic orbits; near the singular periodic orbits the billiard flow is non-smooth and the standard theory is not valid. However, it is of interest to study the behavior near a singular periodic orbit for $\epsilon > 0$. We consider this problem in the next section for the case of so-called *scattering* billiards. Here, the billiard flow is hyperbolic whence all non-singular periodic orbits are hyperbolic. We, nevertheless, show that the singular periodic orbits give rise to stable (elliptic) periodic orbits in the Hamiltonian systems (2.5) limiting to the scattering billiards.

3 Appearance of elliptic islands in the smooth Hamiltonian approximation of scattering billiards

Consider scattering billiards - namely billiards which are composed of concave arcs with the curvature bounded away from zero, and non-zero angles between the arcs at the corner points. The corresponding billiard flows are hyperbolic and exhibit strong ergodic properties (they are K-systems) [21, 1, 11]. In particular, almost every orbit covers the whole phase space densely. In this section we examine how these properties may be lost by the approximating smooth Hamiltonian flows for arbitrarily small positive ϵ values. We propose two mechanisms for the appearance of elliptic islands which destroy these properties: one mechanism is controlled by the existence, in the billiard flow, of a singular periodic orbit and another mechanism is controlled by the existence of a singular homoclinic orbit. To be specific, from here on, we consider only *simple singular* orbits; i.e., those for which the corresponding trajectories in the billiard domain have exactly one tangency to the billiard boundary and do not approach corner points.

First, we study the phase space structure of the local Poincaré map near such orbits, showing that locally these create a "sharp" horseshoe which, embedded in a one parameter family of billiard maps, unravels as the parameter γ varies (see figure 3.3). Then, using theorem 3, we establish that the two parameter family of Hamiltonian flows $h_t(\epsilon; \gamma)$ which approach the family of billiards as $\epsilon \to 0$ undergoes, for sufficiently small ϵ, a series of bifurcations associated with the disappearance of a Smale's horseshoe. It is well established that in this process elliptic islands are created. Thus, it follows that for each sufficiently small ϵ there exist intervals of γ values for which elliptic islands exist.

We end the section with some conjectures on the genericity of the phenomena mentioned above: we expect that singular homoclinic and periodic orbits are, in fact, unavoidable in scattering billiards. Apparently, systems possessing simple singular homoclinic and periodic orbits are dense among all scattering billiards. We provide a numerical example which supports such a conjecture regarding the density of billiards with singular homoclinic orbits. A proof of this conjecture combined with the results presented here would imply that *for any given scattering billiard on a plane, there exists a nearby Hamiltonian flow possessing elliptic islands.*

3.1 Singular periodic orbits.

The hyperbolic structure of the scattering billiards plays a crucial role in the understanding of the behavior near a singular periodic orbit. For the billiard map B (the map relating two consecutive collision points; see section 2.1), the presence of hyperbolic structure implies that for almost every point $P(s,\phi)$ in the phase space there exist stable and unstable directions E_P^u and E_P^s, depending continuously on P. The system of stable and unstable directions is invariant with respect to the linearized map: $d_P B E^{s(u)} = E_{BP}^{s(u)}$, which is uniformly expanding along the unstable direction and uniformly contracting along the stable direction: if $v \in E^u$ ($v \in E^s$), then $||d_P B v|| \geq e^{\lambda\tau}||v||$ (resp., $||d_P B v|| \leq e^{-\lambda\tau}||v||$) in a suitable norm; here, τ is the flight time from P to PB, the uniformity means that the value $\lambda > 0$ is independent of P (see details in [3]).

Equivalently, there is an invariant family of stable and unstable cones: the unstable cone at a point P is taken by the linearized map $d_P B$ into the unstable cone at the point BP; the image is stretched in the unstable direction and shrinks in the stable direction. Similar behavior appears for the stable cone under backward iterations. There is an explicit geometrical description of these cones for scattering billiards [26]. Consider a point (s,ϕ) in the phase space and a small curve passing through this point. Taking two points on this curve defines two inward directed rays emanating from the billiard boundary near s (see figure 3.1). If these rays intersect, then the tangent direction to this curve belongs to the stable cone of (s,ϕ); otherwise, it belongs to the unstable cone (in other words, the unstable cones are given by $ds \cdot d\phi > 0$ and the stable cones by $ds \cdot d\phi < 0$). Moreover, it can also be shown that if the intersection of the rays with each other occurs before the first intersection of the rays with the billiard boundary, then the tangent direction to the forward image of the small curve under consideration belongs to the unstable cone of the image of (s,ϕ).

It follows from the simple geometry above that the tangents to a line of singularity at any point lies in the stable cone, and the tangent to any iteration of the singularity line by the billiard map lies in the corresponding unstable cone. In particular, this implies that intersections of the singularity lines with their images are always transverse.

Next, we find the normal form of the first return map of the billiard map near a simple singular periodic orbit (a periodic orbit with only one tangency). More precisely, consider a periodic orbit L with the corresponding sequence of collision points $P_i(s_i, \phi_i)$ ($i = 0, \ldots n - 1$): $P_{i+1} = BP_i$ where $P_n = P_0$. Since L is a simple singular periodic orbit, assume that $P \equiv P_0$ belongs to the singular set (so $|\phi_1| = \frac{\pi}{2}$). Take a small neighborhood U of P and denote as Σ the line of singular points in U (it is the line composed of the points whose trajectories are tangent to the billiard boundary near s_1). Then, we prove the following proposition:

Proposition 3.1 *Given a simple singular periodic orbit L as above, the local return map near P_0 may be reduced to the form:*

$$\begin{cases} \bar{u} = v \\ \bar{v} = \xi(v - \sqrt{\max(v,0)}) - u + \ldots \end{cases} \tag{3.1}$$

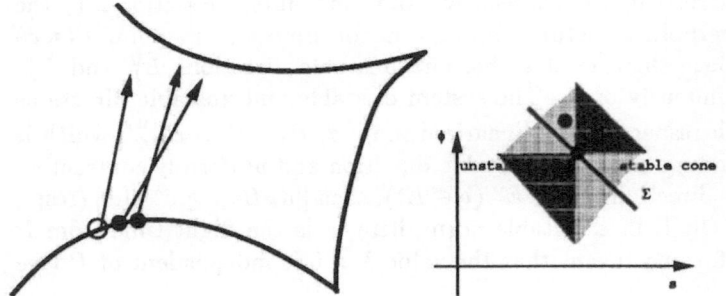

Fig. 3.1. Hyperbolic structure - the stable and unstable cones

a) Geometrical interpretation of stable/unstable directions
b) Phase space structure

where $v = 0$ gives the singularity line, $u = 0$ is its image, and $|\xi| > 2$.

As will be apparent by the proof, it is useful to define an auxiliary billiard $B^{(r)}$, for which the boundary arc by which the tangency of the periodic orbit occurs (i.e. near s_0) is pushed slightly backwards so that the singular periodic orbit becomes a regular orbit for the auxiliary system. The quantity ξ in (3.1) is simply the trace of the linearization matrix of the first return map of the auxiliary billiard about the periodic orbit. Since the auxiliary billiard is scattering, its regular periodic orbits are hyperbolic, hence $|\xi| > 2$.

Proof of Proposition 3.1 Consider the local structure in U, near the singularity line Σ. The line Σ divides U into two parts, U_r and U_s; the orbits starting on U_r (e.g. P_0'' in figure 3.2) do not hit the boundary near s_1 and approach it near the point s_2, the orbits starting on U_s (e.g. P_0' in figure 3.2) have a nearly tangent collision with the boundary in a neighborhood of s_1. Without loss of generality we assume that Σ is locally a straight line $(s - s_0) + k(\phi - \phi_0) = 0$, where $k > 0$ because Σ must lie in the stable cone $(s - s_0)(\phi - \phi_0) < 0$, and that U_r is given by $(s - s_0) + k(\phi - \phi_0) < 0$ and U_s by $(s - s_0) + k(\phi - \phi_0) \geq 0$.

Consider *the first return map* $\bar{B}$ defined on U. The map $\bar{B}$ equals $B_{n-1} \ldots B_2 B_1 B_0$ on U_s and $B_{n-1} \ldots B_2 B_0$ on U_r where B_i is a restriction of the billiard map on a small neighborhood of P_i. According to section 2.1.1, $\bar{B}$ is a continuous map but it loses smoothness on Σ. Namely, the restriction B_{0s} of B_0 on U_s exhibits the square root singularity described in section 2.1.1 whereas the map $B|_{U_r}$ is regular and it can be continued onto the whole U as a smooth map B_{0r}: erasing a small piece of the boundary containing the tangency point s_1, B_{0r} will simply be the billiard map from U to a small neighborhood of P_2 (see the action of B_{0r} on P_0' in figure 3.2). Obviously, $B_{0r}\Sigma = B_1 B_{0s}\Sigma$, therefore the first return map $\bar{B}$ is continuous. One may represent the map $\bar{B}$ as a superposition of regular and singular maps:

$$\bar{B} = B^{(r)} \cdot B^{(s)}$$

37

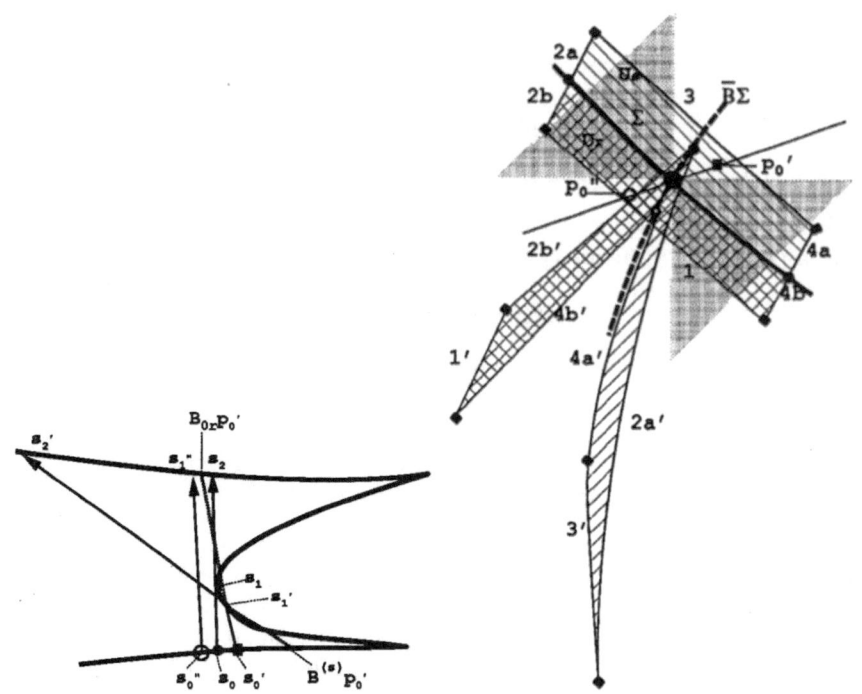

Fig. 3.2. Structure near singular periodic orbit

a) Action of billiard map near a singular segment of trajectory
b) Phase space structure near singular periodic orbit: 1234 is mapped onto $1'2'3'4'$

where

$$B^{(r)} = B_{n-1} \ldots B_2 B_{0r}$$

and

$$B^{(s)} = \begin{cases} id & \text{on } U_r \\ B_{0r}^{-1} B_1 B_{0s} & \text{on } U_s \end{cases}$$

The singular part $B^{(s)} : U \to U$ may be obtained by inverted reflection near the tangency point s_1 (see the action of $B^{(s)}$ on P_0' in figure 3.2). It is not hard to calculate that $B^{(s)}$ is given by

$$\begin{cases} S' = S + k\sqrt{\max(S + k\Phi, 0)} + \ldots \\ \Phi' = \Phi - \sqrt{\max(S + k\Phi, 0)} + \ldots \end{cases}$$

where $S = s - s_0$, $\Phi = \phi - \phi_0$ are coordinates in U, and the dots stand for the quantities infinitely small in comparison with S, Φ or $\sqrt{\max(S + k\Phi, 0)}$ as $S, \Phi \to 0$.

The regular part $B^{(r)}$ is, by definition, the first return map for the auxiliary billiard obtained by pushing the boundary near the tangency point s_1

slightly aside the trajectory of L. The point P is a fixed point for $B^{(r)}$ (as well as for the map $\bar{B}$). Since the auxiliary billiard is still scattering, the point P is a hyperbolic fixed point for $B^{(r)}$. Moreover, the unstable cone $S \cdot \Phi \geq 0$ must be mapped inside itself by the linearization of $B^{(r)}$ at P. If $\begin{pmatrix} b_{11} & b_{12} \\ b_{21} & b_{22} \end{pmatrix}$ is the corresponding linearization matrix, the last condition is equivalent to the requirement that all b_{ij} are of same sign. Recall that $B^{(r)}$ is an area-preserving diffeomorphism, so

$$b_{11}b_{22} - b_{12}b_{21} = 1.$$

Superposition of $B^{(r)}$ and $B^{(s)}$ gives, to leading order in S, Φ and $\sqrt{\max(S + k\Phi, 0)}$, the following formula for the map $\bar{B}$:

$$\begin{cases} \bar{S} = b_{11}S + b_{12}\Phi - (b_{12} - b_{11}k)\sqrt{\max(S + k\Phi, 0)} + \cdots \\ \bar{\Phi} = b_{21}S + b_{22}\Phi - (b_{22} - b_{21}k)\sqrt{\max(S + k\Phi, 0)} + \cdots \end{cases} \quad (3.2)$$

Provided inequalities 3.5 are satisfied, as proved in the lemma below, the normal form 3.1 is obtained from the above expression by changing to the new coordinates u, v where u is aligned with the singularity line ($v \propto S + k\Phi$) and v is aligned with its image. From the calculation, it follows that the quantity ξ is $(b_{11} + b_{22})$, namely the sum of eigenvalues of the linearization of the regular part $B^{(r)}$ of $\bar{B}$ at P. Since the product of the eigenvalues equals to 1 and since they do not lie on the unit circle, it follows that

$$|\xi| > 2, \quad (3.3)$$

as indicated in the Proposition. □.

Lemma 3.1: *The coefficients b_{ij} in (3.2) obey the inequalities:*

$$(b_{12} - b_{11}k)(b_{22} - b_{21}k) > 0 \quad (3.4)$$

$$|b_{12}| < |b_{11}|k \quad (3.5)$$

$$|b_{22}| < |b_{21}|k. \quad (3.6)$$

Proof: Since the image $\bar{B}\Sigma$ of the singularity line $S + k\Phi = 0$ must lie in the unstable cone $\bar{S} \cdot \bar{\Phi} > 0$, it follows from 3.2 that the first inequality $(b_{12} - b_{11}k)(b_{22} - b_{21}k) > 0$ holds. Moreover, it is geometrically evident that for a small piece l of a straight line through P which lies in the unstable cone, i.e., for which the increase of s is followed with the increase of ϕ (see figure 3.2 - imagine a line going through P_0'', P_0, P_0') the image of $l \cap U_r$ by B_0 and the image of $l \cap U_s$ by $B_1 B_0$ lie both to one side of the point P_2 (or s_2 when projected to the configuration plane). In other words, these images belong both to the same half of the unstable cone of P_2 corresponding to a definite sign of $(s - s_2)$. Since the linearization of each of the maps B_i preserves the decomposition into the stable and unstable cones, it follows that the image of l by $\bar{B}$ is a folded line with the vertex at P which divides $\bar{B}l$ in two parts belonging both to the same half of the unstable cone of P; i.e., $\bar{S}$ and $\bar{\Phi}$ have the same sign on $\bar{B}(l \cap U_r)$ and $\bar{B}(l \cap U_s)$. By (3.2), it is equivalent to the condition that the sign of $(b_{12} - b_{11}k)$ is opposite to the sign of b_{12} and b_{11} and the sign of $(b_{22} - b_{21}k)$ is opposite to the sign of b_{22} and b_{21} (recall that all b_{ij} are of same sign). Thus, the second and third inequalities $|b_{12}| < |b_{11}|k$ and $|b_{22}| < |b_{21}|k$ hold. $\quad\Box$.

Now, embed the billiard under consideration in a one parameter family of scattering billiards $b_t(\cdot\,; \gamma)$ for which all arcs depend smoothly on the parameter γ, while the corner points are held fixed; we suppose that the billiard with the simple singular periodic orbit L is realized at $\gamma = 0$. The regular part $B^{(r)}$ of the first return map of U depends smoothly on γ, hence its hyperbolic fixed point $P_\gamma^{(r)}$ is also a smooth function of γ. The same is valid for the position of the singularity line Σ_γ. For a *general* family of billiards, the parameterization by γ may be chosen so that the distance between $P_\gamma^{(r)}$ and Σ_γ is proportional to γ (it is true if, for instance, one changes the billiard boundary locally, near the tangency point s_1 only: such a perturbation moves the singularity line but the map $B^{(r)}$ and the position of its fixed point remain unchanged). Assume, with no loss of generality, that $P_\gamma^{(r)} \in U_r$ for $\gamma > 0$ and that $P_\gamma^{(r)} \in U_s$ for $\gamma < 0$. Therefore, by the definition of $B^{(r)}$, its fixed point is a fixed point of $\bar{B}$ for $\gamma > 0$, and its fixed point is imaginary when $\gamma < 0$.

Thus, for such a family of billiards, the normal form (3.1) of the first return map $\bar{B}$ is now rewritten as

$$\begin{cases} \bar{u} = v \\ \bar{v} = \xi(\gamma + v - \sqrt{\max(v,0)}) - u + \ldots \end{cases} \qquad (3.7)$$

In this form, the map $\bar{B}_\gamma$ looks similar to the well-known Hénon map but it has another type of nonlinearity. In fact we show below: **Proposition 3.2** *Consider the map (3.7). For a small fixed neighborhood U of the origin, let Ω_γ be the set of all orbits of $\bar{B}_\gamma$ which never leave U. Then there exist $\gamma^\pm$ values such that $\Omega_\gamma = \emptyset$ for $\gamma < \gamma^- < 0$, and if $\gamma > \gamma^+ > 0$ and small, then Ω_γ is in one-to-one correspondence with the set of all sequences composed of two symbols (r, s): "r" corresponds to entering U_r and "s" corresponds to entering U_s.*

Proof: Indeed, take a small $\delta > 0$ and let the neighborhood U be a rectangle $\{-\delta < u < \kappa\delta, \ -\delta < v < \kappa\delta\}$ where $\kappa = \frac{1}{2}(\frac{1}{2}|\xi| - 1) > 0$ (recall that $|\xi| > 2$). Let $\gamma^+ = (\frac{1}{2} - \frac{1}{\xi})\delta > 0$ and $\gamma^- = -\frac{2}{|\xi|}\delta$. Then, for sufficiently small δ, one may check that for the given choice of U the map (3.7) takes the horizontal boundaries of U (marked 1 and 3 in figure 3.3) on a finite distance of U for all $\gamma \in [\gamma^-, \gamma^+]$. The images of the vertical boundaries 2 and 4 which intersect the singularity line, fold as indicated in figure 3.3: the

segments 2a,4a are mapped to 2a',4a' and the segments 2b,4b are mapped to
2b',4b'. The folded lines 2',4' may intersect U but they lie on a finite distance
of their preimages (the boundaries 2 and 4) for all $\gamma \in [\gamma^-, \gamma^+]$. Thus, the
image of U by $\bar{B}_\gamma$ has a specific shape of a sharp horseshoe. Changing γ shifts
the horseshoe along the v-axis, so at $\gamma = \gamma^+$ the intersection of the horseshoe
with U consists of two distinct connected components (figure 3.3b). On each
component the map $\bar{B}_\gamma$ is smooth and hyperbolic. The statement regarding
the one-to-one correspondence to Bernoulli shift on two symbols follows as
in the standard construction of the horseshoe map [23, 18]. *In particular, it
implies that each of the two components has a hyperbolic fixed point.* On the
other hand, at $\gamma = \gamma^-$ the intersection of $\bar{B}_\gamma U$ with U is empty (figure 3.3c)
and no fixed points may exist in U. □.

Notice the following three important conclusions from the proof of the
above proposition: first that there exist $\gamma^\pm$ values such that for γ^+ two hyperbolic fixed points exist and for γ^- no fixed points exist in the square region
U near the intersection of the singularity line with its image. Second that $\gamma^\pm$
may be chosen arbitrarily small (by taking smaller U). Third, no fixed points
can pass through the boundary of U as γ varies from γ^- to γ^+ because the
image of the horizontal boundaries of U never intersects the boundary of U
and the image of the vertical boundaries U may intersect only the horizontal
parts of the boundary.

Now, take a *two-parameter* family of Hamiltonians $H(\cdot; \epsilon, \gamma)$ which approach the family of billiard flows $b_t(\cdot; \gamma)$ as $\epsilon \to 0$, in the sense that conditions **I-V** are satisfied uniformly with respect to γ. Note that for the billiard
flow, the structure of the Poincaré map of an arbitrary small cross-section
ω through an *inner* point on the simple singular periodic orbit L is absolutely the same as described above (because the map $\bar{B}$ is a particular case
of the Poincaré map, corresponding to the cross-section made of collision
points, and different Poincaré maps are smoothly conjugate near L; see section 2.1.1). Due to the C^0-closeness result of theorem 3, it follows that for ϵ
sufficiently small the corresponding Poincaré map $\Pi_{\epsilon\gamma}$ for the Hamiltonian
system transforms a rectangle $U' \subset \omega$ (analogous to the rectangle U) to a
horseshoe shape (which is now smooth because the Hamiltonian system is
smooth at all $\epsilon > 0$). At $\gamma = \gamma^-$ the intersection $\Pi_{\epsilon\gamma} U' \cap U'$ is empty for
small ϵ whence $\Pi_{\epsilon\gamma}$- has no fixed points in U'. Moreover, no fixed points
can pass through the boundary of U' as γ varies from γ^- to γ^+ because the
fixed points of the first return billiard map stay on a finite distance from the
boundary of U' for all $\gamma \in [\gamma^-, \gamma^+]$.

The two fixed points of the Poincaré map of the billiard flow which exist
at $\gamma = \gamma^+$ are hyperbolic and do not belong to the singularity line. Thus, by
the corollary 1 to theorem 3, each of these hyperbolic fixed points exists for
the map $\Pi_{\epsilon\gamma^+}$ at all sufficiently small ϵ. Now, fixing any ϵ small enough, a
fixed point of $\Pi_{\epsilon\gamma^+}$ changes continuously as γ decreases, until it merges with
some other fixed point (as we argued, the fixed point must disappear before

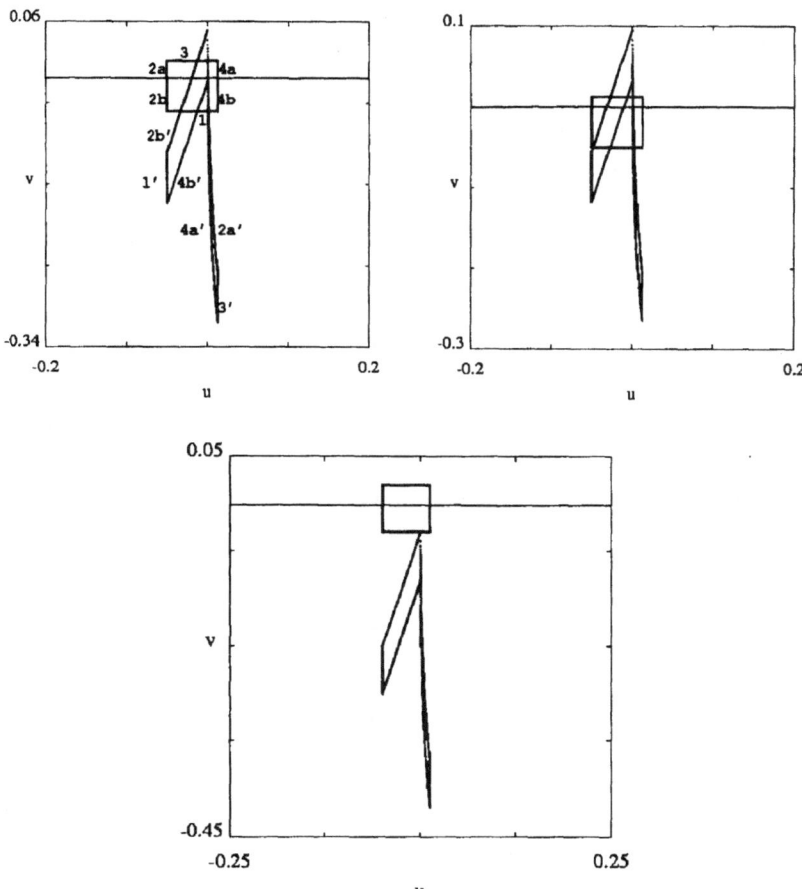

Fig. 3.3. Sharp horseshoe bifurcation near singular periodic orbit

One iterate of the indicated box by the truncation
$(\bar{u} = v, \bar{v} = \xi(\gamma + v - \sqrt{\max(v,0)}) - u)$ of the normal form (3.7)
in all figures $\xi = 3$, $\delta = 0.05$.
a) $\gamma = 0$ b) $\gamma = 0.015 > \gamma^+ = 1/160$ c) $\gamma = \gamma^- = -1/30$.
$\bullet$ - period n point.

$\gamma = \gamma^-$ and it can not leave U' via crossing the boundary). In a *general*
family of sufficiently smooth Hamiltonian systems, one of the merging fixed
points is necessary saddle and another is elliptic. Thus, we have established
that
*generically, for each ϵ small enough, there exists an interval of values of γ
for which the smooth Hamiltonian system possesses an elliptic periodic orbit.*

Without genericity assumptions, we may conclude the following. **Theorem**

4: *If a scattering billiard has a simple singular periodic orbit L, then there exists a one-parameter family of smooth Hamiltonian flows $h_t(\epsilon)$ limiting to the billiard flow as $\epsilon \to 0$ (i.e. satisfying conditions* **I-V***) and for which there exists a sequence of intervals of ϵ values converging to 0 on which elliptic periodic orbits L_ϵ exist in the energy level of L. These elliptic periodic orbits limit to the singular periodic orbit as $\epsilon \to 0$.*

3.2 Singular homoclinic orbits

Consider a *non-singular* hyperbolic periodic orbit L_0 of the billiard flow. Suppose, its stable and unstable manifolds intersect along some orbit Γ. This is a *homoclinic* orbit; i.e., it asymptotes L_0 exponentially as $t \to \pm\infty$. Assume that Γ is *simple singular* which means that its trajectory has one point of tangency with the billiard's boundary (see figure 1.1 b).

Let $P(s, \phi)$ and $\bar{P}(\bar{s}, \bar{\phi})$ be collision points on Γ: P is the last before the tangency and $\bar{P}$ is the first after the tangency. By definition, $\bar{P} = B^2 P$ where B is the billiard map. Consider, in the (s, ϕ) plane, the local segment W^u of the unstable manifold of L_0 to which P belongs. Since the tangent to W^u at P belongs to the unstable cone, it must intersect the singularity line transversely at P. Thus, as explained in the proof of lemma 3.1, the image of W^u in a neighborhood of $\bar{P}$ under the billiard map folds with a sharp square root singularity at $\bar{P}$, see figure 3.4. Now, the point $\bar{P}$ belongs to the stable manifold as well. Since the tangent to W^s belongs to the stable cone, it follows that the folded image of W^u lies to one side of W^s, so a sharp homoclinic tangency is created at $\bar{P}$, as shown in figure 3.4.

In a general family of scattering billiards (as in section 3.1), two transverse homoclinic intersections appear at $\gamma > 0$ and none at $\gamma < 0$. For the corresponding two-parameter Hamiltonian family, arguments analogous to those in the proof of theorem 4 show that
generically, for any ϵ sufficiently small there exists $\gamma^*(\epsilon)$ for which a quadratic homoclinic tangency occurs.

Recall that the occurrence of homoclinic tangencies is a well-known mechanism for the creation of elliptic islands [20]. Thus we have established:

Theorem 5: *If a scattering billiard has a simple singular homoclinic orbit Γ, then there exists a one-parameter family of smooth Hamiltonian flows $h_t(\epsilon)$ satisfying conditions* **I-V***, which limits to the billiard flow as $\epsilon \to 0$ and for which there exist a sequence of intervals of ϵ values converging to zero for which elliptic periodic orbits exist in the energy level of Γ.*

The period of the elliptic periodic orbits mentioned in Theorem 5 goes to infinity as $\epsilon \to 0$. In fact, in the two-parameter family of smooth Hamiltonians elliptic periodic orbits of bounded period limit, as $\epsilon \to 0$, to singular periodic orbits corresponding to $\gamma \neq 0$. Thus Theorems 5 and 4 are very much related. Indeed, like the appearance of stable periodic orbits near a homoclinic tangency is proved in smooth situation (see [12, 20, 13]), one may

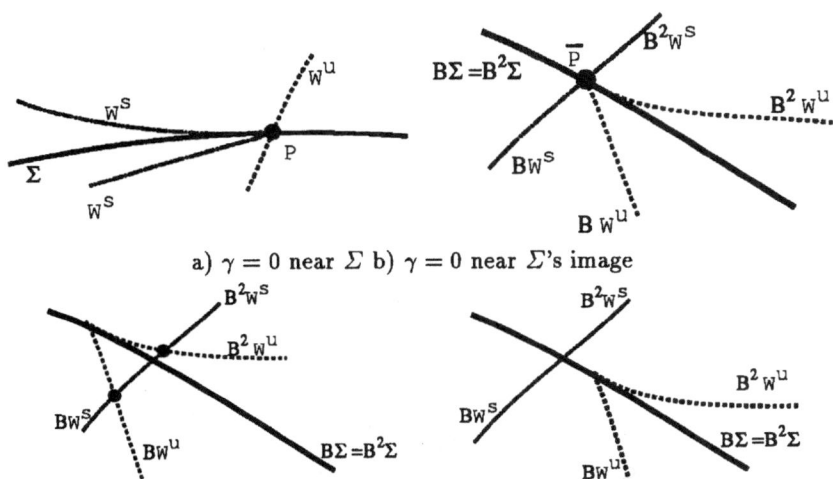

Fig. 3.4. Bifurcation of singular homoclinic orbit

c) $\gamma > 0$ d) $\gamma < 0$, near Σ's image
● - homoclinic points.

show that
in a general family of scattering billiards having a sharp homoclinic tangency at $\gamma = 0$ there is a sequence of values of γ accumulating at $\gamma = 0$ for which singular periodic orbits exist.
Now the reference to theorem 4 gives another proof of theorem 5.

3.3 On the genericity of the elliptic islands creation

It is well known [17, 2, 3] that for scattering billiards the hyperbolic non-singular periodic orbits are dense in the phase space. The stable/unstable manifolds of such orbits cover the phase space densely and the orbits of their homoclinic intersections also form a dense set.

It follows that the periodic orbits and the homoclinic orbits get arbitrarily close to the singularity set. It seems thus intuitively clear that for any scattering billiard very small smooth perturbations may be applied to place a specific periodic orbit or a specific homoclinic orbit exactly on the singularity line, so that Theorem 4 and 5 may be applied. Proving these intuitive statements turns out to be quite a delicate issue, thus we formulate these as conjectures:

Conjecture 1: *Any scattering billiard may be slightly perturbed to a scattering billiard for which a singular (tangent) periodic orbit exists.*

Conjecture 2: *Any scattering billiard may be slightly perturbed to a scattering billiard for which there exists a non-singular hyperbolic periodic orbit which has a singular homoclinic orbit.*

3.4 Numerically produced singular homoclinic orbits

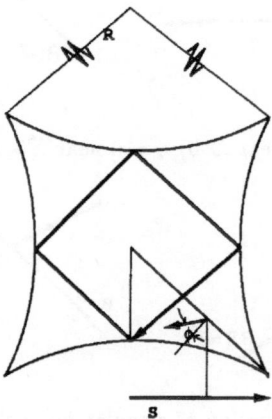

Fig. 3.5. Billiard between four disks

To examine the appearance of singular homoclinic orbits we consider the
billiard in a domain bounded by four symmetrical circles

$$x^2 + (y \pm \frac{1}{\gamma})^2 = R^2; \ (x \pm \frac{1}{\gamma})^2 + y^2 = R^2$$

where $R^2 = 1 + (1 - \frac{1}{\gamma})^2$. The quantity γ (which is, approximately, the
curvature of the circles) serves as the free parameter for unfolding the singularity. We find explicitly the corresponding billiard map, and using DSTOOL
package[10], we find numerically hyperbolic periodic orbits of this mapping
and their stable and unstable manifolds. The billiard map is found on the
fundamental domain of the billiard - a triangular region cut by an arc as
shown in the figure 3.5. We find the return map to the slanted side of
the triangle, which is parameterized by s, the horizontal coordinate, and
by ϕ, the outgoing angle to the normal vector $(-1, -1)$, see figure 3.5. We
choose an arbitrary value of γ and the simplest hyperbolic non-singular periodic orbit, as shown in the figure (the fixed point of the return map to
the slanted side of the reduced domain). Then, we construct the stable and
unstable manifolds for this periodic orbit. We examine how these manifolds
vary by small variation of γ, until we find a value of γ for which singular homoclinic orbit appears. The success (see figure 3.6 and figure 3.7) of
the very crude search for such a delicate phenomena, near every γ value
we have chosen, supports conjecture 2 regarding the density of systems for
which such orbits exist. In fact we have found, by such a search near $\gamma_i = i * 0.05, i = 1, \ldots, 10$, eleven sharp homoclinics to this specific periodic orbit

(at $\gamma = 0.0837, 0.10165, 0.1018, 0.153, 0.2077, 0.2552, 0.29245, 0.3329, 0.3832,$ $0.4143, 0.4692$).

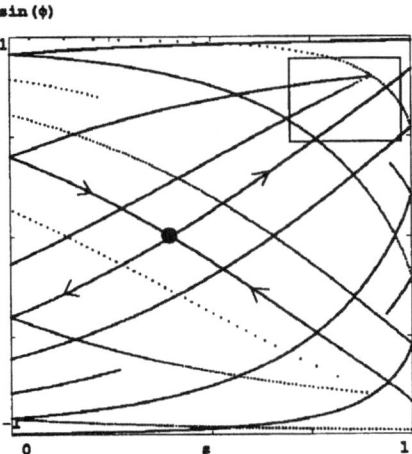

sin (φ)

Fig. 3.6. Numerically produced sharp homoclinics

4 Conclusions

There are two main results in this paper; First, we have found sufficient conditions for establishing that a family of smooth Hamiltonian flows limits to the singular billiard flow (see Theorem 1, 2 and 3). These conditions are fulfilled by smooth Hamiltonians with potentials approaching a step function in almost arbitrary way (see section 2.3); they fail, nevertheless, when the potentials are highly oscillatory (i.e., condition IV or V fails).

Second, we have established that if a scattering billiard (we use the particular hyperbolic structure associated with such billiards) has a singular periodic orbit or a singular homoclinic orbit, then there exist arbitrarily close to it smooth Hamiltonian flows which possess elliptic islands, hence these are not ergodic (Theorem 4 and 5). Finally, we have conjectured, and have provided numerical support to these conjectures, that in fact scattering billiards with singular periodic orbits and singular homoclinic orbits are dense among scattering billiards (conjectures 1 and 2 of section 3.3). If these conjectures are correct, then theorems 4 and 5 will imply that arbitrarily close to any scattering billiard there exists a family of non-ergodic smooth Hamiltonian flows.

Such statements imply that *ergodicity and mixing results concerning two-dimensional non-smooth systems cannot be directly applied to the smooth dy-*

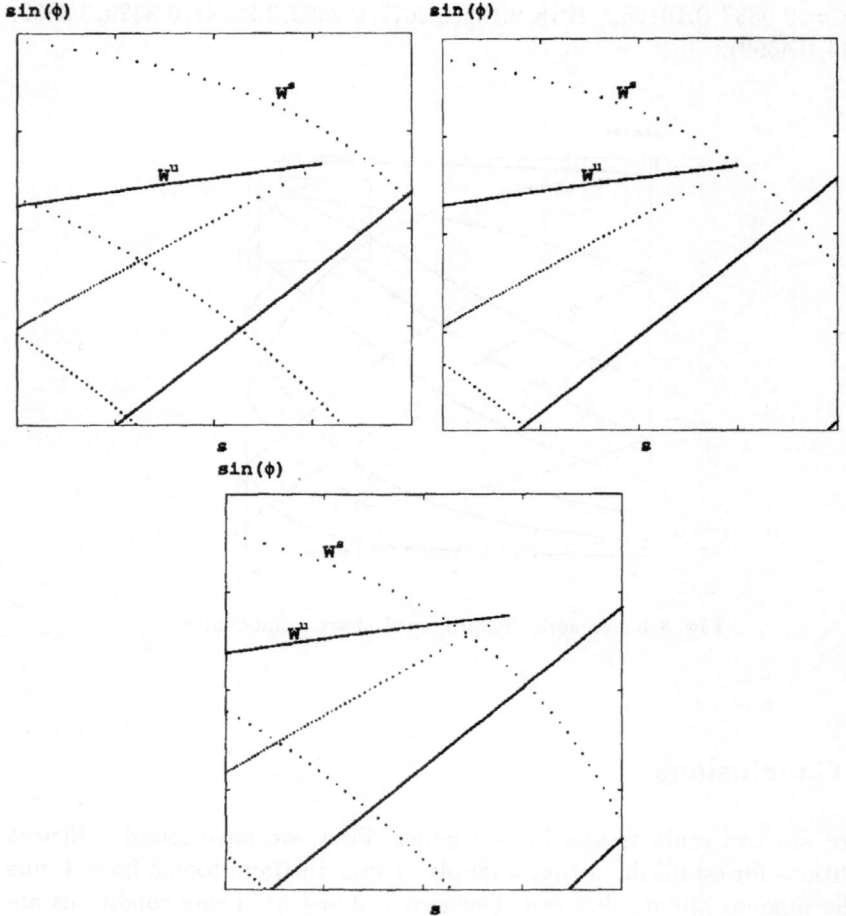

Fig. 3.7. Magnification near numerically produced sharp homoclinics

a) $\gamma = 0.28$ b) $\gamma = 0.29245$ c) $\gamma = 0.31$

namics they model. Whether the same holds for higher dimensional systems, e.g. three-dimensional billiards or multi-particle billiards, is yet to be studied.

On the other hand, eventhough stability islands may appear in smooth billiard-like problems, the size of an individual island is expected to be very small. Thus, with no doubt, while the smooth flow may be non-ergodic, it will "seem" to be ergodic for a very long time; Statistics (e.g. correlation function) which are based upon finite time realizations may appear to behave as in the scattering billiards (e.g. fall off quasi-exponentially [5]). Whether longer realizations will reveal very different statistical properties, depends on the number of elliptic islands, the total area they cover in the phase space and on the "typical" period of the islands. Thus, estimates of the islands sizes, their periods, and of the real potential steepness (the "physical ϵ") are necessary to supply estimates on the time scale for which the mixing property will appear to hold.

We may try to estimate the periodicity of the elliptic periodic orbits of smooth flows approaching generic scattering billiards, by very naive arguments. Indeed, since stable periodic orbits are generated from singular periodic orbits of the billiard, one may expect (if conjecture 1 is correct) that the least period of stable periodic orbits of a smooth Hamiltonian system which is ϵ-close to the billiard is of the order of the Poincaré return time to an ϵ-neighborhood of the singularity surface for the billiard flow. Notice that the billiard flow is a hyperbolic system; therefore, the return time in the billiard and, correspondingly, the typical period of the stable periodic motions in its smooth approximation must, essentially, be *logarithmic* in ϵ and not of a power-law type. Namely, very small ϵ values, corresponding to very steep potentials, may still produce stability islands which are observable on physical time-scales.

Acknowledgement

This work has been supported by MINERVA Foundation, Munich/Germany. We greatly acknowledge discussions with L. Bunimovich, E. Gutkin, A. Kirillov, J. Marsden, L.P. Shilnikov and U. Smilansky.

A   An example of smooth Hamiltonian approximation of the circular billiard with non-reflecting trajectories

Consider the Hamiltonian

$$H = \frac{1}{2}p_x^2 + \frac{1}{2}p_y^2 + \epsilon V(1 - x^2 - y^2) \tag{A.1}$$

where the potential V is given by

$$V(Q) = \frac{1}{Q}\exp\left(-AQ + \int_{1/Q}^{\infty}\frac{\sin z}{z}dz\right), \quad Q > 0, \tag{A.2}$$

with some positive constant A. The potential is of the form $\epsilon V(Q)$ where the pattern function is defined by $Q(x, y; \epsilon) = 1 - x^2 - y^2$ for all ϵ. As $\epsilon \to 0$, the above Hamiltonian satisfies conditions **I** - **III**, which garuantee that near the boundary, $x^2 + y^2 = r^2 = 1$, the correct elastic reflection rules are approached. Thus one may expect that the motion described by (A.1,A.2) limits to the billiard in the unit circle. We show that this is not the case; there exist initial conditions inside the unit circle for which the orbits of the Hamiltonian system (A.1,A.2) stick to the circle boundary for infinitely long time at arbitrarily small ϵ. Notice that condition **IV** is violated.

The specific choice of V is not too important. Essentially, we use that

$$\liminf_{u \to 0} |\frac{V'}{V}| = A < \infty \quad \text{whereas} \quad \limsup_{u \to 0} |\frac{V'}{V}| = \infty. \quad (A.3)$$

Hamiltonian (A.1) is rotationally invariant, thus the particle's angular momentum Ω:

$$\Omega = r^2 \dot{\theta} \quad (A.4)$$

is preserved. It follows that the system is integrable and may be easily analyzed. In polar coordinates ($x = r \cos \theta, y = r \sin \theta$) the equations of motion are of the form

$$\ddot{r} = r(\dot{\theta}^2 + 2\epsilon V'(1 - r^2)) = r(\frac{\Omega^2}{r^4} + 2\epsilon V'(1 - r^2))$$
$$\ddot{\theta} = -2\frac{\dot{r}\dot{\theta}}{r}. \quad (A.5)$$

The radial motion decouples, and is governed by the Hamiltonian:

$$H = \frac{1}{2}\dot{r}^2 + \frac{1}{2}\frac{\Omega^2}{r^2} + \epsilon V(1 - r^2) = \frac{1}{2}\dot{r}^2 + V_{eff}(r; \Omega, \epsilon) \quad (A.6)$$

The maximal polar radius, r^*, reached by an initial condition $(r_0, \dot{r}_0)$ with $\Omega \neq 0$ is found from:

$$V_{eff}(r^*; \Omega, \epsilon) = \epsilon V(1 - r^{*2}) + \frac{1}{2}\frac{\Omega^2}{r^{*2}} = \frac{1}{2}\dot{r}_0^2 + \frac{1}{2}\frac{\Omega^2}{r_0^2} + \epsilon V(1 - r_0^2) = h \quad (A.7)$$

As $\epsilon \to 0$, the value of r^* tends to 1. The time spent by the orbit near $r = r^*$ is given by

$$2 \int_{r^*}^{r} \frac{ds}{\sqrt{h - V_{eff}(s; \Omega, \epsilon)}}, \quad (A.8)$$

thus it is infinite if:

$$V'_{eff}(r^*; \Omega, \epsilon) = -r^*(\frac{\Omega^2}{r^{*4}} + 2\epsilon V'(1 - r^{*2})) = 0 \quad (A.9)$$

(i.e. if $\ddot{r} = 0$ at $r = r^*$). It follows, that if there exist $(r^*(r_0, \dot{r}_0, \Omega; \epsilon) > r_0, \epsilon)$ solving (A.7) and (A.9) simultaneously, then, the phase point will move for infinitely long time close to the unit circle with non-zero angular velocity ($\lim_{t \to \infty} \dot{\theta} = (r_0/r^*)^2 \dot{\theta}_0$).

Next, we show that such a solution exist for many initial condition and for a sequence of $\epsilon \to 0$ values. First, since $V(Q)$ is a monotonic function, for any $r^* > r_0$ one may find ϵ such that (A.7) is satisfied; moreover, $\epsilon \to 0$ as $r^* \to 1$. Resolving (A.7) with respect to ϵ and plugging the result in (A.9) we get

$$r^{*2} \frac{|V'(1 - r^{*2})|}{V(1 - r^{*2}) - V(1 - r_0^2)} = \frac{1}{(r^*/r_0)^2(\frac{\dot{r}_0}{r_0 \dot{\theta}_0})^2 + (r^*/r_0)^2 - 1}. \quad (A.10)$$

According to (A.3), this equation is solved by an infinite number of values of r^* (with their corresponding $\epsilon(r^*; r_0, \dot{r}_0, \dot{\theta}_0)$) limiting to $r^* = 1$, provided

$$(\frac{\dot{r}_0}{r_0 \dot{\theta}_0})^2 + 1 < r_0^2(1 + \frac{1}{A}). \tag{A.11}$$

Clearly, for any given $A > 0$, and for any $r_0 < \sqrt{\frac{A}{1+A}} < 1$ such initial conditions exist. Summarizing: if the initial conditions satisfy (A.11), then there exist an infinite number of values of ϵ, approaching $\epsilon = 0$, for which the orbit sticks to the boundary for infinitly long time.

References

1. L. Bunimovich and Y. Sinai. The fundamental theorem of the theory of scattering billiards. *(Russian) Mat. Sb. (N.S.)*, 90(132):415–431, 1973.
2. L. Bunimovich and Y. Sinai. Markov partitions for dispersed billiards. *Comm. Math. Phys.*, 78(2):247–280, 1980/81.
3. L. Bunimovich, Y. Sinai, and N. Chernov. Markov partitions for two-dimensional hyperbolic billiards. *Russian Math. Surveys*, 45(3):105–152, 1990. Translation of Uspekhi Mat. Nauk, 45, 3(273), 97–134, 221.
4. L. Bunimovich, Y. Sinai, and N. Chernov. Statistical properties of two-dimensional hyperbolic billiards. *Uspekhi Mat. Nauk*, 46(4(280)):43–92, 192, 1991. In Russian. Translation in *Russian Math. Surveys* **46**(4) (1991) 47–106.
5. L. A. Bunimovich. Decay of correlations in dynamical systems with chaotic behavior. *Zh. Eksp. Teor. Fiz*, 89:1452–1471, 1985. In Russian. Translation in Sov Phys. JETP **62** (4), 842-852.
6. P. Constantin, E. Grossman, and M. Mungan. Inelastic collisions of three particles on a line as a two-dimensional billiard. *Physica D*, 83(4):409–420, 1995.
7. I. P. Cornfeld, S. V. Fomin, and Y. G. Sinai. *Ergodic theory*. Number 245 in Fundamental Principles of Mathematical Scienc. Springer-Verlag, New York-Berlin, 1982. Translated from the Russian by A. B. Sosinskiĭ.
8. A. Delshams and R. Ramírez-Ros. Poincaré-melnikov-arnold method for analytic planar maps. *Nonlinearity*, 9:1–26, 1996.
9. V. Donnay. Elliptic islands in generalized sinai billiards. *Ergod. Th. & Dynam. Sys.*, 16:975–1010, 1996.
10. DSTOOLS. Computer program. Cornell university, center of applied mathematics, Ithaca, NY 14853.
11. G. Gallavotti and D. Ornstein. Billiards and Bernoulli schemes. *Comm. Math. Phys.*, 38:83–101, 1974.
12. N. Gavrilov and L. Shilnikov. On three dimensional dynamical systems close to systems with a structurally unstable homoclinic curve i. *Math. USSR Sb.*, 88(4):467–485, 1972.
13. S. V. Gonchenko, L. P. Shilnikov, and D. V. Turaev. Dynamical phenomena in systems with structurally unstable Poincaré homoclinic orbits. *Chaos*, 6(1):15–31, 1996.
14. M. Gutzwiller. *Chaos in Classical and Quantum mechanic*. Springer-Verlag, New York, NY, 1990.

15. K. Hansen. Bifurcations and complete chaos for the diamagnetic kepler problem. *Phys. Rev. E.*, 51(3):1838–1844, 1995.

16. M. Jakobson. Absolutely continuous invariant measures for one parameter families of one-dimensional maps. *Comm. Math. Phys.*, 81:39–88, 1981.

17. Katok, A.B. and Strelcyn, J.M. and Ledrappier, F. and F. Przytycki. *Invariant Manifolds, Entropy and Billiards; Smooth Maps with Singularities*, volume 1222. Springer-Verlag, 1986.

18. L.P.Shilnikov. On a Poincaré-Birkhoff problem. *Math. USSR Sbornik*, 74, 1967.

19. J. Marsden. Generalized Hamiltonian mechanics; a mathematical exposition of non-smooth dynamical systems and classical Hamiltonian mechanic. *Arch. for Rational Mech. and Anal.*, 28(5):323–361, 1968.

20. S. Newhouse. Quasi-elliptic periodic points in conservative dynamical systems. *Amer. J. Of Math.*, 99(5):1061–1087, 1977.

21. Y. Sinai. Dynamical systems with elastic reflections: Ergodic properties of scattering billiards. *Russian Math. Sur.*, 25(1):137–189, 1970.

22. Y. Sinai and N. Chernov. Ergodic properties of some systems of two-dimensional disks and three-dimensional balls. *Uspekhi Mat. Nauk*, 42(3(255)):153–174, 256, 1987. In Russian.

23. S. Smale. Diffeomorphisms with many periodic points. In C. Cairns, editor, *Differential and combinatorial topology*, pages 63–80. Princeton University Press, Princeton, 1963.

24. D. Szász. Boltzmann's ergodic hypothesis, a conjecture for centuries? *Studia Sci. Math. Hungar.*, 31(1–3):299–322, 1996.

25. D. Turaev and V. Rom-Kedar. Islands appearing in near-ergodic flows. *Nonlinearity*, 11(3):575–600, 1998. to appear.

26. M. Wojtkowski. Principles for the design of billiards with nonvanishing lyapunov exponents. *Comm. Math. Phys.*, 105(3):391–414, 1986.

27. G. Zaslavsky and M. Edelman. Maxwell's demon as a dynamical model. *Phys. Rev. E*, 56(5):5310–5320, 1997.

Strong Variation of Global-Transport Properties in Chaotic Ensembles

Itzhack Dana and Tamir Horesh

Department of Physics, Bar-Ilan University, Ramat-Gan 52900, Israel

Abstract. Chaotic transport is studied for Hamiltonians H in which one coordinate, say q, is *cyclic* (i.e., it does not appear in H), leading to the conservation of the conjugate coordinate ("momentum" p). It is assumed that the dynamics depends *nontrivially* on the "parameter" p in H. As a consequence, one expects to observe a variation of the global-transport properties, both normal and anomalous, in a generic chaotic ensemble that exhibits all values of p. By considering the realistic model system of charged particles interacting with an electrostatic wave-packet in a uniform magnetic field, it is shown that this variation can be actually quite strong. This finding may have applications to "filtering" sub-ensembles with well-defined values of p.

Hamiltonian chaos (see, e.g., MacKay and Meiss 1987 and references therein) is a unique phenomenon in that it generically appears interleaved with or-dered/stable motions on all scales of phase space (Meiss 1986; Umberger and Farmer 1985), leading to long-time correlations (Karney 1983; Meiss and Ott 1986) and quasiregularity (Dana 1993) in the chaotic motion. A fundamental question is then to what extent the transport due to the deterministic chaos resembles that associated with a truly probabilistic random process, such as Brownian motion (Chirikov 1979). This question has been investigated exten-sively during the last two decades, mainly for systems which can be described by area-preserving maps. A globally diffusive transport, $\langle R^2 \rangle = 2Dt$ ($\langle \ \rangle$ de-notes initial-ensemble average, $\mathbf{R}$ is some radius vector in the phase space, and D is the diffusion coefficient), is often observed numerically (see, e.g., Chirikov 1979; Dana and Fishman 1985) but occurs rigorously only in very special cases (Cary and Meiss 1981). The self-similar islands-around-islands hierarchy in phase space [Meiss 1986; Zaslavsky et al. (1997)] should be responsible to the anoma-lous global diffusion, $\langle R^2 \rangle \propto t^\mu$ ($0 < \mu < 2$) [Shlesinger et al. 1993; Zumofen and Klafter 1994; Zaslavsky et al. (1997); Afraimovich and Zaslavsky (1997)], which may be described by Lévy random-walk processes (Shlesinger et al. 1993; Zumofen and Klafter 1994).

Because of the complex phase-space structure of a generic Hamiltonian system, chaotic transport is usually quite inhomogeneous *locally* (Karney 1983; MacKay et al. 1984; Dana et al. 1989; Afanasiev et al. 1991). In this paper, we show that one can also observe a high inhomogeneity in the *global*-transport properties due to the following simple scenario. Consider a Hamiltonian H in which one coordinate, say q, is *cyclic*, i.e., it does not appear in H. The conjugate coordinate ("momentum"), p, is then a constant of the motion and appears in H as a "parameter", $H = H(\mathbf{R}, t; p)$. Here $\mathbf{R}$ denotes all the other phase-space coordinates and, for the sake of generality, a dependence on time t is included. Our crucial assumption is that the dynamics in the $\mathbf{R}$ phase space depends *nontrivially* on the "parameter" p. Now, since p is actually a coordinate, a generic, realistic ensemble of particles will exhibit all values of p. Such an ensemble can be divided into sub-ensembles characterized by well-defined values of p. The assumption above then implies that different sub-ensembles will be characterized by different global-transport properties, e.g., a normal-diffusion coefficient $D(p)$ or an anomalous-diffusion exponent $\mu(p)$. As a result, a variation of these properties throughout the entire ensemble will be observed.

We show here that this variation can be actually quite strong by considering the realistic model system of charged particles interacting with an electrostatic wave-packet in a uniform magnetic field. This system is described by the Hamiltonian

$$H = \Pi^2/(2M) + KV(kx, t) , \tag{1}$$

where $\Pi = \mathbf{p} - e\mathbf{A}/c$ is the kinetic momentum of a particle with charge e and mass M in a uniform magnetic field $\mathbf{B}$ (along the z-axis), K is a parameter, $\mathbf{k}$ is the wave-vector (in the x-direction), and V is a general function describing the electrostatic wave-packet. This function is periodic in both kx (with period 2π) and time t (with period T). Without loss of generality, the values of M and k will be both set to 1 from now on.

To see that (1) is a Hamiltonian of the kind described above, let us express it using the natural degrees of freedom in a magnetic field. These are given by the conjugate pairs (x_c, y_c) (coordinates of the center of a cyclotron orbit) and (Π_x, Π_y), see Johnson and Lippmann (1949). Defining $u = \Pi_x/|\omega|$, $v = \Pi_y/\omega$, where $\omega = eB/c$ is the cyclotron frequency, and using the relation $x_c = x + \Pi_y/\omega = x + v$ (easily derivable from simple geometry), (1) can be rewritten as follows

$$H = \omega^2(u^2 + v^2)/2 + KV(x_c - v, t) . \tag{2}$$

It is now clear that y_c is cyclic in H, so that it corresponds to the coordinate q above. The conserved "momentum" p is then x_c.

In what follows, we shall assume the simple wave-packet

$$V(x, t) = -\cos x \sum_{s=-\infty}^{\infty} \delta(t - sT) ,$$

reducing (2) to the Hamiltonian of a kicked harmonic oscillator. The latter system has been investigated extensively by Zaslavsky et al. (1986) (see the review article by Zaslavsky 1991) who assumed, however, the very specific value $x_c = 0$ in (2). These investigations have led to the discovery of the well-known properties of this system. Since the harmonic oscillator is degenerate (linear in the action), the nonlinear perturbation in (2) is strong (in the sense of KAM theory) for all values of K, especially under resonance conditions, $\omega T = 2\pi m/n$ (m and n are coprime integers). One then expects, on the basis of general arguments, that unbounded chaotic motion of (u, v) should exist for arbitrarily small values of K in the resonance case. This motion is observed to take place diffusively on a "stochastic web" [see Fig. 1(a)], analogous in some aspects to the Arnol'd web. For $n = 3$, 4, 6, the web has crystalline symmetry (triangular, square, hexagonal), while for all other values of $n > 4$ it has quasicrystalline symmetry.

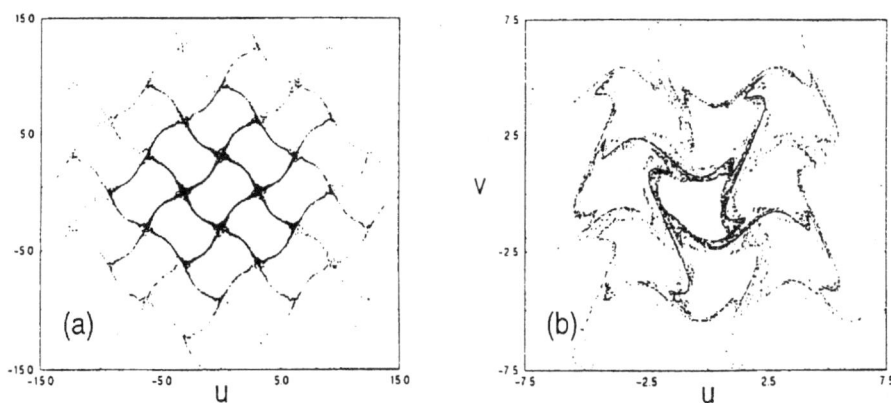

Fig. 1. Portions of the stochastic webs for $m/n = 1/4$, $K = 1.4$, and (a) $x_c = 0$, (b) $x_c = \pi/2$. Each plot contains 40 000 points of chaotic orbits, generated by iterating 100 times an ensemble of 20×20 initial conditions near the origin with the map corresponding to (2). Notice that the diffusion rate in case (b) is slower than in case (a). Without loss of generality, the value of ω in (2) is set to 1 in this paper.

The need to consider general values of x_c has been pointed out only recently by Dana and Amit (1995), who developed a general formalism for calculating the normal-diffusion coefficient $D(x_c)$ for Hamiltonian (2) as a function of x_c. Here $D(x_c)$ is defined, under resonance conditions $\omega T = 2\pi m/n$, by

$$D(x_c) = \lim_{s \to \infty} \frac{1}{2sn} \left\langle R_{sn}^2 \right\rangle_{E(x_c)} , \qquad R_s^2 \equiv (u_s - u_0)^2 + (v_s - v_0)^2 \qquad (3)$$

(assuming the limit exists), where (u_s, v_s), s integer, are the values of (u, v) at times $sT - 0$ and the average $\langle\ \rangle$ is taken over a sufficiently large sub-ensemble $E(x_c)$ of initial conditions (u_0, v_0) at fixed x_c. As already emphasized by Dana and Amit (1995), the average D_{av} of $D(x_c)$ over x_c is of practical importance, since in an experiment one usually measures the average diffusion rate of a generic ensemble, exhibiting all the values of x_c. Here we shall focus on the dependence of the global-transport properties on x_c. An impressive example showing this dependence was given, apparently for the first time, by Dana (1994) in a quantum-chaos context: for the $n = 4$ web (square crystalline symmetry), and for small K, the diffusion rate for $x_c = \pi/2$ is much slower than that for $x_c = 0$. Traces of this phenomenon can be observed already for K not very small, as shown in Fig. 1. Later, Pekarsky and Rom-Kedar (1997) have shown that for small K the $n = 4$ web undergoes a dramatic structural change, mediated by a sequence of bifurcations, as x_c is varied from $x_c = 0$ to $x_c = \pi/2$ (this can also be seen in Fig. 1). They showed that the width of the stochastic layer of the web is proportional to $\exp(-\pi^2/K^\epsilon)$, where $\epsilon = 1$ for $x_c = 0$ and $\epsilon = 2$ for $x_c = \pi/2$. This explains the strong difference in the diffusion rate in the two cases for small K, observed by Dana (1994).

Analytical expressions approximating $D(x_c)$ to high accuracy for K sufficiently large can be obtained using the formalism of Dana and Amit (1995). For example, for the $n = 4$ web we find

$$D(x_c) \approx K^2 \left\{ \frac{1}{4} + \frac{J_0(K)}{2} \cos(2x_c) + \frac{1}{2} \sum_{r=-\infty}^{\infty} \exp(-2irx_c) \times \right.$$
$$\left. [(-1)^r J_0(rK) J_r^2(K) - J_2(rK) J_r^2(K) \cos(4x_c)] \right\} , \tag{4}$$

where $J_r(K)$ is a Bessel function. For K sufficiently large, the expression in (4) can be simplified by identifying the dominant terms in the sum over r.

A variation of the global-transport properties, which is much stronger than that in the normal-diffusion case [e.g., $D(x_c)$ in (4)], can be observed when anomalous diffusion is present,

$$\left\langle R_{sn}^2 \right\rangle_{E(x_c)} \propto s^{\mu(x_c)} , \tag{5}$$

where R_s^2 is defined by (3) and $\mu(x_c)$ is the anomalous-diffusion exponent, $\mu(x_c) \neq 1$. "Superdiffusion", with $1 < \mu(x_c) < 2$, can be observed for sufficiently large values of K in the case of the crystalline webs ($n = 3, 4, 6$). In this case, the translational symmetry allows for the existence of generalized periodic orbits, the "accelerator modes". Their defining equations are

$$u_{sn+ln} = u_{sn} + 2\pi j_1 , \quad v_{sn+ln} = v_{sn} + 2\pi j_2 , \tag{6}$$

for all integers s, where l is the minimal period and $2\pi(j_1, j_2)$ is a lattice vector characterizing the accelerator mode. If sn is replaced by $sn + r$, $r = 1, ..., ln - 1$ (corresponding to the "other" points of the periodic orbit), Eqs. (6) will be satisfied with (j_1, j_2) replaced by $O^r(j_1, j_2)$, where O is a rotation by an angle $\alpha = 2\pi m/n$. If the accelerator mode is linearly stable, each point of it is usually surrounded by a stability island. All the points within an island move essentially (i.e., on a sufficiently large scale) according to Eqs. (6), leading to "acceleration",

$R^2_{sln} \propto s^2$ (i.e., $\mu = 2$). On the other hand, points in the chaotic region (stochastic layer of the web) will "stick" near the boundaries of the islands, following their accelerating motion for a long time interval, and are ejected afterwards back inside the chaotic region. After some time, they will eventually stick again near the boundaries of the islands. This process explains figuratively the origin of the global superdiffusion with an exponent μ taking values between $\mu = 1$ (corresponding to the normal diffusion expected in a strongly-chaotic regime or in the absence of accelerator islands) and $\mu = 2$ (corresponding to acceleration within the islands). A quantitative explanation of superdiffusion and a general relation between μ and the self-similarity properties of accelerator islands have been given recently by Afraimovich and Zaslavsky (1997) [see also the recent review article by Zaslavsky et al. (1997)].

In the case of our system, the crucial observation is that, for a given value of K, accelerator islands may exist only in some intervals of x_c. In these intervals, the characteristics of the islands usually vary strongly with x_c. This is shown in Fig. 2 for the $n = 4$ web at $K = 3.25$. For this value of K, we were able to find only accelerator modes of minimal period $l = 1$ with $(j_1 = 1, j_2 = 0)$ and $(j_1 = 1, j_2 = 1)$ [recall the definition (6)]. These modes exist only in the x_c-intervals covered by the several curves in Fig. 2. The modes are linearly stable and give rise, usually, to accelerator islands only if the trace of their linearity-stability matrix is between -2 and 2 (the two horizontal dashed lines in Fig. 2). Fig. 3 shows an enlargement of Fig. 2 in the main interval of x_c where the

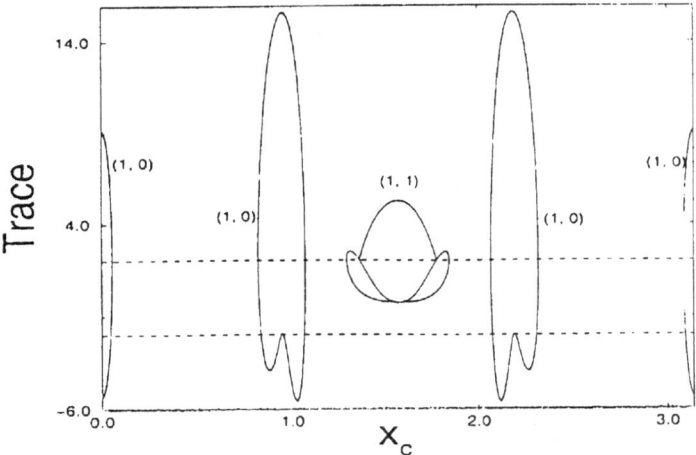

Fig. 2. Trace of the linear-stability matrix as a function of x_c for the accelerator modes with minimal period $l = 1$ in the case of $n = 4$ and $K = 3.25$. These modes exist only in the intervals of x_c covered by the several curves, and are linearly stable only if the trace is between -2 and 2 (the two horizontal dashed lines). The label (1, 0) or (1, 1) near each curve is the type (j_1, j_2) of the corresponding mode.

(1, 1) accelerator mode exists. We also plot here a properly normalized area $S(x_c)$ of the corresponding accelerator island as a function of x_c. Obviously, $S(x_c)$ vanishes for x_c outside the interval.

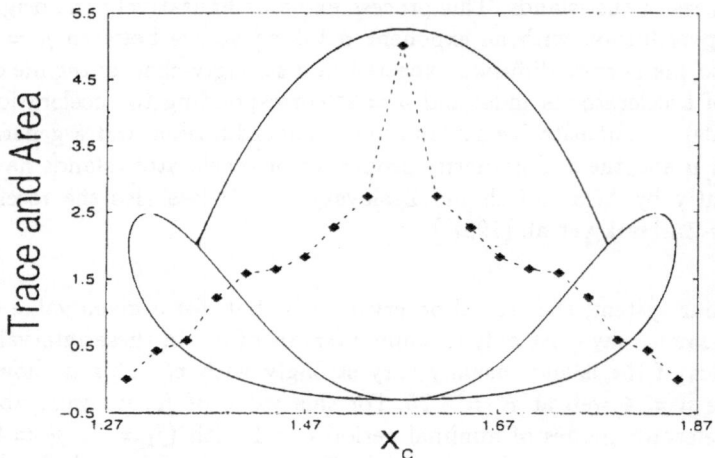

Fig. 3. An enlargement of Fig. 2 in the main x_c-interval of existence of the (1, 1) accelerator mode. The dashed line with diamond symbols gives the area of the corresponding accelerator island, in units such that the maximum value of the area, at $x_c = \pi/2$, is 5. The area was actually computed only for $x_c \leq \pi/2$, and the reflection symmetry around $x_c = \pi/2$ was used to complete the plot for $x_c > \pi/2$.

We have performed an accurate calculation of the anomalous-diffusion exponent $\mu(x_c)$ ($n = 4$, $K = 3.25$) for the same values of x_c used to plot the curve of $S(x_c)$ in Fig. 3. This calculation was made as follows. For a given value of x_c, a large ensemble of 400×400 initial conditions, uniformly distributed in the $2\pi \times 2\pi$ unit cell of the web, was iterated 1219680 times with the map corresponding to (2). Initial conditions inside accelerator islands were easily identified by their accelerating motion, and were removed from the ensemble. The remaining ensemble, $E(x_c)$, should consist then entirely of initial conditions inside the chaotic region. Indeed, we have found that for times $t = sn < 1219680$ the ensemble $E(x_c)$ evolves reasonably well according to the anomalous-diffusion law (5). The anomalous-diffusion exponent $\mu(x_c)$ was determined from the best fit of the function $f(s) = Bs^\mu$ to $\langle R^2_{sn} \rangle_{E(x_c)}$. The results are shown in Fig. 4. The strong oscillatory variation of μ with x_c, from $\mu \approx 1$ (i.e., nearly normal diffusion) to $\mu \approx 1.5$, is quite remarkable! Notice that the oscillatory behavior of $\mu(x_c)$ is quite different from the monotonous one of $S(x_c)$ (the area of the accelerator island in Fig. 3). In particular, the maximal value of $\mu(x_c)$ is not attained at $x_c = \pi/2$, as in the case of $S(x_c)$. In fact, $\mu(x_c)$ is really determined not by $S(x_c)$

but by the self-similarity properties of the accelerator islands [Afraimovich and Zaslavsky (1997); Zaslavsky et al. (1997)].

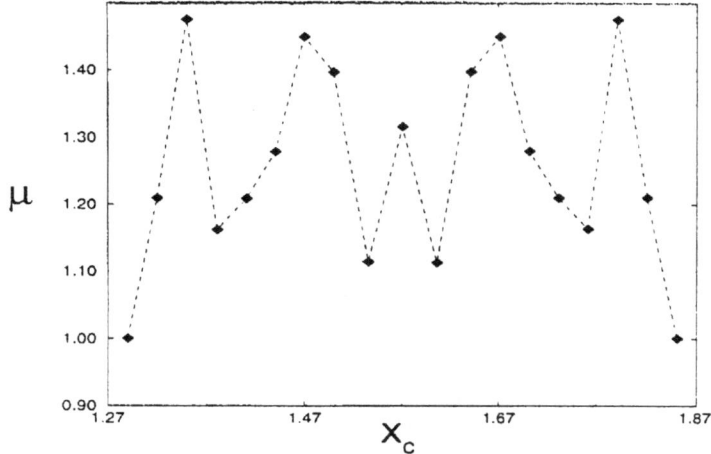

Fig. 4. Anomalous-diffusion exponent $\mu(x_c)$ for $n = 4$ and $K = 3.25$, calculated as explained in the text. As for the area curve in Fig. 3, $\mu(x_c)$ was actually computed only for $x_c \leq \pi/2$, and the reflection symmetry around $x_c = \pi/2$ was used to complete the plot for $x_c > \pi/2$.

In conclusion, we have shown that global-transport properties, such as the normal-diffusion coefficient D and the anomalous-diffusion exponent μ, can vary throughout a chaotic ensemble due to a general and simple scenario. This has been illustrated by a realistic model system of charged particles interacting with an electrostatic wave-packet in a uniform magnetic field. For this system, we have found that the variation of the global-transport properties, mainly the anomalous ones, can be remarkably strong. We expect that this finding should have experimental applications to "filtering" or preparing sub-ensembles characterized by well-defined values of the conserved momentum, e.g., x_c. This can be easily accomplished, for example, by considering electrostatic wave-packets depending on a "phase" ϕ, i.e., $V = V(x - \phi, t)$, and by adjusting ϕ so the maximal transport rate is attained at the desired value of x_c. Other theoretical aspects of the problem considered in this paper will be studied in future publications.

We would like to thank V. Rom-Kedar, G. M. Zaslavsky, F. Skiff, and J. Klafter for useful comments and discussions. This work was partially supported by the Israel Science Foundation administered by the Israel Academy of Sciences and Humanities.

References

Afanasiev, V. V., Sagdeev, R. Z., Zaslavsky, G. M. (1991): CHAOS 1, 143.
Afraimovich, V., Zaslavsky, G. M. (1997): Phys. Rev. E 55, 5418.
Cary, J. R., Meiss, J. D. (1981): Phys. Rev. A 24, 2664.
Chirikov, B. V. (1979): Phys. Rep. 52, 263.
Dana, I. (1993): Phys. Rev. Lett. 70, 2387.
Dana, I. (1994): Phys. Rev. Lett. 73, 1609.
Dana, I., Amit, M. (1995): Phys. Rev. E 51, R2731.
Dana, I., Fishman, S. (1985): Physica (Amsterdam) 17D, 63.
Dana, I., Murray, N. W., Percival, I. C. (1989): Phys. Rev. Lett. 62, 233.
Johnson, M. H., Lippmann, B. A. (1949): Phys. Rev. 76, 828.
Karney, C. F. F. (1983): Physica (Amsterdam) 8D, 360.
MacKay, R. S., Meiss, J. D. (1987): *Hamiltonian Dynamical Systems* (Adam Hilger, Bristol).
MacKay, R. S., Meiss, J. D., Percival, I. C. (1984): Physica (Amsterdam) 13D, 55.
Meiss, J. D. (1986): Phys. Rev. A 34, 2375.
Meiss, J. D., Ott, E. (1986): Physica (Amsterdam) 20D, 387.
Pekarsky, S., Rom-Kedar, V. (1997): Phys. Lett. A 225, 274.
Umberger, D. K., Farmer, J. D. (1985): Phys. Rev. Lett. 55, 661.
Shlesinger, M. F., Zaslavsky, G. M., Klafter J. (1993): Nature 363, 31.
Zaslavsky, G. M. (1991): CHAOS 1, 1.
Zaslavsky, G. M., Edelman, M., Niyazov, B. A. (1997): CHAOS 7, 159.
Zaslavsky, G. M., Zakharov, M. Yu., Sagdeev, R. Z., Usikov, D. A., Chernikov, A. A. (1986): Sov. Phys. JETP 64, 294.
Zumofen, G., Klafter, J. (1994): Europhys. Lett. 25, 565.

Sticky Orbits
of Chaotic Hamiltonian Dynamics

V. Afraimovich[1], G. M. Zaslavsky[2,3]

[1]National Tsing Hua University, Department of Mathematics,
Hsinchu, Taiwan 30043, Republic of China.

[2]Courant Institute of Mathematical Sciences, New York University,
251 Mercer Street New York, New York 10012.

[3]Department of Physics, New York University, 2-4 Washington Place,
New York, New York 10003.

Abstract. Nonuniformity of the phase space of chaotic Hamiltonian dynamics can result from the existence of a sticky set called "Sticky Riddle" (SR) imbedded into the phase space. Fractal and multifractal properties of SR can be described for some simplified situations. Existence of SR imposes similar stickiness for chaotic orbits when they approach the vicinity of SR. As a result, the orbits reveal behavior with power-like tails in the distribution of Poincaré recurrences and exit times, which is unusual for hyperbolic systems. We exploit the generalized fractal dimension to describe the set of recurrences.

Keywords. Chaos, fractals, dimensions, Poincaré recurrences

Introduction

Direct observation of motion of chaotic particles in the phase space displays a certain nonuniformity of orbits distributions. More precisely, a typical point visits different parts of invariant sets with different frequencies. The nonuniformity can manifests itself in different ways: there may be holes in the phase space (islands) which can not be penetrated by an orbit from the outside space (stochastic sea), and there may be a concentration domains inside the stochastic sea due to known or unknown reasons. Numerous simulations show a fairly rich collection of realizations of the nonuniformity. For Hamiltonian systems, which will be discussed in this article, the nonuniformity of the invariant set can occur because of islands, cantori, unstable isolated periodic orbits, boundaries, and other reasons. Observations and descriptions of some of the mentioned properties of orbits can be found in [1,2]. In fact, occurrence of the nonuniformity of the invariant set, or stationary distribution function (using physical terminology), is at the heart of our understanding of the Hamiltonian chaos and its different applications. First, nonuniformity should be related in some way to special, say, singular properties of the dynamics in the phase space. The singularities do not yet have a rigorous and/or convenient classificiation despite the fact that they can easily be observed from the simulation. One can mention a border between chaotic and nonchaotic motion as the most typical example of a singularity. Singular points or sets can be imbedded into the stochastic sea. Secondly, presence of singularities imposes specific large scale space-time asymptotics of distribution functions and correlators. Third, nonuniformity and singularities lead to absence of the familiar Gibbs microcanonical distribution, i.e. to significant changes in statistical and thermodynamical properties of system with chaotic dynamics (see for discussions [3]).

In the light of the above described "unusual" properties of chaotic dynamics, we should seek new approaches which could automatically reveal the absence of characteristic space-time scales. Fractal or multifractal features of systems with different powerwise distributions can be considered using special methods of analysis, which became routine after a set of pioneering publications [4-7]. These results have been followed by rigorous description of the famous $f(\alpha)$ spectrum of dimensions [8-10].

Invariant sets of sticky chaotic dynamics should be considered in accordance with the existence of infinite moments. As a good (and important) example one can mention the distribution of Poincaré recurrences. It was shown by simulation and on the physical level of consideration that for some situations there exists a power-type tail of the Poincaré recurrences distribution with an exponent connected to an exponent of the macroscopic transport [11-13]. Fractal and multifractal analysis can be applied to the Poincaré recurrences [14,15], providing a new tool to study chaotic dynamics.

In this article, we discuss the existence of limit sets, called "sticky riddles" (SR), for Hamiltonian chaotic dynamics which is not uniformly hyperbolic. We apply space-time scaling properties of SR to describe invariant distributions of the Poincaré recurrences.

1 Bad Orbits

In this section, we would like to attract the attention to the existence of zero measure of special type orbits which are bad in the sense that they do not possess a typical hyperbolic property that is valid for all typical orbits of a Hamiltonian system.

Assume that μ is an invariant probability measure and A is the support of μ, i.e. A is a closed invariant set of our dynamical system with $\mu(A) = 1$. Given $x \in X$, one can consider the **upper** and **lower** pointwise dimensions at x

$$\bar{d}_\mu(x) = \limsup_{\epsilon \to 0} \frac{\log \mu(B(x,\epsilon))}{\log \epsilon},$$

$$\underline{d}_\mu(x) = \liminf_{\epsilon \to 0} \frac{\log \mu(B(x,\epsilon))}{\log \epsilon} \tag{1.1}$$

where $B(x, \epsilon)$ is the ball of radius ϵ centered at the point x. If

$$\underline{d}_\mu(x) = \bar{d}_\mu(x) = d_\mu(x) \tag{1.2}$$

then $d_\mu(x)$ is called the pointwise dimension at x ([16]). Thus,

$$\mu(B(x, \epsilon)) \sim \epsilon^{d_\mu(x)} \tag{1.3}$$

at the point x. For ergodic measures with nonzero Lyapunov exponents, $d_\mu(x) = d$ for μ-almost every x , i.e. it might be that $d_\mu(x)$ does not equal d only on a set of zero measure. This result has recently been proved in [17] although it has been known for a long time as the Eckmann-Ruelle conjecture [18].

Let

$$K_\alpha = \{x \in A | d_\mu(x) = \alpha\}, \quad \alpha \in \mathbf{R}, \tag{1.4}$$

then the $f(\alpha)$ spectrum for dimensions is defined by $f(\alpha) = \dim_H K_\alpha$ where $\dim_H K_\alpha$ denotes the Hausdorff dimension of the set K_α. Thus, $f(\alpha)$ is nontrivial if there are many sets of zero measure but nonzero dimension with scaling

$$\mu(B(x, \epsilon)) \sim \epsilon^\alpha, \quad \alpha \neq d . \tag{1.5}$$

The set A can be decomposed now as

$$A = \bigcup_{-\infty < \alpha < \infty} K_\alpha \cup S \tag{1.6}$$

where $S = \{x | d_\mu(x) \text{ does not exist }\}$ is the so called Shereshevsky set [19]. It was shown in [19] that for a class of C^2 surface diffeomorphisms with Smale horseshoe A, having different rates of contraction and expansion at different points, the set S is dense in A and has positive Hausdorff dimension for any good enough (equilibrium) measure μ. Of course, $\mu(S) = 0$ for any good ergodic measure μ. Thus, even for dynamical systems with hyperbolic invariant set we see that:

(1) the existence of zero measure sets K_α with different scaling constants implies nontrivial $f(\alpha)$ spectrum (and, in fact, nontrivial Hentschel-Procaccia spectrum [4]);

(2) the existence of zero measure set S is the reason for many mathematical difficulties in the study of multifractal properties of dynamics.

Bad, irregular orbits appear also in the problem of calculating Lyapunov exponents, in performing of time (Birkhoff) averages, in evaluation of local entropies etc. − see [10, 20]. It is unclear if, the irregular sets are physically negligible even for "*uniformly hyperbolic*" system. Nevertheless, for systems with nonuniformly hyperbolic orbits (such as orbits in chaotic sea of Hamiltonian system), bad orbits become physically recognizable.

For Hamiltonian systems with chaotic dynamics, the chaotic sea does not cover the full phase space, and one needs to extract a set of islands and boundary layers to obtain a region with chaotic behavior. The behavior of orbits near the island boundary layer was studied in [13,21-23] as a fractal object which imposes powerwise distribution in the large time asymptotics of chaotic kinetics. More specifically, the island boundary is sticky, the subisland boundary is stickier and so on. This is the major cause of the fractal (multifractal) space-time behavior of the orbits. As numerical experiments show, sticky orbits "*attract*" a typical orbit in chaotic sea, forcing it to behave nonchaotically for a long time. It is possible to extract some quantitative information if we know scaling laws in a hierarhy of islands of different generations. For that, the hierarhy has to be infinite.

A simple, one-dimensional (non-Hamiltonian) analog of this situation is an infinitely renormalizable map in the family of quadratic maps

$$\bar{x} = ax(1 - x) = f_a(x), \ a \in [0, 4] \ . \tag{1.7}$$

Let us remind that f_a is infinitely renormalizable [24] if there are infinitely many numbers $n_1, n_2, \cdots$ and infinitely many subintervals $I_1, I_2, \cdots$ such that $f^{n_i} I_i \subset I_i$, f^{n_i} has the only one critical point on $I_i = [x_i, y_i]$ and $f^{n_i} x_i = x_i$. If so, then $f^{n_i}|I_i$ can be treated as $f|[0, 1]$ (after some rescalling). It is possible to define a "*sticky*" set

$$F = \bigcap_{i=1}^{\infty} \bigcup_{k=1}^{m_k} f_a^k I_i \tag{1.8}$$

which is really sticky since it attracts almost every orbit on $[0, 1]$. The dynamics $f_a|F$ is nonchaotic - the topological entropy of $f_a|F$ equals zero. There are infinitely many values of $a \in [0, 4]$ for which f_a is infinitely renormalizable. The well-known Feigenbaum attractor is realized for an infinitely renormalizable situation. It is known how length (I_i) behaves as $i \to \infty$ (scaling law is determined by eigenvalues of the period-doubling transformation), therefore it is possible to find the fractal dimension and other characteristics of the set F (see for instance [24], [25]).

In order to obtain information in the Hamiltonian case, one must first define a situation similar to the infinitely renormalizable one for one-dimensional families. We do it in the Sec. 3 by introducing a notion of sticky riddles (SR). Then, one must transform a knowledge of scaling laws into the information about the behavior of sticky orbits and orbits around islands. For that, we use a generalized Carathéodory construciton (Sec. 5), and notions of dimension and capacities for Poincaré recurrences (Secs. 6 and 7).

2   Examples of Sticky Orbits

In this section a few examples will be given of bad orbits that occur in different physical models.

(a) Island-around-islands structure in the standard and web maps.

A typical Hamiltonian system has a rich set of islands in phase space, with a regular dynamics inside the islands and with narrow stochastic layers isolated from the main stochastic sea domain. As an example, one can consider the web map

$$\bar{u} = v, \quad \bar{v} = -u - K \sin v \qquad (2.1)$$

or the standard map

$$\bar{p} = p - K \sin x, \quad \bar{x} = x + \bar{p}, \qquad (2.2)$$

with a fairly well known island structure which will be discussed more below. The islands-around-islands structure was observed and described in different situations [1,21,13,26]. The dynamics near the islands boundary is singular due to the phenomena of stickines, and it can dominate in the large time asymptotics. This influences almost all important probability distributions, such as the distribution of distances, exit times, recurrences, transit times, etc. The main feature of all such distributions is that they may not correspond to either Gaussian or Poissonian (or similar) processes with all finite moments. This is due to the presence of powerlike tails in the asymptotical limits of large space-time scales. More precisely, one must admit that simulations show a possibility of existence of infinite moments of the recurrences distribution for some values of the parameter K, while a rigorous theory of that property does not exist.

Being more specific, one can say that power-wise tailed distributions are a consequence of a (multi-) fractal singular scattering zone near the island boundaries. More precisely, there are different sets of islands with different asymptotics (different powers of distribution tails) and different scales for where and when the asymptotics work. Different intermediate asymptotics is a crucial characteristic of the anomalous transport, as was mentioned in the problem of advection [27], and the problem of charged particle motion in an electromagnetic field [28]. This is the basis for introducing a multifractal description of some distributions in chaotic dynamics.

In numerical experiments, we can not recognize sticky orbits directly and special methods of visualization should be used. Usually, the power-like tails in distributions indicate the presence of sticky orbits. Nevertheless, there are geometric criteria of the existence of such kind of orbits. One of them is an infinite hierarchy of islands of nonvanishing prolifiration numbers q (see below). For example, for the map (2.2) one can find the value $K = K_8 = 6.908745\ldots$ for which an eight-island chain (i.e. $q = 8$) persists in subislands of three successive generation. (See Fig. 1 from [15].) We expect the existence of such a value K_8 for which the sequence of multiplication of islands of smaller and smaller sizes is infinite with a coefficient (proliferation number) $q = 8$ excluding the initial 3-island figure. One can find more such examples in [13,29].

64

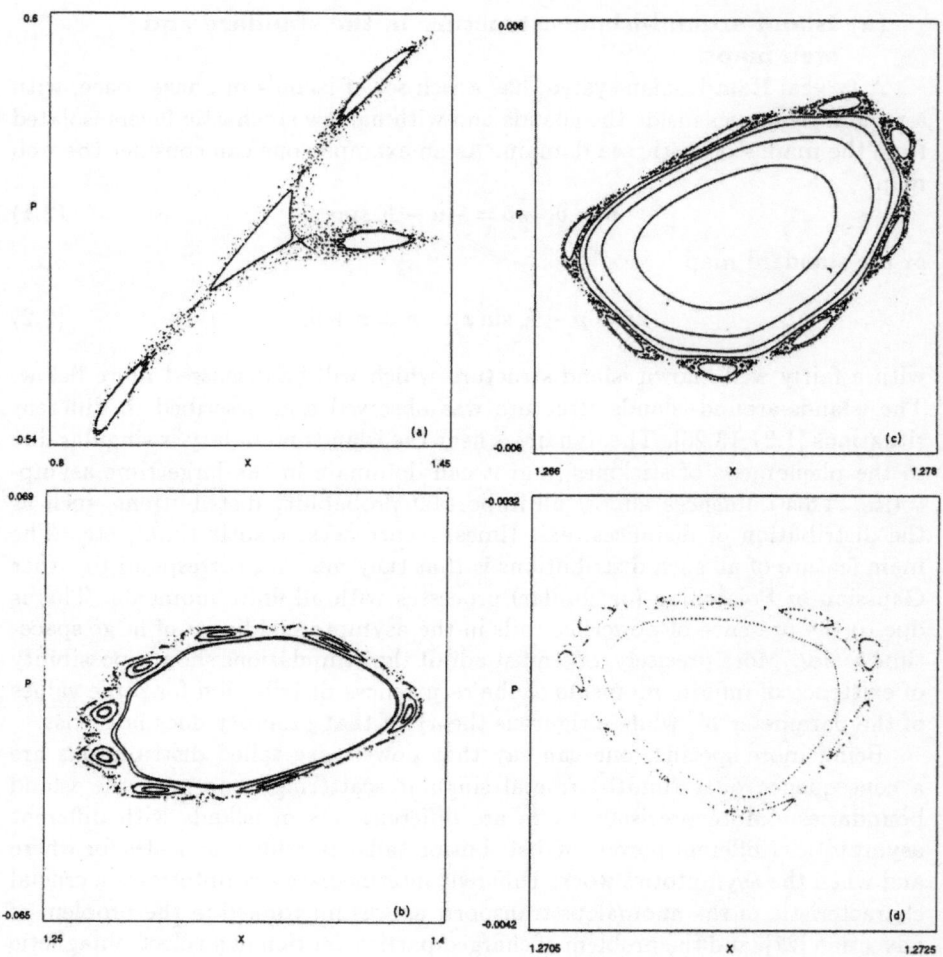

Fig. 1. Example of the island hierarchies for the standard map [13]. (a) The main
island and first generation of a three-island chain. (b) Magnification of the
right island in (a). (c) Magnification of the left island in (b). (d) Magnification
of the bottom island in (c).

In the simplest situation, the islands hierarchy satisfies the exact renormalization conditions:

$$q_n = \lambda_q^n q_0 , \qquad (\lambda_q \geq 3)$$

$$S_n = \lambda_S^n S_0 , \qquad (\lambda_S < 1) \qquad (2.3)$$

$$T_n = \lambda_T^n T_0 , \qquad (\lambda_T > 1)$$

where q_n, S_n, and T_n are, correspondingly, the number of islands, area of islands, and period of the last invariant curve inside islands of the n-th generation. More complicated sequences can be shown for the equations (2.1),(2.2).

(b) Billiards

A particle in billiards of different shapes is another example of a dynamical system with bad orbits. For example, Sinai billiard (with so-called infinite horizon) has infinitely long bouncing orbits. Because of such orbits, there exists a possibility for any orbit to stick for an arbitrarily long time to the bouncing regime. An example of such an orbit is shown in Fig. 2 [30]. In the corresponding Poincaré section in Fig. 2(b), s's (dark lines) correspond to bounces and reflections with $\pi/4$ angle without scatterings. The measure of bad orbits is zero. Nevertheless, any hyperbolic orbit approaches the s's, which leads to strong effects for the Poincaré recurrences distribution [31-33] and kinetics which seems non-Gaussian [30].

3   Criteria of the Existence of Sticky Orbits

It is well-known that Hamiltonian dynamical system generally contains homoclinic orbits [34] belonging to transversal intersections of stable and unstable manifolds of periodic orbits. Therefore (see, for example, [35]), it contains hyperbolic subsets which consist of "*good*", hyperbolic orbits. Nevertheless, the characteristics of hyperbolicity for the hyperbolic subsets intimately depend on the "*orders of the resonances*" of the basic periodic orbits, i.e. on their periods. Numerical simulations show that there are many values of parameters, say K as in (2.1),(2.2), for which stickiness is originated by a fractal (or multifractal) set of islands displayed in Fig. 1. In a rough way, such a fractal is shown in Fig. 3. From another point of view, stickiness means a decreasing of the Lyapunov exponents and angles between stable and unstable manifolds when the islands become smaller, i.e. when the order of generation of islands becomes larger. This effect can be qualitatively explained using Figs. 1 and 3. The boundary layer has a finite width. Infinite proliferation of the number of islands, i.e. $q_n \to \infty$ in (2.3), yields additional scaling property for the angles of intersection of islands' separatrices

$$\vartheta_n = (c\lambda_\perp/\lambda_\|)^n \vartheta_0 = \lambda_\vartheta^n \vartheta_0, \qquad (\lambda_\vartheta = c\lambda_\perp/\lambda_\| < 1) \qquad (3.1)$$

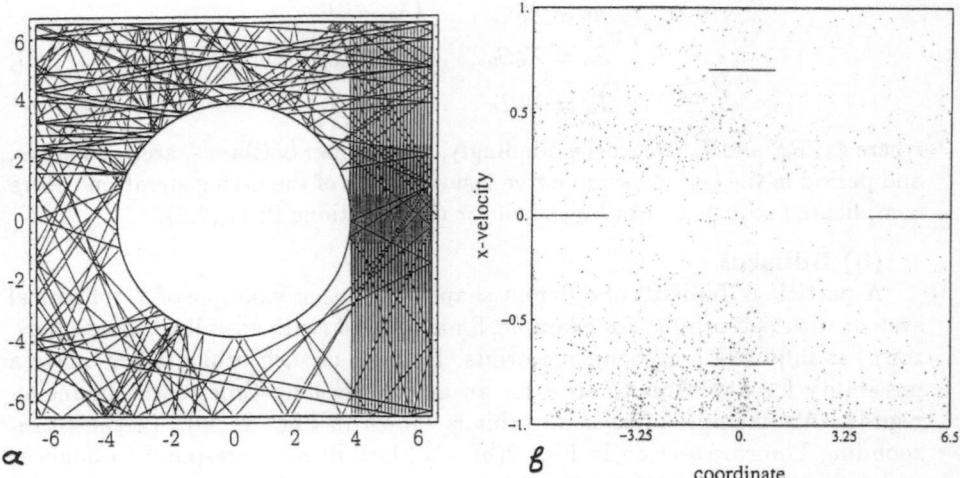

Fig. 2. Sinai billiard with a trajectory that makes many bounces (a) and the corresponding Poincaré section (b).

Fig. 3. Example of the phase space partitioning
for islands-around-islands structure.

with a constant c of order one and scaling constants $\lambda_\perp$, $\lambda_\parallel$ for the islands sizes. Let us note that

$$\lambda_S = \lambda_\perp \lambda_\parallel \tag{3.2}$$

It is possible to estimate that the local Lyapunov exponent is proportional to the angle ϑ of the separatrices intersection at the main saddles of islands, or to the inverse period T_n, and it tends to zero due to (3.1)(3.2) as $n \to \infty$. The same is true for orbits in hyperbolic sets in the neighborhood of homoclinic orbits to these basic periodic trajectories. In other words, the hyperbolicity of orbits becomes weaker if we move from the "*heart*" of chaotic sea to its "*boundary*" and we can expect that it disappears at on the boundary, and that the orbits very close to the boundary layers are not chaotic at all.

Let us attach a mathematical sense to this statement. First of all, give a definition of sticky orbits and sets. For the sake of simplicity we consider only the case of area preserving maps of the plane (i.e. the systems with 1.5 or 2 degrees of freedom). Let $g : \mathbf{R}^2 \to \mathbf{R}^2$ be an area preserving map. A piece of plane u homeomorphic to the disk is said to be an island of stability if $g^q u = u$ for some integer q. In fact, q is the minimal period of periodic points in u.

Definition. *An infinite set* $M = \left\{ u_{i_0 \cdots i_{n-1}} \right\}$, $n = 1, 2, \cdots$, $1 \leq i_s \leq k_s < \infty$, *of islands of stability is said to be the sticky riddle (SR) if*

(1) for any island $u_{i_0, \cdots i_{n-1}} \in M$ there is an island $u_{j_0 \cdots j_{n-1}} \in M$ such that $g\left(u_{i_0 \cdots i_{n-1}} \right) = u_{j_0 \cdots j_{n-1}}$;

(2) if $g\left(u_{i_0 \cdots i_{n-1}} \right) = u_{j_0 \cdots j_{n-1}}$ then for any i, $0 \leq i \leq k_n$, there is j, $0 \leq j \leq k_n$, such that $g\left(u_{i_0 \cdots i_{n-1} i} \right) = u_{j_0 \cdots j_{n-1} j}$;

(3) diam $u_{i_0 \cdots i_{n-1}} \to 0$, therefore area$\left(u_{i_0 \cdots i_{n-1}} \right) \to 0$ as $n \to \infty$;

(4) for any sequence of points $x_n \in u_{i_0 \cdots i_{n-1}}$, the $\lim\limits_{n \to \infty} x_n$ exists;

(5) If $x_n \in u_{i_0 \cdots i_{n-1}} \subset M$, $n \in \mathbf{Z}_+$, $y_n \in v_{j_0 \cdots j_{m-1}} \in M$, $m \in \mathbf{Z}_+$, and $v_{j_0 \cdots j_{k-1}} \neq u_{i_0 \cdots i_{k-1}}$ at least for one k, then $\lim\limits_{n \to \infty} x_n \neq \lim\limits_{m \to \infty} y_m$.

Definition. *Let M be a SR. The set*
$$S = \left\{ x \mid x = \lim_{n \to \infty} x_n, x_n \in u_{i_0 \cdots i_{n-1}} \right\}$$ is called the sticky set, and orbits in S are called sticky orbits.

Let us emphasize that an islands-around-islands structure satisfies the definition of SR. Indeed: (1) an island of the n-th generation is mapped into an island of the same generation; (2) if an island $u_{i_0 \cdots i_{n-1} i}$ lies in a vicinity of the island $u_{i_0 \cdots i_{n-1}}$ then its image $u_{j_0 \cdots j_{n-1} j}$ lies in a vicinity of the image of $u_{i_0 \cdots i_{n-1}}$ i.e. dynamics on the $(n+1)$-th level is consistent with the dynamics of the n-th level; (3) this property is trivially understandable; (4) there is the only point of accumulation of islands $u_{i_0\cdots i_{n-1}}$ along any infinite branch $(i_0 i_1 \cdots)$ of the *"tree"* $\{(i_0 \cdots i_{n-1})\}_{n=1}^{\infty}$; (5) as numerical simulations show, in the islands-around-islands structure the distance between any pair of neighboring islands of a fixed, n-th, generation roughly a constant (idependently of the choice of particular neighboring pair) and depends only on n. It follows that for different branches $(i_0 i_1 \cdots)$, the points of accumulation are different.

Thus, the existence of SR, and in particular the existence of infinite islands-around-islands hierarchy is a criterion of the existence of sticky orbits.

The following questions immediately become very important. Let us formulate them for the maps (2.1), (2.2).

- Is it true that sticky orbits exist for many values of the parameter K ?

- Is it true that for some value K_S of K, the set of sticky orbits has nonzero Lebesgue measure? nonzero Hausdorff dimension?

If true, we may introduce some parameters to measure the difference between good and sticky orbits. (We can treat orbits having positive Lyapunov exponent as good orbits). For example, one can consider the ratio

$$\eta = \frac{mes(S)}{mes(G)}$$

where $S(G)$ is the set of all sticky (good) orbits and *mes* is the Lebesgue measure. Definitely, η should be a function of the parameter $K : \eta = \eta(K)$ whose behavior must reflect essential properties of the systems (2.1), (2.2). Or we can study the Hausdorff dimension of the set S (if it has zero Lebesgue measure). The dependence $\eta(K)$ can be of the devil's-stair-case type.

4   Symbolic Dynamics on the Set of Sticky Orbits

A sticky set S is the topological limit of the sets $u_{i_0\cdots i_{n-1}}$ in the SR. The properties (1)-(5) of the SR allow us to reproduce the behavior of $g|S$. It is useful to consider some symbolic model. Denote by Ω the set of infinite sequences $\underline{w} = (i_0, i_1, \cdots)$, $1 \le i_j \le k_j < \infty$ with a metrics

$$dist\,(\underline{w}', \underline{w}'') = \sum_{k=0}^{\infty} \frac{|i_k' - i_k''|}{a^k}, \quad a > 1. \tag{4.1}$$

If $k_j \le C < \infty$ for some constant C then Ω becomes a compact set homeomorphic to the standard Cantor set (if $k_j > 1$, $j = 1, 2, \cdots$). In order to define a dynamics on Ω, we have to take into account the property (1) in the definition of SR. We define the desired map $T : \Omega \to \Omega$ as follows. For any $\underline{w} = (i_0, i_1, \cdots) \in \Omega$ and any $n = 1, 2, \cdots$, set

$$T\,([i_0, \cdots, i_{n-1}]) = ([j_0, \cdots, j_{n-1}]) \quad \text{if} \quad g\,(u_{i_0\cdots i_{n-1}}) = u_{j_0\cdots j_{n-1}} \tag{4.2}$$

Thus, we have the nested sequence of "*cylinders*" $[j_0, \cdots, j_{n-1}]$, $n = 1, 2, \cdots$, and define $T\underline{w} = \bigcap\limits_{n=1}^{\infty} [j_0, \cdots, j_{n-1}]$.

It follows from the definition that the map T is well-defined and continuous due one-to-one. The map T maps cylinders to cylinders. Therefore it is very similar to the so called dyadic adding machine [24]. Despite the fact that the model looks artificial, it is the simplest symbolic model of the type of dynamics occurring in SR. It is useful to remind the definition of the model.

Dyadic adding machine: Let Ω_2^+ be a set of all one-sided infinite sequences from "0" and "1" with the usual distance. Denote by $\delta : \Omega_2^+ \to \Omega_2^+$ the following map:

- if $\underline{w} = (1, 1, \cdots)$ then $\delta \underline{w} = (0, 0, \cdots)$;
- if $\underline{w} = (i_0, i_1, \cdots)$, $i_k = 1$, $k = 0, \cdots, j-1$, $i_j = 0$, then $(\delta \underline{w})_k = 0$, $k = 0, \cdots, j-1$, $(\delta \underline{w})_j = 1$, $(\delta \underline{w})_{j+s} = i_{j+s}$, $s \geq 1$,

- if $\underline{w} = (0, i_1, \cdots)$ then $\delta \underline{w} = (1, i_1, \cdots)$.

Then the map δ is one-to-one. Indeed,

- $\delta^{-1}(0, 0, \cdots) = (1, 1, \cdots)$,

- if $\underline{w} = (i_0, i_1 \cdots)$, $i_k = 0$, $k = 0, \cdots, j-1$, $i_j = 1$, then $(\delta^{-1} \underline{w})_k = 1$, $k = 1, \cdots, j-1$, $(\delta^{-1} \underline{w})_j = 0$, $(\delta^{-1} \underline{w})_{j+s} = i_{j+s}$, $s \geq 1$,

- if $\underline{w} = (1, i_1, \cdots)$ then $\delta^{-1} \underline{w} = (0, i_1, \cdots)$.

It is clear that δ is continuous. The dynamical system (δ^n, Ω_2^+) is called the dyadic adding machine. The reason for that can be understood if we consider the Abelian group of 2-adic integers $\{k = i_0 + i_1 2 + i_2 2^2 + \cdots + i_{n-1} 2^{n-1}\}$. The transformation that adds the element $\bar{e} = (1, 0, 0, \cdots)$ to the number k corresponds to the map δ generated by our dynamical system. The map δ maps cylinders to cylinders of the same length, for example

$$(1, 1, 0) \xrightarrow{\delta} (0, 0, 1) \longrightarrow (1, 0, 1) \longrightarrow (0, 1, 1) \longrightarrow$$

$$(1, 1, 1) \longrightarrow (0, 0, 0) \longrightarrow (1, 0, 0) \longrightarrow (0, 1, 0) \longrightarrow (1, 1, 0).$$

Introduce the following metrics on Ω_2^+: for $\underline{w} = (i_0, i_1, \cdots)$, $\underline{w}' = (i_0', i_1', \cdots)$, let

$$\text{dist}\,(\underline{w}', \underline{w}) = \frac{1}{n} \quad \text{if} \quad i_0 = i_0', \cdots, i_{n-1}' = i_{n-1}, i_n' \neq i_n \ . \tag{4.3}$$

Then the map δ is isometry because of the property above: $(\delta \underline{w})_k = (\delta \underline{w}')_k$, $k = 0, 1, \cdots, n-1$, and $1/n = \text{dist}_{\text{new}}\,(\delta \underline{w}, \delta \underline{w}') = \text{dist}_{\text{new}}\,(\underline{w}, \underline{w}')$. Therefore, the topological entropy of δ is zero.

Generally, the map $T|\Omega$ can be an m-adic adding machine or a generalized adding machine (see, for instance [24]) with zero topological entropy.

The following theorem shows us that the maps $g|S$ and $T|\Omega$ have the same dynamics.

Theorem 1. *The maps $T|\Omega$ and $g|S$ are topologically conjugated*

PROOF. We prove this for the case when $T|\Omega$ is just a dyadic adding machine. For the general case, the proof is basically the same with nonessential technical complications.

Fix $n \in \mathbf{Z}_+$ and let $u_{0\cdots 0} = \triangle_n^0$, $\triangle_n^j = g^j \triangle_n^0$, $j = 1, \cdots, k_n$. Assume that sets $\triangle_n^j$ are ordered in such a way that if $x \in S$, $x = \lim\limits_{n \to \infty} x_n$, $x_n \in \triangle_n^{j(n)}$, $n \in \mathbf{Z}_+$, then $j(n) = i_0 + i_1 2 + \cdots + i_{n-1} 2^n$ where $i_k \in \{0, 1\}$ and independent of n. Thus, a map $h : x \to (i_0, i_1, \cdots)$ is well-defined. Since $\triangle_n^j$ are disjoint closed sets, we have that h is continuous. It is one-to-one, thanks to the property (5) in the definition of SR. Show that h is a conjugacy. Indeed, we have $g\triangle_n^{j_n} = \triangle_n^{j_n+1}$ for $j_n < 2^n - 1$ and $g\triangle_n^{2^n-1} = \triangle_n^0$. Then, if the sequence $j_n = i_0 + i_1 2 + \cdots + i_n 2^{n-1}$ corresponds to the point $\underline{w} = (i_0, i_1 \cdots)$ and $x \in S$, it follows that the sequence of numbers $\{j_n + 1\}$ for the point gx corresponds to the sequence $\delta \underline{w}$. Thus, $h \circ g = \delta \circ h$. □

This theorem shows that on the set S, area preserving map g in some new "*symbolic*" coordinates is represented as generalized adding machine. In other words, there is a sequence of subsets $h([i_0 \cdots i_{n-1}]) = W_{i_0 \cdots i_{n-1}}$ of different levels, $n = 1, 2, \cdots$, of the sticky set S such that the map g acts as a permutation on every fixed level $n = 1, 2, \cdots$.

Remark. *From a topological viewpoint, the dynamics of the map g on the set of sticky orbits is similar to the dynamics of a dissipative map on the Feigenbaum attractor $\mathcal{F}_\infty$ [24]. Nevertheless, scaling properties of the set S can be different from those for $\mathcal{F}_1$. The main difficulty here is an unfamiliar type of inductive processes for sticky sets. In standard geometric constructions (Moranlike, [10] for example), the sets of the $(n + 1)$-th level are contained inside the sets of the n-th level. In our situation, the sets of the next level do not belong to the sets of the previous level; they are just associated with sticky sets. The study of the fractal properties of sticky sets is therefore an interesting new problem.*

5 Generalized Dimensions

As was mentioned above, temporal characteristics of sticky orbits are as important as spatial ones. To study their fractal properties in space and time we apply a general approach developed by Pesin [36] on the basis of classical Carathéodory results [37].

Assume that X is a metric space with a distance ρ, and $\mathcal{F}$ is a collection of open subsets of X, or $\mathcal{F}$ is the collection of all balls of all diameters. Consider functions $\xi(u), \eta(u)$ of subsets $u \in \mathcal{F}$ satisfying the following properties:

(1) $\eta(u) > 0$ if $u \neq \phi$, $\xi(u) \geq 0$.

(2) For any $\delta > 0$ one can find $\epsilon > 0$ such that $\eta(u) \leq \delta$ for any $u \in \mathcal{F}$ with diam $u \leq \epsilon$ (diam $u_i = \sup_{x,y \in u_i} \rho(x,y)$).

The collection $\mathcal{F}$ and the function $\xi(u), \eta(u)$ of sets in $\mathcal{F}$ form a *Carathéodory structure* [36]. Fix some Carathéodory structure and consider a finite or countable cover $G = \{u_i\}$ of X by sets u_i with diam $u_i \leq \epsilon$. Then introduce the sum

$$M(\alpha, \epsilon, G) = \sum_i \xi(u_i) \eta(u_i)^\alpha \qquad (5.1)$$

and consider its infimum

$$M(\alpha, \epsilon) = \inf_G \sum_i \xi(u_i) \eta(u_i)^\alpha,$$

where the infimum is taken over all finite or countable covers G with diam $u_1 \leq \epsilon$, $u_j \in G$. The quantity $M(\alpha, \epsilon)$ is a monotone function with respect to ϵ, therefore, there exists a limit

$$m(\alpha) = \lim_{\epsilon \to 0} M(\alpha, \epsilon).$$

It was shown in [36] that there exists a critical value $\alpha_c \in [-\infty, \infty]$ such that

$$m(\alpha) = 0, \quad \alpha > \alpha_c, \quad \alpha_c \neq +\infty,$$

$$m(\alpha) = \infty, \quad \alpha < \alpha_c, \quad \alpha_c \neq -\infty.$$

The number α_c is said to be the Carathéodory dimension relative to the structure $(\mathcal{F}, \xi, \eta)$.

For example, if $\mathcal{F}$ is a collection of balls $\{B(x, \epsilon)\}$ of all diameters $\epsilon > 0$, centered at all points $x \in X$, $\xi(B(x, \epsilon)) \equiv 1$, $\eta(B(x, \epsilon)) = \epsilon$, then

$$M(\alpha, \epsilon, G) = \sum_{i=1}^{N} [diam B(x_i, \epsilon_i)]^\alpha, \quad \epsilon_i \leq \epsilon,$$

and $\alpha_c = dim_H X$ is the Hausdorff dimension. If the Hausdorff measure $m(\alpha_c)$ is positive and finite then the average

$$< [diam B(x_1, \epsilon_1)]^{\alpha_c} > \sim \frac{m_c}{N}, \quad \epsilon \gg 1.$$

Consider another structure $(\mathcal{F}, \xi, \eta)$ on the same set X, and let α_c be the corresponding dimension. If $0 < m(\alpha_c) = m_c < \infty$ then the average

$$< \xi(u_1)\eta(u_i)^{\alpha_c} >= \frac{1}{N} \sum_i \xi(u_i)\eta(u_i)^{\alpha_c}$$

also behaves as m_c/N if $\epsilon \ll 1$; here N is the number of sets u_i in a cover with diam $u_i \leq \epsilon$.

Now introduce the sum

$$R(\alpha, \epsilon, G) = \sum_i \xi(u_i)\eta(u_i)^\alpha, \qquad (5.2)$$

where $\{u_i\} = G$ is a cover of X by sets u_i with diam $u_i = \epsilon$ (not $\leq \epsilon$ as above!). Consider the infimum

$$R(\alpha, \epsilon) = \inf_G R(\alpha, \epsilon, G),$$

where the infimum is taken over all covers $G = \{u_i\}$, with diam $u_i = \epsilon$. We may expect that the limit $\epsilon \to 0$ does not exist. Consider the upper and lower limits

$$\bar{r}(\alpha) = \overline{\lim}_{\epsilon \to 0} R(\alpha, \epsilon)$$

and

$$\underline{r}(\alpha) = \lim_{\epsilon \to 0} R(\alpha, \epsilon).$$

It was shown in [36] that there are critical values $\bar{\alpha}_c \geq \underline{\alpha}_c$ such that

$$\bar{r}(\alpha) = \begin{cases} 0, & \alpha > \bar{\alpha}_c, \quad \bar{\alpha}_c \neq +\infty, \\ \\ \infty, & \alpha < \bar{\alpha}_c, \quad \bar{\alpha}_c \neq -\infty, \end{cases}$$

and

$$\underline{r}(\alpha) = \begin{cases} 0, & \alpha > \underline{\alpha}_c, \quad \underline{\alpha}_c \neq +\infty, \\ \\ \infty, & \alpha < \underline{\alpha}_c, \quad \underline{\alpha}_c \neq -\infty, \end{cases}$$

The number $\bar{\alpha}_c$ is said to be the upper and $\underline{\alpha}_c$ the lower capacity relative to the structure $(\mathcal{F}, \xi, \eta)$.

For example, if $\mathcal{F}$ is the collection of the balls $\{B(x, \epsilon)\}$ of all diameters $\epsilon \geq 0$ centered at all points $x \in X$ and

$$\eta(B(x, \epsilon)) = diam B(x, \epsilon) = \epsilon,$$

then

$$R(\alpha, \epsilon) = \inf_G \sum_i [diam B(x_i, \epsilon)]^\alpha = N(\epsilon)\epsilon^\alpha,$$

where $N(\epsilon)$ is the minimal number of balls of the diameter ϵ, needed to cover X. Hence,

$$\bar{\alpha}_c = \overline{dim}_B X, \quad \text{the upper box dimension of } X:$$

$$\overline{dim}_B X = \overline{\lim}_{\epsilon \to 0} \frac{\ln N(\epsilon)}{\ln 1/\epsilon}$$

and

$$\underline{\alpha}_c = \underline{dim}_B X, \quad \text{the lower box dimension:}$$

$$\underline{dim}_B X = \underline{\lim}_{\epsilon \to 0} \frac{\ln N(\epsilon)}{\ln 1/\epsilon}$$

If we assume, in addition, that

$$0 < \underline{r}\left(\underline{dim}_B X\right) < \infty, \quad 0 < \bar{r}\left(\overline{dim}_B X\right) < \infty,$$

then

$$N\left(\bar{\epsilon}_k\right) \sim \bar{\epsilon}_k^{-\bar{\alpha}_c}, \quad N\left(\underline{\epsilon}_j\right) \sim \underline{\epsilon}_j^{-\underline{\alpha}_c},$$

where $\{\bar{\epsilon}_k\}$ $\left(\{\underline{\epsilon}_j\}\right)$ is a sequence of values of ϵ such that

$$\overline{\lim}_{\epsilon \to 0} R\left(\bar{\alpha}_c, \epsilon\right) = \lim_{k \to \infty} R\left(\bar{\alpha}_c, \bar{\epsilon}_k\right)$$

and

$$\underline{\lim}_{\epsilon \to 0} R\left(\underline{\alpha}_c, \epsilon\right) = \lim_{k \to \infty} R\left(\underline{\alpha}_c, \underline{\epsilon}_k\right).$$

If the box dimension exists, i.e.,

$$\overline{dim}_B X = \underline{dim}_B X = b,$$

then

$$N(\epsilon) \sim \epsilon^{-b}, \quad \epsilon \ll 1, \tag{5.3}$$

and for an arbitrary structure $(\mathcal{F}, \xi, \eta)$ we have

$$< \xi(u_i)\eta(u_i)^{\beta} > \sim \epsilon^b. \tag{5.4}$$

provided that $\underline{\alpha}_c = \bar{\alpha}_c = b$. Thus, the described construction allows one to estimate the asymptotic behavior of some average values of functions of sets. We use it to study Poincaré recurrences.

6 Dimension and Capacities for Poincaré Recurrences

Fractal behavior of temporal characteristics: Poincaré recurrences, exit times, transit times, etc., is the key indicator of the presence of sticky orbits. As numerical observations show, an orbit in chaotic sea behaves as follows: after some time interval of "*mixing*", it arrives in the vicinity of a sticky set, and for a long time behaves as a piece of a sticky orbit; after that it returns back to the chaotic sea and is subjected to mixing again; then the process is randomly repeated. Taking into account that the process inside the sticky sets is nonchaotic, as it was shown above, one can understand that such an "*intermittency*" leads to an additional memory of the process and to power tails in the distribution of Poincaré recurrences, exit time, etc., instead of the exponential decay [13]. Fractal analysis can be applied to the temporal characteristics and, particularly, to the Poincaré recurrences.

Orbits in Hamiltonian systems with bounded motion possess a property of repetition of their behavior in time. Given an open set U in M and a point $x \in U$, let us denote by $t(x, U)$ the smallest positive integer for which $g^{t(x,U)}x \in U$ again.

Definition. *1. We call* $t(x, U)$ *the Poincaré recurrence (in fact, the first Poincaré recurrence) for the set* U *specified by the point* x *(it can be* ∞*, of course).*

2. The quantity $\tau(U) = \inf_{x \in U} t(x, U)$ *is called the Poincaré recurrence for the set* U.

Thanks to Poincaré recurrence theorem the number $\tau(u) < \infty$ for any open set u (let us remind that g is area-preserving).

A desired characteristics should be an average value of $\tau(u)$. We use the Carathéodory-Pesin construction to introduce it. Consider the following Carathéodory structure: $(\mathcal{F})$ is the collection of all open sets in the phase space M, $\eta(u) = \text{diam } u$, $\xi(u) = \varphi^q[\tau(u)]$ where $\varphi(t)$ is a monotonically decreasing function and $\tau(u)$ is the Poincaré recurrence for the set $u \in \mathcal{F}$. Then, consider the quantities (5.1) in the form

$$M(\alpha, \epsilon, \varphi, q) = \inf_G \sum_i \varphi(\tau(u_i))^q (\text{diam } u_i)^\alpha \tag{6.1}$$

where infimum is taken over all covers $G = \{u_i\}$, diam $u_i \leq \epsilon$, and (5.2) is considered in the form

$$R(\alpha, \epsilon, \varphi, q) = \inf_H \sum_i \varphi(\tau(u_i)^q (\text{diam } u_i)^\alpha \tag{6.2}$$

and infimum is taken over all covers $H = \{u_i\}$, diam $u_i = \epsilon$. Now apply the general construction and obtain the dimensions $\alpha(q)$ and capacities $\underline{\alpha}(q)$ and $\bar{\alpha}(q)$ for values of the parameter q in an interval. These characteristics are said to be the spectrum of dimensions and the spectra of capacities for Poincaré recurrences. It follows from [36] that $\alpha(q) \leq \underline{\alpha}(q) \leq \bar{\alpha}(q)$. Assume that there is a number q_0 ($\bar{q}_0$ or $\underline{q}_0$) such that $\alpha(q_0) = 0$ ($\bar{\alpha}(\bar{q}_0) = 0$ or $\underline{\alpha}(\underline{q}_0) = 0$). Then the value q_0 ($\bar{q}_0$ or $\underline{q}_0$) is called the dimension (the upper or lower capacity) for Poincaré recurrences.

In order to understand the significance of the definition, let us suppose that $q_0 = \bar{q}_0 = \underline{q}_0$ and $\dim_B M = b$ is the box dimension. Then

$$< \varphi(\tau(u_i))^{q_0} >\sim \epsilon^b, \quad \operatorname{diam} u_i = \epsilon. \tag{6.3}$$

It was shown in [14] that in a nonchaotic situation (minimal sets), the function $\varphi(t) = 1/t$ can serve well, thus for the case of minimal sets we have

$$< \frac{1}{\tau(u_i)^{q_0}} >\sim \epsilon^b \tag{6.4}$$

and we can expect that

$$< \tau(u_i) >\sim \epsilon^{-b/q_0} \tag{6.5}$$

Let us manifest that for the map T on the set of sticky orbits, the function $\varphi(t) = 1/t$ may serve well. For the sake of simplicity, consider the dyadic adding machine with the following distance on Ω_2^+:

$$\operatorname{dist}(\underline{w}, \underline{w}') = \sum_{k=0}^{\infty} \frac{|i_k - i_k'|}{a^{k+1}}, \quad a > 2.$$

It is simple to check that for ϵ, $a^{-n} < \epsilon < a^{-n}(1 - 1/a)^{-1}$, each pair of points $\underline{w}, \underline{w}'$ with $\operatorname{dist}(\underline{w}, \underline{w}') < \epsilon$ has the same n first coordinates $i_0, i_1, \cdots, i_{n-1}$. Moreover, for any two points with different k-th coordinate, $0 \leq k \leq n - 1$, the distance is greater than ϵ. Taking into account the definition of the adding machine, we can see that

$$\tau(u_i) = 2^n \simeq \epsilon^{\ln 2/\ln a} \quad \text{i.e.} \quad \frac{b}{q_0} = \frac{\ell n 2}{\ell n a}.$$

This number does not equal ∞ or 0. This means that $\varphi(t) = 1/t$ is a suitable function. (It is clear that in this case the box dimension is $b = \ln 2/\ln a$, i.e. $q_0 = 1$).

The example in the next section shows that in chaotic cases we should use $\varphi = e^{-t}$, i.e. (6.1) becomes

$$< e^{-q_0 \tau(u_i)} >\sim \epsilon^b \tag{6.6}$$

7 Spectrum of Capacities for Transitive Topological Markov Chains

Let A be the matrix of transitions for a topological Markov chain satisfying the condition of mixing ([35]): there exists $n_0 > 0$ such that A^{n_0} has only positive entries. Therefore, we may introduce such a metric on the space of admissible sequences Ω_A for which

$$\text{diam}([i_0 \cdots i_{n-1}]) = a^{-n}, \quad n > n_0, \quad a > 1, \tag{7.1}$$

$[i_0 \cdots i_{n-1}]$ is a cylinder, i.e.

$$[i_0 \cdots i_{n-1}] = \{\underline{w} = (j_0 j_1 \cdots) \in \Omega_A | j_0 = i_0, \cdots, j_{n-1} = i_{n-1}\},$$

the set of all admissible infinite sequences for which the first n coordinates are determined by $[i_0 \cdots i_{n-1}]$.

Remark. *Such topological Markov chains appear, for example, when we describe some repellers of maps with the constant derivative a. The simplest of them is the map $x \to ax$, mod 1, restricted to a set of orbits belonging to $[0, 1]$ and forming a topological Markov chain.*

We shall calculate the capacity, so we may consider only values of $\epsilon = a^{-n}$, $n \in \mathbf{Z}_+$, and we consider covers of Ω_A by cylinders $\{[i_0 \cdots i_{n-1}]\}$ for a fixed n. Then the Eq. (6.2) becomes

$$R_n(\alpha, q) = \sum_{(i_0 \cdots i_{n-1})} e^{-q\tau(i_0 \cdots i_{n-1})} a^{-\alpha n} \tag{7.2}$$

where the sum is taken over all admissible words $(i_0 \cdots i_{n-1})$. The main idea is to rewrite (7.2) in the form

$$\left(P_1 e^{-q} + P_2 e^{-2q} + \cdots + P_{m_n} e^{-m_n q}\right) a^{-\alpha n} \tag{7.3}$$

where P_k is the number of cylinders $[i_0 \cdots i_{n-1}]$ for which $\tau([i_0 \cdots i_{n-1}]) = k$. Of course, for that we need to show first that $m_n < \infty$. The following result holds.

Proposition 7.1. *$R_n(\alpha, q)$ can be represented in the form (7.3) where*

$$m_n \leq n + n_0$$

(n_0 is a constant in the condition (7.1) of mixing).

It means that Poincaré recurrence for a cylinder of the length n cannot be greater than $n + n_0$. The next proposition tells us that the Poincaré recurrence is realized due to the vicinity of the orbit to a corresponding periodic point.

Proposition 7.2. $\tau([i_0 \cdots i_{n-1}]) = k$ *where* k *is the minimal period of all periodic points belonging to* $[i_0 \cdots i_{n-1}]$.

Let $\lambda_{\max} = \max\{\lambda | \lambda \in spec A\}$. The number of k-periodic points behaves asymptotically as $\lambda_{\max}^k$, $k \gg 1$. It allows one to estimate the upper and lower capacities by using Propositions 7.1, 7.2.

Proposition 7.3. $\underline{\bar{\alpha}}(q) < 0$ *if* $\ell n \lambda_{\max} - q < 0$, *and*

$$\underline{\bar{\alpha}}(q) \leq \frac{\ell n \lambda_{\max} - q}{\ell n a} \tag{7.4}$$

if $\ell n \lambda_{\max} - q \geq 0$.

Proposition 7.4. *The upper capacity* $\bar{\alpha}_c(q)$ *satisfies*

$$\bar{\alpha}_c(q) \geq \frac{\ell n \lambda_{\max} - q}{\ell n a} \tag{7.5}$$

Comparing (7.5) and (7.4) we have

Theorem 2. *(1) If* $q \leq h_{\text{top}}$ *then* $\bar{\alpha}(q) = (h_{\text{top}} - q)/\ln a$;

(2) $\bar{q}_0 = h_{\text{top}} = \ell n \lambda_{\max}$.

Remark. *Taking into account that the number of admissible words* $\{[i_0 \cdots i_{n-1}]\}$ *is asymptotically equal to* $\exp(n h_{\text{top}}) = e^{q_0 n}$, *we can write that*

$$< e^{-q_0 \tau([i_0 \cdots i_{n-1}])} > \sim e^{-q_0 n}, \quad \text{or}$$

$$< \tau([i_0 \cdots i_{n-1}]) > \sim n = -\ln \epsilon / \ln a, \tag{7.6}$$

$(\epsilon = a^{-n})$. *Thus,* $\varphi(t) = e^{-t}$ *works well in an ideal chaotic situation.*

In our considerations, the constant a served as a rate of expansion of the shift map σ at every point $\underline{w} = (i_0, i_1, \cdots)$. We saw that the capacity $\bar{q}_0$ is just the topological entropy, it does not carry new information. In a more real hyperbolic situation when the rates of expansion (and contraction) depend on a point, we may expect that the capacities $\bar{q}_0$ and the topological entropy are independent characteristics.

8 Working with Sticky Orbits

As was shown above, the behavior of Poincaré recurrences in a nonchaotic situation is of the following type

$$< \tau(u_i) > \sim \epsilon^{-\beta}, < \operatorname{diam} u_i > = \epsilon \ll 1,$$

and in a chaotic situation, it is completely different:

$$< \tau(u_i) > \sim -\beta_1 \ell n \epsilon, \quad < \operatorname{diam} u_i > = \epsilon \ll 1,$$

where β and β_1 are constants. How could we describe the behavior of those orbits in the chaotic sea which are spending considerable amount of time in the neighborhood of sticky sets? Let us present some results from our work [15] which show that exit times from a vicinity of a sticky set may reflect some properties of the Poincaré recurrences on the set.

We assume that there is a $SR \left\{ u_{i_0 \cdots i_{n-1}} \right\}$ and the corresponding sticky set S. Let us remind that it follows from the symbolic description (Theorem 1) that there is a sequence of subsets $W_{i_0 \cdots i_{n-1}}$, $h\left([i_0 \cdots i_{n-1}]\right) = W_{i_0 \cdots i_{n-1}}$, of the set S. There are some preliminary results which allow us to believe that the following assumption holds for a general situation.

Main conjecture

$$\operatorname{diam} u_{i_0 \cdots i_{n-1}} \sim \operatorname{diam} W_{i_0 \cdots i_{n-1}}, \quad n \gg 1.$$

In other words, we claim that scaling properties of the islands of the n-th generation, $n \gg 1$, are the same that scaling properties of some subsets of S of the n-th generation which form convinient partitions (coverings) of the sticky set S.

Now consider space-time partitioning that was introduced in [22] – see also [13], [15], and resembles a picture similar to Sierpinsky carpet (Fig. 3). Let the central square be an island of a zero-order generation. Surround the island by an annulus which represent a boundary island layer. It consists of g_1 ($g_1 = 8$ in Fig. 3) subislands of the first generation (dashed small islands in Fig. 1). We can partition the annulus into g_1 domains, so that each of them includes exactly one island of the first generation; then we surround each island of the first generation by an annulus of the second generation and repeat the process. On the nth step the structure can be described by a "*word*"

$$w_n = w(g_1, g_2, \cdots, g_n). \tag{8.1}$$

The total number of islands on the nth step is

$$N_n = g_1 \cdots g_n, \tag{8.2}$$

and any island from the nth generation can be labeled by

$$u_i^{(n)} = u_{i_0 \cdots i_{n-1}}, \quad 1 \leq i_j \leq g_j, \quad \forall j. \tag{8.3}$$

Let us now introduce a time that a particle spends in the boundary layer of an island. This time,

$$T_i^{(n)} = T\left(u_i^{(n)}\right), \tag{8.4}$$

carries all information about the nth generation islands (8.1)-(8.3). By introducing a residence time for each island boundary layer, we have a situation comparable to the plain Sierpinsky carpet or plain fractal situaiton because of the nontriviality of the space-time coupling. In fact, we are attaching an additional parameter responsible for the temporal behavior to the simple geometric construction for a Cantor set.

A simplified situation corresponds to the exact self-similarity of the construction described above, i.e., similar to (2.3):

$$S_i^{(n)} = S^{(n)} = \lambda_S^n S^{(0)}, \quad \forall i,$$

$$T_i^{(n)} = T^{(n)} = \lambda_T^n T^{(0)}, \quad \forall i, \tag{8.5}$$

where $S_i^{(n)}$ is the area of an island $u_i^{(n)}$ and $T_i^{(n)}$ is introduced in Eq.(8.4). Expressions in Eq.(8.5) correspond to equal areas and residence times for all islands of the same generation. Two scaling parameters λ_S and λ_T represent the existence of the exact self-similarity in space and time correspondingly. Precisely such a situation was described in [22,23,13] for maps (2.1) and (2.2), with

$$\lambda_S < 1, \quad \lambda_T > 1. \tag{8.6}$$

In addition to Eq.(8.5), there is a sefl-similarity in the islands' proliferation mentioned in (2.3), i.e.,

$$g_n = \lambda_g^n g_0, \quad \lambda_g \geq 3 \tag{8.7}$$

It follows from Eqs. (8.2) and (8.7) that

$$N_n = \lambda_g^n g_0 = \lambda_g^n \tag{8.8}$$

if we start from the only island $(g_0 = 1)$.

Let us study some characteristics of the process.

Consider a cover of the sticky set S by the sets $W_{i_0 \cdots i_{n-1}} = h\left([i_0 \cdots i_{n-1}]\right)$. Thanks to the main conjecture,

$$\operatorname{diam} W_{i_0 \cdots i_{n-1}} = \sqrt{S^{(0)}} \cdot \left(\sqrt{\lambda_s}\right)^n.$$

Moreover, $T_i^{(n)} = \lambda_T^n T^{(0)}$. Therefore, the sum (6.2) for the function $\varphi(t) = 1/t$ becomes

$$\operatorname{const} \cdot \sum_{(i_0 \cdots i_n)} \lambda_T^{-qn} \left(\sqrt{\lambda_s}\right)^{\alpha n} \tag{8.9}$$

It follows that the spectrum of capacities for Poincaré recurrences on S is

$$\underline{\bar{\alpha}}(q) = 2 \frac{q \ln \lambda_T - \ln \lambda_g}{\ln \lambda_S} \tag{8.10}$$

and the capacities

$$\underline{\bar{q}}_0 = \frac{\ln \lambda_g}{\ln \lambda_T} \tag{8.11}$$

If we consider spectrum of capacities and capacities for the exit times from the vicinity of the sticky set, we arrive exactly at the expresison (8.9) and obtain the same formulas (8.10) and (8.11).

Let us remark that under main conjecture we may calculate the Hausdorff and box dimensions of the SR. For example, in the simplest situtaion described above, the sum (5.2) for the function $\xi(u_i) \equiv 1$ becomes

$$\text{const.} \sum_{(i_0,\ldots,i_{n-1})} (\sqrt{\lambda_S})^{\alpha n} = \text{const } \exp n \left(\ln \lambda_g + \frac{1}{2} \alpha \ln \lambda_S \right).$$

Therefore

$$\alpha_0 = \dim_B S = 2 \ln \lambda_g / |\ln \lambda_S| . \tag{8.12}$$

As in [15], we may consider the space-time partitioning (8.3)-(8.4) to obtain the spectrum and capacities for the reduced times (times per unit of area), etc.

9 Conclusions

Numerical simulations clearly expose a new property of chaotic dynamics in systems that often occur in different applications. This propety can be formulated as intermittency, or stickiness, or nonuniformity of the phase space, and so on. A remarkable feature of the stickiness is that it can be referred to as the existence of "Sticky Riddle" (SR) set embedded into the phase space. The SR is a cause of the existence of bad orbits and can be considered as support for a special kind of dynamics. We speculate about the existence of the SR as a topological and dynamical object with fractal or multifractal properties. The latter can be seen from the distributions of the Poincaré recurrences or exit times, which possesses power-like tails instead of having all finite moments.

10 Acknowledgements

V. Afraimovich was supported by the National Science Council, Republic of China Grant NSC86-2115-M-007-018, and by the U.S. Navy Grant N00014-96-1-0055. G.M. Zaslavsky was supported by the U.S. Navy Grant N00014-96-1-0055 and the U.S. Department of Energy Grant DE-FG02-92ER54184.

We are very thankful to M. Edelman for help in preparing figures.

References

[1] J.D. Meiss, Rev. Mod. Phys. **64**, 795 (1992); Phys. Rev. A. **34**, 2375 (1986).

[2] E. Ott, **Chaos in Dynamical Systems**, Cambridge Univ. Press, Cambridge, 1993.

[3] G.M. Zaslavsky, Chaos **5**, 653 (1995); G.M. Zaslavsky and M. Edelman, Phys. Rev. E **56**, No. 5 (1997).

[4] H.G.E. Hentschel and I. Procaccia, Physica D **8**, 435 (1983); P. Grassberger and I. Procaccia *ibid.* **13** 34 (1984).

[5] U. Frisch and G. Parisi, in **Turbulence and Predictability of Geophysical Flows and Climate Dynamics**, edited by M. Ghill, R. Benzi, and G. Parisi (North-Holland, Amsterdam, 1985).

[6] M.H. Jensen, L.P. Kadanoff, A. Libshaber, I. Procaccia, and J. Stavans, Phys. Rev. Lett. **55**, 439 (1985); T.C. Halsey, M. H. Jensen, L.P. Kadanoff, I. Procaccia, and B.I. Schraiman, Phys. Rev. A **33**, 1141 (1986).

[7] G. Paladin and A. Vulpiani, Phys. Rep. **156**, 147 (1987).

[8] P. Collet, J.L. Lebowitz, and A. Porzio, J. Stat. Phys. **47**, 609(1987).

[9] Y. Pesin and H. Weiss, Chaos **7**, 85(1997).

[10] Y. Pesin and H. Weiss, Journal of Statistical Physics **86**, 233(1997).

[11] G.M. Zaslavsky and M. Tippett, Phys. Rev. Lett. **67**, 3251 (1991).

[12] B.V. Chirikov and D. Shepelyansky, Physica D **13**, 395 (1984).

[13] G.M. Zaslavsky, M. Edelman, and B. Niyazov, Chaos **7**, 159 (1997); G.M. Zaslavsky and B.A. Niyazov, Phys. Rep. **283**, 73 (1997).

[14] V. Afraimovich, Chaos, I, 12 (1997).

[15] V. Afraimovich, and G.M. Zaslavsky, Phys. Rev. E, **55**, 5418(1997).

[16] L.S. Young, Ergod. Theory and Dynamical Systems, **2**, 109(1982).

[17] L. Barreira, Y. Pesin, J. Schmeling, Electromic Research Announc. Amer. Math. Soc. **2**, 69 (1996).

[18] J.-P. Eckmann and D. Ruelle, Rev. Modern Phys. **57**, 617 (1985).

[19] G.M. Shereshevsky, Nonlinearity, **4**, 15 (1991).

[20] L. Barreira, J. Schmeling, Sets of "*non-typical*" points have full topological entropy and full Hausdorff dimension, Preprint, (1997).

[21] G. M. Zaslavsky, D.Stevens, and H. Weitzner, Phys. Rev. E **48**, 1683 (1993).

[22] G.M. Zaslavsky, Chaos **4**, 25 (1994).

[23] G.M. Zaslavsky, Physica D **76**, 110 (1994).

[24] W. de Melo, S. Van Strien, One-dimensional Dynamics, Springer-Verlag, Berlin (1993).

[25] E.B. Vul, Ya.G. Sinai, and K.M. Khanin, in **Advanced Series in Nonlinear Dynamics 1: Dynamical Systems** (ed. by Ya.G. Sinai), World Scientific, Singapore, 501 (1991).

[26] V.K. Melnikov, in **Transport, Chaos and Plasma Physics 2** (eds. S. Benkadda, F. Doveil, Y. Elskens) World Scientific, Singapore, 142, (1996).

[27] W. Young, A Pumir, and Y. Pomeau, Phys. Fluids A **1**, 462 (1989).

[28] V.V. Afanas'ev, R.Z. Sagdeev, and G.M. Zaslavsky Chaos **1**, 143 (1991).

[29] S. Benkadda, S. Kassibrakis, R.B. White, and G.M. Zaslavsky, Phys. Rev. E **55**, 4909 (1997).

[30] G.M. Zaslavsky and M. Edelman, Phys. Rev. E **56**, ... (1997).

[31] J. Machta, J. Stat. Phys. **32**, 555 (1983); J. Machta and B. Reinhold, *ibid.* **42**, 949 (1986); J. Machta and R. Zwanzig, Phys. Rev. Lett. **48**, 1959 (1983).

[32] B. Friedman and R.F. Martin, Jr., Phys. Lett. **105A**, 23 (1984).

[33] A. Zacherl, T. Geisel, and J. Nierwetberg, Phys. Lett. **114A**, 317 (1986).

[34] R.C. Robinson, I. Am. J. Math. **22**, 562 (1970); III. Am. J. Math. **22**, 897 (1970).

[35] A. Katok and B. Hassenblatt. **Introduction to the Modern Theory of Dynamical Systems**, Cambridge University Press, Cambridge (1995).

[36] Y. Pesin, **Dimension Theory in Dynamical Systems: Rigorous Results and Applicaiton**, University of Chicago Press, Chicago, (1997).

[37] C. Carathèodory, Nach. Ges. Wiss. Götingen, 406 (1914).

Part 2:

Fluids and Turbulence

Turbulence: Beyond Phenomenology

A. Tsinober

Faculty of Engineering, Tel-Aviv University,
Tel-Aviv 69978, Israel

> *Correlations after experiments done is bloody bad.*
> *Only prediction is science.*
> **Fred Hoyle**, 1957 *The Black Cloud*, Harper, N-Y.

Abstract. Following a critical overview of the phenomenological aspects a selection of issues which are essentially beyond phenomenology are presented with the emphasis on issues which can be effectively addressed *quantitatively* via *geometrical statistics*. In particular, it is argued that regions with concentrated vorticity (tubes-filaments-worms) are not that important as it has been thought before and do not seem to play a special role in the overall dynamics of turbulent flows: these regions are more the consequence rather than the dominating factor of the turbulence dynamics. The 'random sea'/background, in which are embedded the strongest filaments, appears to be strongly non-Gaussian, not passive and possessing distinct structure. Moreover, apart of enstrophy dominated regions and the background turbulent flows contain other *dynamically* more important regions. These are the strain dominated regions with the following subregions of special interest: *i* - regions responsible for the highest enstrophy generation and its rate, and associated with high values of the largest eigenvalue of the rate of strain tensor Λ_1, and finite curvature of vortex lines, alignment between vorticity ω and the corresponding eigenvector λ_1, *ii* - regions, which are wrapped around the enstrophy dominated regions and associated with alignment between ω and λ_2 and mostly positive values of the intermediate eigenvalue Λ_2, and *iii* - regions with large magnitude of the smallest eigenvalue Λ_3, alignment between ω and λ_3, large curvature of vortex lines and most of vortex compressing, tilting and folding. Among other issues are reduction of nonlinearity, non-Gaussian nature of turbulence and 'kinematic' effects, and nonlocality.

Keywords. Turbulence, phenomenology, geometrical statistics, reduction of nonlinearity, nonlocality, unresolved issues.

This paper is based in part on the lectures delivered by the author in *Ecole Normale Supérieure, Université Paris VII* in May 1995 and in *Laboratoire Modélisation en Mécanique, Université Paris VI* in February 1996 and on the latest work performed since then.

86

1   Introductory notes and background

John von Neumann wrote in his report to the Office of Naval Research in 1949
[179]: *Turbulence is a phenomenon which sets in in a viscous fluid for small
values of the viscosity coefficient ν (reckoning ν in significant units, that is, as
the reciprocal Reynolds' number $1/Re$ [2]), hence its purest, limiting form may
be interpreted as the asymptotic, limiting behavior of a viscous fluid for $\nu \to 0$.*
This asymptotic (long time) behaviour at large enough but finite (not necessarily very large) Reynolds numbers is usually referred as *fully developed (strong)
turbulence* (FDT).[3]
It is commonly believed that FDT in incompressible fluid is described by the
Navier-Stokes equations (NSE) supplemented by initial (IC) and boundary conditions (BC)

$$\partial_t \mathbf{u} + \boldsymbol{\omega} \times \mathbf{u} = -\boldsymbol{\nabla}\left(p/\rho + u^2/2\right) + \nu\nabla^2\,\mathbf{u}, \qquad \boldsymbol{\nabla}\cdot\mathbf{u} = 0, \qquad (1,2)$$

where, $\partial_t \equiv \partial/\partial t$, $\mathbf{u}(\mathbf{x},t)$ and $p(\mathbf{x},t)$ are the fluid velocity and pressure, $\boldsymbol{\omega} =$
curl $\mathbf{u}$, ρ is the density and ν is the kinematic molecular viscosity (both con-
stant).
Though there exist a set of deterministic differential equations probably containing (almost) all of turbulence, most of our knowledge about turbulence comes
from experiment (laboratory, field and numerical). This was understood long ago
by A.N.Kolmogorov:...*I soon understood that there was little hope of developing
a pure, closed theory, and because of absence of such a theory the investigation
must be based on hypotheses obtained on processing experimental data* [236] [4].
Much later he wrote that *the observational material is so large, that it allows to
foresee rather subtle mathematical results, which would be very interesting to
prove* [144].
Indeed the heaviest and the most ambitious armory from theoretical physics and
mathematics was tried for more than fifty years, but without much success [5] -
FDT, as a physical and mathematical problem remains unsolved. This state of
matters is reflected in the characterisitic feature of reviews of turbulence research
which to a large extent deal with methods (and their failures) rather than with
results, showing that turbulence remains among the fields with overproduction
of publications without any real breakthrough in understanding. In addition
to this mathematical/theoretical 'deadlock'[179] one of the main difficulties in
turbulence research in general, and in all the applications in particular, is that
high enough values of Reynolds numbers are inaccessible in the forseeable future

---

[2]The Reynolds number is defined as $Re = LU/\nu$, where L, U are typical length and velocity
scales and ν is the kinematic molecular viscosity of the fluid.

[3]The emphasis in the discussion that follows is mainly on strictly three-dimensional fluid-
dynamical turbulence (again FDT) in incompressible flows.

[4]Therefore, the importance of experimental research in turbulence goes far beyond the view
of those who think of experimentalist as a superior kind of professional fixer knowing how to
turn nuts and bolts into a confirmation of their theories. This is the main reason that this
paper is biased experimentally.

[5]For an up-to-date critical review of these attempts see Chapter 9 in [90].

neither in laboratory nor via direct numerical simulations [171]. There are also many 'technical'difficulties, such as methods of analysis of huge amounts of data on turbulent flows. Ironically turbulence is the most important field in fluid dynamics and in a vast variety of applications. Not accidentally the most sagacious 'practitioners'of turbulence claim that the best thing which can happen in *applied* fluid mechanics would be the creation of a basic theory of developed turbulence.

In the sense of the epigraph above turbulence *'is bloody bad'*: the number of predictions made in the field is limited by the number of fingers on one hand - the rest are *correlations after experiments done*, i.e. 'postdictions'.

In spite of the absence of a sound theory, a hundred years of systematic study resulted in a considerable amount of mostly phenomenological knowledge on *qualitative* manifestations and *quantitative* properties of turbulence, so turbulence *can no longer be viewed as incomprehensible* [61] and *is no longer a complete mystery* [107].

The *qualitative* properties/features of all turbulent flows at high enough Reynolds numbers are essentially the same, i.e. they are *universal*. It is natural to use this *qualitative universality* for a more detailed identificiation of FDT.

Major qualitative properties/features of FDT [174], [235]

● - Intrinsic spatio-temporal randomness, irregularity. FDT is definitely chaos. However, vice versa, generally, is not true: many chaotic flow regimes are not turbulent (e.g. Lagrangian/kinematic chaos, laminar "turbulent" flows).

● - Extremely wide range of strongly interacting scales, i.e. turbulent flows are **large** systems. In atmospheric flows relevant scales range from hundreds km to parts of a mm, i.e. it possess $\sim 10^{18}$ excited degrees of freedom. Hence extreme complexity of FDT.

● - Highly dissipative. A source of energy is required to maintain turbulence. Continuous energy flux from large to small scales: the energy supply is at mostly large scales [6], its dissipation is at small ones. Statistical irreversibility.

● - Three-dimensional and rotational! It is a "random" field of vorticity with predominant vortex stretching (!), i.e. continuous net production of enstrophy by inertial nonlinear processes, which is dissipated by viscosity. Random potential flows are not FDT.

● - Strongly diffusive (random waves are not), i.e. it exhibits strongly enhanced transport processes of momentum, energy, passive objects (scalars, e.g. heat, salt; vectors, e.g. material lines, magnetic field). Laminar 'hyperbolic' flows exibit enhanced transport of passive objects only.

Whereas these *qualitative* features of FDT are *universal* (there are more more *specific* for FDT, see below the main text) and generally characterise FDT flows as a whole, the *quantitaive* properties vary largely with the range of scales of interest. The *large* scale (LS) properties of FDT depend on *particular* mechanisms

---

[6]The common view on turbulence dynamics is via the Richardson-Kolmogorov cascade of energy (the famous poem by Richardson). However, there are numerous examples in which turulence develops from small scales into the larger ones, e.g. in all spatially developing turbulent flows, both free such as turbulent jets, wakes, plumes, and wall bounded such as turbulent boundary layers.

generating turbulence and, generally, are not universal though they possess properties which in some sense are universal. It is the *small scale* (SS) turbulence which, since Kolmogorov, is believed to possess a number of universal properties independent of the large scale flow structure [7]. Small scale FDT is usually considered in a narrow sense as FDT *per se* and is one of the main themes here. This SS turbulence, especially its scaling properties, remains for more than fifty years one of the most active fields of inquiry. Derivation of scaling properties of FDT directly from NSE analytically remains one of the most popular illusive goals of theoretical research. This (scaling) and other phenomenological aspects are extensively reviewed in [90], [228], so only additional notes will be made referring to aspects which deserve updating and comments as a background for the sequel.

It should be stressed that some of the universal properties of FDT are characteristic of much broader class of nonlinear systems, others are *specific* for FDT. Both will be addressed in the sequel with the emphasis on the latter.

2 An Overview of Phenomenology

The term *phenomenology* is defined in the *New Webster's New World Dictionary, College edition (1962)* as follows:
Phenomenology - *The branch of a science that classifies and describes its phenomena without any attempt at explanation.*
It is claimed frequently that in turbulence research phenomenology helps to *explain* some features of turbulent flows. The reader is invited to fashion an opinion of his own whether this is really the case. In any case the combination *phenomenological understanding of tubrulence* sounds somewhat too ambitious.

2.1 Universality, local isotropy and scaling

The first main ingredient of Kolmogorov theory is the hypothesis of local isotropy *'in an arbitrary turbulent flow with sufficiently large Reynolds number $Re = \frac{LU}{\nu}$ in sufficiently small regions G of the four-dimensional space (x_1, x_2, x_3, t) not lying close to the boundaries of the flow or other singularities of it'*[141]. Together with his definition of local isotropy (definition 2 in [141]) this hypothesis postulates at large Reynolds numbers *restoring in the statistical sense of all the symmetries of the Navier-Stokes equations* [8] *locally in time and space* — except of the scaling one. This hypothesis was confirmed in several experiments, the latest and the most impressive ones made by in high-Reynolds-number turbulent flows with mean shear both without and under the influence of large extra mean strain rates in the large wind tunnel of NASA Ames, [211], [212].

---

[7]Though quite an opposite possibility was pointed out: ... *perhaps there is no 'real turbulence problem', but a large number of turbulent flows and our problem is the self imposed and possibly impossible task of fitting many phenomena into the Procrustean bed of a universal turbulence theory* [213]; see also[115], [116].

[8]Space/time translations, Galilean transformations, the full group of rotations, which include reflections.

In order to cope with scale invariance symmetry of NSE at $Re \gg 1$ Kolmogorov introduced his famous *similarity hypotheses*: the first one stating that all the statistical properties of the small scale FDT are *uniquely* defined by *mean* dissipation e[9], $\langle \epsilon \rangle$, and kinematic viscosity, ν , while the second hypothesis referred to the so called inertial range of scales ℓ, $L \gg \ell \gg \eta$, in which the statistical properties of FDT are *uniquely* defined by $\langle \epsilon \rangle$ only. Here $\eta = \nu^{3/4} \langle \epsilon \rangle^{1/4}$ is the Kolmogorov dissipation scale. The second similarity hypothesis implies that at $Re \gg 1$ in the inertial range the scale invariance symmetry of NSE is the same (in the statistical sense) as for the Euler equations (EE) [10] in which the kinematic viscosity $\nu \equiv 0$, and that $\langle \epsilon \rangle$ remains *finite* as $Re \to \infty$. From this Kolmogorov *predicted* his famous 2/3 law, i.e. $h = 1/3$ [141].

$$S_2^{\|}(r) \propto C_2 \langle \epsilon \rangle^{2/3} r^{2/3}, \tag{3}$$

where $S_p^{\|}(r) = \langle (\Delta u_{\|})^p \rangle$, $p = 2$ - is the second order structure function of the longitudinal velocity increment $\Delta u_{\|} \equiv [\mathbf{u}(\mathbf{x} + \mathbf{r}) - \mathbf{u}(\mathbf{x})] \cdot \mathbf{r}/r$, and C_2 is an 'absolute' constant. Another most remarkable *quantitative prediction* made by Kolmogorov is his -4/5 law obtained as a direct consequence from NSE [142]

$$S_3^{\|}(r) = -4/5 \langle \epsilon \rangle r, \tag{4}$$

in which the constant $C_3 = -4/5$. Both predictions have a very solid experimental confirmation (see [90], [174], [175], [211], [212], [228] and references therein). It is noteworthy that in many cases the inertial interval for $S_p^{\|}(r)$ (and/or energy spectra) is considerably longer than that for the structure function $S_p^{\perp}(r)$ of the transverse velocity increments $\Delta \mathbf{u}_{\perp} = \Delta \mathbf{u} - \Delta \mathbf{u}_{\|}$ (see e.g. [129], [158], [175]) – a fact which is definitely beyond phenomenology. This is directly related to the observation that local isotropy occurs in a range of scales much shorter (or even in some cases is not observed at all, see section 3.1) than the range of scales in which (3) and (4) are observed.

Kolmogorov theory raised a number of fundamental issues which dominate to a large extent the contemporary research in FDT. Some of these issues are discussed as a backgroud for the sequel.

The first issue is whether the normalized mean dissipation $\varepsilon = U^3 L^{-1} \langle \epsilon \rangle$ tends really to a finite limit as $Re \to \infty$ or it is Re dependent even at very large Reynolds numbers. There are many speculations on this subject, while the experimental evidence, though favoring the former, is extremely limited. Recent

---

[9]For a Newtonian fluid the local dissipation $\epsilon = 2\nu s_{ik} s_{ik}$, where $2s_{ik} = \partial u_i / \partial x_k + \partial u_k / \partial x_i$ is the rate of strain tensor. The particular expression for ϵ is of no importance for Kolmogorov theory of the inertial range (except of the extent of this range, i.e. its upper cutoff, η, which is dependent on the particular mechanism of dissipation), since viscosity appears in it as a 'subsidiary agent'[179] providing a 'sink'of energy. Therefore any small amount of any kind of this 'subsidiary agent'will do whatever the particular mechanism of energy dissipation is. Moreover, as conjectured by L. Onsager in 1949 [190] and proved recently [67], [79], the role of such a subsidiary agent can be played by chaotic motion on arbitrarily small scales in an inviscid fluid provided the flow field is 'rough'enough at any scale whatever small.

[10]The EE equations are invariant if simultaneously the distance, time and velocity are scaled by λ, λ^{1-h} and λ^h correspondingly, with arbitrary h.

results obtained using glycerol, water and low temperature helium gas [41] allowed to show that $\varepsilon = const$ within the range of Reynolds numbers varying over more than three decades $(3 \cdot 10^3 < Re < 7 \cdot 10^6$, see figure 1) when the flow is forced by 'rough' moving boundaries.[11] However, when the moving boundaries were smooth $\varepsilon(Re) = const$ was a decreasing function of Reynolds number similar to such Re-dependence in other configurations, e.g. in pipes with smooth walls. Nevertheless, the *bulk* of the flow exhibited clear Re-independent behaviour, indicating that the main difference between the two cases is due to the dissimilarity in the coupling between the boundaries and the bulk of the flow. This is in agreement with the results of similar experiments with drag reducing additives which did not exhibit any drag reduction in case of 'rough' boundaries [40].

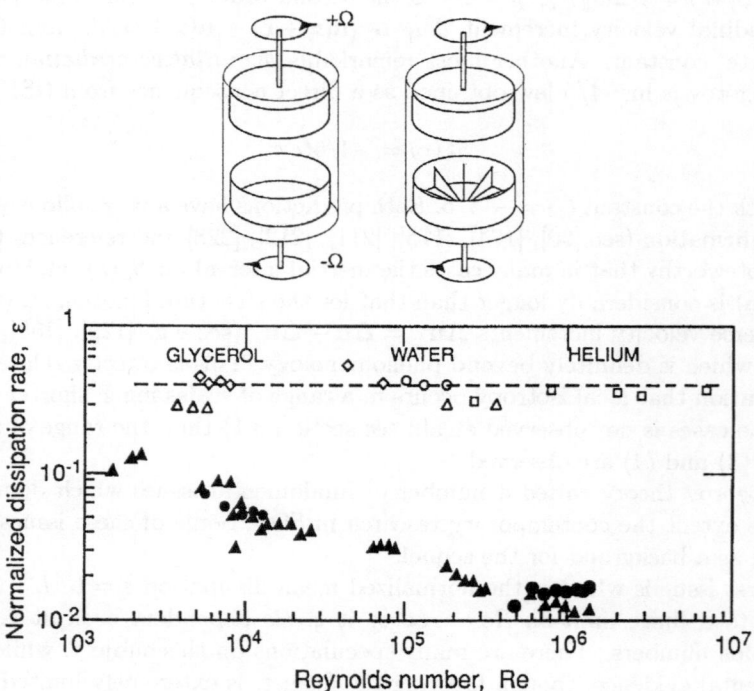

Figure 1. Reynolds number dependence of normalized dissipation rate of energy ε in a turbulent flow in a circular tank forced by counter rotating top and bottom. *Adapted from the Ph. D. Thesis by O. Cadot 1995, Laboratoire de Physique Statistique de l'Ecole Normale Supérieure, Université Paris VII*, see also [41]; top - schematic of water/glycerol and low temperature helium gas facilities [72], [22].

---

[11]It is noteworthy that in the engineering practice this fact has been recognized long ago in a great variety of flow configurations, see e.g. [118].

indicating that drag reduction phenomena in turbulent flows are mainly associated with the specific processes of the coupling between the boundaries and the bulk of the flow.

It is noteworthy that rigorous upper bounds of ε are independent of Reynolds number at large Re (see [39], [136], [180] and references therein), and thereby are consistent with the experimental results. There are some indications from recent DNS of NSE [49] that the statistical properties of inertial range are independent of the dissipation mechanism in conformity with Kolmogorov ideas, see also [179].

The Reynolds independent behaviour of some global characteristics of turbulent flows (such as the total dissipation in the above example) at large Re comprises one of their *quantitative* universal properties. This is distinct from *some* universal (mostly scaling) properties of small scale turbulence presumably independent of the large scale flow structure [177].

2.2   Intermittency

In a broad sense the term *intermittency* refers to an extremely uneven *spatial* distribution of dissipation, though most small scale quantities (such as enstrophy ω^2, here $\boldsymbol{\omega} = curl$ **u** is the vorticity vector and others, see below) exhibit such a spiky behaviour as well. There is no general agreement on the meaning of the term *intermittency,* but it seems unjustified to associate it *solely* or *mostly* with the so called anomalous scaling, just like defining intermittency via r-dependence of the flatness $\langle (\Delta u_{\parallel})^{2n} \rangle / \langle (\Delta u_{\parallel})^2 \rangle^n$ (i.e. associating it with *even* moments *only* [151]). Nevertheless, it is commonly believed that among the manifestations of the small scale intermittency is the experimentally observed deviation of the scaling exponents in relations of the type (3) for structure functions $S_p^{\parallel}(r)$ for $p > 3$ from the values implied by the Kolmogorov theory (i.e. anomalous scaling) which in turn is due to rare strong events. Namely,

$$S_p^{\parallel}(r) \propto C_p \langle \epsilon \rangle^{p/3} r^{\zeta_p^{\parallel}}, \tag{5}$$

where $\zeta_p^{\parallel} = p/3 - \mu_p < p/3$ is a convex nonlinear function of p (see figure 2).

It was claimed recently on the basis of high resolution DNS of NSE and laboratory data [49], [50] that ζ_p *exhibit unambiguous departures from the Kolmogorov 1941 theory* not only for $p \geq 4$ but also for $-1 \leq p \leq 2$, though this departure is less than 5% (see also [261]).

Thus - unlike the other symmetries - the assumed global scale invariance (single scaling exponent h) of NSE at $Re \gg 1$ is broken.

Following the Landau objection to universality [12] Kolmogorov put forward his refined similarity hypothesis (RSH) [143] in which he replaced the mean dissipation $\langle \epsilon \rangle$ by 'local'dissipation ϵ_r averaged over a region of size r. However, 'once the K41 theory was abandoned a Pandorra's box of possibilities is opened'[147].

---

[12]The famous remark by Landau in the first Russian edition of Fluid Mechanics by Landau and Lifshitz 1944 about the role of large scale fluctuations of energy dissipation rate, i.e. nonuniversality of both the scaling exponents ζ_p and the prefactors C_p in (5).

Indeed, the RSH was followed by numerous phenomenological models attempting to describe some aspects of intermittency, such as the multifractal model or its alternative using the statistics of the so called breakdown coefficients (ratios $\epsilon_{r_i}/\epsilon_{r_k}, r_i \leq r_k$) and great many others (for references see [30], [31], [58], [74], [90], [99], [215] which include the most recent ones).

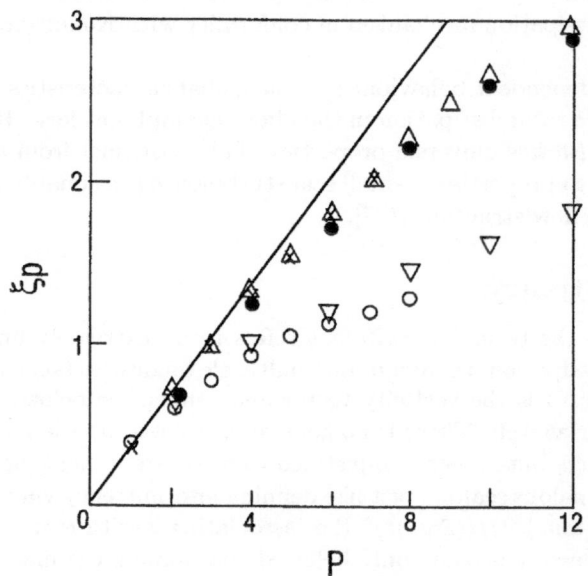

Figure 2. Exponents of structure functions for the longitudinal velocity component ($\triangle, \bullet, \times$) and temperature ($\triangledown, \circ$); $\triangle, \triangledown$ - [4], $\circ$ - [210], $\bullet$ - [259]; $\times$ - exponents of structure functions for the transverse velocity component [185].

All of these models (including the RSH [172]) are in good agreement with the experimental and numerical evidence, e.g. these models exibit the same scaling properties (and some other such as PDFs') as in real turbulence [7], [12], [49], [44], [26], [206], [209], [210], [259]. It is noteworthy that many of these models are based on *qualitatively different* premises/assumptions and with few exceptions have no direct bearing to NSE (for a partial list of recent examples see [20], [25], [44], [30], [31], [33], [38], [48], [74], [77], [99], [172], [178], [218], [215], [261], [277] and numerous references therein).[13] The most common justification for the preoccupation with such models is that (at least some of them) they share

---

[13]Therefore the success of such models can hardly be evaluated on the basis of how well they agree with experiments. For example, there exist a great many theories which produce the $k^{-5/3}$ energy spectrum for *qualitatively* and/or *physically different reasons*. A recent example is a suggestion that the *specrum of fully developed turbulence is determined by the equilibrium statistics of the Euler equations and that a full description of turbulence requires only a perturbation, small in some appropriate metric, of a Gibbsian equilibrium* [62].

the same basic symmetries (perhaps also some hidden symmetries), conservation laws and some other general properties, etc. as the NSE. The general belief is that this - along with the diversity of such systems (there are many having nothing to do with FDT, e.g. granular systems, financial markets, activity in brain) - is the reason for the above mentioned agreement. However, this is not really the case, e.g. in [147] a counter example of a *'dynamical equation is exhibited which has the same essential invariances, symmetries, dimensionality and equilibrium statistical ensembles as the Navier-Stokes equations but which has radically different inertial-range behaviour'*! Many models, such as mostly popular GOY model (for references see [25], [44], [99], [126], [161]), a variant of shell models originated with the systems of hydrodynamical type of Obukhov since 1969 [187], exhibit *temporal* chaos only. Therefore, such and most of other models hardly can be associated with the intermittency of *real* FDT which involves essentially *spatial* chaos as well. Moreover, for the above reasons the agreement between such models and experiments (both laboratory and nimerical) cannot be used for evaluation of the success of such models.

Note that there exists a controversy regarding the exponents $\zeta_p^\perp$ of structure functions for the *transverse* velocity increments ($\Delta \mathbf{u}_\perp = \Delta \mathbf{u} - \Delta \mathbf{u}_\parallel$). According to [42], [46], [45], [127], [185], these are practically indistinguishable from the ones for longitudinal velocity components ($\zeta_p^\parallel$), whereas a substantial diference between the two was found in [31], [32], [58], [70], [102], [105] such that the transverse structure functions are more intermittent than the longitudinal ones both in the sense of deviation of scaling exponents and behaviour of PDFs of velocity increments (especially their *tails*), just like lateral velocity derivatives are more intermittent than the longitudinal ones, and enstrophy is more intermittent than the total strain (see e.g. references in [32] and [57]). It is noteworthy that the discrepancy between $\zeta_p^\perp$ and $\zeta_p^\parallel$ is noticalble for $p > 6$. Therefore, it is not garanteed that it is not an artifact of insufficient number of significantly contributing data points for $p > 6$, i.e. large *total* number of data points may be not sufficient for determination of $\zeta_p^\perp$, since the main contribution to transverse velocity increments comes from enstrophy dominated regions [32], occupying much smaller volume than strain-dominated regions (four times smaller following the results [32]), which - according to [32] - contribute mostly to the longitudinal velocity increments.[14] Note that the deviation of $\zeta_p^\parallel$ from $p/3$ ($\zeta_p^\parallel < p/3$) was much larger when the number of data points was not large enough (see, e.g. figure 14 in [3]), just like the deviation observed for $\zeta_p^\perp$ in [31], [32], [58], [70], [102], [105]. The difference between scaling exponents $\zeta_p^\perp$ and $\zeta_p^\parallel$ for longitudinal and transverse velocity increments - if it is not an artifact - points to a kind of lack of universality (i.e. $\zeta_p^\parallel$ are not that universal). It was proposed, therefore, in

---

[14]However, it was shown in large Reynolds number experiments (up to $Re_\lambda \sim 3 \cdot 10^3$) [198] that the Kolmogorov refined similarity hypothesis (RSH) $\Delta u_r^\parallel = \beta_1 (r\epsilon_r)^{1/3}$ is valid both when the dissipation ϵ_r averaged over scales r was estimated via its conventional surrogate $15\nu(\partial u_1/\partial x_1)$, i. e. based on *longitudinal* velocity component, and the one based on the *lateral (transversal)* velocity component $\frac{15}{2}\nu(\partial u_3/\partial x_1)$. Here β_1 is a stochastic variable independent both of r and Re.

[58] (see also [57]) that *no more than the two sets of independent exponents* (i.e $\zeta_p^{\parallel}$ and $\zeta_p^{\perp}$) *are required for describing the scaling of all small-scale features as longitudinal and transverse velocity increments, dissipation, enstrophy, and circulation.* It should be reminded that there exist other 'universality'proposals involving 'much more'scaling exponents (see e.g. [90], [166], [167] and references therein).

All the above is a clear indication that the question about the origins of intermittency in *real* FDT remains open. Similarly open are the questions on universality of the intermittency manifestations in FDT, though judging by the multitude of models of intermittency there is no universality whatsoever.

Phenomenology and models only will hardly be useful and convincing, since almost any dimensionally correct model, both right or wrong, will lead to correct scaling without appealing to NSE and/or elaborate physics: '*The wonderful thing about scaling is that you can get everything right without understanding anything*'[149]; '*...it is clear that if a result can be derived by dimensional analysis alone... then it can be derived by almost any theory, right or wrong, which is dimensionally correct and uses the right variables*'[37], i.e. scaling laws are not necessarily theories.[15] With all the importance of scaling, turbulence phenomena are infinitely richer than their manifestation in scaling and related things. Most of these manifestations are beyond the reach of phenomenology. Phenomenology is inherently unable to handle the structure of turbulence in general and phase and geometrical relations in particular, to say nothing of dynamical aspects such as build up of *odd* moments, interaction of vorticity and strain resulting in positive net enstrophy generation/predominant vortex stretching (see below). It seems that there is little promise for progress in understanding FDT in going on asking questions about scaling and related matters only [88], [244] without looking into the structure and, where possible, basic mechanisms which are *specific* of turbulent flows. A recent attempt to relate the scaling exponents $\zeta_p^{\parallel}$ and $\zeta_p^{\perp}$ with some aspects of turbulence structure was made in [32] (see also [255]).

3 Beyond phenomenology

3.1 On possible origins of small scale intermittency of FDT

These are roughly of two kinds, kinematic and dynamic.

Direct interaction between large and small scales. One of the manifestations of such interaction is that in many situations the small scales do not forget the anisotropy of the large ones. There exist considerable evidence for

---

[15]Indeed the Kolmogorov theory and many subsequent models used *dissipation* as a basic quantity, i. e. intimately related to *strain*. Several later theories are based on hierarchies of *vorticity* dominated structures. Most of *both* kinds of these theories are in good agreement with experimental results. However, while there is a basic reason, not only on dimensional grounds, for RSH $\Delta u_r^{\parallel} = \beta_1 (r\epsilon_r)^{1/3}$, since it can be seen as a 'local'version of the -4/5 Kolmogorov law $\langle (\Delta u^{\parallel})^3 \rangle = -4/5 \langle \epsilon \rangle r$, a similar claim that $\Delta u^{\perp} = \beta_2 (r\Omega_r)^{1/3}$ $(\Omega = \nu\omega^2)$ [58] remains just one more dimensionally - but not necessarily physically - correct relation (note that $\langle (\Delta u^{\perp})^3 \rangle \equiv 0$). Here, β_1, β_2 are stochastic variables independent of Re and r.

this [5], [75], [129], [130], [200], [266], [275], [276]. Along with deviations from local isotropy there are other manifestations of direct interaction between large and small scales [46], [129], [130], [156], [157], [173], [196], [229] (for a list of earlier references see [241]). These effects seem to occur due to various external constraints like boundaries, initial conditions, forcing (e.g. as in DNS) mean shear/strain, centrifugal forces (rotation), buoyancy, magnetic field, etc., which usually act as an organizing factor, favoring the formation of coherent structures of different kinds (quasi-two-dimensional, helical, hairpins, etc.). These are as a rule large scale features which depend on the *particularities* of a given flow and thus are *not universal*. These structures, especially their edges are believed to be responsible for the contamination of the inertial range.

The following example of a 'simple' turbulent flow provides reliable information on the existence and importance of the direct interactions between the large and small scales in turbulent flows. Consider a unidirectional *in the mean* fully developed turbulent flow such as the flow in a plane channel in which all statistical properties depend on the coordinate normal to the channel boundary, x_2, only. In such a flow a simple *precise* kinematic relation is valid as follows

$$d\langle u_1 u_2 \rangle / dx_2 \equiv \langle \boldsymbol{\omega} \times \mathbf{u} \rangle_1 = \langle \omega_2 u_3 - \omega_3 u_2 \rangle = 0, \tag{6}$$

which is just a consequence of the vector identity $(\mathbf{u} \cdot \boldsymbol{\nabla}) \mathbf{u} \equiv \boldsymbol{\omega} \times \mathbf{u} + \boldsymbol{\nabla}(p/\rho + u^2/2)$ in which incompressibility and $d\langle \cdots \rangle / dx_{1,3} = 0$ where used, and $\langle \cdots \rangle$ means an average in some sense (e.g. time or/and over the planes $x_2 = const$, etc.). Since in these turbulent flows $d\langle u_1 u_2 \rangle / dx_2$ is essentially different from zero at *any arbitrarily large* Reynolds number (see figure 3), one can see from (6) that at least some correlations between velocity and vorticity are essentially different from zero too.

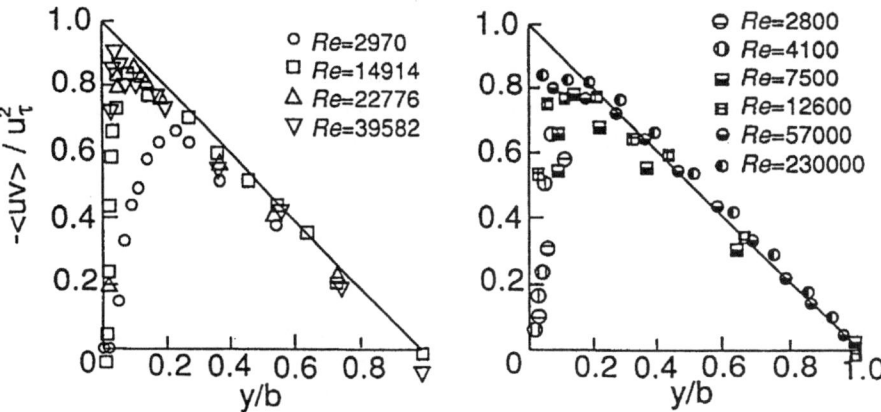

Figure 3. Dependence of the mean Reynolds stress $\langle u_1 u_2 \rangle$ on the distance from the wall in turbulent flows in channels of cross section with large aspect ratio. Adapded from [269].

Note that in isotropic flows $\langle \boldsymbol{\omega} \times \mathbf{u} \rangle \equiv 0$. Since vorticity is essentially a small scale quantity the relation (6) is a clear indication of a dynamically important statistical dependence between the large ($\mathbf{u}$) and small ($\boldsymbol{\omega}$) scales. Without this dependence $d\langle u_1 u_2 \rangle / dx_2 \equiv 0$, which means that the mean flow would not 'know'about its turbulent part at all. It is noteworthy that both correlation co-efficients $\frac{\langle \omega_2 u_3 \rangle}{\langle \omega_2^2 \rangle^{1/2} \langle u_3^2 \rangle^{1/2}}$, $\frac{\langle \omega_3 u_2 \rangle}{\langle \omega_3^2 \rangle^{1/2} \langle u_2^2 \rangle^{1/2}}$ (and many other statistical characteristics, e.g. measures of anisotropy) are of order 10^{-2} even at rather small Reynolds numbers. Nevertheless, in view of the dynamical importance of interaction be-tween velocity and vorticity in turbulent shear flows (the relation (6) is approx-imately valid in almost many important turbulent flow such as boundary layers, wakes, jets, etc. in which $d\langle \cdots \rangle / dx_{1,3} \ll d\langle \cdots \rangle / dx_2$) such 'small'correlation by no means does not imply absence of dynamicaly important statistical de-pendence and direct interaction between large and small scales. In fact, the above result is related to the simple kinematic fact that after all $\mathbf{u} = curl\ \boldsymbol{\omega}$. In this respect it belongs to origins of intermittency of kinematic nature, but not $d\langle u_1 u_2 \rangle / dx_2 = 0$, which is definitely a dynamical effect. The direct interaction between large and small scales similar to the one in the above example may exist in a much broader class of turbulent flows and regions in these flows, e.g. with appropriate scale (in time and space) separation such as vorticity 'pancakes'[36]. **Multiplicative noise intermittency of passive objects in random me-dia.** It is known for about thirty years that passive scalars exhibit anomalous scaling behaviour and strong intermittency (see figure 2) even in pure Gaussian random velocity field (see [4], [56], [146], [151], [152], [153], [201], [220], [221], [272], [273], [274] and references therein). Similar behaviour is exhibited by a passive vector in a random Gaussian field [154], [205], [256]. Though these are dynamically *linear* systems with the so called multiplicative noise (i.e. statisti-cally they are 'nonlinear') and in this sense are kinematic in respect with real FDT, they may reflect the contribution of kinematic nature in real turbulent flows as it was demonstrated in [223]. It is noteworthy that the intermittency ef-fects in such linear systems are *stronger* than in FDT [4], [154], [210] and exhibit anomalous scaling, which, generally, is *nonuniversal* [59]. On other aspects of passive scalar 'misbehaviour'in turbulent flows see [230], [110], [200], [220], [227], [242] and references therein. In view of the recent progress in this field it was claimed that *investigation of the statistics of the passive scalar field advected by random flow is interesting for the insight it offers into the origin of intermittency and anomalous scaling of turbulent fluctuations* [201]. More precisely it offers an insight into the origin of intermittency and anomalous scaling of fluctuations in random media generally and *independently* of the nature of the random motion [272], [273], [274], i.e. it gives some insight into the contributions of *kinematic nature*, but does not offer much regarding the *specific dynamical* aspects of strong turbulence as is FDT. Moreover, anomalous diffusion (including scaling) of pas-sive objects occurs in purely laminar flows in Eulerian sense (E-laminar flows) as a result of Lagrangian chaos (L-turbulent flows). For examples, see [10], [11], [51], [268], [271] and references therein (see section 8).

'Near'singularities. It is not known for sure whether Euler equations (EE) at large Reynolds numbers and/or even NSE develop a genuine singularity in finite time, though there is some evidence that at least for EE this may be true [36], [65], [90], [101], [135], [184], [188]. Whatever is the real situation it seems a reasonable speculation that these 'near'singularities *trigger topological change and large dissipation events* (for NSE); *their presence is felt at the dissipation scales and is perhaps the source of small scale intermittency* [65], though it does not help to understand the *inertial range* intermittency without invoking the *reacting back* of the dissipation range on the inertial range. This is possible due to mentioned above direct coupling between the large and small scales and other nonlocal effects [80], e.g. the 'bottleneck phenomenon'[84] in which viscosity leads to a 'pileup'of energy in the inertial range of scales. The experimentally observed phenomenon of strong drag reduction in turbulent flows of dilute polymer solutions and other drag reducing additives [104] is another example of such a 'reacting back'effect.

Near singular objects associated with non-integer values of the energy spectrum scaling exponents are thought to be closely related with some structure(s) and, consequently, with intermittency of turbulent flows [255].

In any case the 'near'singular objects may be among the origins of intermittency of dynamical nature.

Instantons. This is a recently proposed approach seeking the dynamical origin of intermittency in the so called *instantons* [85], [86], which are path integral analogy of, e.g. solitons associated with the separatrix in the phase space of a pendulum (unstable equilibria). The instanton is both a dynamical and a statistical object - a kind of average characterising the very intense events of the kind under consideration and containig information on their prehistory.

The most popular view is that intermittency *specifically* in FDT is associated mostly with some aspects of its spatio-temporal structure, especially the spatial one. Hence, the close relation between the origin(s) and meaning of intermittency and structure of turbulence. Just like there is no general agreement on the origin and meaning of the former there is no concensus regarding what are the origin(s) and what turbulence structure(s) really mean. What is definite that turbulent flows have lots of structure(s) [16].

3.2 On the structure of turbulence

3.2.1 On the origins of structure(s) of/in turbulence

This question - in some sense - is a 'philosophical'one. But its importance is in direct relation to even more important questions about the origin of turbulence itself. The difficulties of definition what the structure(s) of turbulence are (mean) are of the same nature as the definition of turbulence itself. So before and in

---

[16]The term *structure(s)* is used here deliberately in order to emphasize the duality (or even multiplicity) of the meaning of the underlying problem. The first is about how turbulence 'looks like'. The second implies existence of some entities.

order to 'see'or 'measure'the structure(s) of turbulence one encounters the most difficult questions such as what is (say, dynamically relevant) structure?, structure of what? which quantities possess structure in turbulence?, can structure exist in 'structureless'(artificial) pure random Gaussian field and which? All this - like many other issues - are intimately related to the skill/art to ask the right and correctly posed questions.

Instability. The most commonly accepted view on the origin of turbulence is the (continous state of [238]) flow instability. An additional aspect is that instability is considered as one of the origins of structure(s) in/of turbulence. However, this latter veiw requires to admit pretty long 'memory'of turbulence or, alternatively, presense in the 'purely'turblulent flow regime (i.e. at large enough Reynolds numbers) of instability mechanisms similar to those existing in the process of transition form laminar to turbulent flow state.

Emergence. Another less known view is that structure(s) emerge in large Reynolds number turbulence out of 'purely random structureless'background, e.g. via the so-called inverse cascades. Among the spectacular examples, are the 'geophysical vortices'in the atmosphere and in the ocean. Another example is the emergence of coherent objects (vortex filaments/worms) out of initially random Gaussian velocity field via the NSE dynamics (*cf.*, e.g. figure 1a and figure 3 in [216]).[17]

It "just exists" or do flows become turbulent or they are "just"such? *'To the flows observed in the long run after the influence of the initial conditions has died down there correspond certain solutions of the Navier- Stokes equations. These solutions constitute a certain manifold $\mathcal{M} = \mathcal{M}(\mu)$ (or $\mathcal{M} = \mathcal{M}(Re)$) in phase space invariant under phase flow'*[111].*'Kolmogorov's scenario was based on the complexity of the dynamics along the atractor rather than its stability'*[13] (see also [97], [131], [132]). This view is a reflection of one of the modern *beliefs* that the structure(s) of turbulence - as we observe it in *physical space* - is (are) the manifestation of the generic structural properties of mathematical objects (*in phase space*) which are called (strange) attractors, which are invariant in some sense. In other words here the structure(s) is assumed to be 'built in'the turbulence independenly of its origin (hence universality).

Whatever the origin (both of turbulence and its structure(s)) at large enough Reynolds numbers it is extremely complex — apparently random/stochasic. For instance, the so called "coherent structures"of different scales and shapes appear randomly in space/time and many of their properties change randomly as well.[18]

---

[17]Recall P. W. Anderson with the emphasis on *the concept of 'broken symmetry'*, *the ability of a large collection of simple objects to ababdon its own symmetry as well as the symmetries of the forces governing it and to exhibit the 'emergent property'of a new symmetry* [2]. One of the difficulties in turbulence research is that no objects simple enough and such that a collection of these objects would *adequately* represent turbulent flows were found so far.

[18]The term *scales* is used everywhere in its simple geometrical meaning without any other implication. Speaking about scales of some individual structures in turbulent flows requires to keep in mind that even the 'simplest'most popular structures like vortex filaments/worms have at least *two* very different scales: their length can be of order of the *integral* scale, whereas its cross section is of the order of the *Kolmogorov* scale.

3.2.2 How structure(s) of turbulence 'look(s) like'?

Until recently very little was known what this structure is or how the structures look like (in physical space). The structure in question is the so called fine structure and not the one which is promoted by various external factors and/or constraints like boundaries, mean shear, centrifugal forces (rotation), buoyancy, magnetic field, etc., which usually act as an organizing factor, favoring the formation of coherent structures of different kinds (quasi-two-dimensional, helical, hairpins, etc.). These are as a rule large scale features which depend on the particularities of a given flow and thus are not universal. It is noteworthy that the statement that turbulence has structure is in a sense trivial: to say that turbulent flow is 'completely random'would define turbulence out of existence ([238], p. 295) - after all turbulent flows seem to obey the Navier-Stokes equations.

Since the first DNS simulations [225] a number of other computations were performed (see [32], [112], , [121], [124], [125], [193], [216], [217], [232], [260], and references therein), which demonstrated clearly that fluid-dynamical turbulence which is 'homogeneous'and 'isotropic'has structure(s), i.e. contains a variety of strongly localized events. The primary evidence is related to spatial localization of subregions with large enstrophy (i.e. intense vorticity) which are organized in long, thin tubes-filaments-worms. Such filaments were observed also directly in laboratory experiments (the ones mentioned in figure 1) employing the property of intense vorticity to be strongly correlated with regions of low pressure and using small air bubbles for visualization of these regions ([72], [258] and also [119], [214] and references therein). This follows from the Poisson-like equation for pressure $2\nabla^2 p/\rho = \omega^2 - 2s_{ij}s_{ij}$. There is some evidence that in regions with moderate magnitude of vorticity it is organized in sheet-like structures [134], [216], [232]. Much less is known about regions with large strain, $s_{ij}s_{ij}$, i.e. dissipation. They were tentatively identified as layered vortex sheets in [214], which was not confirmed by other observations or computations so far. Most common observations at Reynolds numbers accessible in DNS showed that isosurfaces of high strain are wrapped around the regions strong enstrophy [35], [121], [137], [193], [207], [208], [217], [232]. However, in [232] and in recent computations, [32] isosurfaces of large strain were observed as sheet-like objects with very sharp edges (razors/flakes). In fact such objects were observed already in [224] (see there figure 21). This does not mean that vorticity field in these regions is simple and is necessarily sheet-like too. Some examples of the results mentioned are shown in figure 4.

The relatively *simple* appearance of the observed structures as shown above prompted a rather popular view that turbulence structrure(s) is (are) simple in some sense and that essential aspects of turbulence structure and its dynamics may be adequately represented by a random distribution of simple (weakly interacting) objects, such as straight strained (Burgers-like) vortices (see [246] and references therein). In particular, it is commonly believed that *most* of the structure of turbulence is associated with and is due to various strongly localized intense events/structures, e.g. mostly regions of concentrated vorticity so that *'turbulent flow is dominated by vortex tubes of small cross-section and bounded eccentricity*'([61], p. 95) and that these events are mainly responsible for the phenomenon of intermittency ([21], [90], [129], [178] and references therein).

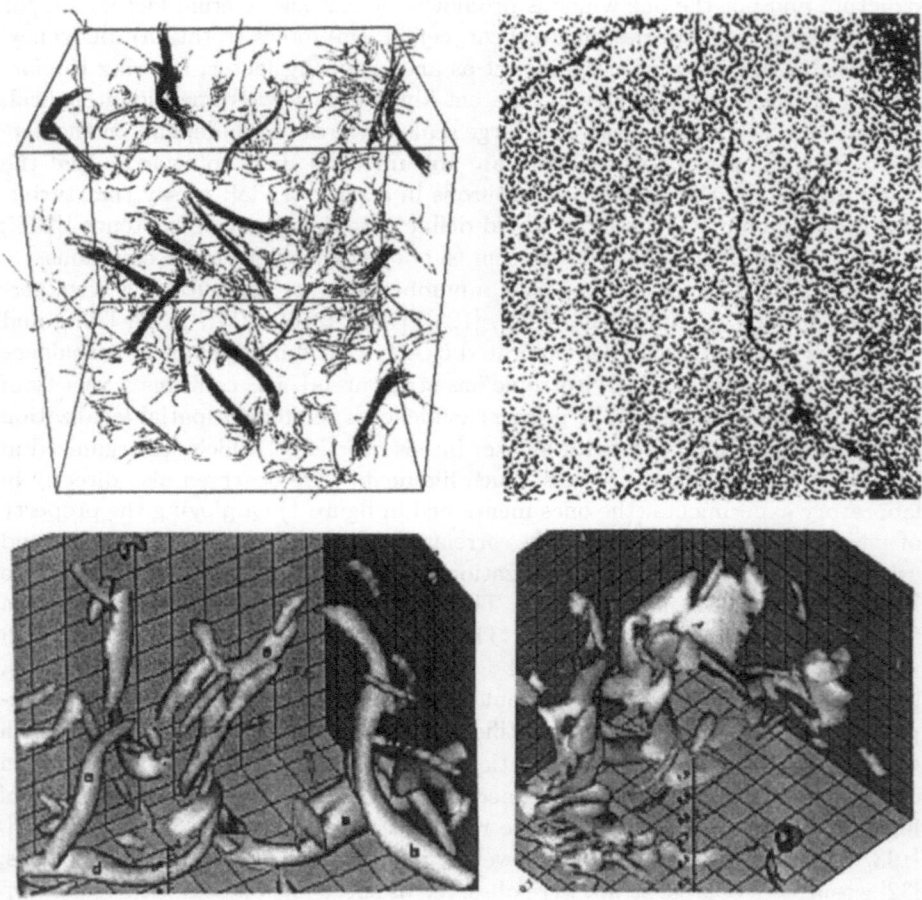

Figure 4. Vortex filaments in DNS [217] (top left) and laboratory [72] (top right). Isosurfaces of the second invariant of the velocty derivatives tensor $Q = \omega^2 - s_{ij}s_{ij}$ (bottom left) at $2rms$ positive level, i.e. vorticity dominated regions, and isosurfaces of strain $s_{ij}s_{ij}$ (bottom right) at $2\ rms$ level [32] , i.e. strain dominated regions. The two bottom pictures were not included in the paper [32], but are available at *http://www.eng.uci.edu/˜boratav/* and are used here by permission of the authors.

It is argued in [246] that such views are inadequate and that – though important – regions of concentrated vorticity are not that important as commonly believed. Namely, regions *other* than concentrated vorticity such as: *i* – 'structureless' background, *ii* – regions of strong vorticity/strain (self) interaction and

largest enstrophy generation, and *iii* – regions with negative enstrophy production are dynamically significant (in some important respects more significant than those with concentrated vorticity) strongly non-Gaussian, and possess
structure. Due to strong nonlocality of turbulence in physical space all the regions are in continuous interaction and are strongly correlated.

These conclusions are the outcome of use of *quantitative* manifestaions of turbulence structure (see sections 4 – 8).

3.2.3 On 'simple'*quantitative* manifestations of turbulence structure

On a *qualitative* level it is widely recognized that fluid-dynamical turbulence
(even 'homogeneous'and 'isotropic') has 'structure(s)', i.e. contains a variety of
strongly localized events, which are believed to influence essentially the properties of turbulent flows. Being extremely useful the individual observations of such
events are inherently limited as compared with statistical information. Indeed,
pure Gaussian velocity field has some structure too (see figure 3 in [216], showing that such field has also some 'vortex filaments', though much less in number
and less intense than in a real flow). In other words, though what we see is
real – the problem is in intepretation. This requires to employ the quantitative
manifestations of turbulence structure. In order to proceed to the *quantitative*
aspects of the problem it is not sufficient to look at pictures (whatever beautiful)
and one has to turn to numbers and quantitative relations such as in mentioned
above anomalous scaling, which is one of many other more specific quantitative
manifestations of turbulence structure.

Speaking about 'structure(s)'in turbulence the implication is that there exist
something 'structureless', e.g. Gaussian random field as a repereentative of
full/complete disorder. It is frequently claimed that '*Kolmogorov's work on the
fine-scale properties ignores any structure which may be present in the flow*'([90],
p. 182) and that it is associated with *near-Gaussian statistics* [217]. The main argument overruling the above claims is the Kolmogorov 4/5 law (4) [19], which is the
first strong indication of presense of structure [20], showing that non-Gaussianity
and structure of turbuence are directly related to the dissipative nature of FDT.
It is remarkable that the title of this paper by Kolmogorov [142] is *Dissipation
of energy in the locally isotropic turbulence*. Likewise the structure functions
of higher odd orders $S_p^{\parallel}(r) = \langle (\Delta u_{\parallel})^p \rangle$ are essentially different from zero (see
references in [28], [228], [246]). An interesting consequence of (4) is that non-
Gaussianity *increases* with scale in contradistinction with the rather common
opposite view. The 4/5 law (4) belongs to the most prominent and distinctive
features of turbulent flows of utmost dynamical significance – *the build up of odd
moments* both in large and small scales which among other things means phase
and geometrical coherency, i.e. structure. Indeed, as was discovered by Taylor
[233], [234] (for later references see [28], [228], [246]) another most important

---

[19]This relation has attracted considerable attention recently (see [7], [90], [106], [130], [163],
[228], [254] and references therein).

[20]In the inertial range, $L \gg \ell \gg \eta$. In the dissipation range $\ell \sim \eta$ turbulence is strongly
non-Gaussian and intermittent too [145], [182].

odd moment associated with small scales – the enstrophy generation $\langle \omega_i \omega_j s_{ij} \rangle$ – is essentially positive, which reflects one of the most basic *specific* properties of three-dimensional turbulent flows – the (prevalence of) vortex stretching process. Hence the particular emphasis on the *odd* moments. In addition, the non-Gaussianity found experimentally both in large and small scales is exhibited not only in the nonzero odd moments, but also in strong deviations of even moments from their Gaussian values. Thus both the large and small scales differ essentially from Gaussian indicating that both possess structure.

A special aspect of non-Gaussian behaviour is related to strong fluctuations of velocity: it was observed in laboratory [185] and numerical experiments [123], [154], [259] that the single point PDF of velocity at large amplitudes of velocity fluctuations is *sub-Gaussian* and recently was confirmed theoretically using the instanton formalism [86].

A number of more subtle issues related to *quantitative aspects* of structure of turbulence can be effectively addressed via what is denoted in the sequel by the term *'geometrical statistics'* (see [64], [68], [245], [252] and references therein).

4 Geometrical statistics

The widely known example of the utmost importance of geometrical relations in turbulence is the *qualitative* difference between the dynamics of 3D and 2D turbulence. This is seen immediately from the equations for vorticity ω_i and enstrophy ω^2

$$D_t \omega_i = \omega_j s_{ij} + \nu \nabla^2 \omega_i, \qquad D_t(\omega^2/2) = \omega_i \omega_j s_{ij} + \nu \omega_i \nabla^2 \omega_i, \qquad (7,8)$$

where $D_t \equiv \partial/\partial t + u_k(\partial/\partial x_k)$. The nonlinear terms $\omega_j s_{ij}$ and $\omega_i \omega_j s_{ij}$ are known to be responsible for the so called vortex stretching (**VS**) and enstrophy generation (**EG**). In other words the essential dynamics of 3D-turbulence is contained in the *interaction* between vorticity $\boldsymbol{\omega}$ and the rate of strain tensor s_{ij}. Both $\omega_j s_{ij}$ and $\omega_i \omega_j s_{ij}$ vanish identically for 2-D flows. Among other things **VS** responsible for the enhanced dissipation in turbulent flows [21] and it is one of the primary mechanisms for formation of structures.

So far there have been given no theoretical arguments in favor of positiveness of $\langle \omega_i \omega_j s_{ij} \rangle$. The argument that the reason is the (approximate) balance between the enstrophy generation and enstrophy dissipation is misleading and puts the consequences before the reasons, since it is known that for Euler equations the enstrophy generation increases with time very fast (apparently without limit) [36], [61], [87], [100], [135], [188]. Another rather common view that the prevalence of vortex stretching is due to the predominance of stretching of material

---

[21]Taylor [234] was the first to realize this fundamental importance of the vortex stretching process and its prevalence over vortex compressing. He demonstrated experimentally the essential positiveness of $\langle \omega_i \omega_j s_{ij} \rangle$ in a turbulent grid flow using the relations imposed by isotropy. Later this has been shown directly for several flows via measurements of all the nine velocity derivatives [249] and also by direct numerical simulations [223].

lines is - at best - true in part only, since, there exist several *qualitative* differences between the two processes.

Vortex stretching versus stretching of material lines[243] • - The equation for a material line element l is a linear one and the vector l is passive, i.e. the fluid flow doesn't 'know' anything whatsoever about l. In other words the vector l (as any passive vector) doesn't exert any influence on the fluid flow. The material element is stretched (compressed) locally at an exponential rate proportional to the rate of strain along the direction of l, since the strain is independent of l.

• - On the contrary the equation for vorticity is a nonlinear partial differential equation and the vector ω is an active one - it 'reacts back' on the fluid flow (vorticity and rate of strain tensor are composed of derivatives of the same velocity field: $\quad = rot\ \mathbf{u}$, $2s_{ij} = \frac{\partial u_i}{\partial x_j} + \frac{\partial u_j}{\partial x_i}$, i.e. the strain does depend in a nonlocal manner on $\quad$ and vice versa). In other words the rate of vortex stretching is a nonlocal quantity, whereas the rate of stretching of material lines is a local one. Therefore the rate of vortex stretching (compressing) is different from the exponential one (and is unknown). Also, it is noteworthy that there are much 'less' vorticity lines than the material ones - at each point there is typically only one vorex line, but infitely many of material lines. This leads to differences in the statistical properties of the two fields.

• - Consequently while a material element ℓ tends to be aligned with the largest (positive) eigenvector of s_{ij}, vorticity $\quad$ tends to be aligned with the intermediate (mostly positive) eigenvector of s_{ij} : the eigenframe of s_{ij} rotates with an angular velocity Ω_s of the order of vorticity $\quad$ [73].

• - For a Gaussian isotropic velocity field the enstrophy generation is identically zero, $\langle \omega_i \omega_j s_{ij} \rangle \equiv 0$ [223], whereas the mean rate of stretching of material lines is essentially positive [17], [63], [191]. The same is true of the mean rate of vortex stretching $\langle \omega_i \omega_j s_{ij} \rangle |\omega|^{-2}$ (for other counter-examples see below) and for purely two-dimensional flows. This means that one can expect that in turbulent flows the mean growth rate of material lines is larger than the one of vorticity [191]. Recently this was really observed in decaying DNS turbulence [114]. In other words the nature of vortex stretching process is dynamical and not a kinematic one as is the stretching of material lines.

• - An additional difference due to viscosity becomes essential for regions with concentrated vorticity, in which there is an approximate balance between enstrophy generation and its reduction. Vortex reconnection is allowed by nonzero viscosity. No such phenomena exist for material lines.

As mentioned above in 3-D turbulence $\langle \omega_i \omega_j s_{ij} \rangle$ is an essentially positive quantity - the PDF of $\omega_i \omega_j s_{ij}$ is strongly positively skewed. This reflects one of the most basic *specific* properties of three-dimensional turbulent flows - the (prevalence of) the vortex stretching process. The enstrophy generation $\omega_i \omega_j s_{ij}$ is an oustanding nonzero *odd* moment of utmost dynamical importance in turbulence. Indeed, in the hypothetical case of absence of VS and EG or even in case when only $\langle \omega_i \omega_j s_{ij} \rangle = 0$ - as assumed by Karman [128] - the three-dimensional turbulence, as we observe it, would not exist.

4.1 Geometrical invariants versus their surrogates

Quantities like $\omega_i \omega_j s_{ij}$ are geometrical invariants (see also section 7), e.g. they remain invariant under the full group of rotations in contradistinction with other *noninvariant* combinations of velocity derivatives. For this reason the geometrical invariants are mostly appropriate for studying physical processes in turbulent flows, their structure and universal properties (see [29], [181], [243], [249] and references threrein).

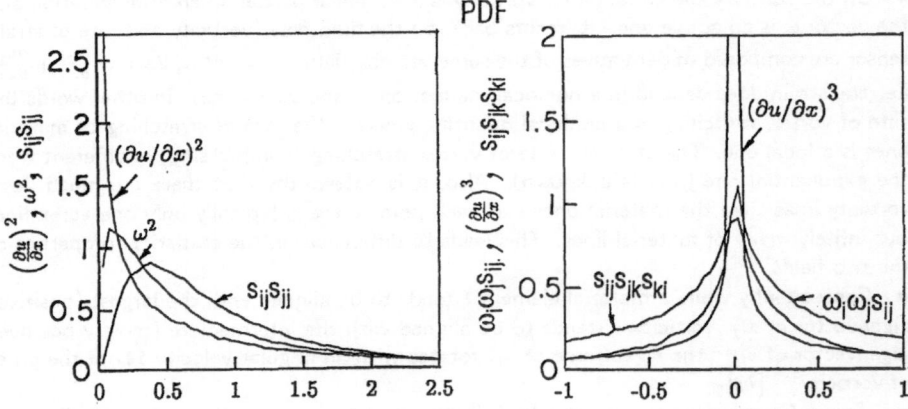

Figure 5. PDF's of enstrophy ω^2, total strain $s_{ij}s_{ij}$ (dissipation) and their surrogate $15(\partial u_1/\partial x_1)^2$ (left column) and enstrophy generation $\omega_i \omega_j s_{ij}$, $s_{ij}s_{jk}s_{ki}$ and their surrogate $17.5(\partial u_1/\partial x_1)^3$ (right column).

Surrogates of the type $(\partial u_1/\partial x_1)^n$ reperesent adequately only the *means* of the true quantities such as mean disspation. Other properties (spectral, fractal, scaling, etc.) of the surrogates and of the true quantities (invariants) are generally different (see, [228], [243], [245], [249] and references therein; for some recent results on such differences see [113], [267]). Likewise there exist qualitative differences between the flow regions dominated by strain and those by vorticity (e.g. [29], [32], [57], [58], [117], [217], [223], [274] and references therein). A new aspect of such a difference is addressed in sections 4.3, 4.4, 10.2.1. An example of the differences between the true quantities and their surrogates is shown in figures 5 and 6 for a DNS of NSE in a cube at Taylor microscale Reynolds number $\mathrm{Re}_\lambda \approx 75$ [250], [252].

A usual phenomenological argument [90] results in the estimate $\omega_i \omega_j s_{ij} \sim \omega^3$, whereas in reality it is only $\omega_i \omega_j s_{ij} \sim \omega^{7/3}$ in slots of ω, but $\omega_i \omega_j s_{ij} \sim \omega^3$ in slots of s (see [246] and references therein), showing the importance of taking into account the mutual *orientation* of vorticity $\boldsymbol{\omega}$ and the eigenframe $\boldsymbol{\lambda}_i$ ($i = 1, 2, 3$) of the rate of strain tensor s_{ij}. In other words the essential dynamics of 3D-turbulence contained in the interaction between vorticity $\boldsymbol{\omega}$ and the rate of strain tensor s_{ij} depends strongly not only on the magnitude of vorticity and strain but also on the geometry of the field of velocity derivatives, in particular on the mutual orientation of vorticity $\boldsymbol{\omega}$ and the eigenframe $\boldsymbol{\lambda}_i$ of the rate of strain tensor s_{ij}.

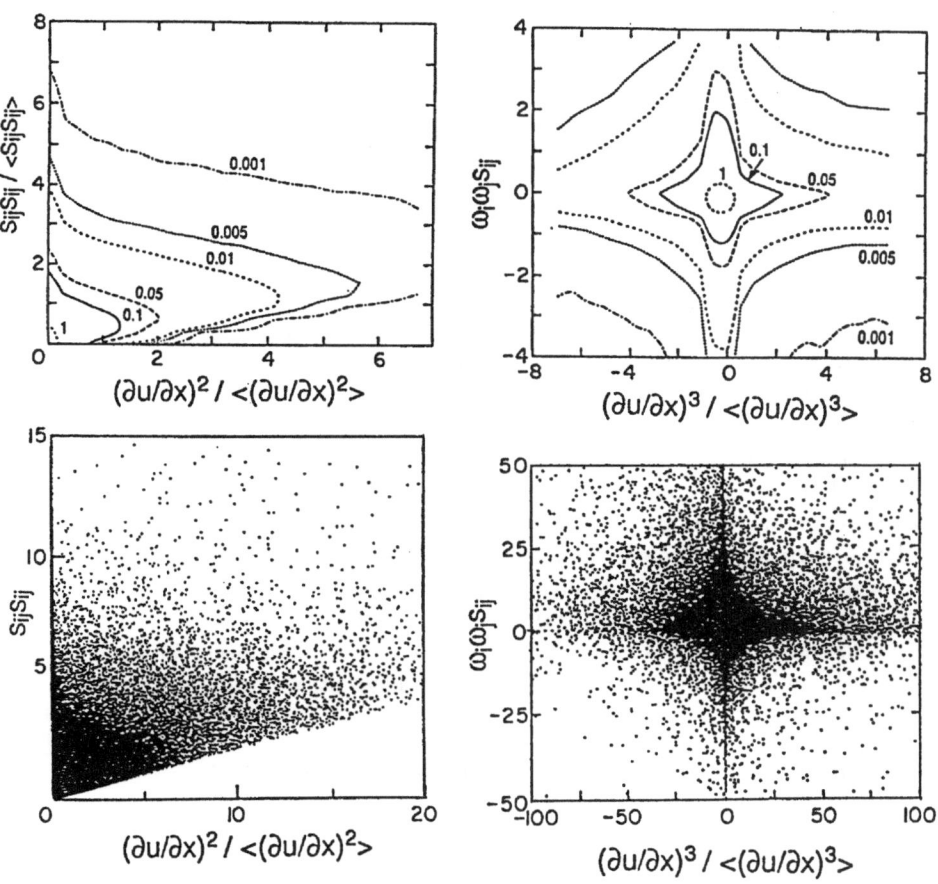

Figure 6. Joint PDF's and scatter plots of enstrophy ω^2, total strain $s_{ij}s_{ij}$ (dissipation) and their surrogate $15(\partial u_1/\partial x_1)^2$ (left column), and enstrophy generation $\omega_i\omega_j s_{ij}$, $s_{ij}s_{jk}s_{ki}$ and their surrogate $17.5(\partial u_1/\partial x_1)^3$ (right column).

One of the simplest means to characterize *quantitatively* this geometrical aspect of turbulence dynamics is to look at the alignments between vorticity and strain, e. g. the PDFs of the cosine of the angle between vorticity $\boldsymbol{\omega}$ and the eigenvectors $\boldsymbol{\lambda}_i$.

4.2 Alignments

Various alignments comprise an important simple *geometrical* characteristics and manifestation of the *dynamics* and *structure* of turbulence. For example, there is a distinct qualitative difference between the PDFs of $cos(\boldsymbol{\omega}, \boldsymbol{\lambda}_i)$ [22] for a real turbulent flow and a random Gaussian velocity field. In the last case *all* these PDFs are precisely flat! [223]. An example of special dynamical importance is the strict alignment between vorticity and the vortex stretching vector $W_i \equiv \omega_j s_{ij}$. In real turbulent flows it is strongly asymmetric [223], [249], [246], [252] in full conformity with the prevalence of vortex stretching over vortex compressing, i.e. positiveness of $\langle \omega_i \omega_j s_{ij} \rangle \equiv \langle \boldsymbol{\omega} \cdot \mathbf{W} \rangle$, whereas it is symmetric for a random Gaussian field [223]. Thus the very existence of alignments such as mentioned above points to the presence of internal organization of flow at various scales, i.e. alignments belong to the rare *quantitative statistical* manifestation of the existence of structure in turbulence. They are the simplest representative of a much broader class of geometrical statistics in turbulent flows [245]. It is noteworthy that while the above mentioned (an some others) alignments are intimately related to the *dynamics* of turbulent flows, there are alignments which are mostly of kinematic nature, e.g. alignment between the Lamb vector $\boldsymbol{\omega} \times \mathbf{u}$ and its potential part (pressure gradient), the alignment between velocity and the eigenvectors of rate of strain tensor and some others [246], [252].

Alignments by their very definition are suitable for events of *any* magnitude, since they do not contain the amplitude of the quantities involved. Finally, alignments are *invariant* in the sense that they are independent of the system of reference and therefore, along with other invariant quantities are the most appropriate in studying of physical processes generally and in particular for characterization of the structural nature of turbulent flows.

Due to these properties using of alignments enables to answer in a simple and reliable way a number of questions on turbulence structure [252].

5 The geometry of vortex stretching

In order to address this issue let us remind some simple relations for the key quantities of turbulence dynamics - the vortex stretching vector $W_i = \omega_j s_{ij}$ and enstrophy generation $\omega_i \omega_j s_{ij}$ and some related quantities (see also section 6).

$$\omega_i \omega_j s_{ij} = \omega_i^2 \Lambda_i \cos^2(\boldsymbol{\omega}, \boldsymbol{\lambda}_i) = \alpha \omega^2; \quad W^2 = \omega_i^2 \Lambda_i^2 \cos^2(\boldsymbol{\omega}, \boldsymbol{\lambda}_i), \quad (9, 10)$$

Here $\alpha = \Lambda_i \cos(\boldsymbol{\omega}, \boldsymbol{\lambda}_i)$ — is the rate of enstrophy generation. It is seen from the relaltions (8,9) that indeed - as mentioned above - the essential dynamics of 3D-turbulence contained in the interaction between vorticity $\boldsymbol{\omega}$ and the rate of strain tensor s_{ij} depends strongly not only on the magnitude of vorticity and strain but also on the geometry of the field of velocity derivatives, in particular on the

---

[22]The eigenvalues of the rate of strain tensor are denoted as Λ_i and $\Lambda_1 > \Lambda_2 > \Lambda_3$, $\Lambda_1 + \Lambda_2 + \Lambda_3 = 0$ due to incompressibility, so that $\Lambda_1 > 0$ and $\Lambda_3 < 0$. It is known from experiments - both numerical [14] and laboratory [249] - that $\langle \Lambda_2 \rangle > 0$.

mutual orientation of vorticity $\boldsymbol{\omega}$ and the eigenframe $\boldsymbol{\lambda}_i$ of the rate of strain tensor s_{ij}. This is true especially regarding the *rate* of enstrophy generation $\omega_i \omega_j s_{ij}$ (i.e. $\alpha = \Lambda_i \cos(\boldsymbol{\omega}, \boldsymbol{\lambda}_i)$) and similar quantity for W^2 (i.e. $W^2/\omega^2 = \Lambda_i^2 \cos(\boldsymbol{\omega}, \boldsymbol{\lambda}_i)$), which depend explicitly only on the orientation of vorticity, but not on its magnitude.

In view of the importance of the predominant vortex stretching and positive net enstrophy generation, i.e. $\langle \omega_i \omega_j s_{ij} \rangle > 0$ it is useful to introduce an angle between $\boldsymbol{\omega}$ and $\mathbf{W}$, since $\omega_i \omega_j s_{ij} \equiv \boldsymbol{\omega} \cdot \mathbf{W}$. It is easy to see from the simple relation

$$\cos(\boldsymbol{\omega}, \mathbf{W}) = \frac{\Lambda_i \cos^2(\boldsymbol{\omega}, \boldsymbol{\lambda}_i)}{\{\Lambda_i^2 \cos^2(\boldsymbol{\omega}, \boldsymbol{\lambda}_i)\}^{1/2}}. \tag{11}$$

that the alignment between $\boldsymbol{\omega}$ and $\mathbf{W}$ (i.e. positive $\omega_i \omega_j s_{ij}$) is realized in two situations [252]: i - $\boldsymbol{\omega}$ is aligned with $\boldsymbol{\lambda}_1$ ($\Lambda_1 > 0$) and ii - $\boldsymbol{\omega}$ is aligned with $\boldsymbol{\lambda}_2$ (Λ_2 assumes both positive and negative values, but is positively skewed [14], [249]) . Indeed, the contributions both to $\sigma \equiv \omega_i \omega_j s_{ij}$ and α associated with Λ_1 and Λ_2 are positive (see table 1).

	$\langle \omega^2 \Lambda_1 \cos^2(\boldsymbol{\omega}, \boldsymbol{\lambda}_1) \rangle$	$\langle \omega^2 \Lambda_2 \cos^2(\boldsymbol{\omega}, \boldsymbol{\lambda}_2) \rangle$	$\langle \omega^2 \Lambda_3 \cos(\boldsymbol{\omega}, \boldsymbol{\lambda}_3) \rangle$
DNS	1.06	0.51	− 0.57
Grid	1.17	0.39	− 0.56
	$\langle \Lambda_1 \cos^2(\boldsymbol{\omega}, \boldsymbol{\lambda}_1) \rangle$	$\langle \Lambda_2 \cos^2(\boldsymbol{\omega}, \boldsymbol{\lambda}_2) \rangle$	$\langle \Lambda_3 \cos^2(\boldsymbol{\omega}, \boldsymbol{\lambda}_3) \rangle$
DNS	1.47	0.49	− 0.97
Grid	1.17	0.46	− 0.63

Table 1. Contribution to the total mean of enstrophy generation $\langle \sigma \rangle \equiv \langle \omega^2 \Lambda_i \cos^2(\omega, \lambda_i) \rangle$ and its rate $\langle \alpha \rangle \equiv \langle \Lambda_i \cos^2(\omega, \lambda_i) \rangle$ from the terms corresponding to the eigenvalues Λ_i of the rate of strain tensor s_{ij}. Grid turbulence and DNS, $Re_\lambda \approx 75$ [252].

It is surprising, at first sight, that the largest contribution to σ and α comes from the regions associated with the *largest* eigenvalue Λ_1 of the rate of strain tensor s_{ij} and not from the ones associated with the *intermediate* eigenvalue Λ_2, since it is known that there exists a strong tendency of alignment between $\boldsymbol{\omega}$ and $\boldsymbol{\lambda}_2$ (as shown in figure 7) ([14], [246] and references therein). This apparent contradiction will become clear in the sequel. Meanwhile we note that the alignments between $\boldsymbol{\omega}$ and $\boldsymbol{\lambda}_1$ and between $\boldsymbol{\omega}$ and $\boldsymbol{\lambda}_2$ correspond to two qualitatively different in several respects regions of turbulent flow [246], [252].

5.1 Strained vortical (Burgers-like) objects

We start from regions with concentrated vorticity, which constutue a subset of much larger regions in which there is a tendency for alignment between $\boldsymbol{\omega}$ and $\boldsymbol{\lambda}_2$. This is clearly seen from the figure 7. Indeed, regions corresponding to $\cos(\boldsymbol{\omega}, \boldsymbol{\lambda}_2) > 0.9$ occupy about 20% of the total flow volume, whereas the set of points with concentrated vorticity (say, $\omega^2 > 3\langle \omega^2 \rangle$) is comprised of less 6% of the total flow volume [252].

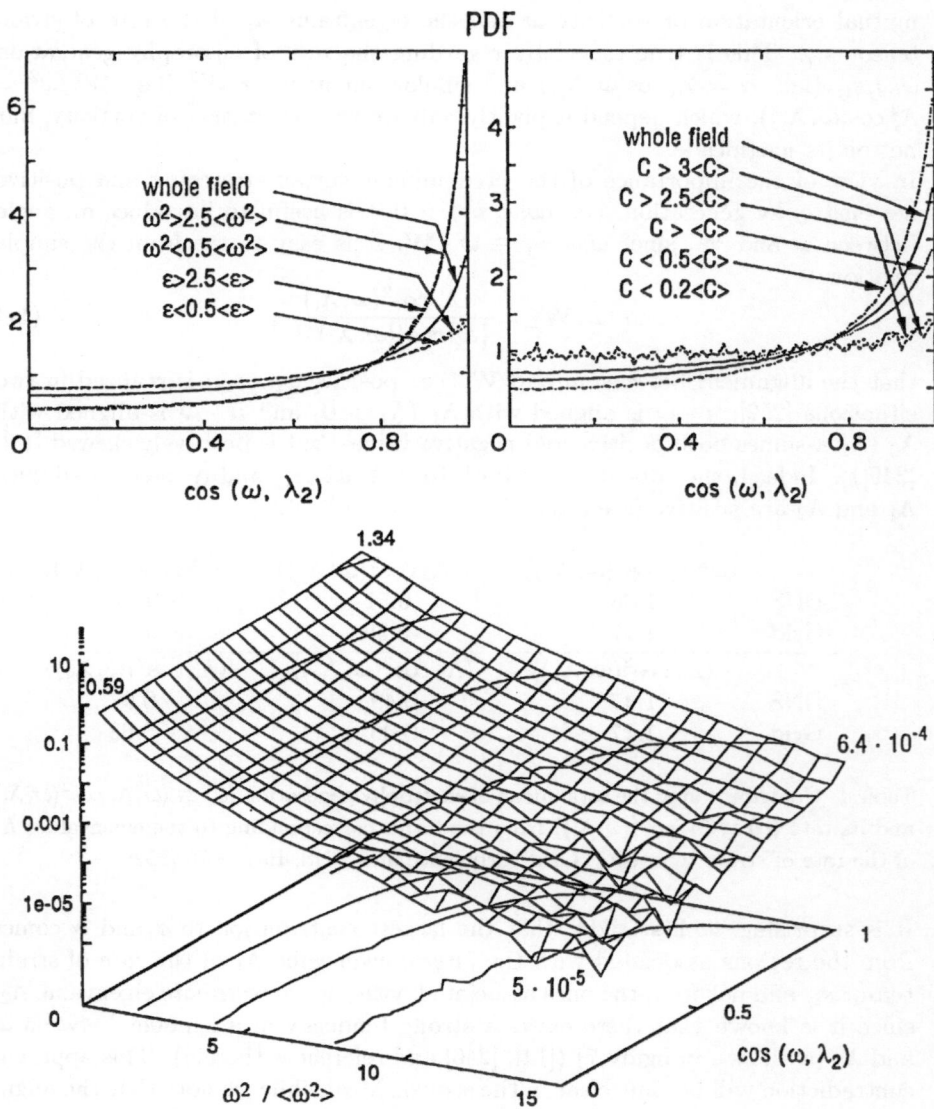

Figure 7. PDFs of $\cos(\omega, \lambda_2)$, DNS, $Re_\lambda \approx 75$ [246], [252]. Top left - conditioned on enstrophy ω^2 and s^2, top right - conditioned on curvature C of vortex lines. Note that the tendency for alignment between ω and λ_2 exists *both* in regions of large ω^2 and large s^2 (see also [87], [217]). Note that for a Gaussian velocity field these PDFs are precisely flat. Bottom - joint PDF of $\cos(\omega, \lambda_2)$ and ω^2 (the joint PDF of $\cos(\omega, \lambda_2)$ and s^2 is similar to the one shown in this figure [252]). It is seen that the maximum of joint PDF of $\cos(\omega, \lambda_2)$ and ω^2 (and similarly of $\cos(\omega, \lambda_2)$ and s^2) takes place at $\cos(\omega, \lambda_2) \approx 1$ and $\omega^2 \approx 0$, i.e. at the points with *weakest* vorticity and *strongest* alignment between ω and λ_2.

The main feature and shortcoming of these objects (straight strained vortices) is that they possess *one*-dimensional vorticity and therefore zero curvature of vortex lines. Though the relation between vorticity and strain is essentially non-local *'the presence of a strained vortex itself modifies the local strain field'*([159] p. 242) – after all both are composed of derivatives of the same velocity field. However, the special feature of the *straight* strained vortices is that they are impotent in the sense that they do not change that part of the strain by which they are strained themselves: this part of strain is prescibed *a priori*, i.e. it is independent decoupled from their vorticity. These vortices do change only that part of their strain which is not reacting back on their vorticity. In other words there is only one way interaction: the vorticity is strained by that part of strain which does not 'know'anything about the vorticity. In this sense such vortices are passive: the essential ingredient of nonlinearity, the main feature of true genuine nonlinear interaction — the self-amplification — is absent in these objects. In this sense the nonlinearity is reduced in these objects. This property is directly related to zero curvature of vorex lines in straight strained vortices (see figure 4, top and bottom left and figure 7, top left) — the genuine nonlinearity is present only in regions with nonvanishing curvature. This is what is observed when looking for (apparent) singularities of Euler equtions ([36], [100], [101], [135], [184], [188], [202]) and vortex reconnection ([87], [138]).

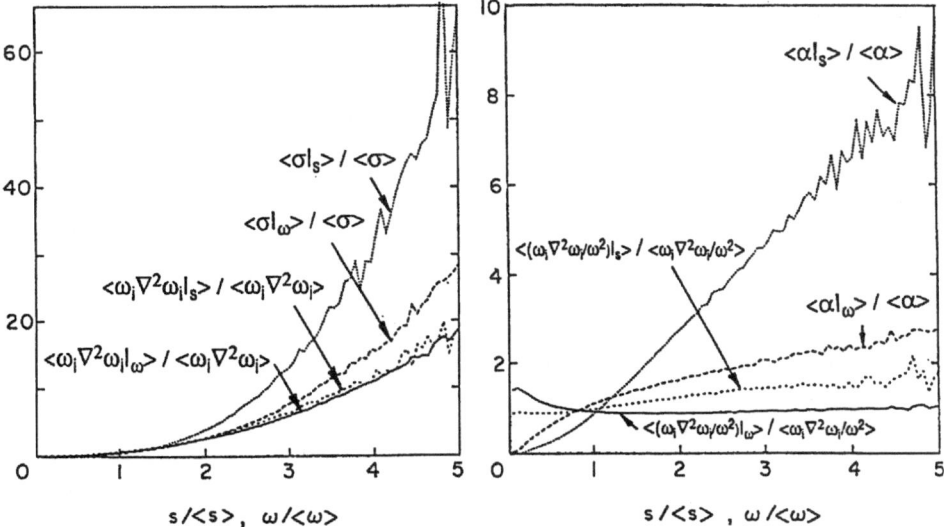

Figure 8. Comparison of enstrophy generation $\sigma \equiv \omega_i\omega_k s_{ik}$ (left) and its rate $\alpha \equiv \omega_i\omega_k s_{ik}/\omega^2$ (right) with their viscous reduction $\nu\omega_i\nabla^2\omega_i$ and $\nu\omega_i\nabla^2\omega_i/\omega^2$ in slots of ω and s. DNS, $Re_\lambda \approx 75$ [251].

In other words regions with concentrated vorticity with small curvature in real turbulent flows seem to be mostly the result, the consequence rather than dominating factor of the turbulence dynamics. Possessing (almost) maximal enstrophy they are in an approximate equilibrium in the sense that their fairly large (but not largest!, see section 5.3) enstrophy generation is approximately balanced by the viscous reduction and in this sense they are less active than the strain dominated regions possessing much larger (apparently maximal) enstrophy generation which is considerably larger than its viscous reduction. This is seen from the comparison of the rate enstrophy generation $\alpha \equiv \omega_i \omega_k s_{ik}/\omega^2$ and its viscous reduction $\nu \omega_i \boldsymbol{\nabla}^2 \omega_i/\omega^2$ in slots of ω and s as shown in figure 8 [251]. Indeed, the imbalance between stretching and viscous terms in slots of s is much *larger* than in slots of ω. This difference is especialy large at large values of of ω and s. This means that the time scale estimated from the imbalance of stretching and viscous terms $\omega^2 \{D_t(\omega^2/2)\}^{-1} \approx \{\omega_i \omega_k s_{ik}/\omega^2 + \nu \omega_i \boldsymbol{\nabla}^2 \omega_i/\omega^2\}^{-1}$ in slots of ω is much larger than such time scale in slots of s, i.e. the life time of regions with concentrated vorticity is large comparing to that of the regions with large strain, i.e. large rate of energy dissipation. This explains – at least in part – the observability of the regions with conce..'rated vorticity and the difficulties in observing the regions with large dissipation (but see [214]) and points to the importance of studying more carefully the regions of turbulent flows with strong imbalance between vortex stretching and viscous destruction of vorticity. It is noteworthy that the Burgers-like objects in real turbulent flows possess essentially nonvanishing curvature [95], [251], so that the self-amplification of their vorticity is not vanishing as in perfectly straight ones.

5.2 Turbulence background – not stuctureless random sea

Use of alignments allowed to show that - contrary to the common view - the so called 'background'is strongly non-Gaussian, is dynamically not passive and is not structureless (figures 7, 9, 10) [246], [252].
Though the strongest tendency for alignment between $\boldsymbol{\omega}$ and $\boldsymbol{\lambda}_2$ is observed for large ω^2 this alignment is still significant (see bottom of figure 7) in the 'background'(say $\omega^2 < \langle \omega^2 \rangle$) especially taking into account that the background is occupying about 70% of the flow volume (*cf.* with the volume occupied by strong vorticity, say $\omega^2 > 3\langle \omega^2 \rangle$, which is only about 6% of the flow volume). Note that this does not contradict the mostly known result about the tendency of alignment between $\boldsymbol{\omega}$ and $\boldsymbol{\lambda}_2$ in regions of concentrated vorticity: the regions with such an alignment are an order of magnitude larger than those with concentrated vorticity only [250], [252].
The significance of the background is stressed more clearly in figure 9, showing the normalized enstrophy generation $\omega_i \omega_j S_{ij} |\omega|^{-1} |W|^{-1} = \cos(\boldsymbol{\omega}, \mathbf{W})$. Just like in the case of $\cos(\boldsymbol{\omega}, \boldsymbol{\lambda}_2)$ the tendency for alignment between $\boldsymbol{\omega}$ and $\mathbf{W}$ exists *both* in regions of large ω^2 and s^2. However, it is much stronger for large strain s^2 (see section 5.3). It is seen that the maximum of joint PDF of $\cos(\boldsymbol{\omega}, \mathbf{W})$ and ω^2 (and of $\cos(\boldsymbol{\omega}, \mathbf{W})$ and s^2) takes place at $\cos(\boldsymbol{\omega}, \mathbf{W}) \approx 1$ and $\omega^2 \approx 0$, i.e. at the points with *weakest* vorticity and *strongest* alignment between $\boldsymbol{\omega}$ and $\mathbf{W}$.

111

PDF

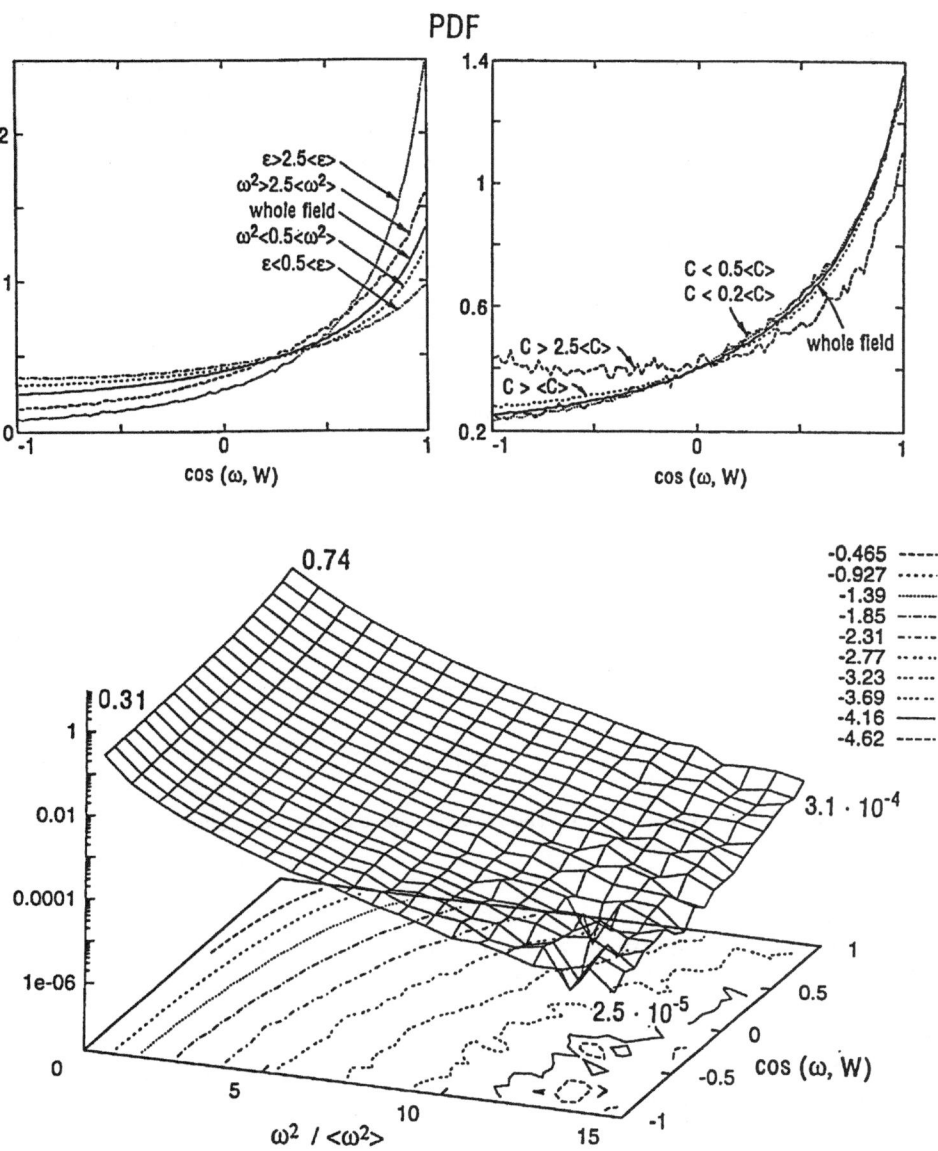

Figure 9. PDFs of $\cos(\omega, W)$, $W_i = \omega_i s_{ij}$; DNS, $\text{Re}_\lambda \approx 75$ [246], [252]. Top left - conditioned on enstrophy ω^2 and s^2, top right - conditioned on curvature of vortex lines; bottom - joint PDF of $\cos(\omega, \lambda_2)$ and ω^2 (the joint PDF of $\cos(\omega, \lambda_2)$ and s^2 is similar to the one shown in this figure [252].

Note also the strong asymmetry of its PDF for the background $\omega^2 < \langle \omega^2 \rangle$, which is almost the same as for the whole field. This asymmetry remains significant even for $\omega^2 < 0.1\langle \omega^2 \rangle$, and becomes *stronger* for $\omega^2 < \langle \omega^2 \rangle$ and $\cos(\boldsymbol{\omega}, \boldsymbol{\lambda}_2) > 0.9$ (not shown, see [250], [252]). Moreover, this asymmetry remains significant for *both* small ω^2 and s^2 (see figure 10). We remind that for Gaussian velocity field the PDF of $\cos(\boldsymbol{\omega}, \mathbf{W})$ is symmetric.

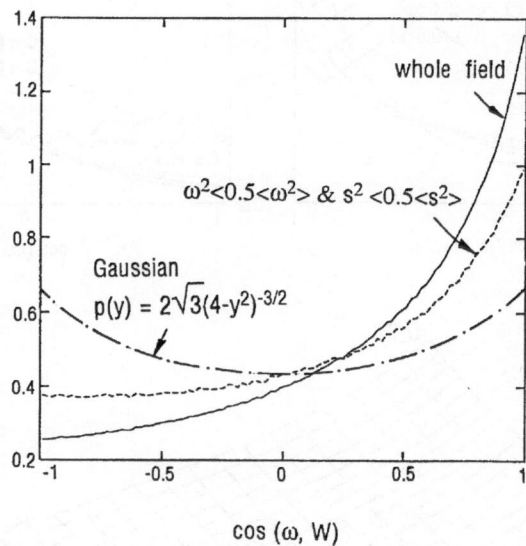

Figure 10. PDFs of $\cos(\omega, W)$ for the 'weakest'part of turbulent flow; DNS, $Re_\lambda \approx 75$ [252].

One can see from figures 7, 9 and 10 that the maxima of the joint PDFs of both $\cos(\boldsymbol{\omega}, \boldsymbol{\lambda}_2)$ and $\cos(\boldsymbol{\omega}, \mathbf{W})$ are located at *weakest* enstrophy and *strongest* alignment between $\boldsymbol{\omega}$ and $\boldsymbol{\lambda}_2$, and $\boldsymbol{\omega}$ and $\mathbf{W}$. The same is true for a variety of joint PDFs of other quantities [250], [251], [252].

The above results show clearly that the background is strongly non-Gaussian, not structureless and not passive.

5.3 Regions of strongest vorticity/strain interaction

As mentiond above the important point is that at least in quasi-isotropic flows the largest contribution to the enstrophy generation $\omega_i \omega_j s_{ij} = \omega_i^2 \Lambda_i \cos^2(\boldsymbol{\omega}, \boldsymbol{\lambda}_i)$ comes from the regions associated with the *largest* eigenvalue Λ_1 of the rate of strain tensor s_{ij} ([249], [250], [251], [252], [260]) and not from the ones associated with the *intermediate* eigenvalue Λ_2 to which belong the regions of concentrated vorticity. Namely the ratio of $\langle \omega^2 \Lambda_1 \cos^2(\boldsymbol{\omega}, \boldsymbol{\lambda}_1) \rangle$ to $\langle \omega^2 \Lambda_2 \cos^2(\boldsymbol{\omega}, \boldsymbol{\lambda}_2) \rangle$ is roughly $2 : 1$. The same is true of $\alpha = \Lambda_i \cos^2(\boldsymbol{\omega}, \boldsymbol{\lambda}_i)$ (see Table 1).

This shows that there exist regions (intense and weak — both structured and dynamicaly active) other than concentrated vorticity regions, which at least in the above sense are dynamicaly more important [250], [251], [252].

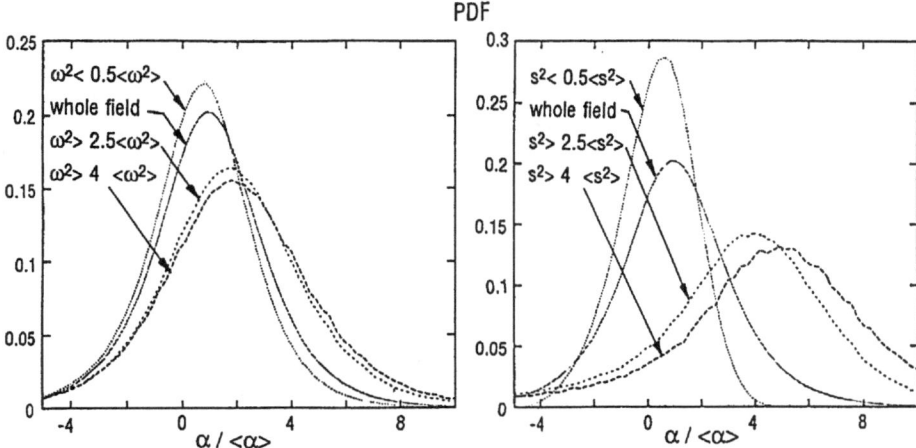

Figure 11. PDF's of the enstrophy generation rate α for the whole field and conditioned on ω^2 (left) and s^2 (right). DNS, $\text{Re}_\lambda \approx 75$ [251]. It is seen that there exist considerable regions with vortex compression (i.e. $\alpha < 0$) also for large enstrophy (see also [125]), whereas in regions with large strain the rate of enstrophy generation α is mostly positive.

These regions are associated mainly with largest strain rather than enstrophy [208], [250], [251], strong tendency of alignment between ω and λ_1 [250], [252], and fairly large curvature of vorticity line [251]. These regions are characterised by largest, apparently maximal, enstrophy generation and its rate (as shown in figure 8), which are much larger than their viscous reduction as discussed in section 5.3. This is consistent with the PDFs of α conditioned on ω and s (see figure 11) and with the results of [66] that the dominating contribution to α comes from the local (self) interaction of vorticity ω and strain s_{ij}, which is absent in Burgers-like objects. The behaviour of W^2 and W^2/ω^2 in slots of ω and s is essentially the same (see figure 16 below).

As implied by the results shown in table 1 these regions are associated with strong tendency for alignment between ω and the largest eigenvector λ_1 (corresponding to the largest eigenvalue Λ_1) of the rate of strain tensor s_{ij} as illustrated in figure 12 [250], [251], [252].

Similarly the dependence of enstrophy generation $\sigma \equiv \omega_i \omega_j s_{ij}$ and its rate $\alpha \equiv \Lambda_i \cos^2(\omega, \lambda_i)$ on ω and on $s \equiv (s_{ij} s_{ij})^{1/2}$ is qualitatively different for small and large curvature of vortex lines in such a way that the nonlinearity is manifested stronger in regions of large curvature [251]. In particular, the disparity in the behaviour of σ and α in slots of ω and s becomes larger at small curvature, whereas at large curvature the dependence of σ and α on ω and s is very similar. This last fact is a reflection of stronger interaction of vorticity and strain in regions with *large* curvature and *positive* α and consequently with non-negligible vortex folding and tilting (see the next section).

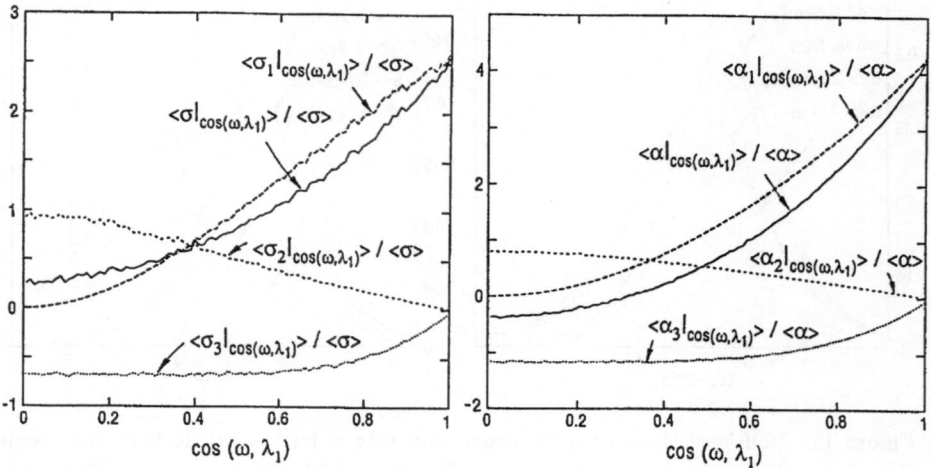

Figure 12. Conditional averages of enstrophy generation σ (left) and its rate α (right) in slots of $\cos(\omega, \lambda_1)$. DNS, $\text{Re}_\lambda \approx 75$.

The regions just discussed comprise a *subset* of larger regions dominated by strain. Namley, these are the regions with large vortex lines curvature. There exist at least two other kinds of strain dominated regions: those with small curvature of vortex lines, which wrap around the vorticity dominated regions (tubes/worms), which contribute mostly to the alignment of $\boldsymbol{\omega}$ and $\boldsymbol{\lambda_2}$ as shown in figure 7, and regions with large magnitude of Λ_3 and large negative α, in which most of vortex *compressing, tilting and folding* occur.

5.4 Vortex compression, tilting, folding and curvature

The most basic phenomenon in turbulence – the predominant vortex stretching, i.e. predominant enstrophy generation $\sigma \equiv \omega_i \omega_j s_{ij}$ so that $\langle \sigma \rangle > 0$ - cannot occur in a finite volume without its concomittants – vortex compressing ($\sigma < 0$) and folding ([60], [61] and references therein; the term folding was introduced by Reynolds in 1894 [204] in the context of folding of material lines). Hence, the importance of looking at properties of turbulent flow in regions with large curvature and $\sigma < 0$, which typically occupy about $1/3$ of the whole flow volume [251], and for the evidence and characterisation of the vortex folding in three-dimensional turbulence. These regions with σ, $\alpha < 0$ play an important role in the dynamics of turbulence. For example, these regions make a positive contribution to the magnitude of the vortex stretching vector $W_i \equiv \omega_j s_{ij}$ in (6). Indeed, $W^2 = \omega^2 \Lambda_i^2 \cos^2(\boldsymbol{\omega}, \boldsymbol{\lambda}_i)$ and $W^2/\omega^2 \equiv \Lambda_i^2 \cos^2(\boldsymbol{\omega}, \boldsymbol{\lambda}_i)$ are large for large $\Lambda_3^2 \cos^2(\boldsymbol{\omega}, \boldsymbol{\lambda}_3)$, see table 2, for which the enstrophy generation $\sigma \equiv \omega_i \omega_j s_{ij} = \omega^2 \Lambda_i \cos^2(\boldsymbol{\omega}, \boldsymbol{\lambda}_i)$ and its rate $\alpha = \Lambda_i \cos^2(\boldsymbol{\omega}, \boldsymbol{\lambda}_i)$ are *negative*.

$\langle \omega^2 \Lambda_1^2 \cos^2(\boldsymbol{\omega}, \boldsymbol{\lambda}_1) \rangle$	$\langle \omega^2 \Lambda_2^2 \cos^2(\boldsymbol{\omega}, \boldsymbol{\lambda}_2) \rangle$	$\langle \omega^2 \Lambda_3^2 \cos^2(\boldsymbol{\omega}, \boldsymbol{\lambda}_3) \rangle$
0.53	0.15	0.32
$\langle \Lambda_1^2 \cos^2(\boldsymbol{\omega}, \boldsymbol{\lambda}_1) \rangle$	$\langle \Lambda_2^2 \cos^2(\boldsymbol{\omega}, \boldsymbol{\lambda}_2) \rangle$	$\langle \Lambda_3^2 \cos^2(\boldsymbol{\omega}, \boldsymbol{\lambda}_3) \rangle$
0.51	0.11	0.38

Table 2. Contribution to the total mean of the magnitude of vortex stretching vector $\langle W^2 \rangle \equiv \langle \omega^2 \Lambda_i^2 \cos^2(\omega, \lambda_i) \rangle$ and its rate $\langle W^2/\omega^2 \rangle \equiv \langle \Lambda_i^2 \cos^2(\omega, \lambda_i) \rangle$ from the terms corresponding to the eigenvalues Λ_i of the rate of strain tensor s_{ij}. DNS, $\mathrm{Re}_\lambda \approx 75$.

Similarly, enstrophy generation (and α) can be small, whereas W^2 (and W^2/ω^2) can be large.

A closely related process is the vortex tilting, which is characterized by the rate of change of direction of vorticty. This rate is obtained from the equations (11), (12) for the magnitude of vorticity ω and its unit vector $\boldsymbol{\varpi}$ which are equivalent to equations (6), (7)

$$D_t \omega = \alpha \omega + vt, \quad D_t \varpi_i = s_{ij} \varpi_j - \alpha \varpi_i + vt, \tag{12, 13}$$

where and vt stands for viscous terms.

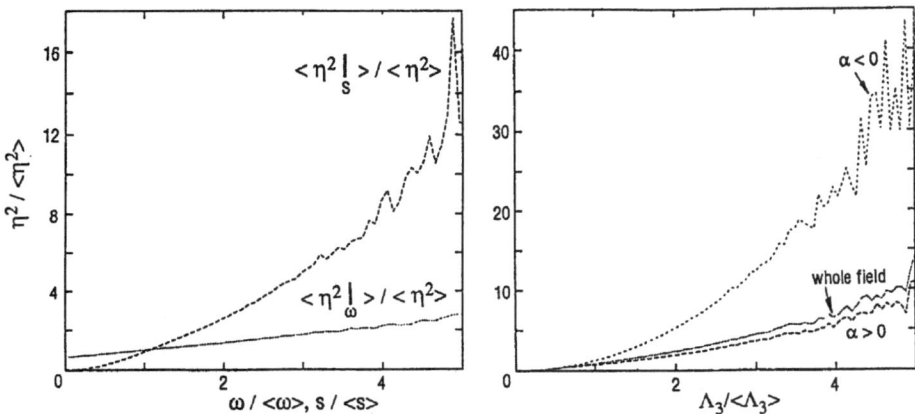

Figure 13. Conditional averages of the magnitude of the rate of change of vorticity direction $\eta^2 = W^2/\omega^2 - \alpha^2 = \Lambda_i^2 \cos^2(\omega, \lambda_i) - \{\Lambda_i \cos^2(\omega, \lambda_i)\}^2$ [251]. Left - in slots of ω and s, from which it is seen that the direction of vorticity is changing much stronger in strain dominated regions. Right - in slots of Λ_3, showing that this rate of change is (apparently) largest in (sub)regions of vortex compression with large magnitude of Λ_3 . Note that η^2 is increasing in slots of Λ_1 and Λ_2 too but at slower rates (not shown).

The vector $\eta_i = s_{ij}\varpi_j - \alpha\varpi_i = W_i/\omega - \alpha\omega_i/\omega$ is the inviscid rate of change of the unit vector ϖ along the direction of vorticity ω, and is responsible for the rate of change of its *direction* [64], and $\eta \perp \omega$, i.e. vector η is associated with vorticity *tilting*. Its magnitude is $\eta^2 = W^2/\omega^2 - \alpha^2 = \Lambda_i^2\cos^2(\omega,\lambda_i) - \{\Lambda_i\cos^2(\omega,\lambda_i)\}^2$. From this it is seen that in regions with negative σ, α the rate of change η of the unit vector ϖ can be large, since, as mentioned, these regions make a positive contribution to the magnitude of the vortex stretching vector $W_i \equiv \omega_j s_{ij}$, so that $W^2/\omega^2 = \Lambda_i^2\cos^2(\omega,\lambda_i)$ can be large and $\alpha^2 = \{\Lambda_i\cos^2(\omega,\lambda_i)\}^2$ can be small. This happens in regions associated with large magnitudes of Λ_3 as is seen from figure 13.

It is reasonable to associate the above process with large curvature of vortex lines and similar quantities, which should reflect their folding and tilting - at least the resulting aspect of these processes. Hence among the questions of interest is about the properties of curvature and relation between curvature and dynamically relevant quantities such as enstrophy ω^2, enstrophy generation σ, rate of enstrophy generation $\alpha \equiv \sigma/\omega^2$ and relations such as various alignments. Of course, the ultimate clarification of such relations can be obtained from looking at global properties. One can hope that some insights can be gained from local analysis, i.e. from working with point quantities at a particular time moment. We bring few typical results relevant to the theme of this paper. More details are given in [251].

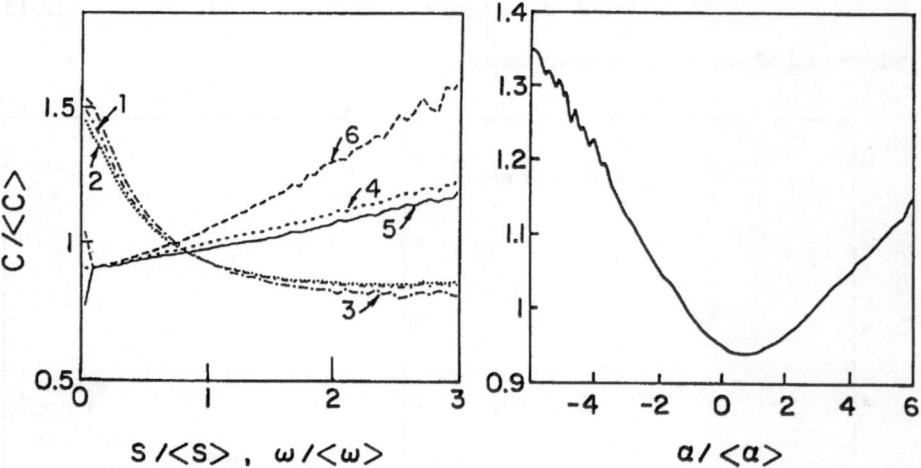

Figure 14. Conditional averages of curvature C of vortex lines [251], DNS, $Re_\lambda \approx 75$. Left - in slots of ω and s for the whole field and for positive and negative rate of enstrophy generation α: $1 - \langle c|_\omega\rangle/\langle c\rangle$, $2 - \langle c|_\omega\rangle/\langle c\rangle$ and $\alpha > 0$, $3 - \langle c|_\omega\rangle/\langle c\rangle$ and $\alpha < 0$, $4 - \langle c|_s\rangle/\langle c\rangle$, $5 - \langle c|_s\rangle/\langle c\rangle$ and $\alpha > 0$, $6 - \langle c|_\omega\rangle/\langle c\rangle$ and $\alpha < 0$. Note the *qualitatively* different behaviour of curvature in slots of ω (decreasing) and in slots of s (increasing). Right - in slots of rate of enstrophy generation α. $Re_\lambda \approx 75$. Note that the curvature of vortex lines *increases* with $|\alpha|$ *both* for negative and positive α.

In a simplified form the logic is that strong stretching results in strong vorticity: indeed regions with strong vorticity are known to be tube-like with small curvature as observed visually in a number of numerical simulations cited above (see also [95]). However, more close inspection shows that the matters are much more complicated (figure 14) due to a number qualitative differences between material and vortex lines (see section 4). One can see that indeed, the curvature decreases in slots of ω. However, this behaviour is practically the same for the whole field, for positive and for *negative* rate of enstrophy generation α (the reader is reminded again that typically regions with $\alpha > 0$ occupy about 2/3 of the turbulent flow field, and regions with $\alpha < 0$ comprise about a 1/3 of the whole flow volume). This last fact, i.e. the behaviour of curvature C versus ω for *negative* rate of stretching $\alpha < 0$ and strong increase of curvature with strain (again for the whole field and both for $\alpha > 0$ and for $\alpha < 0$) undermines the simple analogy with the behaviour of material lines in turbulent flows (see section 4). Similarly, as is expected the curvature of vortex lines is *increasing* with $|\alpha|$ for $\alpha < 0$ due to folding of vortex lines, but again, most interestingly the same behaviour of C is observed for $\alpha > 0$ due to self-induction unlike the case of material lines. This is consistent with the results on the comparison of dependence of enstrophy generation $\omega_i\omega_k s_{ik}$ and its viscous reduction $\nu\omega_i\nabla^2\omega_i$ on ω^2 and s^2 (figure 8) and curvature C . Namely, the preferential alignment between ω and λ_2 is correlated with small curvature and there is no preferential alignment between ω and λ_2 at large curvature (figure 7, top right).

The above shows that the 'most nonlinear'are the regions with large curvature, dissipation, i.e. strain, and preferable alignment between ω and λ_1, and not the regions of concentrated vorticity with small curvature and preferable alignment between ω and λ_2, such as the filaments observed in direct numerical simluations of Navier-Stokes equations and laboratory experiments [23]. This brings us to next issue.

6 Reduction of nonlinearity

This notion has several aspects all of them directly related to geometrical statistics. One of the simplest aspects concerns the magnitude of the vortex stretching and enstrophy generation terms, i.e. $W \equiv |\omega_j s_{ij}|$ and $\omega_i\omega_j s_{ij}$ in (1) and (2). Their magnitude is expected to be smaller than ω^2 and ω^3 respectively due to reduction of nonlinearity in long, thin tubes-filaments-worms which are believed to be in some sense locally quasi-one-dimensional [90], i.e. that nonlinearity is stronger *outside* of these structures. Hence the term depletion (expulsion) of nonlinearity. Following this line one would expect that in regions with strong alignment between vorticity ω and the intermediate eigenvector λ_2 vortex stretching and enstrophy generation should decrease as $|cos(\omega, \lambda_2)|$ increases. Indeed W

---

[23]It is noteworthy that regions of concentrated vorticity are not free of vortex compression in the same proportion as in the whole turbulent field [124], [125] which is possibly associated with Kelvin waves along the worms [257], and is consistent with the results shown in figures 11 and 14.

and its rate are decreasing but remain essentially finite. However, contrary to the above expectation the enstrophy generation and its rate *increase* in slots of $|\cos(\boldsymbol{\omega}, \boldsymbol{\lambda}_2)|$ and becomes *maximal* at $|\cos(\boldsymbol{\omega}, \boldsymbol{\lambda}_2)| \sim 1$ (figure 15).

In other words in these regions the rate of creation of enstrophy ω^2 is the largest and in this sense the nonlinearity is *stronger* and not weaker than in, at least, some of their background using $|\cos(\boldsymbol{\omega}, \boldsymbol{\lambda}_2)|$ as a criterion. The above tendencies are stronger in regions with strong vorticity and survive in the background, e.g. regions of weak enstrophy [248], [252].

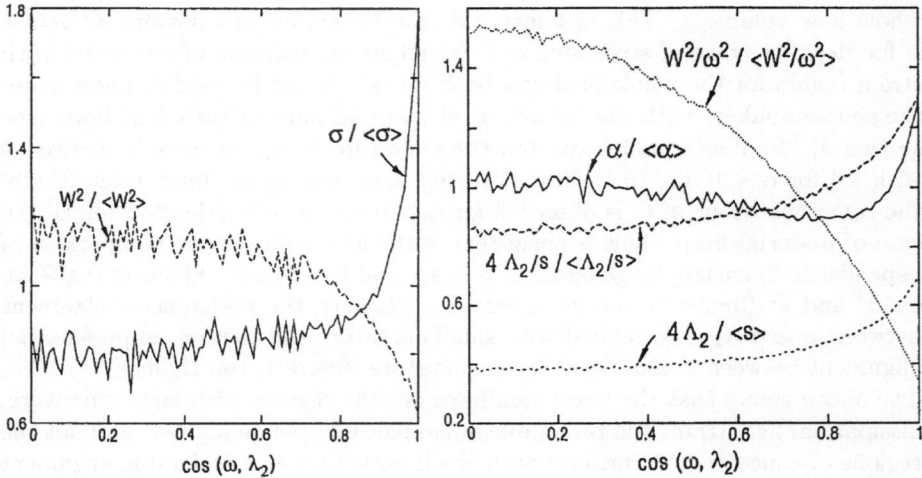

Figure 15. Conditional averages of: left - enstrophy generation σ, vortex stretcing W^2, right - rates of enstrophy generation α, vortex stretcing W^2/ω^2, intermediate eigenvalue of the rate of strain tensor Λ_2 and the ratio Λ_2/s in slots of $\cos(\omega, \lambda_2)$. DNS, Re$_\lambda \approx 75$ [251].

Note that none of the quantities $\omega_i \omega_j s_{ij}$, W, Λ_2 and Λ_2/s become small for $|\cos(\boldsymbol{\omega}, \boldsymbol{\lambda}_2)| \sim 1$ indicating that the flow does not become locally two-dimensional. In particular, it is important that in these regions the intermediate strain (i.e. Λ_2) is positive and is increasing too with $|cos(\boldsymbol{\omega}, \boldsymbol{\lambda}_2)|$, which correspond to strong straining in these regions (cf. with pure two-dimensional flow in which $\Lambda_2 \equiv 0$). Thus one can speculate that there is a tendency to 'localization of nonlinearity'in space which, somewhat paradoxically, is sustained by nonlocal effects due nonlocal relation between strain and vorticity and due to pressure ('nonlocal localization'), see section 8. Note, that the claim on 'localization of nonlinearity'is supported by the behaviour of Λ_2/s in slots of $|cos(\boldsymbol{\omega}, \boldsymbol{\lambda}_2)|$, which is similar to the one of $\Lambda_2/\langle s \rangle$ as shown in figure 15. In order to get more insight it is necessary to look into more subtle aspects of geometrical statistics than just single space/time point alignments. For the moment is it is clear that 'simple'structures in three-dimensional turbulence are qualitatively different from those in pure two-dimensional turbulence in which the nonlinearity is really depleted in such structures — however, in three-dimensional turbulence such structures do not seem to be the best candidates to look for depletion of nonlinearity [250], [248], [252]. Nevertheless, taking the enstrophy generation $\omega_i \omega_j s_{ij}$ as a measure of nonlinearity the objects with strong alignment between $\boldsymbol{\omega}$ and $\boldsymbol{\lambda}_2$ appear to be not the most nonlinear, since their enstrophy generation $\omega_i \omega_j s_{ij}$ comes mostly from the nonlocal effects and not from self-stretching ([123], [250], [252]). Indeed, as shown in figure 8 and section 5.3 the enstrophy generation and its rate are much larger in strain dominated regions (than that in enstrophy dominated ones) with finite curvature of vortex lines and associated with largest eigenvalue Λ_1 of the rate of strain tensor and alignment between $\boldsymbol{\omega}$ and $\boldsymbol{\lambda}_1$. The main contribution to vortex stretching is these regions comes from local effects associated with the (self) interaction of $\boldsymbol{\omega}$ and s_{ij} [66] in contrast with the enstrophy dominated regions in which the vortex stretching is sustained mostly by nonlocal effects (see section 7). Similarly other nonlinear dynamically relevant quantities (vortex stretching W^2 and its rate W^2/ω^2 as shown in figure 16; for η^2 see figure 13 left) also are strongly reduced in regions of concentrated vorticity as compared to their values in strain dominated regions.[24]

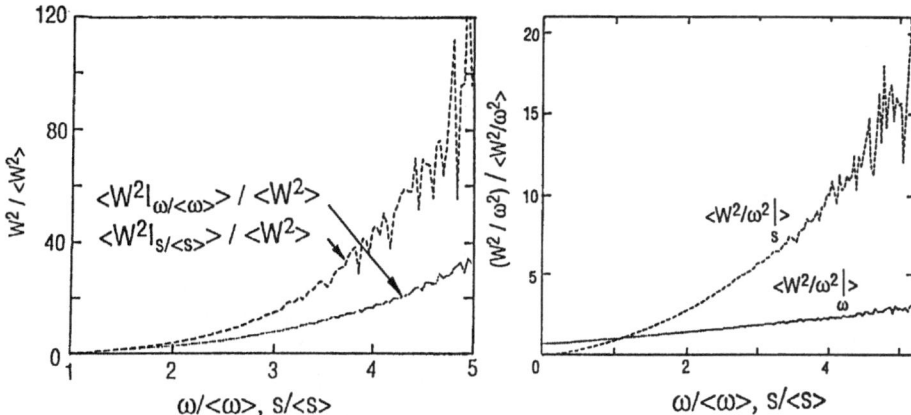

Figure 16. Conditional averages of vortex stretching W^2 (left) and its rate W^2/ω^2 (right) in slots of ω and s. DNS, $\mathrm{Re}_\lambda \approx 75$.

There are also other dynamically important quantities which behaviour in slots of ω and s is *qualitatively* different, see figure 17.

Among other aspects of the problem of reduction of nonlinearity is the compar-

---

[24]It should be emphasized that though the above results are likely to be true at large Reynolds numbers they cannot be seen as an indication that NSE may not develop a singularity in finite time [19], since these results reflect statistical tendencies and there exist, e.g. (small) regions with very large enstrophy, enstrophy generation and alignment between ω and λ_1.

ison of nonlinearities in real turbulent flows with their Gaussian counterparts [155], which is meaningful for even moments only. For example, $\langle |\mathbf{u} \times \boldsymbol{\omega} - \boldsymbol{\nabla}(p + \frac{1}{2}u^2)| \rangle / \langle |\mathbf{u} \times \boldsymbol{\omega} - \boldsymbol{\nabla}(p + \frac{1}{2}u^2)| \rangle_G < 1$ ($\sim 0.5 \div 0.6$) [55], [155]; $\langle W^2 \rangle / \langle W^2 \rangle_G < 1$ ($\sim 0.7 \div 0.8$) [224], [134], [249]. In this sense nonlinearity is reduced. However, in the sense of odd moments the real nonlinearity is 'infinitely'larger, since for a Gaussian velocity field the odd moments vanish identically, e.g. the longitudinal velocity structure functions of odd order $S_{2n+1}(r) = \langle \{[\mathbf{u}(\mathbf{x} + \mathbf{r}) - \mathbf{u}(\mathbf{x})] \cdot \mathbf{r}/r\}^{2n+1} \rangle$, enstrophy generation $\langle \omega_i \omega_k s_{ik} \rangle$ and its rate $\langle \omega_i \omega_k s_{ik}/\omega^2 \rangle$ and many others. In other words build up of *odd* moments such as $S_3(r)$, $\langle \omega_i \omega_k s_{ik} \rangle > 0$ is an important *specific* manifestation of nonlinearity of turbulence along with being the manifestation of its structure. The nonzero $\langle \omega_i \omega_k s_{ik} \rangle$ is associated with the strict alignment between $\boldsymbol{\omega}$ and $\mathbf{W}$ and in this sense this alignment is enhancing the nonlinearity. As is seen from figures 9, 10 this alignment is significant throughout *all* the regions of turbulent flow. On the other hand, the alignment between $\mathbf{u}$ and $rot\ \omega$ [253] is reducing $\langle \omega_i \omega_k s_{ik} \rangle$. Indeed, since $\langle \omega_i \omega_k s_{ik} \rangle = \langle \boldsymbol{\omega} \cdot rot\ (\mathbf{u} \times \boldsymbol{\omega}) \rangle = \langle rot\ \boldsymbol{\omega} \cdot (\mathbf{u} \times \boldsymbol{\omega}) \rangle = -\langle \boldsymbol{\omega} \cdot (\mathbf{u} \times rot\ \boldsymbol{\omega}) \rangle$, and since $-\langle \mathbf{u} \cdot rot\ \boldsymbol{\omega} \rangle \equiv \langle \omega \rangle^2 > 0$ there is a tendency of (anti-)alignment between $\mathbf{u}$ and $rot\ \boldsymbol{\omega}$ reducing the magnitude of $\mathbf{u} \times rot\ \boldsymbol{\omega}$ and thereby of $\langle \omega_i \omega_k s_{ik} \rangle$. Though this is a purely kinematic effect it is directly related to the dissipative nature of turbulent flows, since the mean dissipation $\langle \epsilon \rangle \simeq \nu \langle \omega \rangle^2$.

On other aspects of reduction of nonlinearity, such as resulting from the so called Beltramization related to alignment between $\mathbf{u}$ and $\boldsymbol{\omega}$, etc., see [169], [222], [239], [248], [250] and references therein.

7 Nonlocality

This is one of the main reasons the problem of turbulence is so difficult. As mentioned in section 7 the localization of vorticity in vortex filaments is mostly sustained by the nonlocal effects in physical space, which keep all the regions in turbulent flow in continuous interaction and mutual transformation.

The well known property of nonlocality of NSE in physical space is two-fold. On one hand, it is due to pressure ('dynamic'nonlocality), since $\rho^{-1}\nabla^2 p = \omega^2 - 2s_{ij}s_{ij}$, so that pressure is nonlocal due to nonlocality of the operator ∇^{-2}. The nonlocality is strongly associated with essentially non-Lagrangian nature of pressure. For example, replacing in the Euler equations the pressure Hessian $\frac{\partial^2 p}{\partial x_i \partial x_j}$, which is both nonlocal and non-Lagrangian, by a local quantity $\frac{1}{3}\delta_{ij}\nabla^2 p = \frac{\rho}{6}\{\omega^2 - 2s_{ij}s_{ij}\}$ turns the problem into a local one and allows to integrate the equations for the invariants of the tensor of velocity derivatives $\partial u_i/\partial x_j$ in terms of a Lagrangian system of coordinates moving with a particle ([47] and references therein). This means that nonlocality due to presure is essential for sustaining turbulence without external *random* forcing as, e.g. in Burgers turbulence [192], [277].

Similarly the equations for vorticity (1) and enstrophy (2) are nonlocal in $\boldsymbol{\omega}$

since they contain the rate of strain tensor $s_{ij} = \frac{\partial u_i}{\partial x_j} + \frac{\partial u_j}{\partial x_i}$, $\mathbf{u} = rot^{-1}\boldsymbol{\omega}$ due to nonlocality of the operator rot^{-1} ('kinematic'nonlocality). Both aspects are reflected in the equation for the rate of strain tensor s_{ij} [264]

$$D_t s_{ij} = -s_{ik}s_{kj} - \frac{1}{4}(\omega_i\omega_j - \omega^2\delta_{ij}) - \frac{\partial^2 p}{\partial x_i \partial x_j} + vt, \qquad (14)$$

and for the third order quantities, e.g. $\omega_i\omega_j s_{ij}$, α, [189], [243], [250]

$$D_t \omega_i \omega_j s_{ij} = \omega_j s_{ij}\omega_k s_{ik} - \omega_i\omega_j \frac{\partial^2 p}{\partial x_i \partial x_j} + vt, \qquad (15)$$

$$D_t \alpha = -2\alpha^2 + \lambda_i^2 \cos^2(\boldsymbol{\omega}, \boldsymbol{\lambda}_i) - \pi_i \cos^2(\boldsymbol{\omega}, \boldsymbol{\pi}_i) + vt, \qquad (16)$$

where vt stands for viscous terms and π_i, $\boldsymbol{\pi}_i$ are respectively the eigenvalues and the eigenvectors of the pressure hessian $\frac{\partial^2 p}{\partial x_i \partial x_j}$.

It is seen from the equations (5), (6) that the rate of change of enstrophy generation $D_t\sigma$ and $D_t\alpha$ depend on the geometrical relations between vorticity $\boldsymbol{\omega}$ and both the eigenframe of the rate of strain tensor $\boldsymbol{\lambda}_i$ and that of the pressure hessian $\boldsymbol{\pi}_i$ [181], [189], [250]. An important aspect is that the equation (5) and a similar one for $s_{ik}s_{kj}s_{ij}$ contain two invariant quantities $\mathcal{I}_5 = \omega_i\omega_j \frac{\partial^2 p}{\partial x_i \partial x_j}$; $\mathcal{I}_6 = s_{ik}s_{kj} \frac{\partial^2 p}{\partial x_i \partial x_j}$ reflecting the nonlocal dynamical effects due to pressure and can be interpreted as interaction between vorticity and pressure and between vorticity and and strain. In particular the equation (5) for the enstrophy generation $\sigma \equiv \omega_i\omega_j s_{ij}$ shows both aspects of nonlocality of vortex stretching process. The first term in (5) is strictly positive $\omega_i s_{ij}\omega_k s_{ki} \equiv W^2 > 0$. This means that the nonlinear processes involving vortex stretching (or direct interaction of vorticity and strain) always tend to increase even the *instantaneous* enstrophy generation. Here also the term $W_3^2 = \omega^2 \lambda_3^2 \cos^2(\boldsymbol{\omega}, \boldsymbol{\lambda}_3)$ associated with the negative eigenvector of the rate of strain tensor Λ_3, i.e. vortex compressing or negative enstrophy production $\omega_3^2 \Lambda_3 \cos^2(\boldsymbol{\omega}, \boldsymbol{\lambda}_3)$, makes a positive (!) contribution to the rate of change of enstrophy generation ($W_i^2 = \omega_i^2 \Lambda_3^2 \cos^2(\boldsymbol{\omega}, \boldsymbol{\lambda}_i)$). It is natural to call the term W^2 (which is just the squared magnitude of the vortex stretcing vector) as the inviscid rate of *increase* of the enstrophy generation term. However, the inviscid rate of *change* of enstrophy generation contains also a second term reflecting the interaction between vorticity and the pressure hessian $\frac{\partial^2 p}{\partial x_i \partial x_j}$. It appears that $\langle \omega_i\omega_j \frac{\partial^2 p}{\partial x_i \partial x_j} \rangle$ is positive and is about $\langle W^2 \rangle / 3$, i.e. in the mean the nonlinearity in (5) is reduced by this nonlocal term, since for a Gaussian velocity field $\langle \omega_i\omega_j \frac{\partial^2 p}{\partial x_i \partial x_j} \rangle \equiv 0$. Consequently the PDF of $\cos(\boldsymbol{\omega}, \mathbf{W}^{\Pi})$, $W_i^{\Pi} = \omega_j \frac{\partial^2 p}{\partial x_i \partial x_j}$, is strongly skewed just like the PDF of $\cos(\boldsymbol{\omega}, \mathbf{W})$ [247], [250]. Another aspect of reduction of nonlinearity is seen clearly from figure 17. Namely, the nonlinearity is strongly reduced in the enstrophy dominated regions (filaments/worms), whereas it is enhanced in strain dominated regions. Note the *qualitatively* different behaviour of $\omega_i\omega_j \frac{\partial^2 p}{\partial x_i \partial x_j}$ (figure 17 top) in enstrophy dominated regions (decreasing with ω) as compared to that in strain dominated regions (increasing with s).

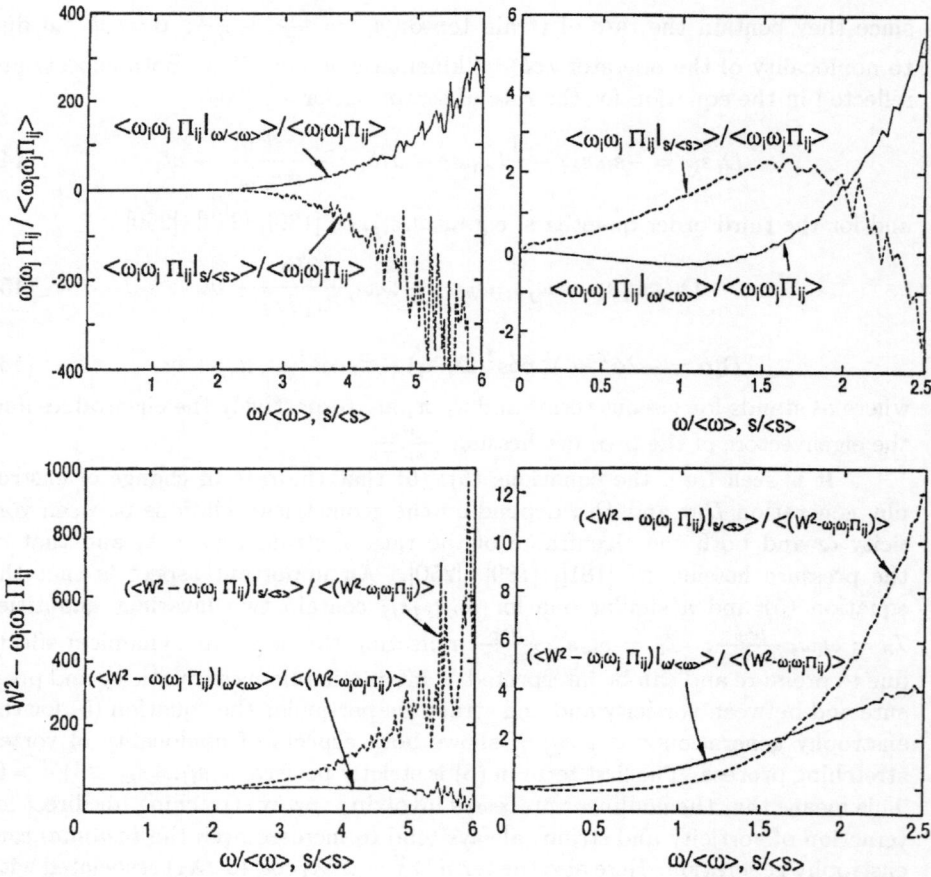

Figure 17. Conditional averages of the nonlocal term $\omega_i\omega_j\frac{\partial^2 p}{\partial x_i \partial x_j}$ (top) and of the inviscid rate of change of enstrophy generation $W^2 - \omega_i\omega_j\frac{\partial^2 p}{\partial x_i \partial x_j}$ (bottom) in slots of ω and s, DNS. $Re_\lambda \approx 75$.

8 Non-Gaussian nature of turbulence and 'kinematic' effects

As mentioned above one of the prominent and distinctive *specific* features of turbulent flows of utmost dynamical significance, such as specific expression of nonlinearity of turbulence and its structure, and the most spectacular manifestations of their non-Gaussian nature is the build up of *odd* moments of various quantities. For this reason the quantity $cos(\boldsymbol{\omega}, \mathbf{W}) = \omega_i\omega_j S_{ij}|\omega|^{-1}|W|^{-1}$ appeared so useful in the diagnostics of the non-Gaussian nature of the 'random' 'structureless' sea in turbulent flows, which appeared to be quite the opposite, i.e. not stuctructureles, dynamically not passive and essentially non-Gaussian.

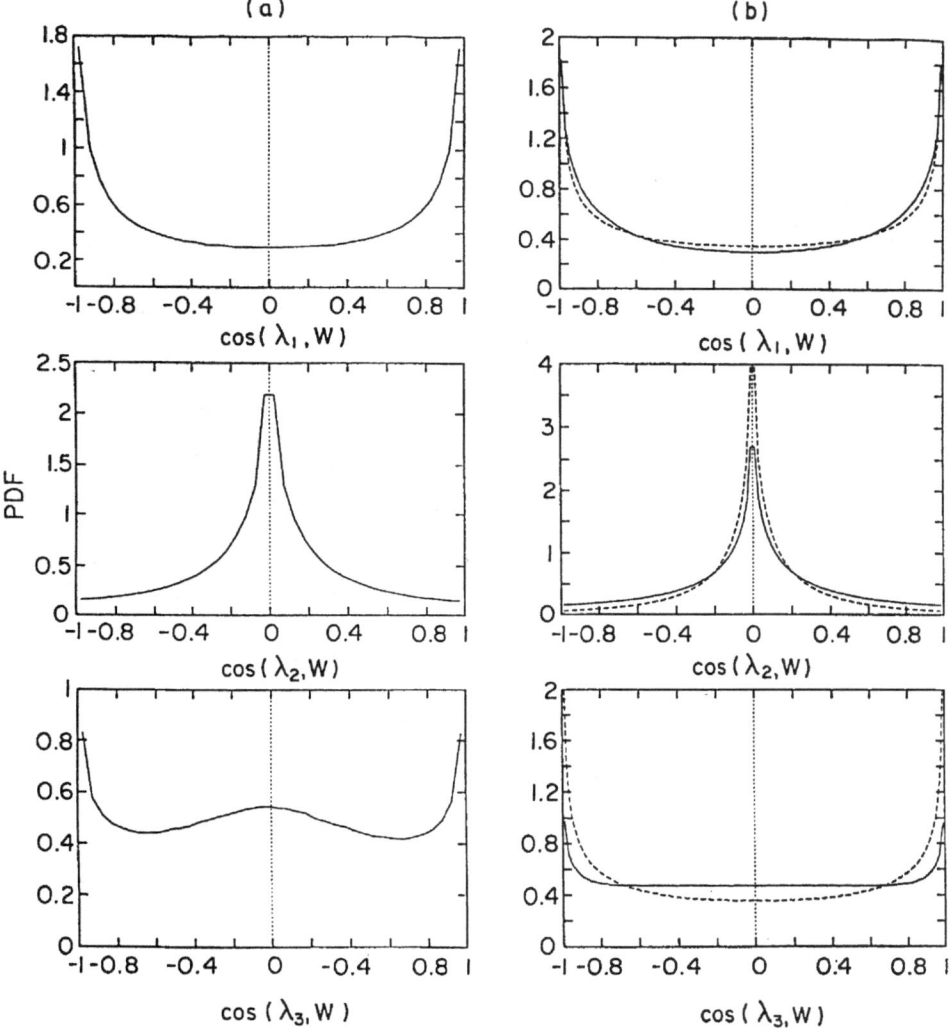

Figure 18. PDFs of the cosine of the angle between vortex stretching vector $W_i = \omega_i s_{ij}$, and the eigenvectors λ_i of the rate of strain tensor;   left – grid turbulence, right – DNS and random Gaussian (– – – –). $Re_\lambda \approx 75$. The behaviour shown in (b) seems to contradict the one shown in figure 7, since there exist a tendency for alignment between ω and $\mathbf{W}$ and between ω and λ_2. However, more close inspection shows that the alignments shown in this figure and in figures 7 are associated mostly with *different* regions in the flow [251].

A simple demonstration how the dynamics of turbulence makes it non-Gausssian can be seen from taking $\langle \ldots \rangle$ from the equation (17) (dropping the viscous term)

$$\frac{D}{Dt}\langle \omega_i \omega_j s_{ij} \rangle = \langle \omega_j s_{ij} \omega_k s_{ik} \rangle - \langle \omega_i \omega_j \frac{\partial^2 p}{\partial x_i \partial x_j} \rangle \qquad (17)$$

For a Gaussian velocity field $\langle \omega_i \omega_j s_{ij} \rangle = 0$, $\langle \omega_i \omega_j \frac{\partial^2 p}{\partial x_i \partial x_j} \rangle = 0$, $\langle \omega_j s_{ij} \omega_k s_{ik} \rangle \equiv \langle W^2 \rangle = \frac{1}{6}\langle \omega^2 \rangle^2$ [223], [250]. If the initial conditions are Gausssian the flow ceases to be gausssian with *finite rate*. In other words it is seen directly from (17) that *turbulence cannot be Gaussian* (see also [183]). In this sense Gaussian initial conditions are not 'good', since *no flow state existing in reality is Gaussian*. It is seen also from (17) that initially Gaussian and *potential* velocity field with small seeding of vorticity will produce - at least for a short time - an essentially positive enstrophy generation.

Turbulence – being essentially non-Gaussian – is such a rich phenomenon that it can 'afford'a number of manifestations (sometimes nontrivial or, at least, not obvious, [223], [239], [262]), which are Gaussian-like. A recent example of such behaviour is shown in figure 18. This example is interesting in that the Gaussian-like behaviour is exhibited by third order quantitites. For other examples see [223].

Many second (and higher) order quantities in turbulent flows exhibit exponential tails in their PDF's. However, precisely the same behaviour is characteristic of purely Gaussian isotropic velocity field. For example, PDF's of ω^2 and s^2 of a Gaussian velocity field have exponential tails [223] and their shape is very similar to that in real turbulent flows like those shown in figure 5. Using the explicit expressions for the PDF's of ω^2 and s^2 from [223] it is staightforward to obtain the values of flatness for ω^2 and s^2 for an arbitrary Gaussian velocity field. Alternatively the same result is obtained directly without invoking the explicit expressions for the PDF's of ω^2 and s^2, but using instead the decomposition rule for forth-order moments. Namely, $F_{\omega^2} = \langle \omega^4 \rangle / \langle \omega^2 \rangle^2 = 5/3$ and $F_{s^2} = \langle s^4 \rangle / \langle s^2 \rangle^2 = 7/5$, i.e. the flatness of enstrophy is *larger* than that of total strain $\langle \omega^4 \rangle / \langle \omega^2 \rangle^2 - \langle s^4 \rangle / \langle s^2 \rangle^2 = 4/15$ [25]. Does one have to conclude from the above result that the enstrophy field is more intermittent than that of total strain in a *Gaussian* velocity field. Definitely not, since by definition Gaussian velocity fields lack any intermittency. This example shows that even moments *only* are not sufficiient for characterisation of the non-Gaussian nature and intermittency of turbulence.

Similarly the PDF of pressure has exponential tails, which in addition is strongly negatively skewed. It is not easy in this case to demonstrate this directly due to nonlocal nature of the ∇^{-2} operator [109]. Much easier is to do this via looking at $\nabla^2 p$. Using the same method as in [223] the PDF $\mathcal{P}(x)$, $x = \frac{\nabla^2 p}{\rho \langle \omega^2 \rangle} = \frac{\omega^2 - 2s_{ij}s_{ij}}{2\langle \omega^2 \rangle}$,

---

[25]This is precisely the extreme of the 'generic'inequality obtained in [54] in somewhat different way using the the decomposition rule for forth-order moments.

is expressed in the following way (Spector 1996, private communication)

$$\mathcal{P}(x) = \{3^{1/2}5^{5/2}\}/(4\pi) \, x^2 \, e^x \, [K_2(4x) - K_1(4x)], \qquad x < 0,$$

which for large $|x|$ has the asymptotics $\sim |x|^{1/2} \, e^{-3|x|}$,
and

$$\mathcal{P}(x) = \{3^{1/2}5^{5/2}\}/(4\pi) \, x^2 \, e^{-|x|}[K_2(4x) + K_1(4x)], \qquad x > 0,$$

which for large x has the asymptotics $\sim x^{3/2} \, e^{-5x}$.

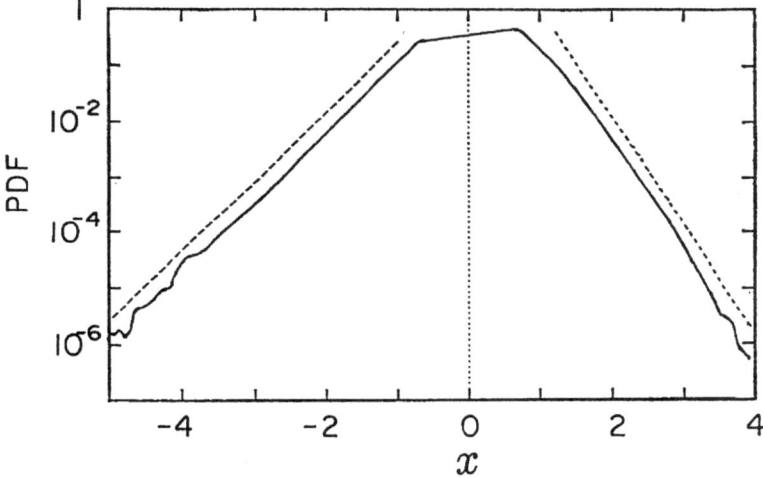

Figure 19. PDF of the laplacian of pressure $x = \frac{\nabla^2 p}{\rho\langle\omega^2\rangle} = \frac{\omega^2 - 2s_{ij}s_{ij}}{2\langle\omega^2\rangle}$. ——— - grid turbulence, $Re_\lambda \approx 75$; - - - - - - slope $+ 4.5$, — — — — - slope $- 2.9$.

So, as expected, the distribution of $\nabla^2 p$ is asymmetric even for a Gaussian velocity field. This result is in good quantitative agreement with the ones from laboratory and DNS experiments (figure 19), showing that these effects are mostly of kinematical nature as many others [223], [239].

In this sense the non-Gaussian strongly intermittent behaviour and anomalous diffusion of passive objects (scalars, vectors) in a Gaussian or any other *a priory* prescribed random velocity field is a kinematic effect, since the Lagrangian velocity field is an extermely complicated non-linear functional of the Eulerian field. How complicated is the issue of the relation between the Lagrangian and Eulerian fields can be seen on the example of the so called Lagrangian (kinematic) chaos or Lagrangian turbulence (chaotic advection) with *a priory* prescribed and *not random* Eulerian velocity field (E-laminar). In such E-laminar, but L-turbulent flows ([10], [11], [51], [103], [203], [268], [271] but see [263] references therein) the statistics of the latter has no counterpart in the former to be related with. The complication is, of course, due to Lagrangian chaos. In genuine E-turbulent flows, which are L-turbulent too, i.e. EL-turbulent, the fluid particle chaos consists of two contributions: Eulerian and Lagrangian.

9   Concluding comments

Turbulence has more unresolved issues than otherwise. In this chapter along with concluding remarks notes on some of these issues are made, partially rather speculative.

9.1   Universality versus nonuniversality

9.1.1   Quantitative versus qualitative

As mentioned in the introduction the qualitative properties/features of all turbulent flows at high enough Reynolds numbers are essentially the same, i.e. it is meaningful to speak about *qualitative univesality* of turbulent flows. In addition to properties/features mentioned in section 1 there are more such as the build up of odd moments (e.g. the predominance of vortex stretching over vortex compressing), a variety of alignments and other aspects of geometrical statistics and there seem to exist many not yet known. The similarity between the large scale properties of (the same) turbulent flows at transitional, very moderate and large values of Reynolds number can be qualified as one of the manifestations of qualitative universality of turbulent flows. The likely reason for this is that the *nonlinear terms...remain active at surprisingly low Reynolds numbers* [168]. Therefore it seems not necessary to 'hunt'very large values of Reynolds number in studying and trying to understand the basic physics of turbulence. Scaling and related matters proved not very useful [88], [244] in *understanding* of basic physics of turbulence so far as to justify, e.g. enormous efforts in accurate measurement of exponents at very large values of Reynolds number. The similarity of large scale properties of turbulent flows at different values of Reynolds number is likely to be among the main reasons for the success (whatever this means) of the most of low dimenional models and their low sensitivity to the details of the subgrid scale models ([38], [93], [107], [122] and references therein). This does not mean that low dimensional models of turbulent flows represent adequately the physics of turbulent flows, just like it is doubtful that things like low dimensional chaos, etc., can be qualified as turbulence (especially FDT).

9.1.2   Large scales versus small

The quantitaive properties vary largely with the range of scales of interest. Though the large scale (LS) properties of FDT depend on particular mechanisms generating turbulence and, generally, are not universal in the sense that they are different for different flows they are universal in the sense that they become independent of Reynolds number for a particular class of flows, e.g. for particular geometry. There is also some evidence that the (statistical) properties of turbulent flows with the same geometry, not homogeneous and at very modest Reynolds numbers, are invariant of the boundary and initial conditions (BC and

IC) at least in some cases. For example, typical DNS computations of NSE of turbulent flows (e.g. in a circular pipe [76] and a plane channel [139] , in a cubic box [123], [252], [260], etc.) involve extensive use of periodic BC. The results of these agree very well with those obtained in laboratory experiments, in which the BC have nothing to do with periodicity [26] and in which the IC were totally different from those in DNS. No explanation of this kind of invariance is known so far, but it is naturally to expect that it is related to some kind of hidden symmetry(ies) of the NSE. If such exist they may be the reason for similarity of results obtained via DNS of NSE in, e.g. periodic boxes by various forcing (different deterministic, random/stochastic, etc.).

On the other hand, the far field statistical properties of free shear turbulent flows (mixing layers, wakes, jets) are known to possess strong memory (sensitive to the conditions at their 'start') with some properties not universal in Reynolds number (see [71], [270] and references therein). These flows develop in space beginning with *small* scales into the *large* ones, in *apparent* contradiction to the Richadrson-Kolmogorov cascade ideas. Another peculiarity of these flows (and other spatially developing flows like boundary layers) is that their turbulent part is confined in a bounded region surrounded by purely *laminar* flow with a *distinct* though very convoluted (fractal?) boundary.

As mentioned in section 1.1 there is reasonable evidence that the normalized mean dissipation $\varepsilon = U^3 L^{-1} \langle \epsilon \rangle$ tends to a finite limit, or at least is bounded by a Reynolds independent bound, as Re $\rightarrow \infty$. Since at large Reynolds numbers $\langle \epsilon \rangle \approx \nu \langle \omega^2 \rangle$ this means that $\langle \omega^2 \rangle \approx \nu^{-1}$ at Re $>> 1$, i.e. it is Reynolds dependent. There exists evidence on Reynolds dependence of less trivial small scale quantities such skewness and flatness of some velocity derivatives and some other [6], [8], [15], [16], [21], [23], [53], [78], [94], [194], [197], [219], [228], [231], [269]. As in the case of large scales there is neither full agreement between the results of different authors nor understanding of the reasons for the Reynolds dependence at large values of the Reynolds number.

At present there are no clear answers to the questions like *How much universal is turbulence (FDT) and in what sense? What is the origin of intermittency in turbulence? Is anomalous scaling necessarily related to some (intermittent) structure of turbulent flow? Are these two terms, i.e. anomalous scaling and intermittency, synonymous?* Answers to such and similar questions cannot be obtained within the frame of phenomenology or similar approaches. In order to make at least some progress one has to look into more details of the turbulence structure and specific aspects of its dynamics.

---

[26]The correlation coefficient between two values of *any* quantity at the opposite such boundaries (i.e. the points separated at *maximal distance* in the flow domain) will be precisely equal to unity and close to unity for the points in the proximity of such boundaries, whereas in any *real* flow the correlation coefficient becomes very small for points separated by a distance of the order of (and larger than) the integral scale of turbulent flow.

9.2 Structure

It is commonly believed that *most* of the structure of turbulence is associated
with and is due to various strongly localized intense events/structures, e .g.
mostly regions of concentrated vorticity so that *'turbulent flow is dominated by
vortex tubes of small cross-section and bounded eccentricity'*([61], p. 95) and
that these events are mainly responsible for the phenomenon of intermittency.
However, it appears that the concentrated vorticity (tubes-filaments-worms) is
not that important as it has been thought before, they *'do not seem to play a
special role in the overall dynamics of turbulent flows'*and *'the effect of removing
the worms is small,...'*[121]. In other words, the 'worms'are more the *consequence
rather than the dominating factor of the turbulence dynamics* (for more details
and references see [246]).
On the contrary all regions in turbulent flow than just those with intense vorticity are spatially structured, since structure in (quasi-homogeneous/isotropic)
turbulent flow is associated with alignments rather than with strong vorticity
only.

9.2.1 Vorticity-dominated versus strain-dominated regions

A useful way of distinguishing between different regions in turbulent flows is
based on inviariants of tensor of velocity derivatives $A_{ij} = \partial u_i/\partial x_j$ and of rate
of strain tensor ([29], [32], [117], [181] and references therein). In particular
it is common to look at enstrophy dominated regions as contrasted to those
dominated by strain. The enstrophy dominated regions are well defined by, say,
high enough enstrophy ω^2 and are tube/filament-like objects. They form a subset of much larger *locally* quasi-two-dimensional regions corresponding to large
$\cos(\omega, \lambda_2)$, i.e alignment between ω and λ_2. However, it is not enough to specify
the magnitude of the total strain s^2 in order to 'visualize'in a unique way the
strain dominated regions, which – as demonstrated above – contain a number of
qualitatively different regions. In particular, the set of strain dominated regions
contains the following three dynamically significant subsets [27] :
i - The largest enstrophy generation occurs in regions of strongest (local) vortic-
ity/strain interaction, which are different from those with concentrated vorticity. These regions contribute most to the enstrophy generation, which is *much
larger* than viscous reduction of enstrophy, and at least in this respect are dynamicaly more important than those of concentrated vorticity. These regions
are associated with alignment between ω and λ_1, large values of Λ_1, fairly large

---

[27]These three strain dominated regions may overlap since large total strain $s^2 = \Lambda_1^2 + \Lambda_2^2 + \Lambda_3^2$
cannot be associated with large value of one of Λ_i *only* due to incompressibily $\Lambda_1 + \Lambda_2 + \Lambda_3 = 0$
($\Lambda_1 > \Lambda_2 > \Lambda_3$; $-2 < \Lambda_2/\Lambda_1 < -1/2$, $-1/2 < \Lambda_2/\Lambda_1 < 1$). However, these regions
mostly are associated also with alignments between ω and corresponding eigenvector λ_i [251].
This excludes overlapping at large enough $\cos^2(\omega, \lambda_i)$. Note that dynamically important
quantities (such as enstrophy generation $\sigma = \omega^2 \Lambda_i \cos^2(\omega, \lambda_i)$, it's rate $\alpha = \Lambda_i \cos^2(\omega, \lambda_i)$,
the magnitude of vortex stretching $W^2 = \omega^2 \Lambda_i^2 \cos^2(\omega, \lambda_i)$ and the rate of change of vorticity
direction $\eta^2 = W^2/\omega^2 - \alpha^2 = \Lambda_i^2 \cos^2(\omega, \lambda_i) - \{\Lambda_i \cos^2(\omega, \lambda_i)\}^2$) contain Λ_i and $\cos^2(\omega, \lambda_i)$
in combinations like $\Lambda_i \cos^2(\omega, \lambda_i)$ and $\Lambda_i^2 \cos^2(\omega, \lambda_i)$ only.

curvature of vortex lines and, obviously, are dynamically most active, strongly non-Gaussian and possess structure.

ii - Regions, which are wrapped around the enstrophy dominated regions and associated with alignment between ω and λ_2 (just like in the enstrophy dominated regions) and mostly positive Λ_2.

iii - Regions with large magnitude of Λ_3, alignment between ω and λ_3, large curvature of vortex lines, vortex compressing (i.e. negative enstrophy generaton and its rate), tilting and folding. In these regions considerable part of curvature of vortex lines is generated along with the other part produced in the regions of largest enstrophy generation via self-induction [251].

9.2.2 Background

The above regions, both enstrophy and strain dominated are surrounded by the so called 'background', which contratry to common beliefs is not stuctructure-less, dynamically not passive and essentially non-Gaussian just like the whole flow field - even with weakest measurable vorticity and/or strain or any other region with 'weak'excitation in some sense. The structure of this apparently random background seems to be rather complicated. The previous qualitative observations (mostly from DNS) on the *'little apparent structure in the low intensity component'*or the *'bulk of the volume'*with *'no particular visible structure'*should be interpreted that no *simple visible* structure has been observed so far in the bulk of the volume in the flow. It is a reflection of our inability to 'see'more intricate aspects of turbulence structure: intricacy and 'randomness'are not synonyms of absence of structure.

The same is true of all the regions other than those with concentrated vorticity. However, the structure is definitely present practically everywhere in the turbulent flow, but it is more complex than just a collection of 'simple'objects such as vortex tubes, though the latter may only seem to be simple. Indeed, even in rather 'simple'nonturbulent configurations (see, e.g. [1], [87], [138], [140], [226] and references therein) the structure of vorticity field is far more complicated than just a collection of simple objects such as vortex tubes, etc. This together with the limited role of the 'simple'objects in turbulent flows leads to the conclusion that it would be somewhat wishfully naive to expect that such a complicated phenomenon like turbulence can be adequately described in terms of collections of such 'simple'(weakly interacting) objects *only*. On the contrary it seems that all the regions — concentrated vorticity, the background, regions of strong vorticity-strain (self) interaction and largest enstrophy generation, and regions with negative enstrophy production — are in continuous interaction and mutual transformation and are strongly correlated due to strong nonlocality of turbulence in physical space. In other words no set of 'simple'weakly interacting objects is known so far, which represent *adequately* the turbulent field.

At this stage little is known about how the structure of the regions different from those with concentrated vorticity looks like. One of the main issues for the future research is to get more insight into this structure and in the dynamics of the interaction and mutual transformation of different regions/structures. This requires much more than single point space-time statistics mostly used in this paper. In particular things like time evolution, Lagrangian statistics (e.g. [114], [181], [217], [265] and references therein) relating the spatial structure and the time dimension have to be used and studied. Some of such work is in progress. All the above results were obtained at rather low Reynolds number ($Re_\lambda \approx 75$). However, a number of very similar results, such as various alignments, etc., were obtained in numerical simulations of Euler equations ([24], [36], [61], [87], [188], [189], [195], [202] and references therein) thereby showing that - at least qualitatively - they should be valid at large Reynolds numbers.

These results seem to be also of importance for further basic research in turbulence and in their implications and consequences for a large number of existing and forthcoming theoretical descriptions of turbulence/turbulent flows, e.g. those based on expansions near a Gaussian field, and those representing the turbulent field as a collection of 'simple'objects.

9.3 On right results not necessarily for the right reasons

Turbulence is an extremely rich phenomenon with a great multitude of manifestations. Therefore, due to the *richness* of turbulence phenomenon correspondence of some 'theory'to some experimental results may occur not necessarily for the right reasons. The correspondence with experimental results is at best only a necessary condition. One of the oldest examples is the mixing length theory and related things like gradient transport, eddy viscosity, etc.,[69], [165], [176]. In a recent example mean velocity distributions were obtained '*from the first principles*'for turbulent Couette and Poiseuille flows, which are in very good agreement with experimental results for real *three*-dimensional flows [27]. The problem is that these theoretical results are based on a *two*-dimensional model lacking any vortex stretching whatsoever. The third example is the popular GOY and similar models (for references see [25], [44], [99], [126], [161]) a variant of shell models originated with the *dynamical* systems of hydrodynamical type of Obukhov since 1969 [187], exhibit *temporal* chaos only. Therefore such models hardly can be associated with the intermittency of *real* FDT which involves essentially *spatial* chaos as well. Therefore a question arises as to whether correspondence of such models (as well as many others [20], [25], [30], [33], [38], [44], [48], [74], [77], [99], [172], [178], [215], [218], [261], [277]) with experimental results in *real* turbulent systems [7], [12] [206], [44], [209], [259] occurs for the right reasons. The fourth example relates to the belief that essential aspects of turbulence fine structure and its dynamics may be adequately represented by a random distribution of strained vortical objects (sheets, filaments, tubes-worms-sinews) - or other '*simple'objects* . This idea goes back to Townsend [237] (see also [18] pp. 159 -161) and it was developed to a high level of sophistication (for references see [246]). However, in this way the turbulent field is represented by elements which comprise - contrary to the common belief - not the most important part of the flow. The main feature and shortcoming of these objects is that they possess single-component vorticity, zero curvature of vortex lines and consequently such objects lack the 'genuine'nonlinearity of turbulent flows —

the self-amplification: they are stretched by the strain which is decoupled from them. In addition these elements are stretched only, whereas in real turbulent flows both vortex *stretching* and vortex *compressing* occur — the last occupying about a third of flow volume. The fifth example is associated with the belief that essential aspects of quasi-two-dimensional turbulent flows (Q2D) can be well described by pure two-dimensional ones (P2D) both globally and locally (e.g. [99], [199]), i.e. ε in $Q2D = P2D + \varepsilon$ is small in some sense. One of the popular beliefs is that *locally* this is true of Q2D regions with concentrated vorticity (vortex filaments). However, it appears that locally quasi-two-dimensional regions corresponding to large $\cos(\boldsymbol{\omega}, \boldsymbol{\lambda}_2)$ – to which belong the regions of concentrated vorticty – are *qualitatively* different from purely two-dimensional ones in that they posses essentially nonvanishing enstrophy generation σ and intermediate eigenvalue Λ_2 of the rate of strain tensor, whch are identically zero in P2D flows [248], [252]. Moreover, in these regions both σ and Λ_2 are *larger* than in the whole field and in this sense ε is *not small* in $Q2D = P2D + \varepsilon$.

In case of *globally* Q2D turbulent flows the matters seem to be even controversial, which is seen from the case of nonhomogeneous flows. For example, the experimental (and recent numerical) results obtained for turbulent MHD flows in channels with large aspect ratio in the presence of an azimuthal magnetic field showed that in such flows (at $\mathrm{Re} \leq 10^4$), - which are Q2D - on one hand, the drag is indistinguishable from its laminar value, and on the other hand, the level of turbulence may be substantially higher than that in the same flow without magnetic field (see e.g. the review [240]). However, the examination of the results of the DNS of plane Poiseulle turbulent flow (which is P2D) at $Re \sim 10^4$ [120] shows that its drag is about twice larger than its purely laminar value and is only twice smaller than its value for the 3D turbulent flow, i.e. P2D plane Poiseulle turbulent flow is not not that low dissipative. Moreover, the Reynolds stresses in this flow are not small either (as was expected before) and contribute about a half to the total stress. These, results were confirmed recently using an essentially different code [164]. The problem of the relation(s) between Q2D and P2D turbulent flows is complicated further by the *multiplicity* of Q2D states: there exist several Q2D flows such as flows in rotating frames, flows with stable density stratification, MHD-flows and some others, which along with being similar kinematically (geometrically) in many respects are very different dynamically. There is little doubt about the qualitative difference between Q2D states produced by physically different processes, e.g. the ones in MHD are of dissipative nature (Joule dissipation), whereas those with rotation are not. Strong anisotropy is a necessary condition only for Q2D and/or low dissipative behavior, e.g. shear turbulent flows with strong shear are both strongly anisotropic and strongly dissipative. Similarly, strong correlations along some direction (i.e. Q2D behavior) do not exclude the possibility of vorticity stretching in this direction [248], [252].

9.4 Final remarks

Due to space limitations many issues which are beyond phenomenology were even not touched [28], so some general remarks are in place.

9.4.1 The closure problem and modeling

Hans Liepmann wrote in 1979: *Turbulent modelling is still on the rise owing to rapid development of computers coupled with the industrial need for management of turbulent flows. I am convinced that much of this huge effort will be of passing interst only. Except for rare critical appraisals...much of this work is never subjected to any kind of critical or comparative judgement. The only encouraging prospect is that current progress in undertsanding turbulence will restrict the freedom of such modelling and guide these efforts toward a more reliable discipline* [162].

Liepman's critisism was directed to (already at that time) great number of publications which used a variety of assumptions all of them *very remote from any physical basis* to say nothing of any rigorous mathematical basis. This state of matters was changed due to recent important rigorous results (see [92] and references therein). Namely, it was proved that the Friedman-Keller chain of equations for the moments [133] (and consequently the Hopf equation) has a unique solution for initial conditions in appropriately chosen functional space. In other words a positive answer was given to the question whether the closure problem has a solution, and an estimate of convergence of approximations for the closure of the infinite chain of equations for moments was given. Thus, it became clear - at least in principle - that turbulence modelling can be put on a rigourus foundation, e.g. [81]-[83], though just like simulation (DNS) by itself does not bring undestanding, neither does modelling of whatever sophistication.

---

[28]For example, the following questions were mentioned in the first announcement on the *Second Monte Verità Colloquium on Basic Problems in Turbulence* to be held in Switzerland in March 22-27, 1998:

Among the main themes at the Monte Verità I were: the implication and relevance of dynamical systems to (fully developed) turbulence, reduction of nonlinearity, some aspects of turbulence structure, fundamental experiments (including the numerical ones), and some others.

In addition to the mentioned above possible themes for discussion include some fundamental mathematical issues, how universal turbulence is? (What are the situations and what are the properties which are invariant of IC and BC?, is the inertial range dynamics independent of viscosity, the type of dissipation, etc.?), is it possibile to put turbulence modelling and the problem of closure on a reasonable/rigorous physical/mathematical foundation?, how realistic/adequate are the attempts to represent turbulence via a collection of 'simple'objects (like 'coherent structures', 'eigensolutions', 'worms', etc.)?, is it possible to construct a kind of 'statistical mechanics'of at least some 'simple'turbulent flows? what can be learnt about the dynamics of turbulence from studies of passive objects (scalars, vectors...) in real and 'synthetic'turbulence? linear versus nonlinear processes in turbulent flows, interplay of kinematics and dynamics, geometrical statistics, to what extent is it possible to control turbulence?, and some others (see also [91], [228] and references therein).

9.4.2 Reduced (low dimensional) representations

Perhaps the biggest fallacy about turbulence is that it can be reliably described (statistically) by a system of equations which is far easier to solve than the full time-dependent three-dimensional Navier-Stokes equations (P. Bradshaw [37]). In spite of this warning there is a general belief that an adequate reduced (low dimensional) description is possible, e.g. via reduction of the huge number of degrees of freedom in FDT by retaining the so called *relevant/important* ones (i.e. *Perhaps large systems boil down to a few degrees of freedom...*[43]), though the meaning of what are the relevant/important modes/degrees of freedom is quite problematic [107], [148]. Most frequently it is argued that these are 'modes'containing most of the energy, but - at least from the physical point of view - the 'modes', e.g. carrying most of the energy dissipation are not less relevant/important in some sense, and 'mixed modes'related to both small and large scales such as eigenfunctions of $\langle u_i \omega_j \rangle$ (note that the vortex stretching vector $W_i = \partial(u_i \omega_j)/\partial x_j$) may appear even more relevant/important (see section 3.1) [241]. Even in such a case there is little hope to obtain a low dimensional approximation representing *adequately* [29], e.g. such a 'simple'turbulent flow as the flow in a plane channel at rather low Reynolds number (Re = 3300) which attractor dimension is estimated to be of the order 10^3 ! [132]. The exceptions are when the flow - though turbulent - at the outset is strongly dominated by some 'low dimensional subsystem', e. g. [108], [160], [186], .

9.4.3 Concluding

Perhaps it is guileless to think that the 'problem of turbulence'would be resolved if one would have a super-hyper computer enabling to 'solve'the NSE or whatever at any Reynolds number. Suppose one can do this and also measure whatever one wants. The real problem is what one is going to do next with the many GB of data. In fact the situation is more serious, which is seen from the following example. There is now available fully resolved data from DNS of NSE on a turbulent flow in a plane channel at rather low Reynolds number (Re = 3300, based on the half-width of the channel and the mean velocity) [139]. While extremely useful in a great variety of aspects this did not lead to any qualitative change in the *understanding* of this flow. Likewise due to DNS at rather moderate Re much more is known about a variety of turbulent flows than before. Nevertheless, little proggress was made in *understanding* of such flows. It seems that for the progress in understanding of basic physics of turbuent flows - at least at present stage - one needs neither very high Reynolds numbers nor precise determination of scaling exponents, etc. at such Reynolds numbers. Turbulent flows at the asymptotic regime at Re $\gg$ 1 ($\to \infty$) - if such exists - do not seem to be simpler than those at very moderate Re, and are definitely much less accessible in every respect.

Quoting Feinmann: *The next great era of awakening of human intellect may*

---

[29]Inadequate (too) low dimensional approximations may lead to *spurious* chaotic behaviour [52], which disappears when the number of the basic functions becomes large enough.

well produce a method of understanding the qualitative content of equations.
To day we cannot. Today we cannot see that the water flow equations contain
such things as the barber pole structure of turbulence... We cannot say that
something beyond it like God is needed or not. And so we can hold strong
opinions either way [89]. Qualitative is the key word, since there is a qualitative difference between being able to measure and/or compute/calculate all one
wants and understanding. Perhaps the efforts of turbulence community should
be somewhat shifted to the qualitative aspects of the problem.

It is remarkable that in spite of - or perhaps just because of - frustrated and
unsuccessful attempts to construct a predictive theory of FDT and generally of
turbulent flows based on the first principles the attraction of the turbulence problem is only growing. This is reflected in disproportionally large to the funding
continuing efforts in most of the areas of the field. And so one can be optimistic
that in the end the glorious enigma of turbulence as a physical phenomenon will
be resolved.

References

[1] S. V. Alekseenko and S. I. Shtork, Russ J. Engn. Thermophys., **2**, 231, (1992).

[2] P.W. Anderson, Science, **177**, 9, (1972); Proc. Natl. Acad. Sci. USA, **192,** 6653, (1995).

[3] F. Anselmet, Y. Gagne, E. G. Hopfinger and R. A. Antonia, J. Fluid Mech., **140,** 63, (1984).

[4] R. A. Antonia, E. G. Hopfinger, Y. Gagne and F. Anselmet, Phys. Rev., **A30,** 2704, (1984).

[5] R. A. Antonia and J. Kim, Phys. Fluids, **6**, 834, (1994).

[6] R. A. Antonia, B. R Satyaprakash and A. J. Chambers, Phys. Fluids, **25**, 29, (1982).

[7] R. A. Antonia, M. Ould-Rouis, Y. Zhu and F. Anselmet, J. Fluid Mech., **332,** 395, (1997).

[8] R. A. Antonia, Boundary Layer Met., **34**, 411, (1986).

[9] H. Aref , in *Whither turbulence? Turbulence at the crossroads*, ed. J. L. Lumley, (Springer, 1990), pp. 258-268.

[10] H. Aref , Phil. Trans. Roy. Soc. Lond., **A333,** 273, (1990).

[11] H. Aref, Phys. Fluids, **A3**, 1009, (1991).

[12] A. Arneodo et al. , Europhys. Lett., **34,** 411, (1996).

[13] V. I. Arnold, Proc. Roy. Soc. London., **A3**, 19, (1991).

[14] Wm. T. Ashurst, A. R. Kerstein, R. A. Kerr and C. H. Gibson, Phys. Fluids, **30**, 2343; **30**, 3293, (1987).

[15] C. W. van Atta and R. A. Antonia, Phys. Fluids, **23**, 252, (1980).

[16] G. I. Barenblatt and A. J. Chorin , Comm. Pure and Appl. Math., **50 (L)**, 381, (1997).

[17] G. K. Batchelor, Proc. Roy. Soc. Lond., **A 213**, 349, (1952).

[18] G. K. Batchelor, *The theory of homogeneous turbulence*, (Cambridge University Press, 1953).

[19] J. T. Beale, T. Kato and A. Majda, Commun. Math. Phys., **94**, 61, (1984); P. Constantin, Commun. Math. Phys., **104**, 311, (1986)

[20] C. Beck, Phys. Rev., **E49**, 3641, (1994).

[21] F. Belin, J. Maurer, P. Tabeling and H. Willame, J. Phys. II France, **6**, 573, (1996).

[22] F. Belin, P. Tabeling and H. Willame, Physica, **D93**, 52, (1996).

[23] F. Belin, J. Maurer, P. Tabeling and H. Willame, Phys. Fluids, **9**, 3843, (1997).

[24] J. B. Bell and D. L. Marcus, Commun. Math. Phys., **147**, 371, (1992).

[25] R. Benzi, R., L. Biferale, R. M. Kerr and E. Travatore, Phys. Rev. , **E53**, 3541, (1996).

[26] R. Benzi, S. Ciliberto, C. Baudet and G. Ruiz-Chavaria 1995, Physica , **D80**, 385, (1995).

[27] V. Berdichevsky, A. Fridlyand and V. Sutyrin, Phys. Rev. Lett., **76**, 3967, (1996).

[28] R. Betchov, Arch. Mech. (Archiwum Mechaniki Stosowanej), **28**, 837, (1976).

[29] M. Blackburn, N. N. Mansour and B. J. Cantwell, J. Fluid Mech., **310**, 269, (1996).

[30] O. N. Boratav, Phys. Fluids, **9**, 1206, (1997).

[31] O. N. Boratav, Phys. Fluids, **9**, 3120, (1997).

[32] O. N. Boratav and R. B. Pelz, Phys. Fluids, **9**, 1400, (1997).

[33] M. S. Borgas, Phys. Fluids, **A4**, 2055, (1992).

[34] M. S. Borgas, and B. L. Sawford, J. Fluid Mech., **279**, 69, (1994).

[35] M. E. Brachet, Fluid Dyn Res., **8**, 1, (1991).

[36] M. E. Brachet, M. Meneguzzi, A. Vincent, H. Politano and P. L. Sulem, Phys. Fluids, **A4**, 2845, (1992).

[37] P. Bradshaw, Experiments in fluids, **16**, 203, (1994).

[38] M. Briscolini and P. Santangelo, J. Fluid Mech, **270**, 199 (1994); Advances in turbulence, **5**, 41, (1995).

[39] F. H. Busse, in *Nonlinear Physics of complex systems*, eds. J. Parisi and S. C. Müller, (Springer, 1996), pp. 1-9,

[40] O. Cadot, D. Bonn and S. Douady , Phys. Fluids, **10**, 426, (1998).

[41] O. Cadot, Y. Couder, A. Daerr, S. Douady and A. Tsinober, Phys. Rev., **E56**, 427, (1997).

[42] R. Camussi and R. Benzi, Phys. Fluids, **9**, 257, (1997).

[43] D. G. Dritschel, J. Fluid Mech, **216**, 657, (1990).

[44] R. Camussi, C. Baudet, R. Benzi and S. Ciliberto, Phys. Fluids, **8**, 1686 (1996).

[45] R. Camussi, D. Barbagallo, G. Guj and F. Stella, Phys. Fluids, **8**, 1181, (1996).

[46] R. Camussi, and G. Guj, J. Fluid Mech, **348**, 177, (1997).

[47] B. J. Cantwell, Phys. Fluids, **A4**, 782 (1992); **A5**, 2008, (1993).

[48] N. Cao and Chen S., Phys. Rev., **E52**, R5757, (1996).

[49] N. Cao, N., Chen S. and She, Z.-S., Phys. Rev. Lett., **76**, 3711, (1996).

[50] N. Cao, S. Chen and K. R. Sreenivasan, Phys. Rev. Lett., **77**, 3799, (1996).

[51] Cardoso O., Glukmann B., Parcolet O. and Tabeling P., Physics of Fluids, **8**, 209, (1996).

[52] J. H. Carry, J. R. Herring, J. Loncaric and S. A. Orszag, J. Fluid Mech., **147**, 1, (1984).

[53] B. Chabaud, A. Naert, J. Peinke, F. Chilla, B. Castaing and B.. Herbal, Phys. Rev. Lett., **73**, 3227, (1994).

[54] H. Chen and S. Chen 1998, Phys. Fluids., **10**, 312, (1998).

[55] H. Chen, J. R. Herring R. M. Kerr and R. Kraichnan, 1989, Phys. Fluids, **A1**, 1844, (1989).

[56] S. Chen and N. Cao, Phys. Rev. Lett., **78**, 3459, (1997).

[57] S. Chen, K. R. Sreenivasan, and M. Nelkin, Phys. Rev. Lett., **79**, 1253, (1997).

[58] S. Chen, K. R. Sreenivasan, M. Nelkin and N. Cao, Phys. Rev. Lett., **79**, 2253, (1997).

[59] M. Chertkov, G. Falkovich and V. Lebedev, Phys. Rev. Lett. , **76**, 3707, (1996).

[60] A. J. Chorin, Comm. Math. Phys., **39**, (special issue), 517, (1982).

[61] A. J. Chorin, *Vorticity and Turbulence.* (Springer, 1994).

[62] A. J. Chorin, Phys. Rev., **54**, 2616, (1996).

[63] W. J. Cocke, Phys. Fluids, **12**, 2488, (1969).

[64] P. Constantin, SIAM Rev., **36**, 73, (1994).

[65] P. Constantin, Lect. Appl. Math., **31**, 219, (1996).

[66] P. Constantin, C. Fefferman and A. J. Majda, Comm. Part. Diff. Eq., **21** (3&4), 559, (1996).

[67] P. Constantin, W. E and E. S. Titi, Comm. Math. Phys., **165**, 207, (1994).

[68] S. Corrsin, in *Statistical models and turbulence,* eds. M. Rosenblatt and C. van Atta, (Springer, 1972), pp.300-316.

[69] S. Corrsin, Adv. Geophys, **18A**, 25, (1974).

[70] B. Dhruva, Y. Tsuji and K. R. Sreenivasan, Phys. Rev., **E 56**, R4928, (1997).

[71] P. E. Dimotakis, Progr. Astron. and Aeron., **137**, 265, (1991).

[72] S. Douady, Y. Couder and M.E. Brachet, Phys. Rev. Lett., **67**, 983, (1991).

[73] E. Dresselhaus and M. Tabor, J. Fluid Mech., **236**, 415, (1991).

[74] B. Dubrulle, J. Phys. II France, **6**, 1825, (1996).

[75] P. A. Durbin and C. G. Speciale, J. Fluids Engn., **113**, 707, (1991).

[76] J. G. M. Eggels, F. Unger, M. H. Weiss, J. Westerweel, R. J. Adrian, R. Friedrich and F. T. M. Nieuwstadt, J. Fluid Mech., **268**, 175, (1994).

[77] J. Eggers and S. Grossmann, Phys. Rev., **A54**, 2360, (1992).

[78] V. Emsellem, L. Kadanoff, D. Lohse, P. Tabeling and Z. J. Wang, 1997, Phys. Rev., **E55**, 2672, (1997).

[79] G. Eyink, Physica., **D78**, 222, (1994).

[80] G. Eyink, Phys. Fluids, **6**, 3063, (1994).

[81] G. Eyink and F. J. Alexander, Phys. Rev. Lett., **78**, 2563, (1997).

[82] G. Eyink, Phys. Rev., **E56**, 5413, (1997).

[83] G. Eyink, "Action principle in statsistical hydrodynamics", in *Proc. XIIth Congr. Math. Phys., Brisbane, Australia, 13-19 July, 1997, in press.*

[84] G. Falkovich, Phys. Fluids, **6**, 1411, (1994).

[85] G. Falkovich, I. Kolokolov, V. Lebedev and A. Migdal, Phys. Rev., **E54**, 4896, (1996).

[86] G. Falkovich and V. Lebedev, Phys. Rev. Lett., **79**, 4159, (1997).

[87] V. M. Fernandez, N. J. Zabusky and V. M. Gryanik, Fluid Mech., **299**, 289, (1995).

[88] M. Feigenbaum, in *Nonlinear Dynamics, Chaotic and Complex Systems*, eds. E. Infeld, R. Zelazny and A. Galkovski, (Springer, 1997), pp. 321-326.

[89] R. Feinmann, *Lectures on Physics*, **2**, (Addison-Wesley, 1964), p. 41-12.

[90] U. Frisch, *Turbulence: The legacy of A. N. Kolmogorov*, (Cambridge University Press, 1995).

[91] U. Frisch and S. A. Orszag, Phys. Today, **9**, 24, (1990).

[92] A. V. Fursikov and O. Yu Emanuilov, Russian Acad. Sci. Sb. Math, **81**, 235, (1995).

[93] C. Fureby, G. Tabor, H. G. Weller and A. D. Gosman, Phys. Fluids, **9**, 1416, (1997).

[94] M. Gad-el-Hak and P. R. Bandyopadhyay, Appl. Mech. Rev., **47**, 307, (1994).

[95] B. Galanti, I. Procaccia and D. Segel, Phys. Rev., **E 54**, 5122, (1996).

[96] S. Garg and Z. Warhaft, Phys. Fluids., **10**, 662, (1998).

[97] W. K. George, in *Forum on turbulent flows*, eds. Bower, W.M., Morris, M.J. and Samimy, M., *FED-Vol.* **94**, (1990), pp.1-11.

[98] S. S. Girimaji and S. B. Pope, J. Fluid Mech., **220**, 427, (1990).

[99] E. Gledzer, E. Villermaux, H. Kahaleras and Y. Gagne, Phys. Fluids, **8**, 3367, (1996).

[100] R. Grauer, T. Sideris, Physica, **D88,** 116, (1995).

[101] J. M. Greene, O. Boratav, Physica, **D107,** 57, (1997).

[102] M. S. Grossmann, D. Lohse and A. Reeh, Phys. Fluids, **9**, 3817, (1997).

[103] A. M.Guzmán and C. H. Amon, J. Fluid Mech., **321,** 25, (1996).

[104] A. Gyr, ed., *Structure of turbulence and drag reduction*, (Springer, 1990); A. Gyr and H.-W. Bewersdorf, *Drag reduction of turbulent flows by additives*, (Kluwer, 1995).

[105] J. Herwejer and W. van de Water, Advances in Turbulence, **5,** 210,(1995); W. van de Water and J. A. Herwejer, Physica Scripta, **T67**, 136, (1996); "High order structure functions of turbulence", J. Fluid Mech., sub judice.

[106] Hill R.J, *J. Fluid Mech.,* **353,** 67, (1998).

[107] P. J. Holmes, G. Berkooz and J. L. Lumley, *Turbulence, coherent structures, dynamical systems and symmetry*, (Cambridge University Press, 1996).

[108] P. J. Holmes, G. Berkooz, J. L. Lumley, J. Mattingly and R. W. Wittenberg, Physics Reports, **287,** 337, (1997).

[109] M. Holzer and E. Siggia, Phys. Fluids, **A5,** 2525, (1993).

[110] M. Holzer and E. Siggia, Phys. Fluids, **6**, 1820, (1994).

[111] E. Hopf, Comm. Pure Appl. Math., **1,** 303, (1948).

[112] I. I. Hosokawa and K. Yamamoto, J. Phys. Soc. Japan, **59**, 401, (1989).

[113] I. I. Hosokawa and S.-I. Ode, Phys. Rev. Lett., **77**, 4548, (1996).

[114] M.-J. Huang, Phys. Fluids, **8**, 2203, (1996).

[115] J. C. R. Hunt and D. J. Carruthers, J. Fluid Mech., **212**, 497, (1990).

[116] J. C. R. Hunt, J. C. Vassilicos and N. K. R. Kevlahan, Progr. Astron. Aeronaut., **162**, 1, (1994).

[117] J. C. R. Hunt, A. A. Wray and P. Moin , in *Studying Turbulence Using Numerical Simulation Databases - II, CTR-S88*, 193, (1988).

[118] I. E. Idelchik, *Handbook of hydraulic resistance,* (Springer, 1996).

[119] J. D. Jacob and Ö. Savas, ZAMP, **S46**, S699, (1995).

[120] J. Jimenez, 1990, Fluid Mech., **218**, 265, (1990).

[121] J. Jimenez, in *New Approaches and Concepts in Turbulence,* ed. T. Dracos and A. Tsinober, (Birkäuser, Basel, 1993), pp. 95-151.

[122] J. Jimenez, *CTR-1995 Ann. Res. Briefs*, 25, (1995).

[123] J. Jimenez, A. A. Wray, P. G. Saffman and R. S. Rogallo, J. Fluid Mech., **255**, 65, (1993).

[124] J. Jimenez and A. A. Wray, Meccanica, **4**, 453, (1994).

[125] J. Jimenez, A. A. Wray, *CTR-1994 Ann. Briefs*, 287, (1994).

[126] L. Kadanoff, D. Lohse and J. Wang, Phys. Fluids, **7**, 617, (1995).

[127] H. Kahaleras, Y. Malecot and Y. Gagne, Advances in Turbulence, **6**, 235, (1996).

[128] Th. von Karman, J. Aeronaut. Sci., **4**, 131, (1937).

[129] G. G. Katul, M. B. Parlange and C. R. Chu, Phys. Fluids, **6**, 2480, (1994).

[130] G. G. Katul, M. B. Parlange, J. D. Albertson and C. R. Chu, Boundary-Layer Meteorology, **72**, 123, (1995).

[131] L. Keefe, in *Near wall turbulence - 1988 Zaric memorial conference,* eds. S. J. Kline and H. N. Afagn, (Hemisphere, 1990), pp. 63-80.

[132] L. Keefe, P. Moin and J. Kim, J. Fluid Mech. , **242**, 1, (1992).

[133] L. V. Keller and A. A. Fridman, in *Proc. First Int. Congr. Appl. Mech.*, ed. C. B. Biezeno and J. M. Burgers, (Waltman, Delft, 1925), pp. 395-405.

[134] R. M. Kerr, J. Fluid Mech., **153**, 31, (1985).

[135] R. M. Kerr, Phys Fluids, **A 5**, 1725, (1993); Nonlinearity, **9**, 271, (1996).

[136] R. R.. Kerswell, Physica, **D100**, 355, (1997).

[137] S. Kida and K. Ohkitani, Phys. Fluids, **A4**, 1018, (1994).

[138] S. Kida and M. Takaoka, Annu. Rev. Fluid Mech., **26**, 169, (1994).

[139] J. Kim, P. Moin and R. J. Moser, J. Fluid Mech., **177**, 133, (1997).

[140] S. Kishiba, K. Ohkitani and S. Kida, J. Phys. Soc. Japan, **63**, 2133, (1994).

[141] A. N. Kolmogorov, Dokl. Akad. Nauk SSSR, **30**, 19 (1941); English translation: *Selected works of A. N. Kolmogorov,* **I,** p. 321, ed. V. M. Tikhomirov, (Kluwer, 1991).

[142] A. N. Kolmogorov , Dokl. Akad. Nauk SSSR, **32**, 19 (1941); English translation: *Selected works of A. N. Kolmogorov,* **I,** p. 324, ed. V. M. Tikhomirov, (Kluwer, 1991).

[143] A. N. Kolmogorov, J. Fluid Mech., **13**, 82, (1962).

[144] A. N. Kolmogorov, "Remarks on statistical solutions of Navier-Stokes equations", Uspekhi Matematicheskikh Nauk, **33**, 124, (1978), in Russian. This section was omitted in the English translation.

[145] R. H. Kraichnan, Phys. Fluids, **10**, 2080, (1967).

[146] R. H. Kraichnan, Phys. Fluids, **11**, 945, (1968).

[147] R. H. Kraichnan, J. Fluid Mech., **62**, 305, (1974).

[148] R. H. Kraichnan, J. Stat. Phys., **51**, 949, (1988).

[149] R. H. Kraichnan as cited by L. P. Kadanoff in *New Perspectives in Turbulence*, ed. L.Sirovich, (Springer, 1990), p. 265.

[150] R. H. Kraichnan , private communication (1994). Along with rare very strong events the even moments can be sinignificantly affected by regions of unusually small excitation.

[151] R. H. Kraichnan, Phys. Rev. Lett., **72**, 1016, (1994).

[152] R. H. Kraichnan, Phys. Rev. Lett., **75**, 240, (1995).

[153] R. H. Kraichnan, Phys. Rev. Lett., **78**, 48, (1997).

[154] R. H. Kraichnan and Y. Kimura, Progr. Astron. Aeronaut., **162**, 19, (1994);

[155] R. H. Kraichnan and R. Panda, Phys. Fluids, **31**, 2395, (1988).

[156] V. R. Kuznetsov, A. A. Praskovsky and V. A. Sabelnikov, J. Fluid Mech., **243**, 595, (1992).

[157] R. Labbe, J.- F. Pinton and S. Fauve, *Phys. Fluids*, **8**, 914; J. Phys. II France, **6**, 1099, (1996).

[158] S. E. Larsen, "Hot-wire measurements of Atmospheric turbulence near the ground ", *Risø-R-233*, Risø National Laboratory, Roskilde, Denmark, (1986), 342 pp.

[159] S. Le Dizes, M. Rossi and H. K. Moffatt, Phys. Fluids, **8**, 2084, (1996).

[160] M. Lesieur and O. Metais, Ann. Rev. Fluid Mech., **28**, 45, (1996).

[161] E. Leveque and Z.-S. She, Phys. Rev. Lett., **75**, 2690, (1995).

[162] H.W. Liepmann, Amer. Scientist, **67**, 221, (1979).

[163] E. Lindborg, J. Fluid Mech., **326**, 343, (1996).

[164] Lomholt S., "Boundary Layer Dynamics in Two-dimensional Flows ", *Risø-I-1063(EN)*, Risø National Laboratory, Roskilde, Denmark, (1996), 112 pp.

[165] J. L. Lumley, in *Advances in Turbulence*, eds. W. K. George and R. Arndt, (Hemisphere & Springer, 1989), pp. 1-10.

[166] V. L'vov and I. Procaccia, Phys. Fluids **8**, 2565, (1996).

[167] V. L'vov, E. Podivilov and I. Procaccia, Phys. Rev Lett. **79**, 2050, (1996).

[168] N. N. Mansour and A. A. Wray, Phys. Fluids, **6**, 808, (1994).

[169] H. K. Moffatt, in *Whither turbulence? Turbulence at the crossroads*, ed. J. L. Lumley, (Springer, 1990), pp. 250-257; in *New Approaches and Concepts in Turbulence*, ed. T. Dracos and A. Tsinober, (Birkäuser, Basel, 1993), pp. 402-404.

[170] H. K. Moffatt and A. Tsinober, *Ann. Rev. Fluid Mech.*, **24**, 281,(1992).

[171] P. Moin and J. Kim, Scientific American, **276**, 1, (1997).

[172] G. M. Molchan, Phys. Fluids, **9**, 2387, (1997).

[173] N. Mordant, J.-F. Pinton and F. Chilla, J. Phys. II France, **7**, 1729, (1997).

[174] A. S. Monin and A. M. Yaglom, *Statistical fluid mechanics, vol.* **1**, (MIT Press, 1971), 2nd *Russian edition*, (Gidrometeoizdat, St. Petersburg, 1992); *vol.* **2**, (MIT Press, 1975), 2nd *Russian edition*, (Gidrometeoizdat, St. Petersburg, 1996).

[175] L. Mydlarski and Z. Warhaft, J. Fluid Mech., **320**, 331, (1996).

[176] R. Narasimha, in *Whither turbulence? Turbulence at the crossroads*, ed. J. L. Lumley, (Springer, 1990), pp. 13-48.

[177] M. Nelkin 1989, J. Stat. Phys., **54**, 1, (1989).

[178] M. Nelkin 1995, Phys. Rev. ,**E52**, R4610, (1995).

[179] J. von Neumann, in *Collected works*, **6** (1963), pp. 437 - 472, ed. A. H. Taub, Pergamon.

[180] R. Nicodemus, S. Grossmann and M. Holthaus, Phys. Rev. Lett., **79**, 4170, (1997).

[181] K. K. Nomura and G. K. Post, "The structure and dynamics of vorticity and rate-of-strain in incompressible homogeneous turbulence ", *J. Fluid Mech.*, sub judice; K. K. Nomura, G. K. Post and P. Diamesis, AIAA 97-1956, (1997), pp. 1-29.

[182] E. A. Novikov, J. Appl. Math. Mech., **27**, 1445, (1963).

[183] E. A. Novikov, Soviet Physics-Doklady, **12**, 1006, (1968).

[184] E. A. Novikov, Fluid Dyn. Res., **16**, 297, (1990).

[185] A. Noullez, G. Wallace, W. Lempert, R. B. Miles and U. Frisch, J. Fluid Mech., **339**, 287, (1997).

[186] J. O'Neil and C. Meneveau, J. Fluid Mech., **349**, 253, (1997).

[187] A. M. Obukhov, Soviet Phys.-Doklady, **14**, 32, (1969).

[188] K. Ohkitani, "Time evolution of the enstrophy of a class of three-dimensional Euler flows: the Taylor-Green vortex", J. Fluid Mech., sub judice.

[189] K. Ohkitani and S. Kishiba, Phys. Fluids , **7**, 411, (1995).

[190] L. Onsager, Nuovo Cimento, **VI**, *Suppl., No.2*, 279-287, (1949).

[191] S. A. Orszag, in *Fluid Dynamics*, eds. R. Balian and J.-L. Peube, (Gordon and Breach, 1977), pp. 235-274.

[192] A. M. Polyakov, Phys. Rev., **E52**, 6183, (1995).

[193] T. Passot, H. Politano, P.- L. Sulem, J. L. Angiella, M. Meneguzzi, J. Fluid Mech. , **282**, 313, (1992).

[194] G. Pedrizzetti, A. A. Novikov and A. A. Praskovsky, Phys. Rev., **E53**, 475, (1996).

[195] D. H. Porter, A. Pouquet and P. R. Woodward, Lect. Notes Phys., **462**, 51, (1995).

[196] A. A. Praskovsky, E. B. Gledzer, M. Yu. Karyakin and Y. Zhou, J. Fluid Mech. , **248**, 493, (1993).

[197] A. A. Praskovsky and S. Onsley, Phys. Fluids, **A6**, 2886, (1994).

[198] A. A. Praskovsky, E. Praskovskaya and T. Horst, Phys. Fluids, **9**, 2465, (1997).

[199] I. Procaccia, and P. Constantin, Phys. Rev. Lett., **70**, 3416, (1993).

[200] A. Pumir and B. I. Shraiman, Phys. Rev. Lett, **75**, 3114, (1995).

[201] A. Pumir, B. I. Shraiman and E. D. Siggia, Phys. Rev., **E55**, R1263, (1997).

[202] A. Pumir and E. D. Siggia, Phys. Fluids, **A2**, 220, (1990).

[203] C. Reyl, T. M. Antonsen and E. Ott, Phys. Rev. Lett, **78**, 2559, ; Physica, **D**, 202, (1998).

[204] O. Reynolds, Nature., **50**, 161, (1894).

[205] I. Rogachevskii and N. Kleorin, Phys. Rev., **E56**, 417, (1997).

[206] G. Ruiz-Chavaria, C. Baudet and S. Ciliberto, Phys. Rev. Lett., **74**, 1986, (1995).

[207] G. R. Ruetsch and M. R. Maxey, Phys. Fluids, **A3**, 1587, (1991).

[208] G. R. Ruetsch and M. R. Maxey, Phys. Fluids, **A4**, 2747, (1992).

[209] G. Ruiz-Chavaria, C. Baudet and S. Ciliberto, J. Phys. II, France, **5**, 485, (1995).

[210] G. Ruiz-Chavaria, C. Baudet and S. Ciliberto, Physica , **D99**, 369 (1996).

[211] S. G. Saddoughi, J. Fluid Mech., **348**, 201, (1997).

[212] S. G. Saddoughi and S. V. Veeravalli, J. Fluid Mech., **268**, 333, (1994).

[213] P. G. Saffman, Lect. Notes Phys., **76**, 274, *(1978)*.

[214] K. W. Schwarz, Phys. Rev. Lett. **64**, 415, (1990).

[215] Z.-S. She, Lect. Notes Phys. **491**, 28, (1997).

[216] Z.-S. She, E. Jackson and S. A. Orszag, Nature, **344**, 226, (1990).

[217] Z.-S. She, E. Jackson and S. A. Orszag, Proc. Roy. Soc. Lond., **A 434**, 101, (1991).

[218] Z.-S. She and E. Leveque, Phys. Rev. Lett., **72**, 336, (1994).

[219] Z.-S. She, S. Chen, G. Doolen, R. Kraichnan and S. A. Orszag, Phys. Rev. Lett, **70**, 3251, (1993).

[220] B. Shraiman and E. Siggia, Phys. Rev. Lett., **77**, 2463, (1996).

[221] B. Shraiman and E. Siggia, "Fluctuations and mixing of a passive scalar in turbulent flow ", *Proceedings of the Cargese 1996 Summer School* Mixing, Chaos and Turbulence, *in press.*

[222] L. Shtilman, Phys. Fluids, **A4**, 197, (1992); L. Shtilman and W. Polifke, Phys.Fluids, **A1**, 778, (1989).

[223] L. Shtilman, M. Spector and A. Tsinober, J. Fluid Mech., **247**, 65, (1993).

[224] E. D. Siggia, J. Fluid Mech., **107**, 375, (1981).

[225] E. D. Siggia, Phys. Fluids, **24**, 1934, (1981).

[226] G. M. Smith and T. Wei, J. Fluid Mech., **259**, 281, (1994).

[227] K. R. Sreenivasan, Phys. Fluids, **A8**, 189, (1996).

[228] K. R. Sreenivasan and R. Antonia, Ann. Rev. Fluid Mech., **29**, 435, (1997).

[229] K. R. Sreenivasan ang G. Stolovitzky, Phys. Rev. Lett., **77**, 2718, (1996).

[230] R. W. Stewart, Radio Sci., **4**, 1269, (1969).

[231] P. Tabeling, G. Zocci F. Belin, J. Maurer and H. Willame, Phys. Rev., **E53**, 1613, (1996).

[232] M. Tanaka and S. Kida, Phys. Fluids **A 5**, 2079–2082, (1993).

[233] G. I. Taylor, J. Aeronaut. Sci., **4**, 311, (1938).

[234] G. I. Taylor, Proc. Roy. Soc., **A164**, 15, (1938).

[235] H. Tennekes and J. L. Lumley, *A first course of turbulence*, (MIT Press, 1972).

[236] V. M. Tikhomirov, ed., 1991 *Selected works of A. N. Kolmogorov*, **I**, p. 487, Kluwer, 1991.

[237] A. Townsend, Proc. Roy. Soc., **A208**, 534, (1951).

[238] D. Tritton, *Physical fluid dynamics*, (Clarendon Press, Oxford, 1988).

[239] A.Tsinober, Phys. Fluids, **A2**, 484, (1990).

[240] A. Tsinober A., Progr. Aeron. Astron., **123**, 327, (1990).

[241] A. Tsinober, in *New Approaches and Concepts in Turbulence*, ed. T. Dracos and A. Tsinober, (Birkäuser, Basel, 1993), pp. 141 -151.

[242] A. Tsinober, Lect. Notes Phys, **450**, 3, (1995).

[243] A. Tsinober, Actes du 12e Congrès Francais de Mécanique , **3**, 409, (1995).

[244] A. Tsinober, J. Fluid Mech., **317**, 407, (1996).

[245] A. Tsinober, Advances in Turbulence, **6**, , 263 (1996).

[246] A. Tsinober, "Is concentrated vorticity that important?", Eur. J. Mech., **B**/Fluids, (1998), in press.

[247] A. Tsinober and L. Shtilman, "Rate of change of enstrophy generation. Interaction of vorticity and pressure hessian ", *3rd Eur. Conf. Fluid Mech*, Götinngen, Sept 15-18, 1997.

[248] A.Tsinober, J. G. M.Eggels and F. T. M. Nieuwstadt, Fluid Dyn. Res., **16**, 297, (1995).

[249] A. Tsinober, E. Kit and T. Dracos, J. Fluid Mech., **242**, 169, (1992).

[250] A. Tsinober, L. Shtilman, A. Sinyavskii and H. Vaisburd, Lect. Notes Phys., **462**, 9, (1995)

[251] A. Tsinober, M. Ortenberg and L. Shtilman, "Geometrical properties of vorticity field in numerical turbulence ", Advances in Turbulence, **7**, in press (1998).

[252] A. Tsinober, L. Shtilman and H. Vaisburd, Fluid Dyn. Res., **21**, 477, (1997).

[253] A. Tsinober and H. Vaisburd, Phys. Lett., **A197**, 293, (1995); Advances in turbulence, **5**, 586, (1995)

[254] S. I. Vainstein, Phys. Rev., **E56**, 447, (1997).

[255] J. C. Vassilicos , in *Eddy structure identification*, ed. J. P. Bonnet, (Springer, 1996), pp. 197-270.

[256] M. Vergassola, Phys. Rev., **E53**, R3021, (1996).

[257] A. Verzicco, J. Jimenez and P. Orlandi, J. Fluid Mech., **299**, 367, (1995).

[258] E. Villermaux, B. Sixou and Y. Gagne, Phys. Fluids, **7**, 2008 (1995).

[259] A. Vincent and M. Meneguzzi, J. Fluid Mech. , **225**, 1, (1991).

[260] A. Vincent and M. Meneguzzi, J. Fluid Mech. , **258**, 245, (1994).

[261] V. Yakhot 1998 , Phys. Rev., **E57**, 1737.

[262] V. Yakhot and S. A. Orszag, Nuclear Physics, **B** (Proc. Suppl) **2**, 417, (1987).

[263] A. N. Yannacopoulos, I. Mezic, G. Rowlands and G. P. King, Phys. Rev., **E57**, 482, (1998).

[264] V. E. Yanitskii, Sov. Phys.-Doklady, **27**, 701, (1982).

[265] P. K. Yeung, Phys. Fluids, **9**, 2981, (1997).

[266] P. K. Yeung, J. Brasseur and Q. Wang, J. Fluid Mech, **283**, 743, (1995).

[267] L.-P. Wang, S. Chen, J. Brasseur and J. Wyngaard, J. Fluid Mech, **309**, 113, (1996).

[268] E. R. Weeks, J. S. Urbach and H. L. Swinney, Physica, **D 97,** 291, (1996).

[269] T. Wei and W. W. Willmarth, J. Fluid Mech., **204**, 57, (1989).

[270] I. J. Wygnanski and R. A. Petersen, AIAA J., **25**, 201, (1987).

[271] G. M. Zaslavsky, M. Edelman and B. A. Niyazov, Chaos, **7,** 159, (1997).

[272] Ya. B. Zel'dovich, S. A. Molchanov, A. A. Ruzmaikin and D. D. Sokolov, Sov. Phys.-Uspekhi., **30,** 353, (1987).

[273] Ya. B. Zel'dovich, S. A. Molchanov, A. A. Ruzmaikin and D. D. Sokolov, Sov. Sci. Rev., **C7,** 1, (1988).

[274] Ya. B. Zel'dovich, S. A. Molchanov, A. A. Ruzmaikin and D. D. Sokolov, *The Almighty Chance*, (World Scientific, 1990).

[275] J.Y. Zhou, P. K. Yeung and J. G. Brasseur, Phys. Rev., **E 53,** 1261, (1996).

[276] Y. Zhu, R. A. Antonia and J. Kim, Phys. Fluids, **8,** 838, (1996).

[277] O. Zikanov, A. Thess and R. Grauer, Phys. Fluids, **9,** 1362, (1997).

Forced and Decaying 2D Turbulence: Experimental Study

P. Tabeling[1], A.E. Hansen[1,2] and J. Paret[1]

[1] LPS / ENS, 24 rue Lhomond, F-75231 Paris, France
[2] CATS, The Niels Bohr Institute, Blegdamsvej 17, DK-2100 Copenhague, Denmark

Abstract. We report experimental results obtained on freely decaying and forced two-dimensional turbulence. The flow is produced in a thin stratified layer of electrolyte, using an electromagnetic forcing. The velocity and vorticity fields are measured using a particle image velocimetry (PIV) technique. The study of the temporal evolution of the system confirms in detail the scaling theory of Carnevale et al. (1991). We further measure the merging time τ, the mean free path λ, and the mean square displacement σ_v^2 of the vortices. We find the following laws : $\tau \sim t^{0.57}$, $\lambda \sim t^{0.45}$, $\sigma_v^2 \sim t^{1.3}$. The statistics of passive particles (albeit virtual) in the system is also studied. They move hyperdiffusively, with an exponent identical to that obtained for the vortex motion. We find the dispersion of the particles is controlled by Lévy flights, produced by the jets formed by the dipoles. We finally underline the close relationship between the decay of turbulence decay and the dispersion phenomena. We further turn to the forced case. We find the energy spectrum displays a clear $k^{-5/3}$ law with a Kolmogorov constant lying in the range 5.5–7.5, which is consistent with the current numerical estimates. The dispersion of pairs of passive particles is found to be controlled by Richardson law, throughout the inertial range of scales revealed by the analysis of the flow field. No evidence for the existence of Levy flights has been found. At variance with the decaying case, it is not clear whether coherent structures may play any role in the control of the main characteristics of the inverse cascade, along with the pair dispersion in the corresponding inertial range of scales.

1 Introduction

Two-dimensional turbulence has been much studied in recent years, because of its applications in astrophysics and geophysics, its relative accessibility to numerical computation, and as a fascinating field in its own right. The formation of coherent structures or vortices has been established both numerically and experimentally as a characteristic feature of 2D turbulent flows. In freely decaying turbulence, the vortices tend to live long compared to their turn-over time. From the time at which the coherent structures have been formed, and until the final dipole state has been reached, the governing dynamical processes are the mutual advection of vortices, and the inelastic merging of like-sign vortices. Assuming a self-similar evolution of the vortex system, Carnevale et al. (1991) proposed the extremum vorticity of the core of the vortices, ω_{ext}, and the total kinetic energy as the two invariants controlling the decaying regime. Assuming further that the vorticity is concentrated in vortices, they obtained the following scaling laws for the density of vortices ρ, the vortex radius a, the mean separation between vortices r, the velocity u of a vortex, the total enstrophy Z, and the kurtosis Ku of the vorticity distribution :

$$\rho \sim \mathcal{L}^{-2}\left(\tfrac{t}{\mathcal{T}}\right)^{-\xi}, \quad a \sim \mathcal{L}\left(\frac{t}{\mathcal{T}}\right)^{\xi/4},$$

$$r \sim \mathcal{L}\left(\tfrac{t}{\mathcal{T}}\right)^{\xi/2}, \quad u \sim \sqrt{E},$$

$$Z \sim \mathcal{T}^{-2}\left(\tfrac{t}{\mathcal{T}}\right)^{-\xi/2}, \, Ku \sim \left(\frac{t}{\mathcal{T}}\right)^{\xi/2}, \tag{1}$$

in which length $\mathcal{L}$ and time scale $\mathcal{T}$ are defined by :

$$\mathcal{L} = \omega_{ext}^{-1}\sqrt{E}, \qquad \mathcal{T} = \omega_{ext}^{-1}. \tag{2}$$

The exponent ξ is not determined by the theory. Numerical studies, both of the full Navier-Stokes equations and of point-vortex models, have consistently given values $\xi = 0.71 - 0.75$ (Carnevale et al. (1991), Weiss and McWilliams (1992)). On the experimental side, the early investigations of the decay regime (Couder (1983), Hopfinger et al. (1983)) confirmed that as time increases, the vortex population becomes depleted and the mean vortex size raises up. However, accurate quantitative analysis of the phenomenon was not successfully achieved. Recently, one author of the present paper made detailed measurements of the decay regime of quasi-two dimensional flows (Cardoso et al. (1994)). However, in the system they explored, three dimensional perturbations were suggested to play an important role. All this means that at the moment, our knowledge of the phenomenon essentially relies on numerical studies.

Theoretically, several attempts have been made to determine ξ (Pomeau (1996), He (1996b), Trizac and Hansen (1996), Huber and Alstrøm (1993)). Based on the scaling laws (1), one of the authors of the paper by Carnevale et al. (1991) has proposed a derivation that yields $\xi = 1$, but argued for lowering corrections (Pomeau (1996)). On the other hand, on the background of a theoretical approach using a probabilistic method to describe the motion of vortices in an external strain-rotation field, it has been suggested that the value of ξ depends on initial conditions (He (1996b)). In a related context, the 2D ballistic agglomeration of hard spheres with a size-mass relation mimicking the energy conservation rule for vortices, the value $\xi = 0.8$ is derived under mean-field assumptions (Trizac and Hansen (1996)). Further, in another possibly related context, that of Ginzburg-Landau vortex turbulence, the value $\xi = 3/4$ has been proposed (Huber and Alstrøm (1993)).

The first part of the paper is thus devoted to the study of freely decaying turbulence, along with the dispersion of passive particles. We will confirm the theory of Carnevale et al, and find an appealing correspondence between the decay and the dispersion problems. Throughout the study, the coherent structures, i.e. the long living vortices with circulation well above the background, will appear to play a central role.

In a second part, we study the forced case. We will show evidences for the existence and stationarity of the inverse cascade ; moreover we report the observation that, within the inertial range of scales, the pair dispersion is governed by Richardson law, i.e. the distance between such pairs increases, in the average, as the time raised to power three. These results will suggest a strong difference between the free decay and the forced case, concerning the role of the structures.

2 Experimental set-up

The experimental set-up we use has been described in a number of papers. The flow is generated in a square PVC cell, 15 cm × 15 cm. The bottom of the cell is made of a thin (1 mm thick) plate of glass and permanent magnets are placed just below. They produce a vertical magnetic field which has a maximum value of about 0.3 T and decay over a typical length of 3 mm. Using different magnet arrangements, we are able to study different spatial structures of the forcing. The cell is filled with two layers of NaCl solution of different densities in a stable configuration, that is the heavier underlying the lighter. Each layer is 3 mm thick so that the aspect ratio of the fluid domain is small. We drive an electric current from one side of the cell to the other, and the interaction of this current with the magnetic field produces forces which drive the flow. The flow is visualized by using clusters of tiny particles placed at the free surface. We record the images of the flow and then process them using particle image velocimetry techniques in order to get the velocity fields. The grids we use for computing the velocity fields are typically 40 × 40 in the decaying case (Sect. 3) and 64 × 64 in the forced case (Sect. 4). The advantages of this method are that it is completely non-intrusive and that it allows us to measure the complete velocity field at any time.

There are several qualitative arguments in favour of the two-dimensionality of the flows generated in our set-up : the aspect ratio, the Froude number (which measure the "strength" of stratification) and the ratio of measured horizontal divergence to maximum vorticity are all very small (a few percents each). More precise experiments, involving measurements both at the free surface and at the inner interface, have recently shown that three-dimensional perturbations were relaxing very quickly and that the assumption of two-dimensionality was justified for our experiments.

3 The decay of turbulence

To study the decay of turbulence, we force the system for a few seconds until the wished flow structure has been obtained, and from a time defined as $t = 0$, let the system decay freely (without any further energy input). A similar method has been used in earlier studies (Cardoso et al. (1994), Tabeling et al. (1991)), however *without* the use of density stratification.

3.1 The effect of bottom friction

On Fig. 1 we plot the evolution of the energy for a typical experiment. In a transient time period of maximum duration 2 seconds after the forcing, the flow field reorganizes into a stationary vertical profile (Paret et al. (1997)). Thereafter, the only effect of tridimensionality is the friction on the bottom of the cell, causing the energy to decay exponentially, $E = E_o exp(-\alpha t)$. The time constant α agrees with the characteristic time for the relaxation of a Poiseuille profile $1/\alpha = 2b^2/\pi^2\nu$, where b is the total thickness of the fluid layer and ν is the kinematic viscosity.

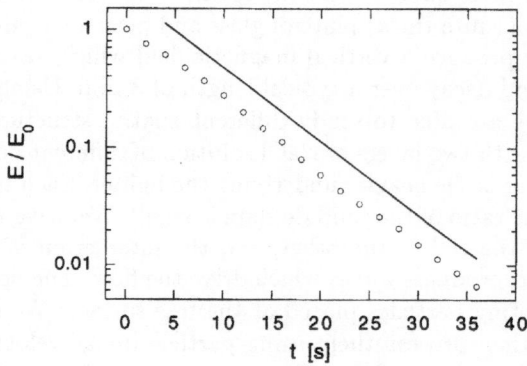

Fig. 1: The evolution of the system energy pr. area. Straight line: $e^{-0.14t}$.

The friction against the bottom can in the case of a stationary velocity profile be represented by a linear term in the 2D Navier-Stokes equations, or in terms of the vorticity $\omega(x, y)$,

$$\partial_t\omega + J(\omega, \psi) = \nu\nabla^2\omega - \alpha\omega \tag{3}$$

where $\omega = -\nabla^2\psi$ and $J(\cdot, \cdot)$ is the Jacobian, and ψ the streamfunction. In order to counterbalance the term $-\alpha\omega$ we apply the transformation

$$\omega(w, y, t) \to \tilde{\omega}(x, y, t)e^{-\alpha t} \tag{4}$$

and rescale the time as $\alpha t^* = 1 - e^{-\alpha t}$. One then arrives at

$$\partial_{t^*}\tilde{\omega} + J(\tilde{\omega}, \tilde{\psi}) = \nu^*\nabla^2\tilde{\omega} \tag{5}$$

where $\nu^* = e^{\alpha t}\nu$. The flow is therefore equivalent to a two-dimensional flow with a time dependent viscosity. In the limit of high Reynolds numbers, it is legitimate to discard the temporal variation of the viscosity and confront the experimental results to pure two dimensional systems using a constant viscosity.

All measurements of temporal properties will hereafter be expressed in terms of the transformed time t^*. Note that under the transformation the maximal observational time is $t_\alpha^* = 1/\alpha$.

3.2 Vortex identification

In Fig. 2 we show a scatter plot of vorticity versus streamfunction for a typical experimental situation. We note, that for large absolute values of ω there is functional relation between ψ and ω, $\omega = f(\psi)$, with $f'(\psi) > 0$. This means that the nonlinear term in the Navier-Stokes equation vanishes, and hence that the structures are stationary.

As a definition of a vortex, we take areas of the flow with large magnitude of the vorticity (Benzi et al. (1987), McWilliams (1990)). So, we search for values $|\omega(x, y)|$ of the vorticity field around a unique, local extrema ω_{ext} such that $\omega_{ext} > |\omega(x, y)| > \omega_s$, where ω_s is a threshold. We use values of the threshold such that the initial number of vortices is correctly determined. In pratice this limits the ratio $\omega_s/\omega_{ext} \sim$ to lie between 0.4 and 0.5. Our method agrees with the *Weiss Criterion* (McWilliams (1984)), associating vortices to the areas of the flow with a negative determinant of the velocity derivatives.

In the numerical study of McWilliams (1990), there were added requirements to the axi-symmetry of the vortices. We find this too restrictive, and on the border of what is meaningful given the experimental resolution.

An example of how our procedure works is shown in Fig. 2. The mean vortex radius a is found from the mean area occupied by the vortices. The mean distance r between the vortices is found by averaging over the distances between nearest neighbours. The position of a vortex is defined by the position of its extremum vorticity.

3.3 The statistics of vortices

Qualitative aspects. The governing dynamical processes observed in the experiments are the mututal advection of vortices, and the merging of like-sign vortices.

The time evolution of the vorticity field in a typical experiment is displayed in Fig. 3. The image borders coincides with the solid rim around the edge of the experiment. We arrange the magnets such that the initial forcing produces an 8×8 array of vortices (upper image). At that point, the forcing is turned off. After some time, like-sign vortices start to merge, almost exclusively with one of their initial nearest neighbours. Fewer and larger structures are thus formed, as one will see turning to the middle image, obtained after 4.2 sec. Both well-formed vortices and pairs in the midst of a merging are visible. The formation of dipoles is in general observed throughout the decay, but they will usually not move very far - either because one of the vortices breaks off and merges with another vortex, or simply due to the constraining action of the field of the surrounding vortices. Finally, at $t^* = 11$ sec. (lower image) the energy is so small

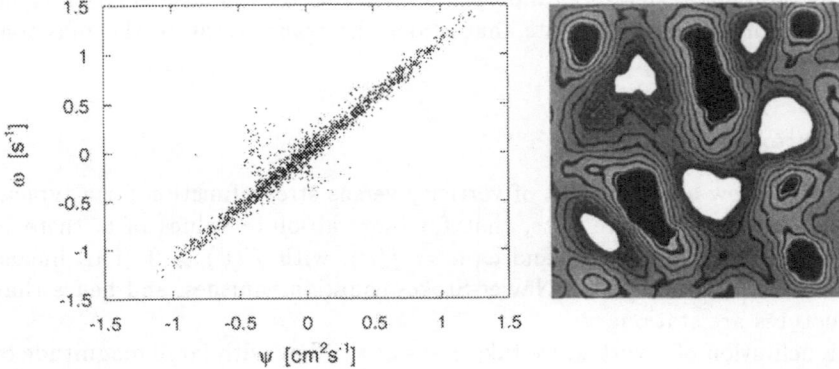

Fig. 2: *Left:* Vorticity ω versus streamfunction ψ, for a typical experiment. *Right:* An example of the procedure used to identify the vortices. The areas colored uniformly white correspond to negative-sign vortices, grey areas to positive-sign vortices.

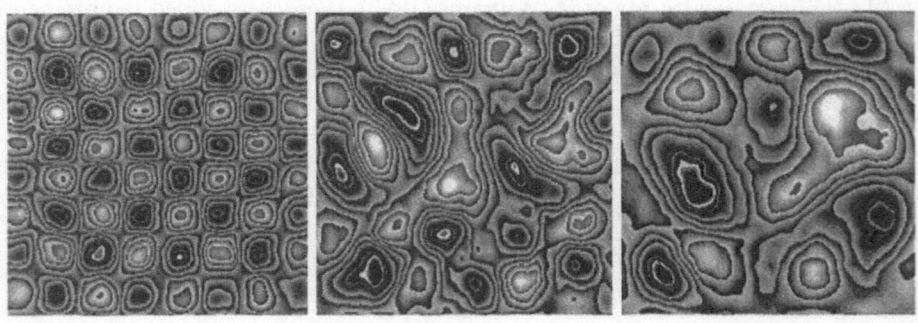

Fig. 3: Examples of calculated vorticity fields, showing the qualitative evolution of the flow from a large number of small vortices, to a smaller number of larger vortices. The overall exponential decline of the energy does not allow the final state to be reached. Left: $t^* = 0$ s (initial field). Middle: $t^* = 4.2$ s. Right: $t^* = 11$ s.

that no further evolution of the vortices can be observed. With the initial large number of vortices, chosen to get good statistics in the decay regime, the limited experimental time (see Sect. 3.1) does not allow us to reach the final state. This is however possible starting with a smaller number of vortices, see Marteau et al. (1995), Marteau (1996).

Measurements. We show the geometrical proporties of the vortices; that is, the number of vortices, their radius and nearest-neighbour separation. We also measure the global dynamical quantities characerizing the system, apart from the energy the enstrophy, extremum vorticity and kurtosis.

In Fig. 4 I show on double-logarithmic scale the time evolution of the number of vortices obtained in several experiments with similar initial conditions. There is a scatter from one realisation to another, but they all have the following charateristics:

For $t^* < 1$ s, the decrease in number is slow or vanishing. As explained in Sect. 3.1, 3D effects might still play a role in the system up to app. 1.5 s. As well, it takes some time (up to a second), before the vortices are liberated from the effect of the initial conditions, and start to move among each other, and merge. For 1 s $< t^* < 10$ s, the number of vortices decreases. For $t^* > 10$ s, the energy has decreased to a few percent of its initial value, and the vortices start to disappear compared to the experimental noise.

Fig. 5 shows the evolution of the number of vortices obtained after ensemble averaging over nine experiments. A least-squares algebraic fit to the time period 1 s $< t^* < 10$ gives an exponent $\xi = -0.70$. Furthermore, plotting the logarithmic slope of $N(t^*)$ (inset in Fig. 5) a plateau appears for the above time period, thus confirming the algebraic decay of the vortex number. If plotting $N(t^*)$ on a semi-logarithmic scale, a curvature is seen that is clearly incompatible with an exponential law.

Summing up, we find the following law for the decay of the vortex number

$$N \sim t^{*-0.70\pm0.1}, \tag{6}$$

where the error bar accounts for the variability in the individual runs.

In Fig. 4 I show as well the vortex radius a, and the mean vortex separation r as a function of the rescaled time. The mean separation is evaluated taking distances between nearest-neighbour vortices. Both quantities increase, showing that the geometry expands as time increases. Algebraic laws are well defined for the period 1 s $< t^* < 10$ s (2 s $< t^* < 10$ s for the radius).

$$a(t^*) \sim t^{*0.21\pm0.05}, \qquad r(t^*) \sim t^{*0.38\pm0.15}. \tag{7}$$

The evolution in vortex radius a deserves a closer inspection. In Fig. 5 we show the evolution of the instantaneous vortex radius distribution, $P(a/ < a >))$. There is a reasonable collapse, and no trend in the time evolution can be seen. This confirms the self-similar behaviour of the system (Weiss and McWilliams (1992)).

The enstrophy and the kurtosis of the system, along with the extremum vorticity, are represented on Fig. 6 (the enstrophy and extremum vorticity are corrected for the exponential decline of the energy). If power laws are assumed, they read

$$\frac{\omega_{ext}}{\sqrt{E}} \sim t^{*-0.09\pm0.04}, \qquad \frac{Z}{E} \sim t^{*-0.18}, \qquad Ku \sim t^{*0.12} \tag{8}$$

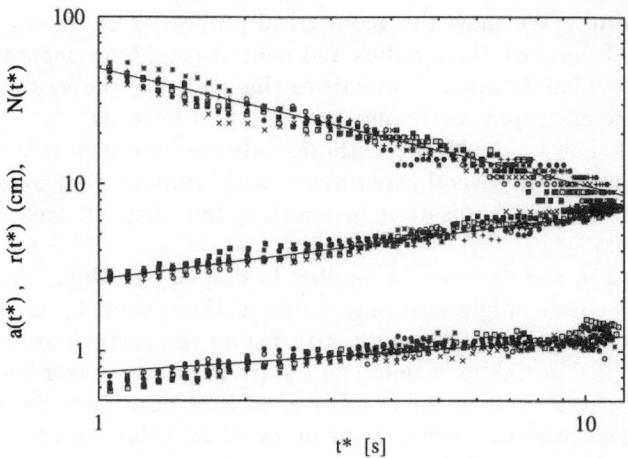

Fig. 4: Number of vortices, their separation and radius versus rescaled time for several experiments, log-log scale. Lines: $t^{*-0.70}$, $t^{*0.21}$, $t^{*0.38}$ (6,7).

The conservation of the extremum vorticity is a basic assumption in the theory (Carnevale et al. (1991)). Here, we find a slight decay, which is probably due to a finite Reynolds number effect (as seen as well the numerical studies by Weiss and McWilliams (1992)).

Discussion. Let us now compare our results to the theory (Carnevale et al. (1991) and equations (1)). If we take 0.70 ± 0.1 as the value defining the exponent ξ, we expect 0.18 ± 0.025 for the increase of the vortex size $a(t)$, 0.35 ± 0.05 for the distance between vortex centers $r(t)$ which agrees well with the experiment.

Another way to test the theory is to whether the internal relations proposed by the theory agree with the experiment. According to the theory, one should have :

$$E = \rho \omega_{ext}^2 a^4 \qquad (9)$$

On Fig. 6 we plot the ratio $E/\rho\omega_{ext}^2 a^4$. Within 10 %, a plateau is found, showing that the above relation hold in the experiment.

The laws for the kurtosis and the enstrophy, while having the right signs, are slower than the predictions (Carnevale et al. (1991) and equations (1)) gives $t^{*0.35}$ for the kurtosis, and $t^{*-0.35}$ for the enstrophy), and the power laws are not well defined. This is probably due to a lack of spatial resolution of the PIV technique at early times, i.e. when the structures have the smallest size. The presence of systematic errors leading to an underestimate of the enstrophy in the first few seconds may explain why the power law behaviour is deteriorated,

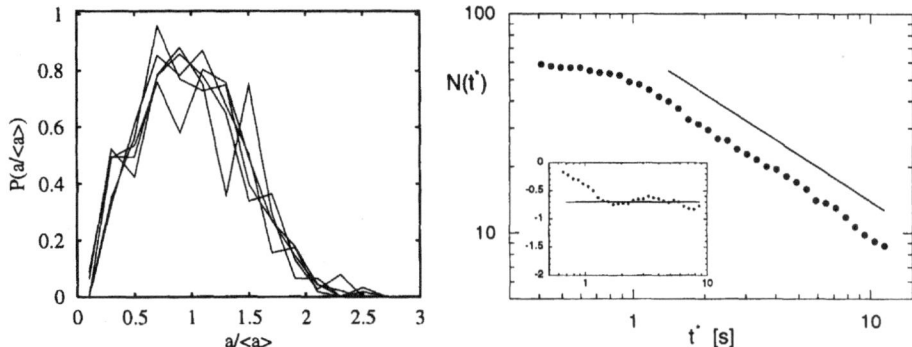

Fig. 5: *Left:* Instantaneous distributions of vortex radii, for times $t^* = 2.2, 3.4,$ 4.8, 6.9, 10.0 s. The values of a are divided by the mean value $< a >$. *Right:* Number of vortices versus rescaled time averaged over 9 experiments on log-log scale. Line: $t^{*-0.70}$. Inset: the logarithmic slope.

and the value of the exponent is below the prediction (in absolute value). A similar argument holds for the kurtosis.

To sum up, we find good agreement between the experimental results and the self-similar decay theory (Carnevale et al. (1991)), the value of ξ being determined to $\xi = 0.70 \pm 0.1$ consistent with numerical estimates.

3.4 The movement of vortices

We now turn the attention to the *dynamical* characteristics of the vortices. In order to continously follow the motion of the vortex centers, the intervals between the calculated vorticity fields correspond to a small movement of the vortices, compared to the inter-vortex distance. We typically calculate 65-75 fields over the whole duration of the experiment, or 4-8 pr. rescaled second. Having tracked all vortices thoughout the experiment, we calculate: the mean collision time τ, the mean free distance λ, the mean velocity u, and the mean squared displacement σ_v^2. The notions will be detailed in the following.

The collision time τ could also be called the 'lifetime' of a vortex. It is the mean time between two subsequent mergings of the same vortex. The time it takes two vortices to merge is 5-10 times smaller than their collision time, and can be neglected.

Correspondingly, the mean free path λ is the mean displacement of the vortex center between two mergings. Additionally, we also measure λ_{tr}, defined as the length of the trajectory followed by the vortex center in the time interval in question.

The vortex velocity u is calculated as a finite difference between vortex positions, using a fixed timestep $dt = 0.5$ s. The timestep is larger than the intervals

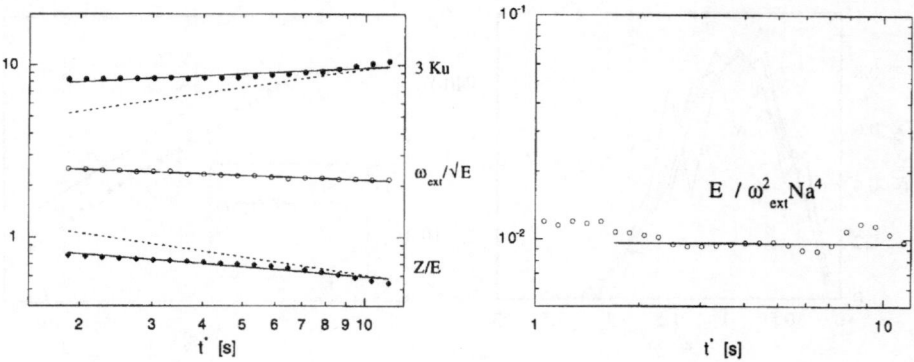

Fig. 6: *Left:* Time evolution of global dynamical quantities: kurtosis, extremum vorticity and enstrophy. Full lines: $t^{*0.12}$, $t^{*-0.09}$, $t^{*-0.18}$. Dotted lines: the expected laws for the enstrophy ($\sim t^{*-0.35}$) and the kurtosis ($\sim t^{*0.35}$). *Right:* Test of the expression $E = \rho\omega_{ext}^2 a^4$ for the system energy. Plotting the ratio a plateau is formed.

between the calculated vorticity fields, to avoid the influence of noise, due to the finite resolution of the vortex positions.

For each time, the mean is obtained by averaging over the properties of all the vortices present in the system.

Finally, we calculate the total mean square displacement σ_v^2 of the vortices that survive through the whole experiment (allowing for a vortex to merge on its way). The 'surviving' vortex in a merging is defined as the one with highest extremum vorticity.

Measurements. In Fig. 7 we plot the quantities λ, λ_{tr} and τ versus time on log-log scale. The data stems from the same nine realisations as in the previous Sect. 3.3. All three quantities *grow* with time in a similar manner.

We remark that at $t^* = 7$ s, the mean lifetime τ is app. 5 s. This means that the measurement will start to be influenced of the finite duration of the experiment (up to $t^* = 11$ s), and the curves saturate.

We have examined the power law behaviour of λ and τ in the (short) time interval 1.5 s $< t^* < 7$ s. Assuming algebraic laws, the results of least-squares fits are the following

$$\tau \sim t^{*0.57\pm0.12}, \qquad \lambda_{tr} \sim t^{*0.49\pm0.09}, \qquad \lambda \sim t^{*0.45\pm0.10}. \qquad (10)$$

In agreement with the qualitative impression, the measured exponents are close. As before, the errorbars are estimated by taking into account the dispersion in the curves from the individual experiments.

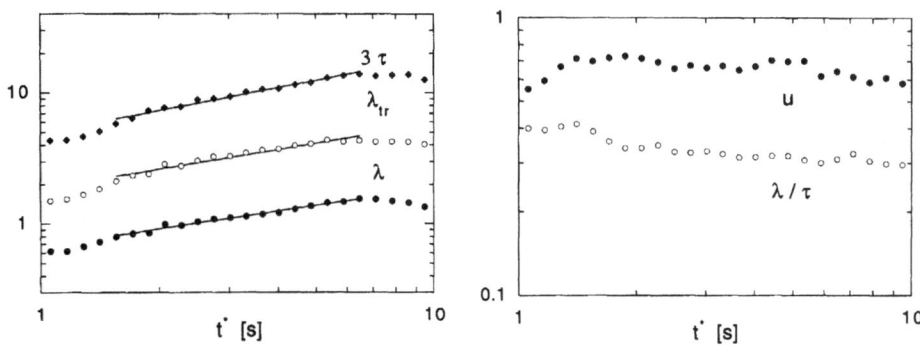

Fig. 7: *Left:* Time evolution of the mean free distance λ, the mean free distance measured by the length of the trajectories between mergings λ_{tr}, and the collision time τ. Straight lines: $t^{*0.49}$, $t^{*0.45}$, $t^{*0.57}$. *Right:* Ratio λ/τ, and the vortex velocity u , and ratio λ/λ_{tr} .

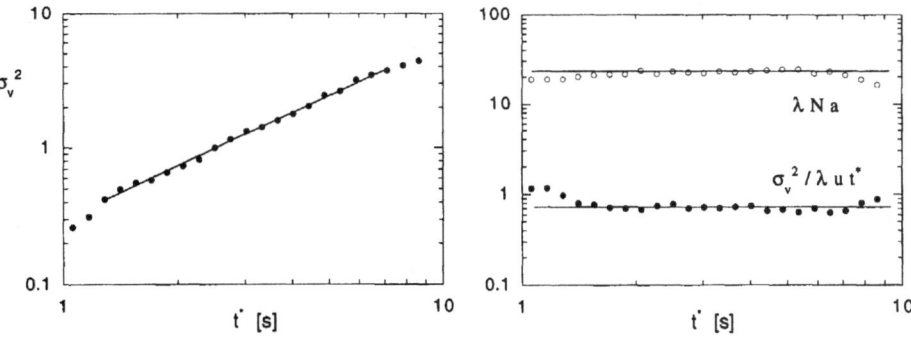

Fig. 8: *Left:* The mean square displacement of the vortices σ_v^2. Straight line: $t^{*1.3}$. *Right:* Invariants: $\lambda N a$, and $\sigma_v^2/\lambda u t^*$. Flat lines are shown for comparison.

In Fig. 8 we plot the mean square displacement σ_v^2, calculated for 34 vortices tracked throughout the whole experimental time. A well-defined power law is observed. For the time interval 1.2 s $< t^* < 7$ s we find

$$\sigma_v^2 \sim t^{*1.3\pm0.1}. \tag{11}$$

The exponent is markedly larger than one, so the vortices move *hyperdiffusively*.

Discussion. The increase in λ, λ_{tr} and τ is not surprising. The system gets more and more *dilute* during the decay; the area occupied by the vortices $N a^2$

decreasing as $t^{*-\xi/2}$. Therefore, the mean distance λ a vortex has to travel before meeting and merging with another vortex will increase. The fact that λ_{tr} is a factor of three larger than λ, is a result of the intricate movement of the vortices, that can be seen in Fig. 9.

Let us try to quantify these ideas using a simple geometrical argument, similar to that classically given in kinetic gas theory. Estimate λ as the distance travelled by a vortex in unit time ($\sim u$, the advection velocity), divided by the probability of suffering a collision in this time interval ($\sim u$ times the collisional cross section times the density of vortices). Then

$$\lambda \sim \frac{u}{u\sigma\rho} \sim \frac{1}{\rho a} \qquad (12)$$

since the collisional cross section is proportional to the radius for a system of circular disks (Melander et al. (1988)). (ρ is the density of vortices, N/L^2). The collision time τ will be $\tau \sim \lambda/u$. Inserting the algebraic laws for the time evolution of a and ρ, one arrives at

$$\lambda \sim \tau \sim t^{\xi}t^{-\frac{1}{4}\xi} \sim t^{\frac{3}{4}\xi}. \qquad (13)$$

With the obtained value $\xi = 0.7$, we should have $\tau \sim \lambda \sim t^{0.5}$, which is in agreement with the observed power laws (10). However we observe an unexpected, slight decrease in the mean vortex velocity u, which seems to cause the exponents to differ from each other (Fig. 7). To further test (12), we have in Fig. 8 plotted the product $\lambda N a$, as given by the data. A clear plateau is observed for times $1.5 \text{ s} < t^* < 7 \text{ s}$, corresponding to the scaling regime of λ. We conclude that the expressions (12) and (13) are confirmed by experiment.

The behaviour of σ_v^2 can be understood as follows: introduce a vortex diffusion coefficient D by

$$\sigma_v^2 \equiv Dt. \qquad (14)$$

Again adapting arguments from kinetic gas theory, D is taken as the mean free path times the vortex speed u . We get

$$D = \lambda u \qquad (15)$$

Thus, the growth in length scale causes D to grow as well. Further, the mean square displacement of the vortices σ_v^2 is now given by :

$$\sigma^2 \sim Dt \sim \lambda ut. \qquad (16)$$

If D had been constant, one would have that $\sigma_v^2 \sim t$, that is, Brownian motion of the vortices. But now D increases with time, and in turn the variance grows faster than t. The proposal (16) can be further tested directly, by plotting $\sigma_v^2/(\lambda ut^*)$ versus time (Fig. 8). A plateau appears for times larger than 1.5 s, so (16) is well verified by experiment.

It is tempting to infer, from the above relations, a formula between ξ and an exponent characterising the temporal evolution of the mean square displacement of the vortex centres. From (12) and (16), on may deduce that if σ_v^2 grows as t^ν, one must have the following relations between ξ and ν :

$$\nu = 1 + \frac{3}{4}\xi \tag{17}$$

By taking $\xi = 0.7$, one should find the mean square displacement of the vortices is characterized by an exponent equal to 1.5. This is a bit larger than the observed exponent which is 1.3; however systematic errors adds so as to violate (17). The main factor is that u is not exactly constant, but slightly decrease with time. In practice, one can say that the relations (12) and (16) are consistent with the experiment, and the corresponding straightforward relation (17) must be taken only as a crude formula expressing the existence of a link between the exponent characterizing the decay, and that characterizing the dispersion of the vortices in the experiment.

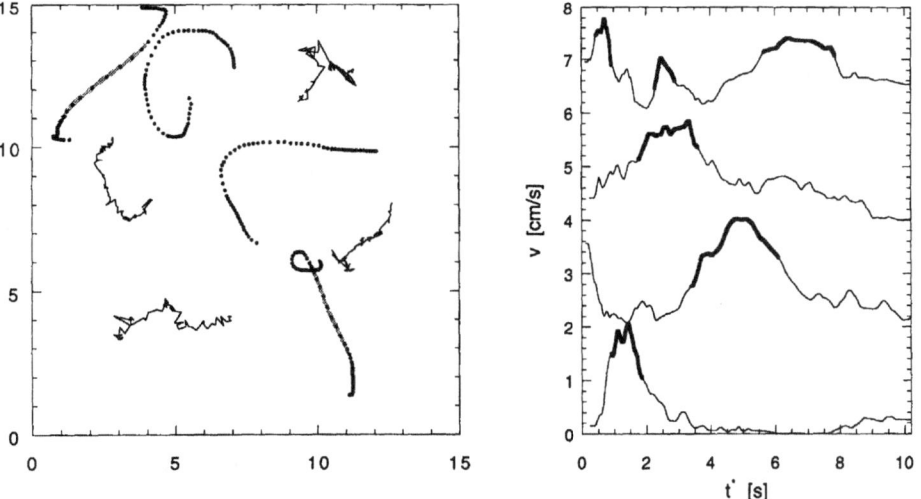

Fig. 9: *Left:* Some trajectories of vortices (lines), and passively advected particles (points, with app. equal time spacings. The analysis uses five times the shown resolution). Parts of particle trajectories defined as a flight is shown with a grey line. *Right:* Examples of flight identification. We show the velocities of single advected particles; regions defined as flights by the procedure described in Section 3.5 are marked with a thick line. (There is added 2, 4 and 6 cm/s to the upper curves in order to separate the curves).

3.5 Dispersion of passive particles

In this section, we present studies of the dispersion of passive, imaginary particles. The trajectories $\mathbf{x}(t)$ is obtained by integrating the equation

$$\frac{d\mathbf{x}(t)}{dt} = \mathbf{v}(\mathbf{x}, t) \tag{18}$$

for chosen initial conditions. The derivatives $\mathbf{v}(\mathbf{x})$ are given by the experimentally determined velocity fields. As above, the velocity fields are found with small, regular time intervals, to ensure that the velocities only change slightly. The 40×40 velocity fields are interpolated in space and time, and the trajectories are calculated using a standard 4th order Runge-Kutta method with adaptive stepsizing.

We find that this is an effective method for extracting statistical quantities from the flow fields. It allows us to obtain a number of trajectories, that would be difficult to achieve experimentally. The imaginary particles are indeed truly passive, so we do not have to consider the question of Stokes drag or similar experimental problems. On the other hand, since there is a limited resolution of the velocity field (0.375 cm between neighbouring vectors); particles cannot correctly sample motions on a much smaller scale. We remark that in any case the quantity we are interested in, namely the mean square displacement σ^2 of the particles, is determined by the large scale properties of the flow (and already after 1 sec. the typical $\sqrt{\sigma^2} > 0.7$ cm).

To demonstrate the qualitative behaviour of the particle motion, we have in Fig. 9 shown some examples of particle trajectories. We note that trapping effects are not visible. Particles tend to get ejected from vortices, both during mergers, and as an effect of the straining of vortices due to the surrounding field. The observation that the vortex cores are characterized by a low tracer density is in qualitative accordance with Elhmaidi et al. (1993). Trapping in vortices and sticking on their periphery is not an effect that seems important for the particle dispersion in out experiment; indeed, well-defined trappings are too rare to justify a detailed analysis. Flights, or parts of the trajectories with a velocity persistently higher than the mean velocity, are on the other hand often observed.

In Fig. 10 we show the mean of the absolute squared displacement σ^2 of the particles. Conditioning the average only to include particles that do not visit high-vorticity regions, does not make any difference. This suggests that the vortices are not important for the dispersion properties of the passive particles. A clear power law emerges for times between 1 s and 7 s. The inset shows the logarithmic derivative of the preceding points. For small times, the exponent decreases from 1.8, while from t = 6 s, the exponent drops to 1; this is in accordance with the classical prediction (Taylor (1921)). However, the change in exponent shows a clear plateau, thus defining a dispersion coefficient for intermediate times, with the value $\sigma^2 \sim t^{1.39}$. As a mean over three experiments, we find

$$\sigma^2 \sim t^{1.4 \pm 0.1} \tag{19}$$

159

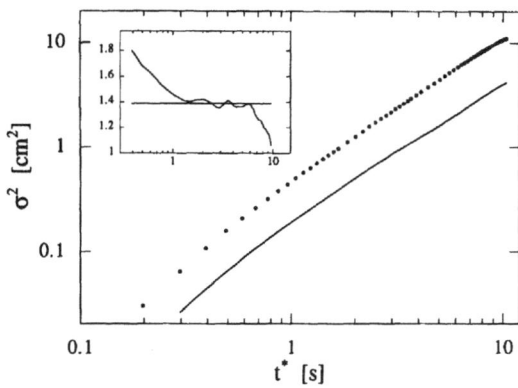

Fig. 10: The mean square displacement of 3200 passive, imaginary particles (circles) and a subset of 2208 particles not entering high-vorticity regions (line). The latter curve has been divided by 2. Inset: the logaritmic slope of the overall mean.

It is remarkable that this exponent is undistinguishable from that corresponding to the mean square displacement of the vortex centers.

Characterization in terms of flights. It turns out that the velocities of the particles vary strikingly, according to which region of the flow the particle sample. To make this observation quantitative, we have developed a procedure to analyze the trajectories for flight events; some examples of the procedure are shown on Fig. 9. The flights are determined by searching for extrema of the velocity above a threshold (taken as 0.80 cm/s, where the square root of the total, constant, system energy per unit area, is 0.71 cm/s). The beginning and end of a flight event is defined by the maximum and minimum in acceleration before and after a velocity extremum. The absolute value of the acceleration is required to be above another threshold (taken as 0.3 cm/s^2), ensuring that the flight corresponds to the time between when the particle enters and exits a flow region with high velocity. The conclusions remain valid for a variation of the above thresholds within $\pm$ 15%.

We stress that there does not exist a universal algorithm to define flights of particles in hydrodynamical flows. We have checked for a large number of trajectories that our procedure correctly identifies the events, that strike an observer 'by eye' as flights.

The flight time distribution on Fig. 11 is a result of this procedure. 5700 particle tracks have been analyzed, giving a total of 4400 flight events. For flights longer than 1.5 seconds and less than 6 seconds, the distribution follows a power law. For long flights, the statistics will be influenced by the finite duration of

160

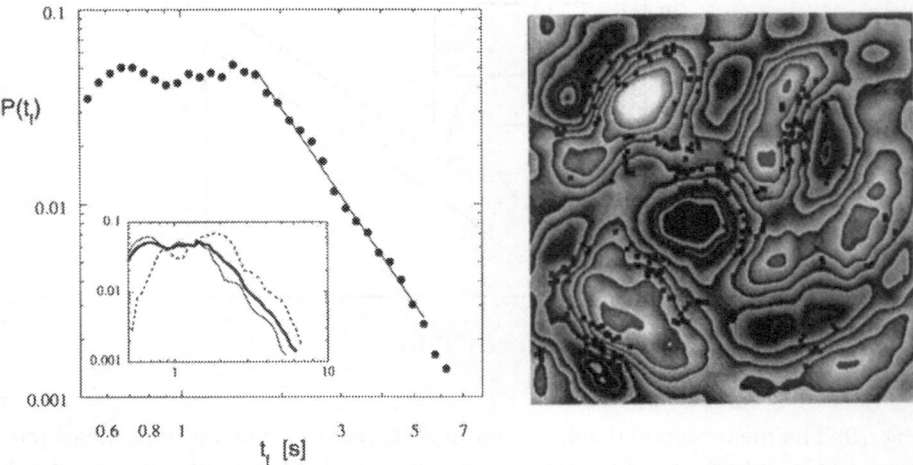

Fig. 11: *Left:* Probability distribution of flight times. Straight line: $t_f^{-2.56}$. Inset: the same distribution(thick line), along with the distribution of flights with extremum velocity occurring between $t^* = 2$ to 5 seconds (thin line), and $t^* = 5$ to 10 seconds (dotted line). *Right:* Vorticity field to t = 8.5 s. The positions of particles undergoing a flight are marked with black squares.

the trajectories (the total duration of this experiment is 11 s); giving a rapid decrease in $P(t_f)$ for $t_f > 6$ s. We conclude that the distribution of flight times has an algebraic tail,

$$P(t_f) \sim t_f^{-2.6\pm0.2}. \qquad (20)$$

We have investigated the temporal evolution of the characteristics of this distribution; this is shown on the inset of Fig. 11. The plot shows the same analysis, but preformed over a smaller range of time, so as to see how the characteristics of the distribution evolves with time. Although the statistics is on the border of being sufficient to draw reliable conclusions, it seems that the tails stay parallel to each other as time increases, so that the slope of the distribution does not vary with time. This means that the distribution are not sensitive to the fact that the system expands.

We have investigated in which regions of the flow the particles move when they are subject to a flight. As demonstrated in Fig. 11, flights predominantly occur for particles located between opposite-sign vortices. This is not surprising, since the regions between two close opposite-sign vortices are characterized by large velocities, forming a jet-like structure. So there is a straightforward physical explanation for the occurrence of flights.

The exponents we find for the flight distribution are consistent with those for the variance. According to Klafter et al. (1987), we effectively have $\sigma^2 \sim t^{4-2.6} \sim t^{1.4}$, in good agreement with the previous result (19). This shows that we can regard the hyperdiffusion for the passive particles as anomalous, that is caused by extreme (flight) events.

3.6   Spectral characteristics

In Fig. 12 we show energy spectra $E(k)$, calculated from the obtained velocity fields at t = 0s and t = 23 s. The spectrum is initially very peaked around the forcing wavenumber, but tends to broaden for the first seconds of the experiment. This is in agreement with the observation that there is a spread in the distribution of vortex sizes for times in the scaling range, as seen qualitatively in Fig. 3, and quantitatively in Fig. 5.

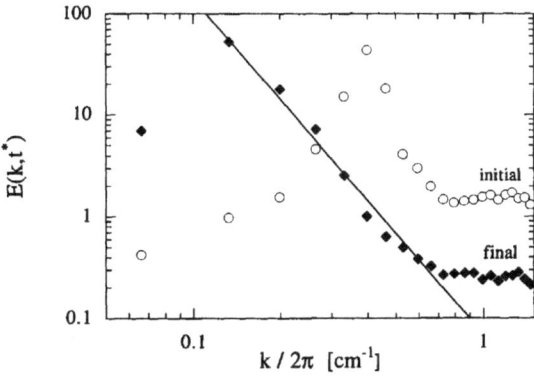

Fig. 12: Energy spectrum at t= 0 s and t = 23 s (the last multiplied by 100 to show the curves at the same scale). Straight line: $\sim k^{-3.3}$.

Regarding the slope of the spectrum for late times in the range $0.1 < k/2\pi < 0.7$, a comparison with a straight line of slope -3.3 shows that a power law is not well-defined, there being turnings on the curve suggesting competing effects. The mean value of the vortex radius a for t = 23 s is 1.2 cm, giving a wavenumber at the order of $k/2\pi \approx 1/(2a) \approx 0.4$ cm $^{-1}$. Thus, the vortices are present at the scales where the systematic decrease in $E(k)$ takes place, suggesting that they account for the spread in energetic wavenumbers.

An intriguing observation is the value of the slope of the spectrum, which evokes the enstrophy cascade. Indeed, the transfer of enstrophy seems to operate towards larger scales, whereas the usual enstrophy cascade proceeds in the other direction. Nevertheless, it cannot be ruled out that this spectrum is influenced by mechanisms close to those involved in the usual enstrophy cascade, thus giving rise to a $k^{-3.3}$- dependence. Further experiments are needed to clarify this point.

4   Forced turbulence - Inverse cascade of energy

We now turn to the issue of forced turbulence. Contrary to the decaying case, the flow is driven during the whole duration of the experiments and there is no time limitation except the one imposed by the fact that for very long times the stratification is lost and the two-dimensionality assumption is no longer justified. We use periodic magnet arrangements in order to input the energy at a given wave-number. This wave-number is chosen large (energy input at small scales) in order to have a wide range of scales available for the cascade of energy toward large scales. Finally, in order to impose a zero mean flow so as to get homogeneous turbulence, we use a random-in-time forcing: the electric forcing is made of a time series of impulses of constant amplitude and random sign. Each impulse has a duration longer than the characteristic time of the vertical transfers. Each experiment has a typical duration of 6 minutes which corresponds approximately to 50 turn-over times. The injection Reynolds number, based on the root-mean-square velocity and the injection scale has a typical value of 100. One should notice that these characteristics are comparable to the largest numerical simulations performed on the subject (Borue (1994), Smith and Yakhot (1994)).

The experiments display two regimes: first, a short transient phase which duration is around 30 seconds, then, a phase during which the properties of the flow do not vary much. A typical stream-function field taken during this second regime is shown in Fig. 13. It can be seen that there are many structures of different sizes. From such a plot, it would be difficult to infer a single characteristic size as it is possible in the decaying case. This shows that the dynamics involved in both cases may be rather different. Since we input energy at small scales, the fact that there are large structures present in the stationary regime is a sign that there is indeed a transfer of energy toward large scales.

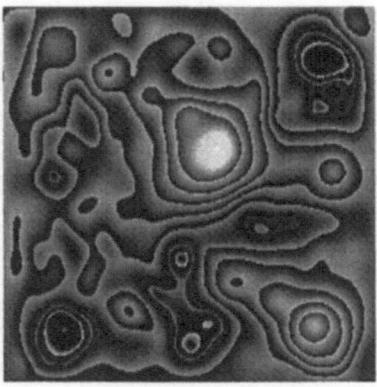

Fig. 13: Typical stream-function for the stationary regime

4.1 Spectral properties

A typical series of energy spectra are displayed in Fig. 14(a): the initial one ($\diamond$),
computed 2 seconds after the current has been switched on, an intermediate
one ($\circ$) averaged over a few fields covering a short time interval centered at t
= 10 s, and the final spectrum ($\bullet$), averaged over 160 fields, well beyond the
transient regime. The initial spectrum ($\diamond$) shows that the injection of energy
is well localized in wave-number space. At later times, in the transient regime,
the energy transfer from large to small wave-numbers is signalled in Fourier
space by the progressive building up of a spectrum with a $k^{-5/3}$ power law,
as illustrated by the transient spectrum ($\circ$). The final spectrum ($\bullet$) displays
a $k^{-5/3}$ behavior over a range slightly narrower than one decade (the black
line corresponds to a calculated $-5/3$ scaling). In Fig. 14(b) the final spectrum
of Fig. 14(a) is displayed compensated by the Kolmogorov scaling, $E(k)k^{5/3}$.
Over the same range of scales as those for which the scaling law is observed in
Fig. 14(a), a clear plateau is observed, which confirms that the spectral exponent
is close to $-5/3$. For the present experiment, the most energetic wave-number is
$k_0 \approx 0.13$ cm^{-1}, which is larger than the wave-number $2\pi/L$ based on the cell
size.

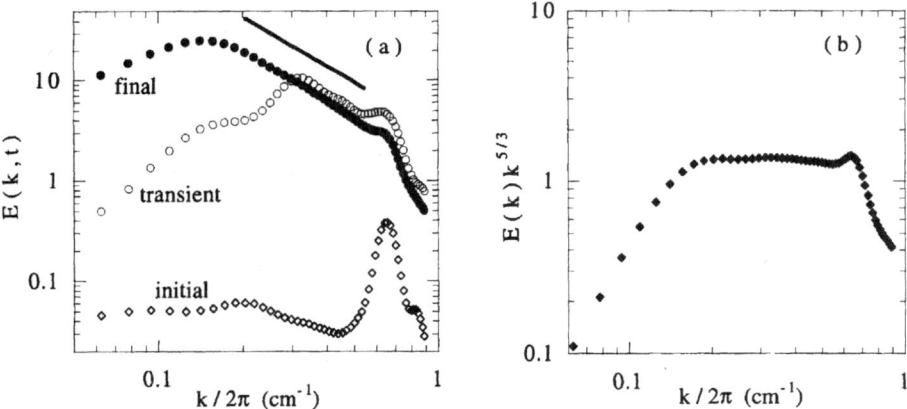

Fig. 14: Energy spectra : (a) temporal evolution, (b) compensated energy spectrum for the stationary regime.

We have checked that our flow is stationary, homogeneous and isotropic.
Stationary is assessed by the fact that both the energy and the enstrophy of
the flow are almost constant during the whole experiment, apart from the short
initial transient. The root-mean-square velocity deviations from the mean flow
(which is approximately zero) display no space dependency in the bulk of the
flow. Of course, because of the no-slip condition on the walls of the cell, these
deviations must fall to zero when we approach the boundaries. Isotropy is less

straightforward: because of the geometry of the cell and the structure of the magnet arrangement, the flow cannot be isotropic at the largest scales and at the forcing scales. However, if we band-pass-filter the velocity fields, just keeping the modes lying in the inertial range, and compute the angular energy spectrum (Fig. 15), we find that the angular dependency of the energy spectrum is very weak and that isotropy is quite well satisfied in the inertial range.

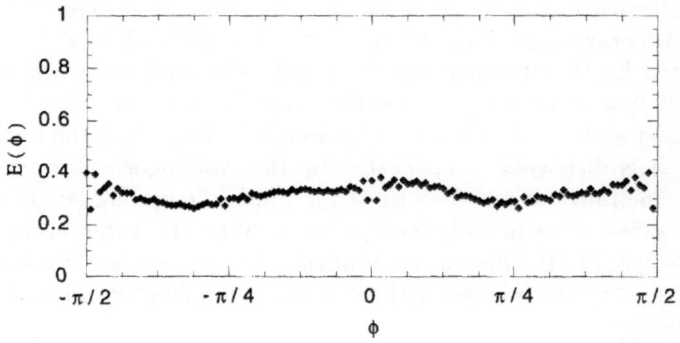

Fig. 15: Angular energy spectrum (inertial range)

We have also determined the Kolmogorov constant in our experiments. First, it must be noticed that, since the injection Reynolds number is not very large, a non-negligeable part of the energy input flows to the viscous scales without taking part in the inverse cascade. Thus, determination of ε from the total dissipation rate overestimates the energy transfer rate in the inverse cascade and yields a very low value for the Kolmogorov constant. Using three different methods, we have determined the true energy transfer rate and we have found $C_k = 6.5 \pm 1$ which is consistent with the estimations from the best numerical simulations (Matrud and Vallis (1991), Smith and Yakhot (1994)).

4.2   Statistics of the vorticity field

We now turn to the physical space characteristics of this inverse energy cascade, and more precisely to its consequences in term of coherent structures and vortices. The first consequence of the shape of the energy spectrum is that the enstrophy spectrum $Z(k)$ has a maximum value when k is equal to the injection wave-number. Actually, we have $Z(k) = k^2 E(k)$ and since $E(k)$ scales as $k^{-5/3}$ for $k \leq k_i$ and as k^{-n} with $n \geq 3$ for $k \geq k_i$, the above conclusion is straightforward. This shape of the enstrophy spectrum implies that, if there are vortices in the flow, they must have a characteristic size of the order of the injection scale. This is confirmed by the distribution of vortex sizes displayed in Fig. 16. The use of log-lin scales makes it clear that this distribution is a decreasing exponential function of the vortex radius r for sizes corresponding to the inverse cascade. This shows that even if there are coherent structures in theses flows, their influence on the dynamics is probably very weak. This is certainly the most striking difference between decaying and forced turbulence.

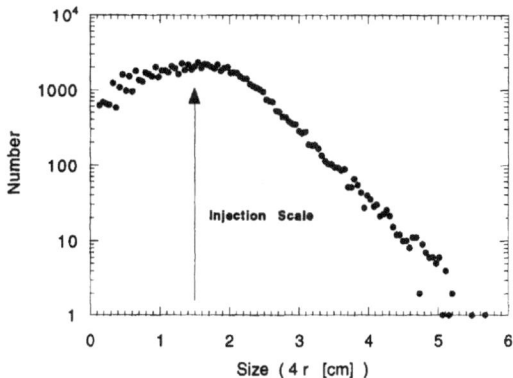

Fig. 16: Distribution of vortex sizes

These characteristics of the vorticity field may also be a hint of the physical mechanism underlying the inverse cascade. The naive approach consists in saying that the inverse cascade is made of a sequence of merging events between same sign vortices as it is the case in decaying turbulence. However, if it was so, the distribution of vortex sizes should be broad and probably not an exponentially decreasing function. The inverse cascade should rather be an aggregation process, that is the formation of large patches of same sign vorticity. This picture is favoured when looking at the *transient* regime of our experiments. Figure 17 displays, at 4 different times in the transient regime, the regions of positive (white) and negative (black) vorticity: it can be seen that, as the cascade builds up, the vorticity tends to seggregate and form larger and larger patches of the same sign.

4.3 Dispersion - Richardson's law

We have studied some dispersion properties of the inverse cascade by looking at the temporal evolution of the separation of pairs of particles initially close to each other, using the technique described in Sect. 3.5. We used 10000 pairs in order to achieve a good statistical convergence. The results are displayed in Fig. 18. It can be seen that there exists a range of time during which the squared separation R^2 scales as t^3. Such a result was originally proposed by Richardson in the twenties (Richardson (1926)). It has been observed in numerical simulations, but we probably provide one of the first experimental observation of Richardson's

Fig. 17: Transient evolution of the vorticity field (t = 2,6,10 and 22 s)

law. Since this result can be inferred from dimensional analysis (assuming that R^2 depends only on t and ε), it is probable that there are no Levy flights in the forced case. Indeed, using many possible definitions of a flight (including of course the definition used in Sect. 3.5) we have not found any probability distribution of flight times with a power-law scaling. Thus, the existence or non-existence of flights in such flows and the existence of deviations from dimensional analysis predictions cannot be decided at the moment. Further experiments using more particles and measurements of higher order moments of the probability density function of the separation R will be the aim of future work.

5   Conclusion

To conclude briefly, the study shows a clear distinction between decaying and forced turbulence. In the former case, long live vortices play a central role,and an approach based on the analysis of the behavior of individual objects, with well defined characteristics, allows to describe correctly the main features of the decaying regime. In the latter case, statistical dimensional arguments, based on the hypothesis that the system is geometrically self similar, provides correct answers to finding out the exponents of the spectral laws and those controlling the dispersion of Richardson pairs. We thus have two examples where different approaches - one structural, and the other statistical - are successful and at the moment alternative ways of tackling the two problems do not seem available. It is clear that in the forced case, the two approaches are at least complementary. We tend to think that it would be interesting to figure out the content of such a

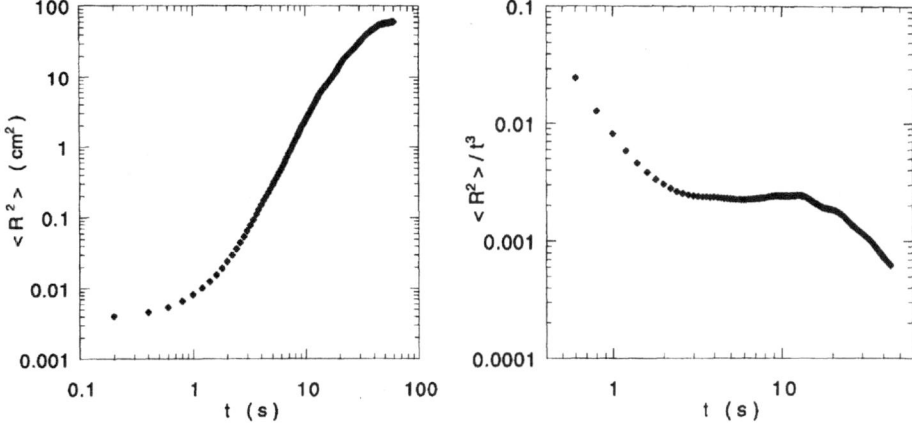

Fig. 18: Temporal evolution of the squared separation R^2 between particles initially close to each other. Left: R^2 vs. t . Right: R^2/t^3 vs. t

complementarity, so as to reach a situation where a complete, consistent, physical description of the inverse cascade is made available.

The authors want to thank the Centre National de la Recherche Scientifique, the Ecole Normale Supérieure and the Universités Paris 6 et Paris 7 for providing support for this work.

References

Benzi, R., Patarnello, S. and Santangelo, P. (1987): Self-similar coherent structures in two-dimensional decaying turbulence, *J. Phys. A: Math. Gen.* **21**, 1221–1237.

Borue, V. (1994): Inverse energy cascade in stationary two-dimensional homogeneous turbulence, *Phys. Rev. Lett.* **72**, 1475–1478.

Cardoso, O., Marteau, D. and Tabeling, P. (1994): Quantitative experimental study of the free decay of quasi-two-dimensional turbulence, *Phys. Rev. E* **49**, 454–461.

Cardoso, O., Gluckmann, B., Parcollet, O. and Tabeling, P. (1996): Dispersion in a quasi-two-dimensional turbulent flow: an experimental study, *Phys. Fluids* **8**, 209–214.

Carnevale, G.F., McWilliams, J.C., Pomeau, Y., Weiss, J.B. and Young, W.R. (1991): Evolution of vortex statistics in two-dimensional turbulence, *Phys. Rev. Lett.* **66**, 2735–2737.

Couder, Y. (1983): Observation expérimentale de la turbulence bidimensionnelle dans un film liquid mince, *C. R. Acad. Sci. Paris* **297**, 641–645.

Elhmaidi, D., Provenzale, A. and Babiano, A. (1993): Elementary topology of two-dimensional turbulence from a Lagrangian viewpoint and single-particle dispersion, *J. Fluid Mech.* **257**, 533–558.

Hansen, A.E., Schröder, E., Alstrøm, P., Andersen, J.S. and Levinsen, M.T.(1997): Fractal particle trajectories in capillary waves: imprint of wavelength, *Phys. Rev. Lett.* **79**, 1845–1848.

He, X. (1996a): Dispersion of a vortex trajectory in 2D turbulence, *Physica D* **95**, 163–166.

He, X. (1996b): On exponent of vortex number in 2D turbulence, in *Advances in Turbulence VI*, 277–278, eds. S. Gavrilakis, L. Machiels & P.A. Monkewitz, Kluwer Academic Publishers, Dordrecht.

Hopfinger, E., Griffiths, R.W. and Mory, M. (1983): The structure of turbulence in homogeneous and stratified rotating fluids, *J. Mec. Theor. App.* **2**, 21–44.

Huber, G. and Alstrøm, P. (1993): Universal decay of vortex density in two dimensions, *Physica A* **195**, 448–456.

Jüttner, B., Marteau, D., Tabeling, P. and Thess, A. (1997): Numerical simulations of experiments on quasi-two-dimensional turbulence, *Phys. Rev. E* **55**, 5479–5488.

Klafter, J., Blumen, A. and Schlesinger, M.F. (1987): Stochastic pathway to anomalous diffusion, *Phys. Rev. A* **35**, 3081–3085.

Maltrud, M.E. and Vallis, G.K. (1991): Energy spectra and coherent structures in forced two-dimensional and beta-plane turbulence, *J. Fluid Mech.* **228**, 321–342.

Marteau, D., Cardoso, O. and Tabeling, P. (1995): Equilibrium states of two-dimensional turbulence: an experimental study, *Phys. Rev. E* **51**, 5124–5127.

Marteau, D. (1996): Etude expérimentale du declin de la turbulence bidimensionelle, Thèse de Doctorat, Université Paris 6, Paris.

McWilliams, J.C. (1984): The emergence of isolated coherent vortices in turbulent flows, *J. Fluid Mech.* **146**, 21–43.

McWilliams, J.C. (1990): The vortices of two-dimensional turbulence, *J. Fluid Mech.* **219**, 361–385.

Melander, M.V., McWilliams, J.C. and Zabusky, N.J (1988) Symmetric vortex merger in two dimensions: causes and conditions, *J. Fluid Mech.* **195**, 303–340.

Paireau, O., Tabeling, P. and Legras, B. (1997): A vortex subjected to a shear: an experimental study, *J. Fluid Mech.* **351**, 1–16.

Paret, J., Marteau, D., Paireau, O. and Tabeling, P. (1997): Are flows electromagnetically forced in thin stratified layers two-dimensional?, *Phys. Fluids* **9**, 3102–3104.

Paret, J. and Tabeling, P. (1997): Experimental observation of the two-dimensional inverse energy cascade, *Phys. Rev. Lett.* **79**, 4162–4165.

Pomeau, Y. (1996): Vortex dynamics in perfect fluids, *J. Plasma Physics* **56**, 407–418.

Richardson, L.F (1926): Atmospheric diffusion shown on a distance-neighbour graph, *Proc. R. Soc. London A* **110**, 709–737.

Smith, L.M. and Yakhot, V. (1994): Finite-size effects in forced two-dimensional turbulence, *J. Fluid Mech.* **274**, 115-138.

Solomon, T.H., Weeks, E.R. and Swinney, H.L. (1994): Chaotic advection in a two-dimensional flow: Lévy flights and anomalous diffusion, *Physica D* **76**, 70–84.

Tabeling, P., Burkhart, S., Cardoso, O. and Willaime, H. (1991): Experimental study of freely decaying two-dimensional turbulence, *Phys. Rev. Lett.* **67**, 3772–3775.

Tabeling, P. (1994): Experiments on 2D turbulence, in *Turbulence: a tentative dictionary, NATO ASI Series* **341**, 13–20, eds. P. Tabeling & O. Cardoso, Plenum Press, New-York.

Taylor, G.I (1921): Diffusion by continous movements, *Proc. London Math. Soc.* **20** 196–211.

Trizac, E. and Hansen, J.P. (1996): Dynamics and growth of particles undergoing balistic coalescence, *J. Stat. Phys.* **82**, 1345–1370.

Weiss, J.B. and McWilliams, J.C. (1992): Temporal scaling behaviour of decaying two-dimensional turbulence, *Phys. Fluids A* **5**, 608–621.

Anomalous diffusion in quasi-geostrophic flow

J. S. Urbach[2], E. R. Weeks[1], and H. L. Swinney[1]

[1] Center for Nonlinear Dynamics and Department of Physics, University of Texas at Austin, Austin, TX 78712, USA
[2] Department of Physics, Georgetown University, Washington, DC 20057, USA

Abstract. We review a series of experimental investigations of anomalous transport in quasi-geostrophic flow. Tracer particles are tracked for long periods of time in two-dimensional flows comprised of chains of vortices generated in a rapidly rotating annular tank. The tracer particles typically follow chaotic trajectories, alternately sticking in vortices and flying long distances in the jets surrounding the vortices. Probability distribution functions (PDFs) are measured for the sticking and flight times. The flight PDFs are found to be power laws for most time-dependent flows with coherent vortices. In many cases the PDFs have a divergent second moment, indicating the presence of Lévy flights. The variance of an ensemble of particles is found to vary in time as $\sigma^2 \sim t^\gamma$, with $\gamma > 1$ (superdiffusion). The dependence of the variance exponent γ on the flight and sticking PDFs is studied and found to be consistent with calculations based on a continuous time random walk model.

1 Introduction

An ensemble of particles in a non-uniform fluid flow will disperse as a consequence of the variations in the fluid velocity as well as the effects of molecular diffusion. In most situations, advection due to fluid motion is much faster than molecular diffusion, and dominates the transport process. Coherent large scale structures, such as vortices and jets, are frequently present in fluid flows and strongly influence particle motion. A quantitative understanding of the effect of coherent structures on transport and mixing in fluids is essential to accurately model such diverse processes as the dispersal of pollutants in the ocean and atmosphere, the persistence of the atmospheric ozone hole, and mixing and chemical reactions in stirred fluids.

Coherent structures typically result in inhomogeneous transport, with particles mixing well in some regions of the flow but isolated from others. The most prominent example of this phenomenon is the maintenance of the ozone hole by the circumpolar night jet. Similarly, Jupiter's Great Red Spot stays red despite the extremely turbulent environment because the existence of a stable vortex inhibits turbulent mixing.

The presence of coherent structures results correlations in particle motion that can persist for long distances and/or times. This may result in the inapplicability of the Central Limit Theorem used to derive the equation for the dispersion of particles in a normal diffusive process, $\sigma^2 = \langle x^2 \rangle - \langle x \rangle^2 \sim t$, often resulting instead in anomalous diffusion, $\sigma^2 \sim t^\gamma, \gamma \neq 1$ [1]. The presence of anomalous diffusion in the atmosphere was recognized in 1926 by Richardson [2], who investigated the separation of weather balloons and found in some circumstances $\sigma^2 \sim t^3$. (At very long times, transport in fluids of finite extent will necessarily be normally diffusive due to Brownian motion [3]. In many realistic flows, however, there are several orders of magnitude between the time scale for mixing due to advection and that due to Brownian motion.)

In this paper we review transport studies performed in a two dimensional (2D) flow. The study of 2D flows is of interest in part because its relative simplicity facilitates comparison between theory and experiment. In addition, most atmospheric and other geophysical flows are predominantly 2D as a result of the effect of planetary rotation, as are some flows of importance in plasma physics from the effects of applied magnetic fields. Finally, the equations of motion for tracer particles in a 2D flow are identical to Hamilton's equations of motion in phase space for dynamical systems [4], so 2D fluid flow provides a unique avenue for investigating Hamiltonian chaos.

The experiments described below were performed in a rotating annulus designed to match the important dimensionless parameters of large scale geophysical flows. Rapid rotation ensures a predominantly 2D flow, as predicted by the Taylor-Proudman theorem [5]. A schematic of the annulus is show in Fig. 1. The annulus is completely filled with fluid. The top and sides are transparent to allow for illumination and visualization. The flow is forced by pumping fluid into and out of the annulus through concentric rings of holes in the base of the annulus. The pumping generates a radial pressure gradient which, through the action of the coriolis force, generates an azimuthal jet, co-rotating when the source ring lies outside the sink ring, and counter-rotating for the opposite configuration. The bottom of the annulus has a slope of 0.1, which mimics the dynamical effect of planetary curvature on atmospheric flows (the beta-effect) [5]. A discussion the design considerations for the annulus can be found in [6].

2 Transport at high Reynolds number

Experiments investigating the dynamics of strongly nonlinear geostrophic flow are described in Refs. [6-8]. In addition to quantitative studies of the dynamical instabilities, qualitative studies of mixing and transport were performed by dye injection. For westward (counter-rotating) jets, large coherent vortices were found to persist in a turbulent background over a wide range of flow parameters. Dye injected into the vortices remained inside the vortex for long periods of time, while dye injected outside of the vortex mixed rapidly throughout the turbulent flow, but would not significantly penetrate the vortices, even after several minutes (the vortex turnover time was about 2 seconds).

For eastward (co-rotating) jets, a narrow wavy jet was found to exist up to the highest accessible forcing. Dye injected into the jet diffused quickly

173

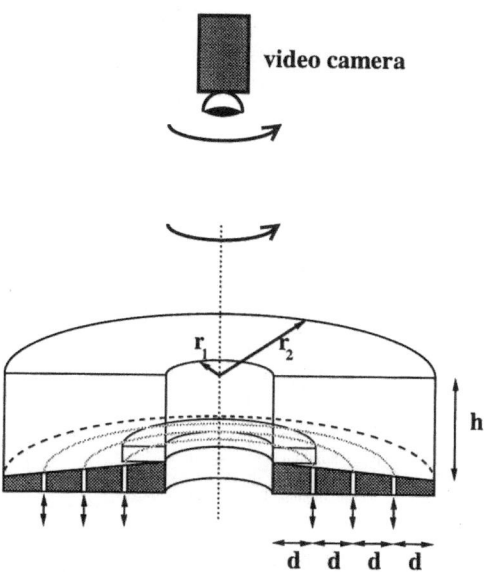

Fig. 1. Schematic of rotating annulus: $r_1 = 10.8$ cm, $r_2 = 43.2$ cm, $d = 8.1$ cm, and $h = 20.3$ cm at r_2. See text for details. The configuration of the annulus was slightly different for the experiments described in section 2 [7], [8], [6].

within the jet, and then slowly filled the region outside of the jet (and, to a lesser extent, inside) through a series of tongues generated from the crests of the traveling wave [6], [8]. Dye injected far from the jet spread uniformly in the region delimited by the jet, but virtually no cross-jet transport was observed, even after 500 rotations of the annulus. This effective dynamical barrier appears to work in much the same way the southern polar night jet acts as a barrier to transport of ozone from lower latitudes into the polar region.

3 Trajectories in vortex chains

For less energetic flows, we have performed detailed measurements of individual particle trajectories in single annular chains of vortices to investigate the role of chaotic advection in particle transport [9-12]. These experiments were performed at Reynolds numbers above the initial instabilities in the axisymmetric flow that exists at very low forcing, but below any indications of turbulent flow. At low pumping rates, the vortex chain rotates at a constant rate, producing a periodic signal on a hot film velocity probe mounted at a fixed position on the annulus (Fig. 2(a)).

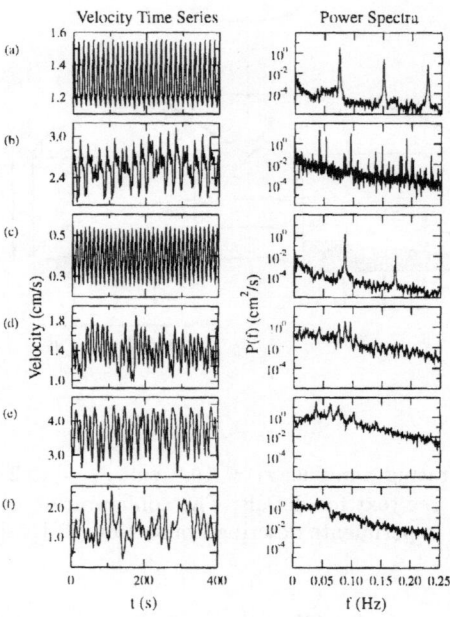

Fig. 2. Velocity time series and power spectral density $P(f)$ obtained from hot film probe measurements of the azimuthal velocity component at $r = 35.1$ cm: (a) *time-independent flow*; (b) *seven-vortex flow* with periodic time dependence in the reference frame co-rotating with the vortex chain (see Fig. 3); (c) *six-vortex flow* with periodic time dependence in the vortex chain reference frame; (d) *five-vortex flow* with chaotic time dependence; (e) *four-vortex flow* with chaotic time dependence (see Fig. 4); (f) *weakly turbulent flow* (see Fig. 5). These data are taken in the *tank* frame of reference, as opposed to the co-moving frame of reference used for the particle pictures in this paper.

Transport is measured by putting several hundred small ($\sim$ 1mm diameter), neutrally buoyant tracer particles into the tank. They are illuminated by light shining through the outer cylinder of the annulus and are viewed through a video camera rotating about the experimental set up. Automated tracking techniques [13] are used to find the trajectories of the individual particles.

3.1 Flows Studied

At the forcing rates used in this experiment, a counter-rotating jet is unstable to a chain of four, five, six, or seven vortices above the outer ring of holes [14]. The instability at the inner shear layer is inhibited by a 6.0 cm tall annular Plexiglas barrier with outer radius of 19.4 cm that is inserted above the inner ring of holes (see Fig. 1). (Without the barrier, the flow would be composed of two vortex chains, one above each forcing ring [14].) The vortex chain rotates relative to the tank at approximately half the speed of the azimuthal jet as seen in the annulus frame of reference (typically 4 cm/s). In a reference frame moving with the vortices, the vortex chain is sandwiched by azimuthal jets going in opposite directions (e.g. Fig. 3).

We study transport in flows generated with six different forcing techniques, using either water (kinematic viscosity $\nu = 0.009$ cm^2/s), or a water-glycerol mixture (38% glycerol by weight; a kinematic viscosity $\nu = 0.03$ cm^2/s). The time-dependence of some of the flows are similar, so in this paper we label some of the flows by their structure (number of vortices). The six flows, listed with the pumping rate, F, tank rotation rate, Ω, and working fluid, are:

1. *Time-independent flow* with six vortices ($F = 45$ cm^3/s, $\Omega = 1.5$ Hz, water-glycerol). The inner (outer) ring of holes acts as a source (sink) through with fluid is pumped into (from) the tank. In the reference frame co-rotating with the vortex chain, the flow is time-independent (Fig. 2(a)). This flow should not have chaotic mixing; tracers should follow the streamlines.

2. *Seven-vortex flow* with quasi-periodic time dependence (see Fig. 3; $F = 45$ cm^3/s, $\Omega = 1.5$ Hz, water-glycerol). The parameters for this flow are the same as the time-independent flow, but the initial conditions were different. In the reference frame co-rotating with the vortex chain, this flow is time-periodic; in the reference frame of the tank, the motion of the vortices around the annulus results in quasi-periodic time dependence (Fig. 2(b)). This flow is termed "modulated wave flow" in Ref. [11].

3. *Six-vortex flow* with quasi-periodic time dependence ($F = 45$ cm^3/s, $\Omega = 1.5$ Hz, water-glycerol). This flow is generated with the same techniques as the time-independent flow, except that the radial forcing has a non-axisymmetric perturbation. The forcing flow through one 60° sector of source and sink holes is restricted to less than half that for the rest of the forcing holes. Thus the vortex chain is perturbed as it moves past this constricted sector, with the period of the perturbation being the time for a vortex to precess around the annulus (70.0 s). In the reference frame of the vortex chain, the flow is time-periodic. In the reference frame of the tank, the flow is also time-periodic (Fig. 2(c)), as the perturbation is stationary with respect to the tank. In all other reference frames, the flow is quasi-periodic in time. This flow is termed "time-periodic flow" in Refs. [10], [11].

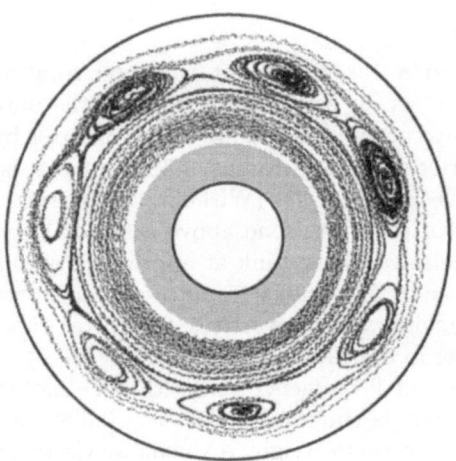

Fig. 3. The *seven-vortex flow* is revealed by the trajectories of 20 particles tracked for 300 s in a reference frame co-rotating with the vortices. In this reference frame, the vortex chain is sandwiched between two azimuthal jets. This flow has periodic time dependence in this reference frame. The inner and outer circles represent the annulus boundaries, and the grey circle indicates the location of the Plexiglas barrier. (Figure from Ref. [11].)

4. *Five-vortex flow* with chaotic time dependence ($F = 45$ cm^3/s, $\Omega = 1.5$ Hz, water-glycerol). This flow is similar to the six-vortex flow, except that the flux through the perturbing sector is completely shut off. There are still well-defined vortices in this flow, but the number of vortices alternates between five and six over long periods of time. This flow has chaotic time-dependence, as can be seen from the hot film probe measurements (Fig. 2(d)). The word chaotic in this case denotes Eulerian chaos, that is, a chaotic velocity field, as distinct from Lagrangian chaos of the particle trajectories. This flow is termed "chaotic flow" in Refs. [10], [11]. We do not actually know this flow is chaotic in the sense of positive lyapunov exponents, but the noise floor shown in Fig. 2(d) is higher than the previous flows, a signature of chaos.

5. *Four-vortex flow* with chaotic time dependence (see Fig. 4; $F = 52$ cm^3/s, $\Omega = 1.0$ Hz, water). Rather than the inner and outer forcing rings, this flow uses the inner and middle forcing rings ($r = 18.9$ cm and 27.0 cm), to allow the vortices to be larger, and prevent an inner jet from forming, as can be seen in Fig. 4.

At this high pumping rate, the motion of the vortices is chaotic, as shown in the velocity power spectrum shown in Fig. 2(e). As with the five-vortex flow, this is Eulerian chaos. This flow was termed "chaotic flow" in Ref. [12]. (Again, the chaos of this flow has not been rigorously

Fig. 4. The *four-vortex flow* is revealed by the trajectories of 12 particles tracked for 100 s in a reference frame co-rotating with the vortices. The inner and outer circles represent the annulus boundaries, and the grey circle indicates the location of the Plexiglas barrier. (Figure from Ref. [12].)

confirmed, but the power spectrum is reasonable evidence of the chaos of the flow, as are the qualitative observations of the vortex motion.)

6. *Weakly turbulent flow* (see Fig. 5; $F = 45$ cm^3/s, $\Omega = 1.5$ Hz, water). This flow was generated using a special forcing configuration. Only the outer ring holes were used ($r = 35.1$ cm). The ring is divided into 60° sectors, alternating between sources and sinks. The resulting flow consists of vortices of both signs, and there are no persistent jets or other structures. Note that the previous flows are all laminar; this is the only velocity field that is turbulent.

The velocity power spectrum consists of broadband noise and no dominant spectral components; see Fig. 2(f). This flow is termed "turbulent flow" in Ref. [10] and "weakly turbulent flow" in Ref. [11].

The flows are summarized in Table 1.

3.2 Analysis techniques

After a typical experimental run of 4 hours, we have tracked typically 5-10 trajectories with duration greater than 20 minutes, 30 with 10-20 minutes duration, and several hundred with 2-10 minutes duration. Statistics for the longer times are improved by repeating the experiments with the same control parameters (but see discussion in Sec. 5).

The transport is analyzed as a one-dimensional process in the azimuthal direction θ. The variance is calculated by the relations

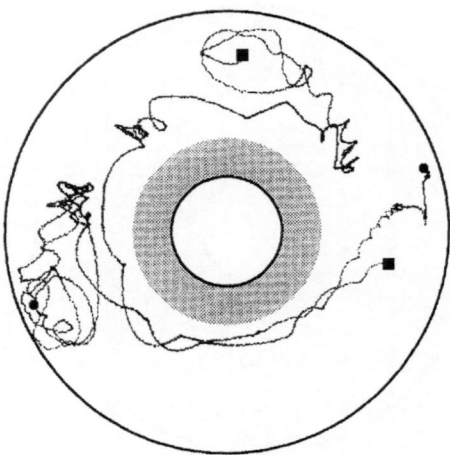

Fig. 5. Two trajectories show the lack of long-lived coherent structures in the *weakly turbulent flow*. The beginning and end of one trajectory is marked with circles, the other with squares; both particles start at the far right. The inner and outer circles represent the annulus boundaries, and the grey circle indicates the location of the Plexiglas barrier. The particles are shown in the reference frame of the annulus. (Figure based on Ref. [11].)

Table 1. Summary of the flows investigated, with kinematic viscosity ν, pump flux F, and dimensionless numbers Ro, Ek, and Re (calculated using $U = 3$ cm/s as the typical velocity for all flows). The rotation rate $\Omega = 1.5$ Hz for all flows (except the four-vortex flow as noted). Time dependence listed is in the reference frame co-rotating with the vortex chain.

Flow name	ν (cm^2/s)	F (cm^3/s)	$Ro \times 10^2$	$Ek \times 10^6$	Re
Time-independent (with six vortices)	0.03	45	4.0	4.0	400
Seven-vortex[1] (time-periodic)	0.03	45	4.0	4.0	400
Six-vortex[2] (time-periodic)	0.03	45	4.0	4.0	400
Five-vortex[3] (Eulerian chaos)	0.03	45	4.0	4.0	400
Four-vortex[4] (Eulerian chaos)	0.009	52	12	2.0	1000
Weakly turbulent[5]	0.009	45	16	1.2	1100

$$\sigma^2(t) = \langle \Delta\theta^2(t,\tau) \rangle - \langle \Delta\theta(t,\tau) \rangle^2 \quad , \tag{1}$$
$$\Delta\theta(t,\tau) = \theta(\tau + t) - \theta(\tau) \quad ,$$

where the ensemble average is over τ for individual trajectories and over the different trajectories in the run. This procedure treats each tracer as though starting from the same angle at the same time. This method is accurate for times greater than typical vortex turnover times (typically 10-20 s) but results in a variance that grows as t^2 for short times [15]. Only those trajectories that display both sticking and flight events are used in the calculation of the variance. The first and last events (sticking or flight) are removed to avoid any biasing. (That is, when a particle is first observed, it is in the middle of an event; we consider the trajectory only after this event has finished, so that all particles are considered at the beginning of a flight or sticking event, rather than in the middle of an event.) Different analysis techniques were examined to insure that the results are not strongly dependent on the biasing effects.

Sticking and flight time probability distribution functions (PDFs) are determined from local extrema of $\theta(t)$; see, e.g., Fig. 8. A flight is identified by an angular deviation $\Delta\theta > \Theta_{\mathrm{vortex}}$ (angular width of a single vortex) between successive extrema, and the sticking events are the intervals between flights. The PDFs are normalized histograms of the durations of these events. Histograms are generated with logarithmic binning, normalized, and plotted on log-log or log-linear scales.

The PDFs are adjusted to correct for biases toward shorter sticking/flight times, due to the finite duration of the measured trajectories. The adjustment is determined by generating long, artificial trajectories numerically with known, ideal power law sticking and flight time distributions. These long trajectories are then chopped randomly into smaller sections with a distribution of durations comparable to those in the experiment. PDFs determined from these chopped trajectories are also biased toward smaller times. The adjustment is determined by comparing the PDFs from the chopped trajectories to the ideal PDFs (both from numerical data); the exponents characterizing the PDFs for the chopped time series are about 0.3 larger than for the original long time series. Note that all reported exponents are the *corrected* values; the values measured directly from the PDFs are reported in footnotes for each PDF figure.

4 Results

4.1 Time-independent flow: no chaotic mixing

Ideally, particle trajectories in a time-independent flow fall on closed streamlines and there is no chaotic advection. While molecular diffusion of the tracer particles is completely negligible on the time scale of the experiments, slight imperfections due to noise, Ekman pumping, and finite-size particle effects

180

can have a noticeable effect on the trajectories. Such imperfections are inevitable in an experiment, even when Fourier spectra indicate that the velocity field is time-independent, as is the case for the flow in Fig. 6. The imperfections allow tracers to wander between neighboring streamlines, apparently filling the interior of a vortex; see Fig. 6(a). The imperfections occasionally lead to the escape of a tracer particle near a separatrix, but we find that in practice tracers remain trapped for long periods of time. Trapping times of 800 s (approximately 40 vortex turnover times) such as the one shown in Fig. 6(a) are common. Similarly, tracers that start in a jet remain in the jet for long times, e.g. Fig. 6(b).

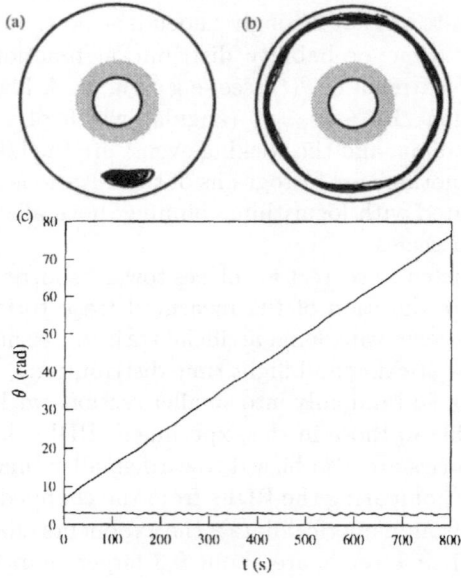

Fig. 6. (a) and (b) Tracer particle trajectories in the time-independent flow, viewed in a reference frame co-rotating with the vortex chain. (c) The azimuthal displacement as a function of time for the particles in (a) and (b); the starting angle $\theta(t = 0)$ is arbitrary. The inner and outer circles represent the annulus boundaries, and the grey circle denotes the Plexiglas barrier. (Figure based on Ref. [10].)

The azimuthal coordinate $\theta(t)$ for a particle in a vortex oscillates about a constant value, while for a particle in a jet with constant velocity, $\theta(t)$ grows linearly with time, as shown in Fig. 6(c). In the absence of noise, the variance of a distribution of particles grows as t^2 (ballistic separation) [16].

4.2   Time-periodic flows: power law flights

Chaotic advection is observed in the seven- and six-vortex flows, the two flows with periodic time dependence in the reference frame co-rotating with the vortex chain. Particles frequently make transitions to and from vortices, as seen in Fig. 7. Instead of being trapped indefinitely, particles have sticking events interspersed with flights in the jet regions.

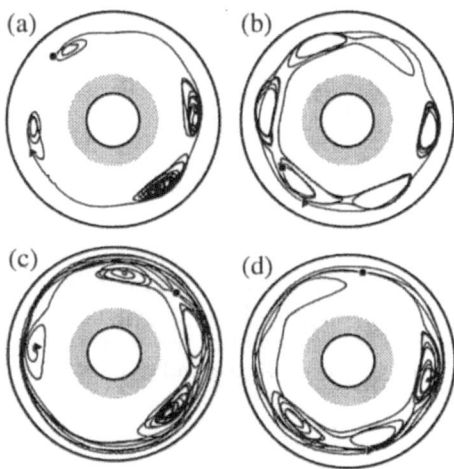

Fig. 7. Chaotic particle trajectories in the six-vortex flow (time-periodic in the reference frame of the vortex chain). Long sticking events can be seen in each case, and flights of length greater than one rotation about the annulus can be seen in (c), (d). Hyperbolic fixed points, near which the particle motion is particularly susceptible to transitions between flights and sticking events, are evident in all of the trajectories. The particle motion is viewed from a reference frame that is co-rotating with the vortex chain, and the beginning of each trajectory is marked by a triangle, the end by a circle. (Note that this is incorrectly labeled in Ref. [10], as can be seen by comparing Fig. 6(a) and Fig. 7(a) in that article. It is also incorrectly labeled in Ref. [9]; compare Fig. 1(b) and Fig. 2(b).) These trajectories are from the six-vortex flow, but are typical in appearance for the seven-vortex and five-vortex flows. (Figure based on Ref. [10].)

This intermittent sticking/flight behavior is apparent in plots of $\theta(t)$, as shown in Fig. 8. The observed sticking times and flight times range from ~10 s to ~600 s. The lower boundary of ~10 s is half a vortex turnover time. (The vortex turnover time, measured by doubling the average time between successive reversals for particles in a vortex, is ~23 s. The five-vortex and six-vortex flows have similar vortex turnover times, ~20 s.)

In Fig. 8 it can be seen that the slopes of the flight segments are approximately constant, indicating that the azimuthal velocity, $\omega = d\theta/dt$, remains

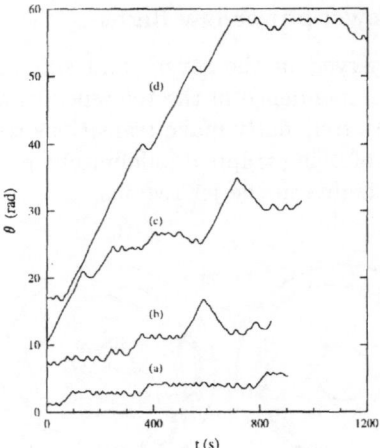

Fig. 8. Azimuthal displacement $\theta(t)$ as a function of time for the particle trajectories in Fig. 7. The oscillations of the tracer particle trajectories correspond to motion around a vortex, and the diagonal lines correspond to flights. The starting angle $\theta(t = 0)$ is arbitrary. These trajectories are from the six-vortex flow, but are typical in appearance for the seven-vortex and five-vortex flows. (Figure from Ref. [10].)

steady during the flights, except when the tracer passes near a hyperbolic point, where both ω and the radial component of velocity can decrease nearly to zero. Some asymmetry is observed in the flight speed and the relative probability of clockwise and counter-clockwise flights, but the PDF exponents appear to be the same. A numerical simulation designed to approximate these flows showed significant asymmetry [17].

To find the PDFs for the flight and sticking events, the trajectories of 1300 particles were analyzed for the seven-vortex flow, and 1700 particles for the six-vortex flow. The cleanest data (of all six flows) were obtained for the quasi-periodic seven-vortex flow, and the results are shown in Fig. 9. The flight PDF shows clear power law decay, $P_F(t) \sim t^{-\mu}$ with $\mu = 3.2 \pm 0.2$. The PDFs for flights in the $+\theta$ and $-\theta$ directions were compared and found to have similar decay exponents. The sticking PDF has a curvature indicating asymptotic behavior steeper than a power law (but does not appear exponential).

The fact that the velocity of the flights are approximately constant, and that the PDF show power law behavior suggests that the results of of the continuous time random walk model (CTRW) developed in Refs. [18], [19], [12] are applicable to this system. The results relevant for this work are summarized in figure 10, which shows the predicted variance exponent γ as a function of the exponents μ and ν of the flight and sticking PDFs, for both

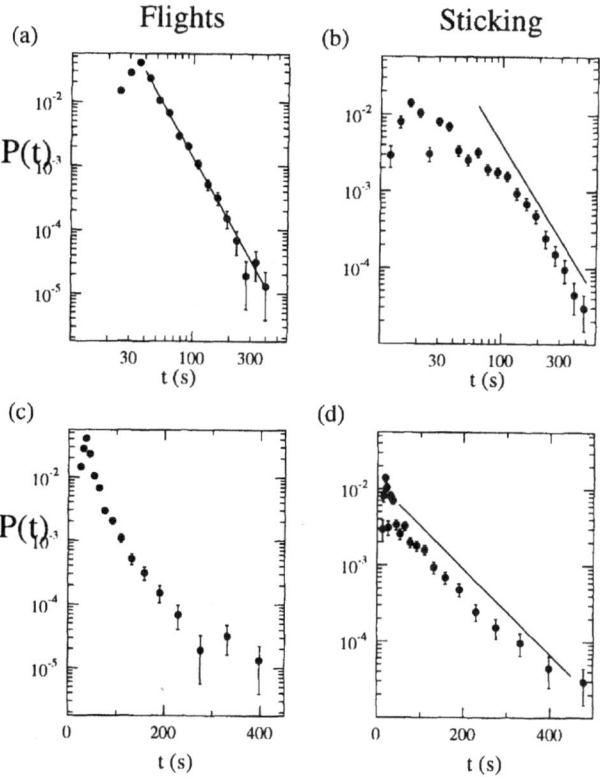

Fig. 9. Seven-vortex flow: (a,c) flight and (b,d) sticking probability distribution functions, shown on (a,b) log-log axes and (c,d) log-linear axes. The error bars show the statistical uncertainty ($\sqrt{N}$). The flight PDF shows power law decay, $P_F \sim t^{-\mu}$; the line drawn in (a) is a least squares fit to the decaying data yielding $\mu = 3.2 \pm 0.2$. The sticking PDF does not show a clear power law decay nor an exponential decay; the straight line (drawn for comparison) in (b) has a slope of -2.55, with the slope obtained from a least squares fit to the last 8 points in the tail. (The uncorrected value of μ is 3.4 ± 0.2; see Sec. 3.2 for details of the correction.)

symmetric and asymmetric random walks. For a detailed discussion of the model, see Ref. [12].

Since $\mu > 3$ for this flow, the Central Limit Theorem predicts normal diffusion ($\sigma^2(t) \sim t^\gamma$ with $\gamma = 1$). We compute the variance as discussed in Sec. 3.2, with the results shown in Fig. 11. The slope of the variance plot is shown in the inset, and suggests that the variance grows superdiffusively. For short times ($t < 10$ s), the variance grows ballistically, $\gamma = 2$. This is because of the vortex turnover time: for times less than $\sim$10 s, particles in flight are indistinguishable from those stuck in a vortex [15]. Particles all appear to be

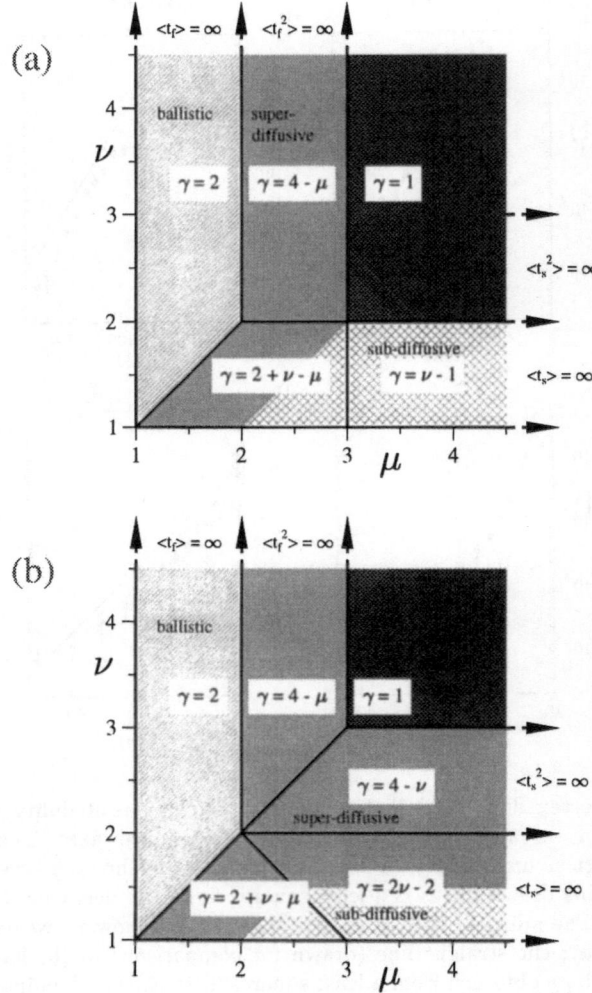

Fig. 10. Phase diagrams for variance of (a) symmetric and (b) asymmetric (or biased) random walks, from Ref. [12]. μ and ν are the exponents controlling the asymptotic power law decay of the flight and sticking PDFs, respectively: $P_F(t_f) \sim t_f^{-\mu}$ and $P_S(t_s) \sim t_s^{-\nu}$, as $t \to \infty$. For each region, bordered by the solid lines, the relationship between the variance exponent γ $[\sigma^2(t) \sim t^\gamma]$ and μ and ν is shown. The shadings indicate areas where the behavior is normally diffusive ($\gamma = 1$), subdiffusive ($\gamma < 1$), superdiffusive ($\gamma > 1$), and ballistic ($\gamma = 2$).

moving with a constant velocity (different for each particle), some in opposite directions, and thus the variance must grow ballistically.

For longer times, γ cannot be determined accurately, most likely due to a lack of trajectories with long durations. It is clear that for our data γ does not ever approach 1, the value expected for normal diffusion. The Berry-Esséen theorem predicts that the time for a (symmetric) random walk to reach normally diffusive behavior scales as [20], [21]

$$ t \sim (\frac{\langle |l|^3 \rangle}{\langle l^2 \rangle^{3/2}})^2 \, , \tag{2} $$

where the moments are for the flight length PDF. For the seven-vortex flow, however, the third moment is infinite, since $\mu < 4$. From Ref. [12], the first two exponents in an asymptotic expansion for the variance at long time, $\sigma^2(t) \sim Ct^\gamma + C't^{\gamma'}$ are $\gamma = 1$ and $\gamma' = 0$ for $\mu > 4$, so that only the leading term grows with time. For $3 < \mu < 4$, however, $\gamma = 1$ and $\gamma' = 4 - \mu$, thus the second order term also grows with time, and a very slow convergence is expected.

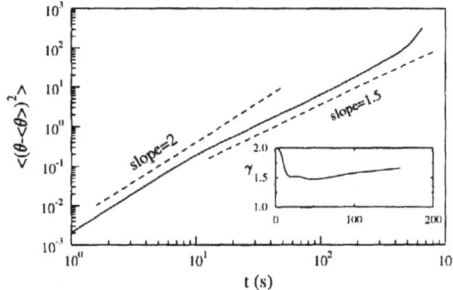

Fig. 11. Variance $\sigma^2(t)$ for the ensemble of tracer particles for the seven-vortex flow (solid line). The slope, shown in the inset, indicates that the variance grows superdiffusively. (Figure based on Ref. [11].)

The flight and sticking PDFs for the six-vortex flow are shown in Fig. 12. Again, the flight PDF shows clear power law decay, with a slope of $\mu = 2.5 \pm 0.2$. The PDFs for leftward and rightward flights separately had similar decay exponents (within their uncertainties). The sticking PDF clearly decays faster than a power law, although it is unclear if the decay is exponential. Note that this interpretation is different from Refs. [9], [10], where it is stated that the sticking-time PDF appears to show power law decay. (The PDFs in those articles were constructed with constant-width bins, so the deviation from power law behavior was less evident.)

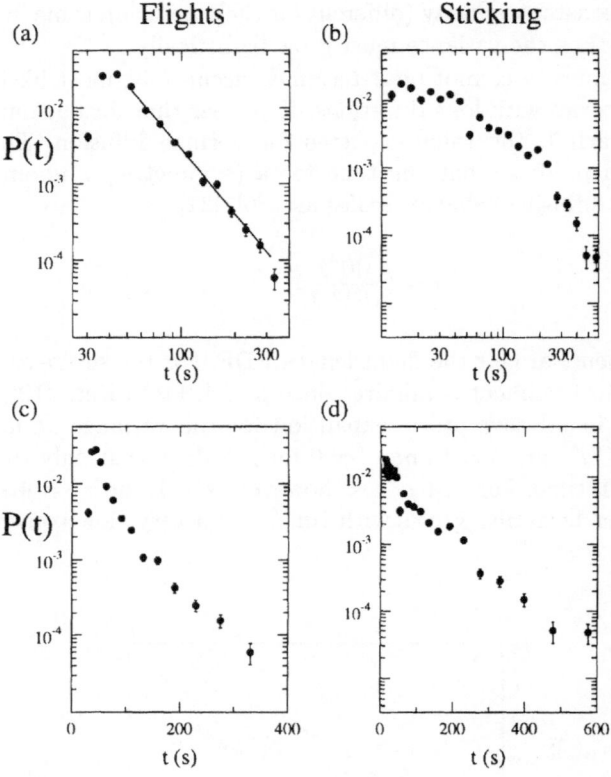

Fig. 12. Six-vortex flow: (a,c) flight and (b,d) sticking probability distribution functions, shown on (a,b) log-log axes and (c,d) log-linear axes. The error bars show the statistical uncertainty ($\sqrt{N}$). The flight PDF shows power law decay, $P_F \sim t^{-\mu}$; the line drawn in (a) is a least squares fit to the decaying data yielding $\mu = 2.5 \pm 0.2$. The sticking PDF does not show a clear power law decay nor an exponential decay. Note that these PDFs are slightly different from those shown in Refs. [9], [10] due to the improvement in binning technique (Sec. 3.2). (The uncorrected value of μ is 2.8 ± 0.2; see Sec. 3.2 for details of the correction.)

Again, the results can be compared with the analysis from the CTRW. The six-vortex flow particles are undergoing an asymmetric random walk with $\mu = 2.5$ and $\nu \to \infty$, suggesting that the variance should grow as $\sigma^2(t) \sim t^\gamma$ with $\gamma = 4 - \mu = 1.5$, that is, superdiffusively (see Fig. 10(b)). Figure 13 shows that for $t > 20$ s, the variance grows with $\gamma = 1.65 \pm 0.15$. Given the uncertainty of μ (± 0.2), the predicted and measured values for γ are in accord. As noted above for the seven-vortex flow, the variance grows ballistically for times shorter than a vortex turnover time.

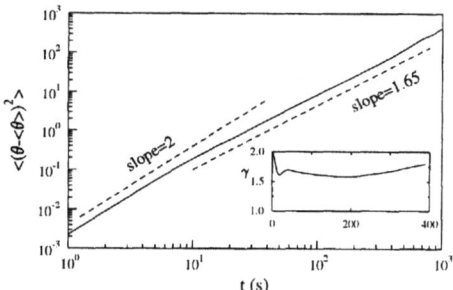

Fig. 13. Variance $\sigma^2(t)$ for the ensemble of tracer particles for the six-vortex flow (solid line). The slope, shown in the inset, indicates that the variance grows superdiffusively, with $\gamma = 1.65 \pm 0.15$. (Figure based on Ref. [9].)

While both the seven-vortex flow and the six-vortex flow have periodic time dependence, it is not surprising that the transport results are different. The seven-vortex flow has naturally arising time-dependence, while the six-vortex flow is perturbed periodically by an artificial change in the forcing (as described in Section refflowss). In the vortex reference frame, the instability of the seven-vortex flow has a frequency of 0.00033 Hz and a mode number of 3 (measured from particle tracking). The mechanical perturbation of the six-vortex flow appears with a frequency of 0.014 Hz (in the vortex reference frame) and is mode number 1.

4.3 Chaotic flows

The two chaotic flows, the five-vortex flow and the four-vortex flow, also exhibited chaotic mixing. Similar to the seven- and six-vortex flows, the difference between the two chaotic flows is the nature of the forcing: the chaotic time dependence of the five-vortex flow is due to the mechanical perturbation, while the chaotic time dependence of the four-vortex flow arises due to natural instabilities.

The trajectories for the five-vortex flow appear similar to those shown in Fig. 7, while typical trajectories of the four-vortex flow are shown in Fig. 14. The four vortices are not stationary but move erratically. (The pictures shown are taken in a frame of reference co-rotating with the *average* speed of the vortex chain, but there is substantial variation in the *instantaneous* speed of each vortex.)

Figure 15 shows the angular position of the particles as a function of time in the four-vortex flow. The oscillatory behaviors correspond to motion when the particle is "sticking" in a vortex, and the longer diagonal lines are flights in the outer jet. Flights are distinguished from sticking motions by examining the azimuthal distance traveled before reversing direction: particles travel in

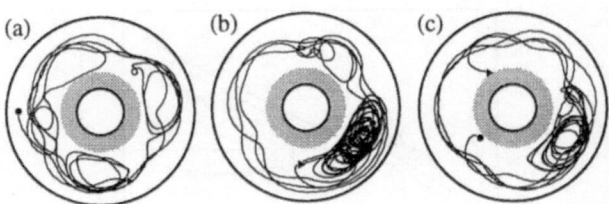

Fig. 14. Chaotic particle trajectories in the four-vortex flow (chaotic time dependence). Nearly all of the flight behavior is in the outer jet; a brief flight in the inside can be seen in (a). The chaotic motion of the four vortices can be seen in (b), where the particle spends most of its time in the same vortex which moves erratically. The beginning of each trajectory is marked by a circle, the end by a triangle. (Figure from Ref. [12].)

a vortex for at most $\pi/2$ radians before changing directions, while a particle that leaves one vortex and enters the next (the minimum flight distance) will move at least $\pi/2$ radians. Unlike the other flows, for the four-vortex flow there is no strong inner jet and particles do not travel long distances on the inner side of the vortex chain. Approximately 10% of the flights seen in the four-vortex flow are short hops on the inner side of the vortex chain, from

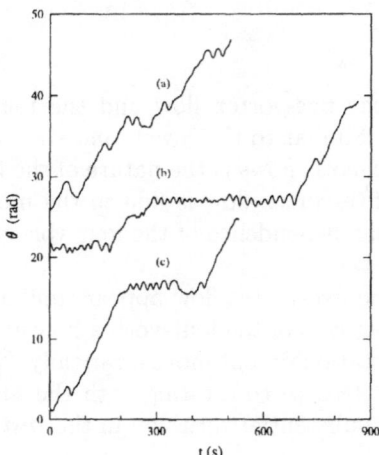

Fig. 15. Angular displacement $\theta(t)$ as a function of time for the trajectories shown in Fig. 14. Diagonal lines indicate flights, while the small oscillations correspond to particle motion within a vortex. Despite the chaotic motion of the vortices, a clear distinction can be made between flight behavior and sticking behavior. (Figure from Ref. [12].)

one vertex to an adjacent vortex; these hops take less than 40 s, and do not contribute to the long-time statistics.

To compile the flight and sticking PDFs, 1100 particles were examined for the five-vortex flow and 210 particles were examined for the four-vortex flow. (It was very difficult to track particles for long times for the four-vortex flow; particles disappeared from the visible area rapidly.)

The flight and sticking PDFs for the five-vortex flow are shown in Fig. 16.

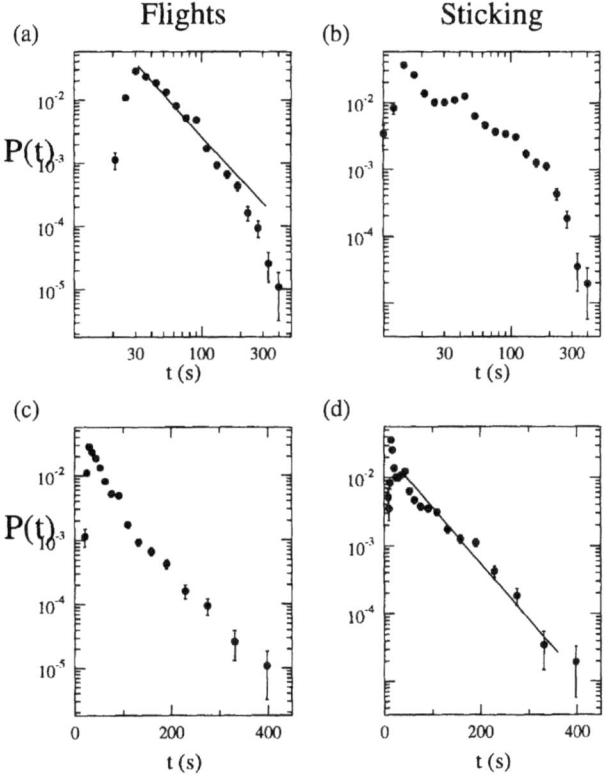

Fig. 16. (a,c) flight and (b,d) sticking probability distribution functions, shown on (a,b) log-log axes and (c,d) log-linear axes, for the five-vortex flow. The error bars show the statistical uncertainty ($\sqrt{N}$). The flight PDF appears to decay faster than a power law; the line drawn for comparison has a slope of -2.2, and is a least squares fit to the data for $t > 30$ s. The sticking PDF does not show a clear power law decay nor an exponential decay, although the data in (d) look roughly linear; a least squares fit line is shown. Note that these PDFs are slightly different from those shown in Ref. [10] due to the improvement in binning technique (Sec. 3.2). (Figure based on Ref. [10].)

Neither PDF shows power law decay, nor do they show convincing exponential decay. Note that this interpretation is different from that given in Ref. [10]. As discussed in Sec. 4.2, this is presumably due to an improvement in the analysis technique.

Given the uncertainty of the decay rate of the flight and sticking PDFs, comparison with the results of the CTRW model is difficult. The most reasonable interpretation of Fig. 16 would be $\mu \to \infty$, $\nu \to \infty$, yielding $\gamma = 1$ by the Central Limit Theorem. The growth of the variance measured from the experiment is shown in Fig. 17, and shows superdiffusive growth with $\gamma = 1.55 \pm 0.15$.

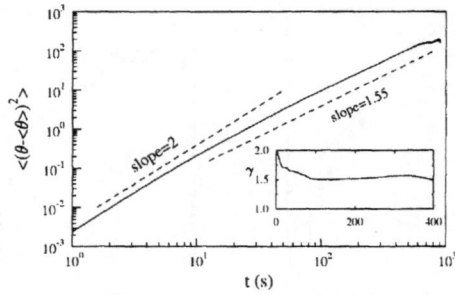

Fig. 17. Variance $\sigma^2(t)$ for the ensemble of tracer particles for the five-vortex flow (solid line). The slope, shown in the inset, indicates that the variance grows superdiffusively with $\gamma = 1.55 \pm 0.15$. (Figure based on Ref. [10].)

The PDFs for the four-vortex flow are shown in Fig. 18. This is the only flow for which both flight and sticking PDFs show power law decay. The decay exponents, adjusted for finite trajectory duration, are $\mu = 2.0 \pm 0.2$ (flight) and $\nu = 1.3 \pm 0.2$ (sticking). (This implies an infinite mean residence time for particles in a vortex, which would violate the incompressibility condition [22]. The PDF presumably falls off at longer times not accessible experimentally.) It is remarkable that these PDFs have a power law form despite the presence of Eulerian chaos. Although the vortices are moving erratically with respect to each other, particle motion still displays the effects of long-time correlations. The model of Refs. [22], [23] exhibits power law flight PDFs for chaotic vortex motion because the outer boundary of the annulus plays the role of an invariant surface.

The long term transport can be deduced from the CTRW for the four-vortex flow. Taking $\nu = 1.3$ and $\mu = 2.0$, the variance should grow as t^γ with $\gamma = 2 + \nu - \mu \approx 1.3$. The experimentally determined variance for this flow is shown in Fig. 19. It is difficult to track particles for long enough times in this flow to gather the statistics necessary to determine the variance accurately;

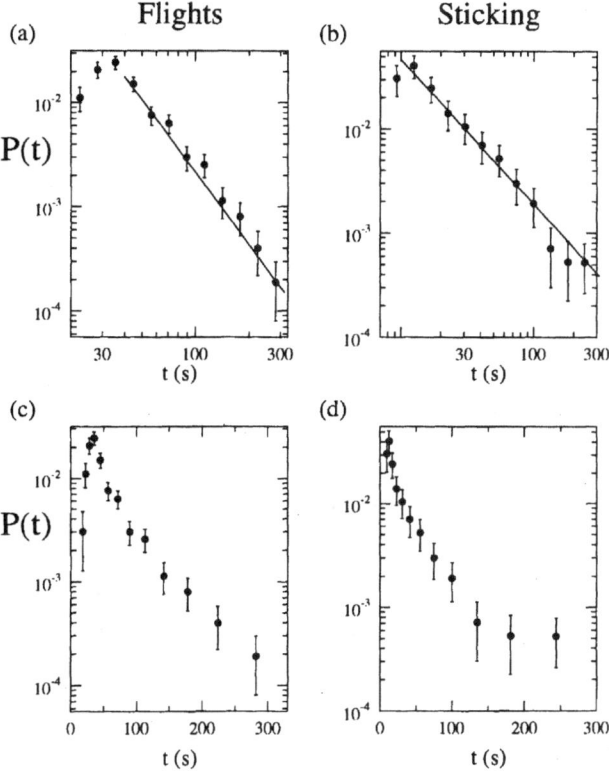

Fig. 18. Four-vortex flow: (a,c) flight and (b,d) sticking probability distribution functions, shown on (a,b) log-log axes and (c,d) log-linear axes. The error bars show the statistical uncertainty ($\sqrt{N}$). The flight PDF decays as a power law, $P_F(t) \sim t^{-\mu}$, with $\mu = 2.0 \pm 0.2$. The sticking PDF also appears to decay as a power law, with a decay exponent of $\nu = 1.3 \pm 0.2$. The error bars for these PDFs are much larger than for Figs. 9, 12, and 16 as this flow had much less data. (The uncorrected value of μ is 2.3 ± 0.2, ν is 1.4 ± 0.2; see Sec. 3.2 for details of the correction.)

hence quantitative comparison with the results of the CTRW is difficult. However, the behavior is appears superdiffusive with an exponent γ between 1.5 and 2.0. At longer times, the exponent drops below 1.5, and the prediction is in the limit $t \to \infty$, so the experimental results appear consistent.

The failure of the variance to reach its asymptotic behavior despite the large number of long time trajectories can be understood from an analysis of crossover times in the CTRW model. The time necessary to approach the asymptotic state can be calculated by retaining lower order terms in the expansion for σ (see Ref. [12] for details). Using the values of $\mu = 1.9, \nu = 1.3$,

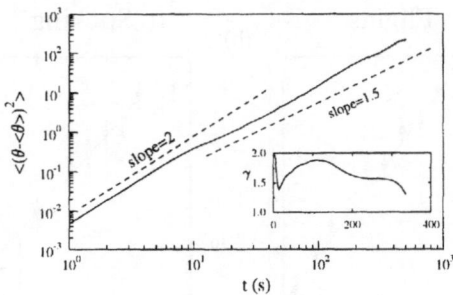

Fig. 19. Variance $\sigma^2(t)$ for the ensemble of tracer particles for the four-vortex flow (solid line). The slope, shown in the inset, indicates that the variance grows superdiffusively. (Figure based on Ref. [12].)

and cutoff times $t_F = 22$ s, $t_S = 10$ s, yields $\sigma \sim 0.055t^{1.4} - 0.10t^{1.1}$. A plot of this function on a log-log scale does not reach a slope of 1.5 until 400 s, and our data only extend to ~500 s. This slow convergence to asymptotic behavior is a generic feature of Lévy processes and complicates analysis in many experimental situations and numerical simulations (see discussion in Sec. 5).

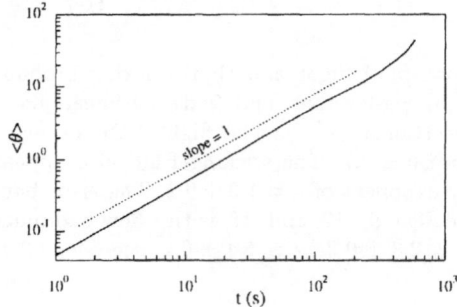

Fig. 20. Mean particle position for four-vortex flow, $\langle \theta(t) \rangle$ (solid line). (Figure from Ref. [12].)

Figure 20 shows that the mean particle position $\langle x \rangle$ for the four-vortex flow grows approximately linearly with time for most of the range. For longer times, $\langle x \rangle$ appears to start growing faster than linearly in time. For times less than a vortex turnover time, linear growth is expected, as all particles are moving with constant velocity (whether in a vortex or in the jet). For longer times, the model of Ref. [12] predicts (for $\mu = 2.0$ and $\nu = 1.3$) that

$\langle x \rangle \sim t^{0.3}$. It is probable that the asymptotic scaling is not reached due to lack of statistics at long times (see [12]).

4.4 Weakly turbulent flow: no long flights

The absence of long-lived vortices and azimuthal jets leads to a behavior in the turbulent regime that contrasts markedly with that in the laminar and chaotic regimes. Tracers in the turbulent flow wander erratically, and there are no well-defined flights (which are dependent on jet regions) or sticking events: compare plots of trajectories in the turbulent flow, Fig. 5, with those for the six- and four-vortex flows, Figs. 7 and 14, and compare plots of azimuthal displacement $\theta(t)$ in Fig. 21 with Figs. 8 and 15.

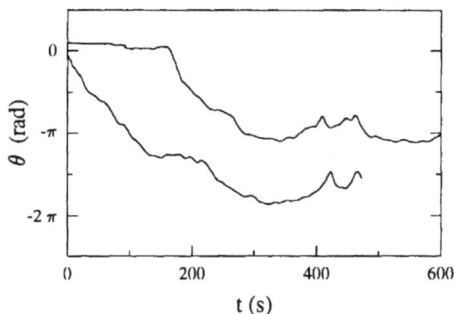

Fig. 21. Angular displacement $\theta(t)$ as a function of time for the trajectories shown in Fig. 5. The upper trace is for the particle marked with circles. (Figure from Ref. [11].)

While there are no flights or sticking events in the turbulent flow, the trajectories can be treated as random walks by defining a step as the time between two successive extrema in $\theta(t)$. We find that the probability distribution function is exponential, $P(t) = Ae^{-t/\tau}$, with $A = 0.158$ and $\tau = 15.2$ s (see Fig. 22), in contrast to the power law PDFs observed for flights in the time-periodic and chaotic regimes.

The slope γ of a log-log plot of the variance $\sigma^2(t)$ (Fig. 23) drops steadily from 2 and appears to approach the value expected for normal diffusion ($\gamma = 1$) at long times; however, we cannot follow particles for long enough times to determine the asymptotic behavior. This is in agreement with the Central Limit Theorem, which predicts $\gamma = 1$ for an exponentially decaying flight PDF. This also agrees with a result derived in 1921 by Taylor [15]. Taylor showed that for very short time scales, a turbulent flow should have ballistic mixing ($\sigma^2(t) \sim t^2$). This ballistic behavior lasts until particle motions be-

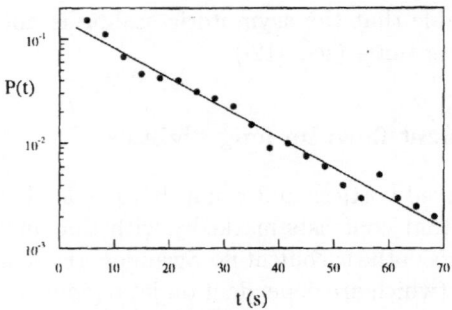

Fig. 22. Probability distribution for azimuthal displacement in the turbulent flow. The distribution is exponential with a decay time of 15.2 s. (Figure from Ref. [10].)

come uncorrelated; for our weakly turbulent flow, this time scale appears to be about 6 s.

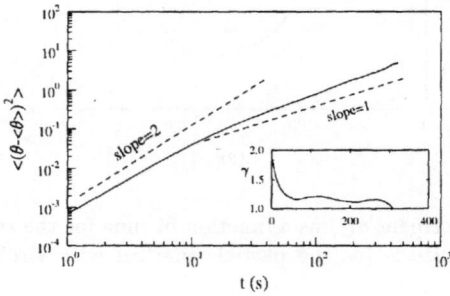

Fig. 23. Variance $\sigma^2(t)$ for the ensemble of tracer particles for the weakly turbulent flow (solid line). The slope, shown in the inset, suggests the long term behavior may be normally diffusive. (Figure based on Ref. [10].)

A diffusion coefficient can be found for the turbulent flow by fitting the variance data, yielding $D = 0.010 \pm 0.003 \ \mathrm{rad}^2/\mathrm{s}$. The data was fit for $t > 10$ s and $t > 100$ s, both giving a similar value. By using $r = 30$ cm as the approximate radial position of the particles, the diffusion coefficient can be written as $D_{\mathrm{eff}} = 9 \ \mathrm{cm}^2/\mathrm{s}$. For particles diffusing purely due to Brownian motion, the Einstein relation for the diffusion coefficient is $D_{\mathrm{Brownian}} = RT/6\pi\eta aN$ with R the universal gas constant, η the dynamic viscosity, a the particle radius, and N Avagadro's number [24]. For our tracer particles this formula yields $D = 4.4 \times 10^{-12}$, a factor of 10^{12} smaller that the measured diffusion coefficient.

5 Discussion

We have found superdiffusion in a variety of flows. The data from the six regimes are summarized in Table 2. Except for the five-vortex flow, all experiments with jets had power law flight behavior. The variance grows super-diffusively for all flows with nontrivial time dependence, except the weakly turbulent flow which appears to approach normal diffusion for very long times, as expected.

Table 2. Exponents ν and μ characterizing the power law decay of probability distribution functions for the sticking and flight times, respectively, and the exponent γ for the power law time dependence of the variance of the azimuthal displacement (measured and predicted). A — entry indicates the exponent is undefined.

Flow name	μ	ν	γ_{expt}	γ_{theory}
Time-independent (with six vortices)	—	—	2	2
Seven-vortex (time-periodic)	3.2 ± 0.2	?	~ 1.5	1
Six-vortex (time-periodic)	2.5 ± 0.2	∞	1.65 ± 0.15	1.5
Five-vortex (Eulerian chaos)	?	?	1.55 ± 0.15	1?
Four-vortex (Eulerian chaos)	2.0 ± 0.2	1.3 ± 0.2	~ 1.5	1.3
Weakly turbulent	∞	—	~ 1.2	1

Precise verification of the CTRW model is not possible given the experimental limitations. The predictions shown in Fig. 10 for the variance are only correct as $t \to \infty$; for finite t, the variance is composed of several terms. Competition between these terms controls the approach to the asymptotic behavior. These higher order terms can cause the variance to grow faster than its asymptotic growth (γ to appear larger at short times).

There are several reasons that particles are not observable for long periods of time. The most significant reason is probably Ekman pumping [5], a boundary layer effect that results in weak flows that are not perpendicular to the axis of rotation. Particles are illuminated only in a narrow horizontal region, and Ekman pumping provides a small vertical velocity which can move particles into and out of this illuminated slice. Additionally, particles that come too close to the edges of the annulus are lost, although they may be tracked as a new particle if the particle returns to the visible region. A final concern is the non-neutral-buoyancy of the particles; centrifugal effects could cause particles to drift out of the illuminated region. If there are any correlations linking the particle behavior to their longevity in the visible region, this could further affect results. For example, if particles stuck in vortices have a faster vertical drift (perhaps due to Ekman pumping which should be stronger in a vortex), then the observations of long-lived particles will be biased towards flights.

Despite these difficulties, the experiments show that power law scaling and Lévy flights are directly observable in fluid transport. In addition, the diffusive process is clearly anomalous for a broad range of times, and is well described the the continuous time random walk model.

Acknowledgments
We are indebted to Tom Solomon for his role in the initial experiments described in this paper, and also acknowledge helpful discussions with J. Klafter, M. F. Shlesinger, and G. M. Zaslavsky. This work was supported by the Office of Naval Research Grant No. N00014-89-J1495. ERW acknowledges the support of an ONR augmentation Award for Science and Engineering Training.

References

[1] For reviews, see J. Klafter, M. F. Shlesinger, G. Zumofen, Beyond Brownian motion, Physics Today **49**, 33 (Feb. 1996); M. F. Shlesinger, G. M. Zaslavsky, J. Klafter, Strange kinetics, Nature **363**, 31 (1993); E. W. Montroll and M. F. Shlesinger, in *Nonequilibrium Phenomena II: From Stochastics to Hydrodynamics*, Studies in Statistical Mechanics, Vol. II, eds. J. L. Lebowitz and E. W. Montroll (North-Holland, Amsterdam, 1984), 1.

[2] L. F. Richardson, Atmospheric diffusion shown on a distance-neighbour graph, Proc. Roy. Soc. (London) Ser. A **110** (1926) 709.

[3] T. H. Solomon and J. P. Gollub, Passive transport in steady Rayleigh-Bénard convection, Phys. Fluids **31**, 1372 (1988).

[4] H. Aref, Stirring by chaotic advection, J. Fluid Mech. **143**, 1 (1984).

[5] J. Pedlosky, *Geophysical Fluid Dynamics*, 2nd ed. (Springer-Verlag, New York, 1987).

[6] J. Sommeria, S. D. Meyers and H. L. Swinney, in *Nonlinear Topics in Ocean Physics*, ed. A. Osborne (North-Holland, Amsterdam, 1991), p. 227.

[7] J. Sommeria, S. D. Meyers, and H. L. Swinney, Laboratory simulation of Jupiter's Great Red Spot, Nature **331** (1988) 689.

[8] J. Sommeria, S. D. Meyers, and H. L. Swinney, Laboratory model of a planetary eastward jet, Nature **337** (1989) 58.

[9] T. H. Solomon, E. R. Weeks, H. L. Swinney, Observation of anomalous diffusion and Lévy flights in a two-dimensional rotating flow, Phys. Rev. Lett. **71**, 3975 (1993).

[10] T. H. Solomon, E. R. Weeks, H. L. Swinney, Chaotic advection in a two-dimensional flow: Lévy flights and anomalous diffusion, Physica D **76**, 70 (1994).

197

[11] E. R. Weeks, T. H. Solomon, J. S. Urbach, H. L. Swinney, Observation of Anomalous Diffusion and Lévy Flights, in: *Lévy Flights and Related Topics in Physics*, eds. M. F. Shlesinger, G. M. Zaslavsky and U. Frisch (Springer-Verlag, Heidelberg, 1995) pp. 51.

[12] . R. Weeks, J. S. Urbach, H. L. Swinney, Anomalous diffusion in asymmetric random walks with a quasi-geostrophic flow example, Physica D **97**, 219 (1996).

[13] M. S. Pervez and T. H. Solomon, Long-term tracking of neutrally buoyant tracer particles in two-dimensional fluid flows, Exp. Fluids **17**, 135 (1994).

[14] T. H. Solomon, W. J. Holloway, H. L. Swinney, Shear flow instabilities and Rossby waves in barotropic flow in a rotating annulus, Phys. Fluids A **5**, 1971 (1993).

[15] G. I. Taylor, Diffusion by continuous movements, Proc. Lon. Math. Soc. 2 **20**, 196 (1921).

[16] I. Mezic and S. Wiggins, On the dynamical origin of asymptotic t^2 dispersion of a nondiffusive tracer in incompressible laminar flows, Phys. Fluids **6**, 2227 (1994).

[17] D. del-Castillo-Negrete, Asymmetric transport and non-Gaussian statistics of passive scalars in vortices in shear, submitted to Phys. Fluids (1997).

[18] M. F. Shlesinger, Asymptotic solutions of continuous-time random walks, J. Stat. Phys. **10**, 421 (1974).

[19] J. Klafter and G. Zumofen, Lévy Statistics in a Hamiltonian System, Phys. Rev. E **49**, 4873 (1994).

[20] W. Feller, *An Introduction to Probability Theory and Its Applications*, (John Wiley & Sons Inc., New York, 1966) Vol. 2, Chap XVI.8, p. 525.

[21] M. F. Shlesinger, Comment on "Stochastic process with ultraslow convergence to a Gaussian: The truncated Lévy flight", Phys. Rev. Lett. **74**, 4959 (1995).

[22] S. Venkataramani, T. M. Antonsen, E. Ott, Anomalous diffusion in bounded temporally irregular flows, Physica D, to appear.

[23] S. Venkataramani, T. M. Antonsen, E. Ott, Lévy flights in fluid flows with no Kolmogorov-Arnold-Moser Surfaces, Phys. Rev. Lett. **78**, 3864 (1997);

[24] J. P. Bouchaud and A. Georges, Anomalous diffusion in disordered media: statistical mechanisms, models and physical applications, Phys. Rep. **195**, 127 (1990).

Chaotic Dynamics of Passive Particles in Three-Vortex System: Dynamical Analysis

Leonid Kuznetsov[2] and George M. Zaslavsky[1,2]

[1] Courant Institute of Mathematical Sciences, New York University, 251 Mercer St., New York, NY 10012, USA

[2] Department of Physics, New York University, 2-4 Washington Place, New York, NY 10003, USA

Abstract. Analytical study of a passive particle advection in three point-vortex system is described in details. We specify two extreme cases of strong and weak chaos depending on the geometry of 3-vortex system. Mappings are derived for both cases and domains of chaotic dynamics are calculated analytically. We discuss the origin of the coherent vortex cores — holes in the stochastic sea filled with KAM orbits, and obtain the expression for the core radius in case of strong chaos. A weak dependence of core radius on geometrical parameters is discovered. For the case of weak chaos the separatrix map is constructed and the stochastic layer width is estimated. Numerical simulations have been performed and it was found that there exists a fine structure of the coherent core boundary layer, which consists of islands and subislands. We also have found the stickiness of the advected particle to the boundaries of vortex cores.

1 Introduction

Motion of small particles in flows, known as advection, is a subject of many recent publications related to such areas as fluid stirring, chemical reactions, pollution of the atmosphere and the ocean, visualization of flows, etc. Speaking about a small particle, we have in mind the particle that does not influence the flow. Other terms are sometimes used, such as tracers, passive particles, or Lagrangian particles. For the cases when the flow is stationary, incompressible, and inviscid, an advected particle trajectory coincides with a streamline of the flow, and the information obtained from the advection, describes the structure of the flow. Numerous articles and reviews are devoted to different aspects of the advection. The general form of the advection equation is

$$\dot{\mathbf{r}} = \mathbf{v}(\mathbf{r}, t) \tag{1.1}$$

where $\mathbf{r}$ is a coordinate vector of the particle, and $\mathbf{v}$ is a given velocity field which defines the flow. Formally (1.1) describes a dynamical system which, generally speaking, possesses chaotic solutions for some parameter values and initial conditions. In this case the advection is known as chaotic. The notion of Lagrangian turbulence is also used for the chaotic solutions of (1.1). Chaotic advection has been observed in numerous experiments and computer simulations, and a number of different theoretical descriptions have been published (see for example works [1]-[10] which represent only a small part of the actual list of important publications).

For a divergence-free velocity field (div $\mathbf{v}=0$) the system (1.1) can be written in the Hamiltonian form. This form is known explicitly for the three-dimensional stationary case $\mathbf{v}=\mathbf{v}(x,y,z)$, and for two-dimensional generic case $\mathbf{v}=\mathbf{v}(x,y,t)$, when (1.1) can be rewritten as

$$\dot{x} = v_x(x,y,t) = \partial\Psi/\partial y$$
$$\dot{y} = v_y(x,y,t) = -\partial\Psi/\partial x \tag{1.2}$$

using the stream function $\Psi = \Psi(x,y,t)$ that plays a role of the Hamiltonian. Both of the above mentioned cases belong to the so-called Hamiltonian systems with 1-1/2 degrees of freedom. Such a Hamiltonian system possesses chaotic advection for some domains of the phase space, which is simply (x,y)-plane for the case (1.2).

The phase space of the Hamiltonian system with 1-1/2 degrees of freedom can be roughly described as domains of chaotic motion (stochastic sea) mingled with islands, inside which stable quasiperiodic motion dominates. The transport is performed along the area of chaotic motion and a structure (topology) of this area is crucial for all properties of transport.

Application of the methods of the dynamical systems theory to the chaotic advection problem has become an established technique. By now chaotization of the tracer motion was observed and studied in a large number of systems, including point vortex flows [11],[12],[13],[14], Rayleigh-Bénard convection [15],[16], blinking vortices [1], oscillating vortex pair [5],[6] and vortex pair perturbed by another point vortices [17], point vortex flows is closed domains [18],[19], etc. We want to emphasize, that correspondence of a trajectory of advected particle in physical (coordinate) space to a phase-space trajectory of a Hamiltonian system implies more than just a fact of appearance of Lagrangian chaos in a generic unsteady velocity field. One should also expect the configuration space of a tracer to contain all the typical structures of Hamiltonian phase space such as islands of regular motion around elliptic points, cantori, island-around-island chains, thin stochastic layers inside the islands, etc.

In fact, such structures were observed in a number of works devoted to the chaotic advection in the time-periodic flows, where the possibility of constructing a Poincare map for the tracer motion provides us with a convenient visualization technique [5],[18],[19],[20],[21]. The importance of this observation is due to the crucial role of some of these structures for the transport process. For Hamiltonian systems with 1 1/2 degrees of freedom one can establish connections between the topology of the phase space and properties of transport [22], which makes the detailed study of the phase space topology more significant.

Point vortex flows, a singular solutions of the 2D Euler equation, are one of the most natural and most frequently addressed examples of velocity fields, generating the Lagrangian chaos of the tracers. Being a traditional subject matter of fluid mechanics for over a hundred years, (see [23],[24],[25] for a review) point vortex systems on one hand share certain features with more involved case of coherent vortex structures in 2D turbulence [26],[27],[28],[29], and on the other hand can be regarded as Hamiltonian systems with few degrees of freedom, so that not only the advection, but the dynamics of the flow itself (in Lagrangian framework), can be attacked by the machinery of the Hamiltonian dynamics.

In the present paper we apply the traditional tools of chaotic dynamics to the problem of the coherent vortex cores – the areas of regular advection around vortices. Appearance of this non-mixing patches is by far not limited to point vortex flows, in fact, they constitute the robust feature of practically any flow with the regions of concentrated vorticity [26],[27],[28],[29],[30],[31],[32],[33]. The comparative numerical study of the advection in point vortex systems and in 2D turbulence carried out in [26] revealed certain common features of advection patterns in these two cases. Point vortex models have enormous advantages for the analytical treatment of the tracer motion, and in particular, for the study of the coherent cores. These advantages stem from the fact that the vortices themselves are advected by the flow and thus, as we already mentioned, their motion can be described via Hamiltonian equations. The appropriate dynamical variable for the problem of the tracer moving coherently with the vortex is the displacement of the tracer from the vortex, in other words, the coordinates of the tracer in the moving vortex reference frame. It is clear, that the equations governing the dynamics of the displacement will be Hamiltonian, since those for both vortex and tracer are.

Our goal is to understand in full detail the advective dynamics of a passive particle in a flow provided by the simplest three-vortex system with regular dynamics. Possibility to study completely the problem can explain the origin of the near-vortex isolated non-penetrable core. Its specific structure possesses a level of universality which can be efficiently used for a tracer dynamics in the multi-vortex system. Existence of the cores as coherent structures can be utilized for speculations on 2D-vortex turbulence, since each vortex can be considered as an advected particle in many-vortex system.

In Section 2 we consider general unbounded planar inviscid incompressible 2D flow with a point vortex in it, and in particular, advection of a tracer particle in the neighborhood of the vortex. We write down the advection equations in the reference frame moving with the vortex, and demonstrate its Hamiltonian structure. When the tracer is close to the vortex, the Hamiltonian system for the dynamics of the displacement is close to an integrable one. We specify the small parameter of the problem and rewrite the stream function in the neighborhood of the vortex in a form, paradigmatic for the theory of near-integrable systems:

$$\Psi(x, y, t) = H_0(x, y) + \epsilon H_1(x, y, t) \tag{1.3}$$

This section serves to prepare an equation to be studied further in a form suitable to apply methods of nonlinear dynamics.

To start a more detailed study of the tracer dynamics, one has to be more specific about the flow. In Section 3 we revisit one of the oldest examples of 2D chaotic advection – the motion of a passive tracer particle in the field of three point vortices, sometimes called the restricted four vortex problem due to its analogy to the restricted three-body problem in celestial mechanics [13]. It is worthwhile to mention here, that non-integrability of the general four-vortex system was established in [34] and further discussed in [35],[36]. The non-integrability of the equations for tracer trajectories in the field of three vortices, and the existence of the chaotic motion in this system was demonstrated numerically in [11], and analytically, using Melnikov method [37], in [12], actually, before the term chaotic advection established itself. In fact, the application of the Melnikov method in [12] was not complete, since the method also allows to obtain the boundary of the region of stochastic motion in case of weak chaos. This work is performed in Section 5, where the separatrix map [38],[10] is constructed for the tracer motion.

Numerical studies in [13],[26] have demonstrated the existence of the coherent cores around the vortices with a sharp border between regular and chaotic regions. In [13], different regimes of advection were simulated, and Poincare sections of the tracer trajectories were constructed to visualize the geometry of the mixing region.

In this paper, we present a more detailed description of the structures in the advection pattern, paying special attention to the boundaries between the areas of chaotic and regular motion. Apart from typical sticky island-around-island structures on the border of islands around elliptic points, we have found that for certain geometries of the vortex motion, the boundaries of the regular cores surrounding vortices contain a complex mingle of small, extremely elongated islands, that create strong stickiness.

The integrability of the underlying vortex motion makes it possible to carry out a detailed analytical study of the advection. In Section 4 we consider the case of strong chaotization of the tracer motion. Using the equations in the moving vortex frame, we construct a mapping for the trajectories of the tracer inside the coherent core. Applying the stochasticity criterion to this map, (see [10],[39] on chaos in area-preserving maps), we obtain the analytic expression for the radius of the coherent core, which is in a good agreement with numerical results.

When the vortex motion is close to steady rotating equilateral triangle configuration, motion of the tracer is chaotic only inside the thin layer around the destructed separatrices of the integrable "equilateral" case. We calculate the width of those layers in Section 5 by constructing corresponding separatrix map and evaluating Melnikov integral.

2 Coherent vortex cores in 2D flow

In this section we will specify the dynamical structure of equations for advection due to a point vortex in a non-stationary flow. Consider a motion of a point vortex in two-dimensional incompressible flow described by the velocity field $\mathbf{v}(x, y, t) \equiv (v_x(x, y, t), v_y(x, y, t))$. Due to the incompressibility condition div $\mathbf{v} = 0$ the velocity field of such flow can always be expressed in terms of a scalar stream function $\Psi(x, y, t)$ as

$$v_x = \frac{\partial \Psi}{\partial y}; \qquad v_y = -\frac{\partial \Psi}{\partial x}. \tag{2.1}$$

We also introduce the scalar vorticity $\omega(x, y, t)$ as

$$\omega \equiv \frac{\partial v_y}{\partial x} - \frac{\partial v_x}{\partial y} = -\frac{\partial^2 \Psi}{\partial x^2} - \frac{\partial^2 \Psi}{\partial y^2} \tag{2.2}$$

A singular localized distribution of vorticity given by:

$$\omega_{vort} = k\delta(x - x_1)\delta(y - y_1) \tag{2.3}$$

is referred to as a point vortex of strength k located at the point $x_1(t), y_1(t)$. By a point vortex in a two-dimensional flow we will imply the following kind of vorticity distribution [40]:

$$\omega = k\delta(x - x_1(t))\delta(y - y_1(t)) + \omega_R(x, y, t) \tag{2.4}$$

where $\omega_R(x, y, t)$ is regular (nonsingular) at $x = x_1(t)$, $y = y_1(t)$. In general, ω_R can possess singularities at some other points, e.g. other point vortices, etc.

Stream function and velocity field in this case will also have singularities at $x = x_1(t)$, $y = y_1(t)$. From (2.2), it follows that the stream function of an unbounded planar flow (at rest at infinity) with a point vortex at $x_1(t), y_1(t)$ has the form:

$$\Psi(x, y, t) = -\frac{k}{4\pi} \ln[(x - x_1(t))^2 + (y - y_1(t))^2] + \psi_R(x, y, t) \tag{2.5}$$

where $\psi_R(x, y, t)$ is regular at $x = x_1(t)$, $y = y_1(t)$ and is related to the regular part of the vorticity by[1]

$$\omega_R = -\frac{\partial^2 \psi_R}{\partial x^2} - \frac{\partial^2 \psi_R}{\partial y^2} \tag{2.6}$$

The components of the velocity field in this case are:

$$v_x = \frac{\partial \Psi}{\partial y} = -\frac{k}{2\pi} \frac{y - y_1}{(x - x_1)^2 + (y - y_1)^2} + \frac{\partial \psi_R}{\partial y}$$
$$v_y = -\frac{\partial \Psi}{\partial x} = \frac{k}{2\pi} \frac{x - x_1}{(x - x_1)^2 + (y - y_1)^2} - \frac{\partial \psi_R}{\partial x} \tag{2.7}$$

If there is no external forces, acting on the fluid, the dynamics of the flow can be obtained from Helmholtz equation:

$$\frac{\partial \omega}{\partial t} + (\mathbf{v}\nabla)\omega = 0 \tag{2.8}$$

---

[1]inverse relation is given by Biot-Savart formula, [40]

which states, that vorticity is freely transported by the flow. The above equation can also be rewritten in terms of stream function Ψ only:

$$\frac{\partial \nabla^2 \Psi}{\partial t} + \frac{\partial \Psi}{\partial y}\frac{\partial \nabla^2 \Psi}{\partial x} - \frac{\partial \Psi}{\partial x}\frac{\partial \nabla^2 \Psi}{\partial y} = 0 \tag{2.9}$$

Equations of motion of the point vortex follow from (2.8) by substitution of the expressions (2.4) and (2.7) for ω and $\mathbf{v}$ (see appendix A for derivation):

$$\dot{x}_1 = \frac{\partial \psi_R}{\partial y}(x_1, y_1); \qquad \dot{y}_1 = -\frac{\partial \psi_R}{\partial x}(x_1, y_1) \tag{2.10}$$

they express the fact, that vortex moves along streamlines of the external flow $\psi_R(x, y, t)$, i.e. it has no direct self-action.

Now we address our attention to the advection of a free passive particle in a flow with the point vortex in it, (2.7). We assume that the solution of (2.8) is known, i.e. $\psi_R(x, y, t)$ and $x_1(t)$, $y_1(t)$ are given functions of time. In this case, a trajectory of a passive fluid particle $\{x(t), y(t)\}$ can be found from

$$\dot{x} = v_x = -\frac{k}{2\pi}\frac{y - y_1}{(x - x_1)^2 + (y - y_1)^2} + \frac{\partial \psi_R}{\partial y}(x, y, t)$$

$$\dot{y} = v_y = \frac{k}{2\pi}\frac{x - x_1}{(x - x_1)^2 + (y - y_1)^2} - \frac{\partial \psi_R}{\partial x}(x, y, t) \tag{2.11}$$

which can be regarded as a Hamiltonian system with Hamiltonian function given by (2.5).

One of our prime goals in this paper is to determine whether a passive particle initially in the neighborhood of the vortex will stay in this neighborhood, or will eventually travel away from it. For this reason we go to the reference frame of moving vortex and define the displacement $\vec{\xi} \equiv (\xi_x, \xi_y)$ of the particle from the vortex:

$$\xi_x \equiv x - x_1(t); \qquad \xi_y \equiv y - y_1(t) \tag{2.12}$$

as the coordinates of the particle in the moving vortex frame. We can easily obtain the equations governing dynamics of the displacement $\vec{\xi}$ by subtracting (2.10) from (2.11):

$$\dot{\xi}_x = -\frac{k}{2\pi}\frac{\xi_y}{\xi_x^2 + \xi_y^2} + \frac{\partial \psi_R}{\partial y}(x_1 + \xi_x, y_1 + \xi_y, t) - \frac{\partial \psi_R}{\partial y}(x_1, y_1, t)$$

$$\dot{\xi}_y = \frac{k}{2\pi}\frac{\xi_x}{\xi_x^2 + \xi_y^2} - \frac{\partial \psi_R}{\partial x}(x_1 + \xi_x, y_1 + \xi_y, t) + \frac{\partial \psi_R}{\partial x}(x_1, y_1, t) \tag{2.13}$$

Another way of arriving at this equations is to make a canonical transformation to the new variables. Generating function of this transformation is:

$$F_2(x, \xi_y) = (x - x_1(t))\xi_y + xy_1(t) \tag{2.14}$$

In this way we see that the Hamiltonian structure of the original system (2.11) is preserved in (2.13), and the new Hamiltonian is

$$\bar{\Psi}(\xi_x, \xi_y, t) = \Psi + \frac{\partial F_2}{\partial t} = \Psi + \xi_x \dot{y}_1(t) - \xi_y \dot{x}_1(t) - x_1(t)\dot{y}_1(t) =$$

$$= -\frac{k}{4\pi} \ln(\xi_x^2 + \xi_y^2) + \psi_R(x_1 + \xi_x, y_1 + \xi_y, t) - \xi_x \frac{\partial \psi_R}{\partial x} - \xi_y \frac{\partial \psi_R}{\partial y} \qquad (2.15)$$

where we have dropped the irrelevant term depending only on time, and expressed the velocity of the vortex $\dot{x}_1(t), \dot{y}_1(t)$ through the regular part of stream function ψ_R using (2.10).

Thus we have established that the displacement of the advected passive particle from the point vortex (which also moves with the flow according to (2.10)), is described by a Hamiltonian system, with time-dependent Hamiltonian function given by (2.15). From the form of (2.15) we can make several conclusions. When the passive particle is close to the vortex, i.e. the distance

$$\xi \equiv (\xi_x^2 + \xi_y^2)^{1/2}$$

between the particle and the vortex is small enough, the first term in (2.15), which has no explicit time dependence,

$$H_0(\xi_x, \xi_y) \equiv -\frac{k}{4\pi} \ln(\xi_x^2 + \xi_y^2) = -\frac{k}{2\pi} \ln \xi \qquad (2.16)$$

is much larger than other terms,

$$H_1(\xi_x, \xi_y, t) \equiv \psi_R(x_1 + \xi_x, y_1 + \xi_y, t) - \xi_x \frac{\partial \psi_R}{\partial x} - \xi_y \frac{\partial \psi_R}{\partial y} \qquad (2.17)$$

which are of order $O(\xi^2)$, and therefore system (2.13) is close in this region of phase space to an integrable system defined by Hamiltonian function $H_0(\xi_x, \xi_y)$. We can estimate the size of this region by comparing the magnitude of different terms in the equations (2.13), which gives

$$\xi \ll (k/8\pi\Psi'')^{1/2} \qquad (2.18)$$

where we have defined

$$\Psi'' \equiv \max\left\{ \left|\frac{\partial^2 \psi_R}{\partial x^2}\right|, \left|\frac{\partial^2 \psi_R}{\partial y^2}\right|, \left|\frac{\partial^2 \psi_R}{\partial x \partial y}\right| \right\} \qquad (2.19)$$

For the moment let's neglect time-dependent terms, $H_1(\xi_x, \xi_y, t)$. In that case $\vec{\xi}$ describes the displacement of a passive particle from single point vortex moving together in an arbitrary uniform (perhaps non-stationary) velocity field. The equations of motion (2.13) reduce to:

$$\dot{\xi}_x = -\frac{k}{2\pi} \frac{\xi_y}{\xi_x^2 + \xi_y^2}; \qquad \dot{\xi}_y = \frac{k}{2\pi} \frac{\xi_x}{\xi_x^2 + \xi_y^2} \qquad (2.20)$$

their solution is:

$$\xi_x = \xi_0 \cos(\nu t + \phi_0), \qquad \xi_y = \xi_0 \sin(\nu t + \phi_0) \qquad (2.21)$$

where

$$\nu = k/2\pi\xi_0^2 \qquad (2.22)$$

and ϕ_0 is an arbitrary constant initial phase. It describes the circular rotation of the passive particle around the vortex, with radius ξ_0 and angular frequency ν. Thus, circles $\xi_x^2 + \xi_y^2 = \xi_0^2 = const.$ are invariant curves of the motion.

Under the influence of perturbation term (2.17), some of this circles may experience a certain deformation of their shape (KAM curves), and some break down and are replaced by areas of chaotic motion. As a next step, we will consider the breaking of the KAM curves near the singularity at $\xi = 0$. Let's look at the structure of the perturbation terms in the neighborhood of one of the invariant circles $\xi = \xi_0$, which lies close enough to the vortex, so that (2.18) is satisfied. Expanding $\psi_R(x_1 + \xi_x, y_1 + \xi_y, t)$ in Taylor series in ξ_x, ξ_y, around the vortex position, we obtain:

$$H_1(\xi_x, \xi_y, t) = \frac{\partial^2 \psi_R}{\partial x^2}\xi_x^2 + \frac{\partial^2 \psi_R}{\partial x \partial y}\xi_x \xi_y + \frac{\partial^2 \psi_R}{\partial y^2}\xi_y^2 + O(\xi^3) \qquad (2.23)$$

Note, that terms, linear in ξ, cancel out, leaving H_1 proportional to $O(\xi^2)$. Now we can write $H_1(\xi_x, \xi_y, t)$ in a form, where the small parameter of the problem,

$$\epsilon \equiv \xi_0^2 \Psi''/k \qquad (2.24)$$

enters explicitly:

$$H_1(\xi_x, \xi_y, t) = \epsilon V(\xi_x, \xi_y, t) \qquad (2.25)$$

where

$$V(\xi_x, \xi_y, t) \equiv \frac{k}{\Psi''\xi_0^2}\left[\frac{\partial^2 \psi_R}{\partial x^2}\xi_x^2 + \frac{\partial^2 \psi_R}{\partial x \partial y}\xi_x \xi_y + \frac{\partial^2 \psi_R}{\partial y^2}\xi_y^2 + O(\xi^3)\right] \qquad (2.26)$$

Note that $V(\xi_x, \xi_y, t)$ is of the same order of magnitude as $H(\xi_0)$. Thus, the problem of the advection in the vicinity of the selected vortex can be reduced to the problem of the time dependent Hamiltonian perturbation of the near-a-core motion of the tracer (2.21). This formulation provides ground for the analysis of the core size in Section 4. It is apparent now that the effect of the external part of the flow on the tracer particle near the vortex can be rendered arbitrarily small (but not completely absent) as we take particles closer to the vortex.

Let the external flow ψ_R have a characteristic frequency ν_1. Unperturbed curves in the vicinity of the vortex satisfy the condition

$$\nu_1 \ll \nu = k/2\pi\xi_0^2 \qquad (2.27)$$

in correspondence to (2.22). It means that for fairly deep orbits the changes of any parameters of the system are always adiabatic, and the invariant curves remain undestroyed. In the extended phase space (ξ_x, ξ_y, t) the invariant cylinders of the unperturbed system (2.20) undergo only a slight (exponentially small) deformation under the perturbation $H_1(\xi_x, \xi_y, t)$ which is due to the external flow with upper bounded frequency. We can introduce cylindrical tubes in the extended phase space which are the invariant manifolds of system (2.13). Using polar coordinates in (ξ_x, ξ_y)-plane,

$$\xi = (\xi_x^2 + \xi_y^2)^{1/2}, \qquad \phi \equiv \arctan \xi_x / \xi_y \qquad (2.28)$$

the equation defining this tubes can be written as:

$$\xi(\phi, t) = \xi_0(1 + f(\epsilon, \phi, t)) \qquad (2.29)$$

where $f(\epsilon, \phi, t)$ is bounded, and $f \to 0$ as $\epsilon \to 0$. Estimation of $f(\epsilon, \phi, t)$ for 3-vortex flow will be given in section 4, more general situation will be considered elsewhere.

Each of the invariant cylinders divides the phase space into two regions. If the motion starts in one of this regions, it will stay there forever, since the trajectory cannot move across the invariant manifold (2.29). The implication for the flow motion is that if at some instant of time $t = t_0$ a passive particle is located at (ξ_0, ϕ_0), then at any other time t it will stay inside the region bounded by the curve (2.29). It is important, that, though the shape of the boundary (2.29) depends on time, it always remains just a slightly perturbed circle, and will never evolve into a filament structure. That means that the part of the fluid occupying this region will never mix with the fluid outside and will be carried together with the vortex as a coherent structure. This core structures are very rigid due to the adiabaticity of the perturbation leading to the exponentially small corrections. We will specify this statement in Section 4.

Thus, the question of the existence of the coherent vortex core in a given flow is reformulated as the question of the existence of the invariant curves of the forced Hamiltonian system with one degree of freedom. For the particles which are close enough to the vortex, the Hamiltonian function (2.15), can be written in the form

$$H(\xi_x, \xi_y, t) = H_0(\xi_x, \xi_y) + \epsilon V(\xi_x, \xi_y, t) \qquad (2.30)$$

with H_0 and V given by (2.16) and (2.25) and the small parameter ϵ (2.24).

3 Equations, Classification, and Examples

Advection of a passive tracer particle in the velocity field of three point vortices provides us with a valuable example, which on one hand demonstrates a variety of different advection patterns, and on the other hand allows a thorough numerical study. This problem can be approached analytically in special cases where the relevant mappings can be constructed to approximate the advected particle motion. The integrability of the equations of motion of three point vortices [41], [42], implies that the stream function (2.5) can be found explicitly in this case, and the periodic character of the motion of vortices in rotating frame means that the flow is periodic in this frame, and gives us a convenient way to monitor the advection numerically by constructing the Poincare map for the motion of the tracer [11], [13].

To proceed with the discussion of the advection problem, we have to take a more detailed look at the motion of vortices themselves. Evolution of planar incompressible inviscid flows, consisting of a finite number of point vortices N, has attracted the attention of physicists for more than hundred years (see [23] and [25] for review of literature). These flows have a particular property – their dynamics can be derived from the autonomous Hamiltonian system with N degrees of freedom, which is essentially the set of equations on the vortex positions (2.10). System (2.10) possesses four independent integrals of motion, and three of those integrals are in involution, independent of what the strengths of the vortices are. As a consequence, motion of three point vortices is always integrable. It was studied extensively by several authors, and by now we have a complete description of motion types for arbitrary vortex strengths [41],[42],[43],[44].

Below, for the convenience of readers, we briefly outline the solution of the equations (2.10) for the case of three identical point vortices following [36]. As above, denote by k the strength of vortices. It is convenient to introduce complex coordinate in the plane: $z = x + iy$, and to specify the positions of the three vortices by means of complex-valued functions of time: $z_m(t) = x_m(t) + iy_m(t)$, $m = 1, 2, 3$.

Stream function of this flow is a sum of stream functions corresponding to each vortex:

$$\Psi(z, z^*, t) = -\frac{k}{4\pi} \sum_{j=1}^{3} \ln |z - z_j(t)|^2 \tag{3.1}$$

so the flow is completely determined by $z_m(t)$, $m = 1, 2, 3$, and its dynamics is reduced to the set of equations for the positions of vortices:

$$\dot{z}_m^* = \frac{k}{2\pi i} \sum_{j \neq m} \frac{1}{z_m - z_j}, \qquad (j, m = 1, 2, 3) \tag{3.2}$$

These equations can be written in Hamiltonian form:

$$\dot{z}_m^* = [z_m^*, H] \tag{3.3}$$

with Hamiltonian function (compare to (2.16),(2.12))

$$H = -\frac{k}{4\pi} \sum_{j \neq m} \ln |z_m - z_j|^2 \tag{3.4}$$

and fundamental Poisson bracket

$$[z_m, z_j] = 0, \quad [z_m, z_j^*] = -2i\delta_{mj} \tag{3.5}$$

The Hamiltonian (3.4) immediately reveals its translational and rotational invariance which allows to construct corresponding integrals of motion:

$$\sum_{j=1}^{3} z_j \equiv Q + iP \qquad (3.6)$$

and

$$\sum_{j=1}^{3} |z_j|^2 \equiv L^2 \qquad (3.7)$$

Using (3.5) one can derive the non-zero Poisson brackets:

$$[Q, P] = 3; \qquad [Q, L^2] = 2P; \qquad [P, L^2] = -2Q \qquad (3.8)$$

with two independent integrals in involution:

$$[Q^2 + P^2, L^2] = 0 \qquad (3.9)$$

Since Hamiltonian has zero Poisson bracket with any combination of Q, P and L, the system has three independent integrals in involution: H, L^2 and $Q^2 + P^2$.

The integrals Q and P defined above are the coordinates of the center of vorticity, so conservation of (3.6) means that the center of vorticity doesn't move, and below we will put $Q = P = 0$, i.e. always place the origin of coordinate system in the center or vorticity. Second integral L^2 defines the spatial scale of the motion and can be made equal to one by an appropriate rescaling of coordinates $z \to z/L$. Provided $Q = P = 0$, this integral is equal to the average squared distance between the vortices:

$$L^2 = \frac{1}{3} \sum_{j>m} |z_j - z_m|^2, \qquad (j, m = 1, 2, 3) \qquad (3.10)$$

Equations (3.2) can be written in a dimensionless form by rescaling the time variable $t \to (k/L^2)\, t$ The only parameter that is left is the value of the Hamiltonian H, which can be related to the geometry of the problem, e. g. to the product of side lengths of the vortex triangle Λ:

$$e^{-2\pi H} = \Lambda \equiv |z_1 - z_2||z_2 - z_3||z_3 - z_1| \qquad (3.11)$$

written in the dimensionless form with $k = L = 1$ (compare to (B.17)).

Another geometrical parameter is a half of the minimum distance between two vortices during their motion, which we denote as a. It is related to H as:

$$3a - 4a^3 = e^{-2\pi H} = \Lambda \qquad (3.12)$$

and will be used in numerical simulations. The range of these parameters is:

$$H \in (0, \infty); \qquad \Lambda \in (0, 1); \qquad a \in (0, 1/2). \qquad (3.13)$$

Solution for the the equations (3.2) can be found in quadratures (see Appendix B) and written in the following form, with expressions for $I(t)$, $\phi_1(t)$ and $\phi_2(t)$ given in the Appendix B:

$$z_m(t) = \frac{L}{\sqrt{6}}e^{i\phi_2/2}\left[(1 - I^{1/2})^{1/2}e^{-2\pi i(m-1)/3}e^{-i\phi_1/2}+\right.$$

$$\left. + (1 + I^{1/2})^{1/2}e^{-4\pi i(m-1)/3}e^{i\phi_1/2}\right], \qquad m = 1, 2, 3 \qquad (3.14)$$

where the first factor contributes only to the overall rotation of the triangle formed by vortices, and the second one describes a relative motion, i.e. a change of the triangle shape. The relative motion of three vortices is periodic (except of a the special case $\Lambda = \Lambda_c \equiv 1/\sqrt{2}$), i. e. after a certain time T_{rel} (see (B.27),(B.29)) the vortex configuration repeats itself.

In the meanwhile (during the period T_{rel}) the triangle spanned by vortices is rotated as a whole by a certain angle $\Theta(\Lambda)$, so the motion of the vortex triangle looks like a superposition of the periodic pulsations and uniform rotation with rotation frequency $\Omega(\Lambda) = \Theta(\Lambda)/T_{rel}$ (see (B.33)). For a general initial configuration, Θ is incommensurate with 2π, and the the frequency of relative motion $\omega_{rel}(\Lambda) = 2\pi/T_{rel}$ is incommensurate with rotation frequency Ω and the overall motion is quasiperiodic. However it is periodic in the reference frame rotating with Ω, so it is convenient to consider the motion in the reference frame co-rotating with vortices. We denote as $\tilde{z}_m(t)$ trajectories of vortices in the co-rotating frame. They are simply related to the laboratory frame coordinates:

$$\tilde{z}_m(t) = z_m(t)e^{-i\Omega t}, \qquad m = 1, 2, 3. \qquad (3.15)$$

The functions $\tilde{z}_m(t)$ are periodic with period T_{rel} and satisfy the equations:

$$\dot{\tilde{z}}_m^* = \frac{k}{2\pi i}\sum_{j\neq m}\frac{1}{\tilde{z}_m - \tilde{z}_j} + i\Omega\tilde{z}_m^* , \qquad (j, m = 1, 2, 3) \qquad (3.16)$$

Depending on the value of the relevant parameter of the system (Λ, H or a), one encounters different types of relative motion of the vortices . Following [41],[42], we give a brief description of what happens when Λ decreases from 1 to 0 (H increases from 0 to ∞), and write down the symmetries of different regimes, important for the advection problem.

For $\Lambda=1$, we have $H = 0$, $a = 0.5$, $I(t) \equiv 1$ (see (B.24)) and there is no relative motion at all. Vortices form an equilateral triangle uniformly rotating around its center with angular velocity $\Omega(1) = 3/2\pi$. Advection problem in this case is integrable. When Λ is close to 1, vortices perform small oscillations around the equilateral triangle, and advection exhibits the onset of chaos in the neighborhood of the separatrices of the steady rotating case $\Lambda = 1$. We consider this case in detail in Section 5.

In the range $1/\sqrt{2} = \Lambda_c < \Lambda < 1$ the shape of the vortex triangle oscillates between the two isosceles triangles - the acute one with the base length $2a$, when the area of the triangle reaches its maximum, and the obtuse one (minimum of the area). During one period of the relative motion, each of these shapes is repeated three times, corresponding to three different cyclic permutations of vortices. This is a manifestation of the particular symmetry of motion in this range:

$$\tilde{z}_m(t + T_{rel}/3) = \tilde{z}_j(t)e^{2\pi i/3}, \quad m = 1,2,3; \quad j - 1 = m \mod 3 \qquad (3.17)$$

in rotating frame, or

$$z_m(t + T_{rel}/3) = z_j(t)e^{i(2\pi + \Theta(\Lambda))/3}, \quad m = 1,2,3; \quad j - 1 = m \mod 3 \quad (3.18)$$

in the laboratory frame. Note that the orientation σ, defined as $\sigma = +1$ for the counterclockwise arrangement of vortices 1,2,3 along the perimeter of the vortex triangle, and $\sigma = -1$ for the clockwise arrangement does not change during the motion.

As Λ approaches its critical value Λ_c, T_{rel} grows logarithmically, and the solution for $\Lambda = \Lambda_c$ is aperiodic (except of the situation when the vortices are perfectly aligned, one exactly in the middle between the other two, which is a hyperbolic equilibrium point of the system). More details concerning this regime are in the Section 4, where we address the advection problem for $\Lambda \approx \Lambda_c$.

For $0 < \Lambda < \Lambda_c$ the relative motion of vortices is periodic again, but it is distinctly different from the case $\Lambda_c < \Lambda < 1$. Now the shape of the vortex triangle oscillates between two congruent isosceles triangles (maximum of the area, base length $2a$) with different orientation, passing through the linear configuration in the midway. The equivalence of all three vortices (in a sense of the symmetry (3.17)) does not exist anymore – one of the vortices becomes special, it always stays in the sharp corner of the triangle, never coming to any of the other two as close as they come to each other. We number this vortex by $m = 1$ (in this case we have to pick a particular solution of (B.30) for ϕ_1). For this case we have a following symmetry with respect to change of the orientation:

$$\tilde{z}_2(t + T_{rel}/2) = \tilde{z}_3(t); \quad \tilde{z}_3(t + T_{rel}/2) = \tilde{z}_2(t); \quad \tilde{z}_1(t + T_{rel}/2) = \tilde{z}_1(t). \quad (3.19)$$

In practice, symmetries (3.17),(3.19) allow to reduce CPU time required for constructing the Poincare map for the tracer motion in three or two times respectively.

Now, let us look at the motion of an advected particle in the co-rotating frame. We denote position of the particle in that frame as $\tilde{z} = \tilde{x} + i\tilde{y}$. Its trajectory satisfies the equation:

$$\dot{\tilde{z}}^* = [\tilde{z}^*, \tilde{\Psi}] = \frac{k}{2\pi i} \sum_{m=1}^{3} \frac{1}{\tilde{z} - \tilde{z}_m} + i\Omega\tilde{z}^* \qquad (3.20)$$

where stream function $\tilde{\Psi}$ is given by (3.1) with additional term, corresponding to rotational "energy":

$$\tilde{\Psi}(\tilde{z}, \tilde{z}^*, t) = -\frac{k}{4\pi} \sum_{j=1}^{3} \ln|\tilde{z} - \tilde{z}_j(t)|^2 + \frac{\Omega}{2}\tilde{z}\tilde{z}^* \qquad (3.21)$$

To estimate size of the coherent structures around vortices, we pick a particular vortex, for example one with $m = p$, and consider dynamics of the displacement of the advected particle

$$\zeta \equiv \tilde{z} - \tilde{z}_p \qquad (3.22)$$

Note that ζ is defined in the rotating frame, and is related to the displacement in the laboratory frame $\xi = \xi_x + i\xi_y$, introduced in (2.12), by

$$\zeta = \xi e^{-i\Omega t}.$$

From (3.20) and (3.16) it follows immediately:

$$\dot{\zeta}^* = \frac{k}{2\pi i}\left[\frac{1}{\zeta} + \sum_{m=2}^{3}\left(\frac{1}{\zeta - \tilde{z}_{mp}} + \frac{1}{\tilde{z}_{mp}}\right)\right] + i\Omega\zeta^* \qquad (3.23)$$

where we have denoted the relative positions of vortices

$$\tilde{z}_{mj} \equiv \tilde{z}_m - \tilde{z}_j \qquad (3.24)$$

The equation (3.23) can be considered as a special case of (2.13) in the rotating reference frame. It can be obtained from a time dependent Hamiltonian (2.15) with a centrifugal term

$$\tilde{\Psi}(\zeta,\zeta^*,t) = -\frac{k}{4\pi}\left[\ln|\zeta|^2 + \sum_{m=2}^{3}\left(\ln|\zeta - \tilde{z}_{mp}|^2 + \frac{\zeta}{\tilde{z}_{mp}} + \frac{\zeta^*}{\tilde{z}_{mp}^*}\right)\right] + \frac{\Omega}{2}\zeta\zeta^* \quad (3.25)$$

We will use this representation in Section 4 to obtain the formula for the radius of the coherent cores around vortices.

To get a general idea about the character of the advection, we have performed numerical simulations of the "three-vortex + tracers" system (3.16),(3.20). To visualize spatial pattern of advection we have constructed Poincare sections of the tracer motion (taking snapshots in the rotating frame with intervals T_{rel}) for various vortex geometries. Figures 1a, 1b correspond to the case $\Lambda \approx 1$, when vortices slightly oscillate near the equilateral triangle configuration. Two separatrix families of the integrable case $\Lambda = 1$ (see Section 5, Fig. 6), split and merge for fairly small deviations of Λ from unity, Fig 1a. Area of chaotic motion grows rapidly as Λ decreases; for $\Lambda = 0.97926\ldots$ (Fig. 1b) there already exists a well-developed stochastic sea — large connected region with strong mixing properties. Similar situation occurs for other vortex geometries when $\Lambda > \Lambda_c$, see Fig. 1c, 1d.

This figures reveal two main types of islands of regular advection inside the mixing region:

1. Vortex cores — smooth, near-circular, robust islands around the vortices, existing for any values of Λ.

2. Elliptic islands — irregularly-shaped islands around elliptic points, extremely sensitive to variations in Λ, each existing only in the limited range of Λ values.

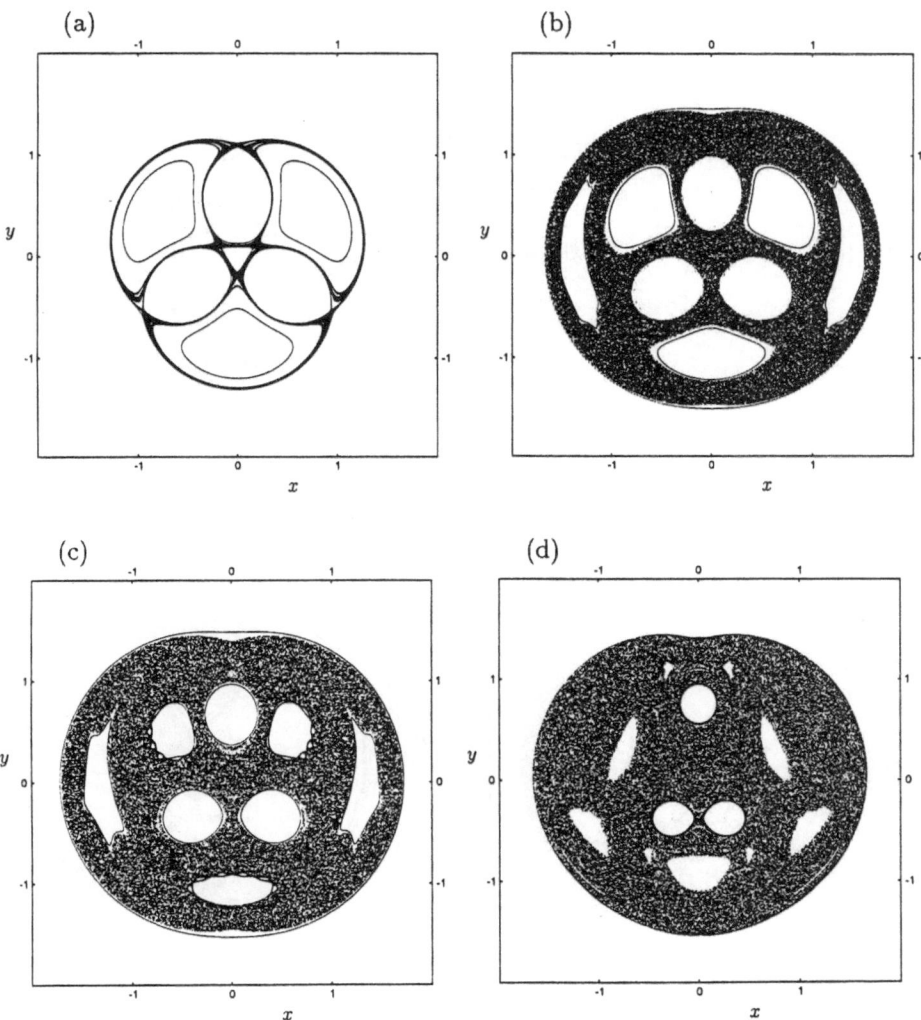

Figure 1: Poincare sections of tracer particle trajectories for different vortex geometries: (a) onset of chaotic advection, $a = 0.4996$, $1 - \Lambda = 9.6 \cdot 10^{-7}$; (b) well developed stochastic sea exists already for $\Lambda = 0.97926$; (c) $\Lambda = 0.94$, elliptic island boundary has a typical island-chain structure, see Fig. 3a for zoom. (d) $\Lambda = 0.71666$, note narrow sticky bands around vortex cores, see zoom in Fig. 3b, 3c, 3d for details of the structure.

214

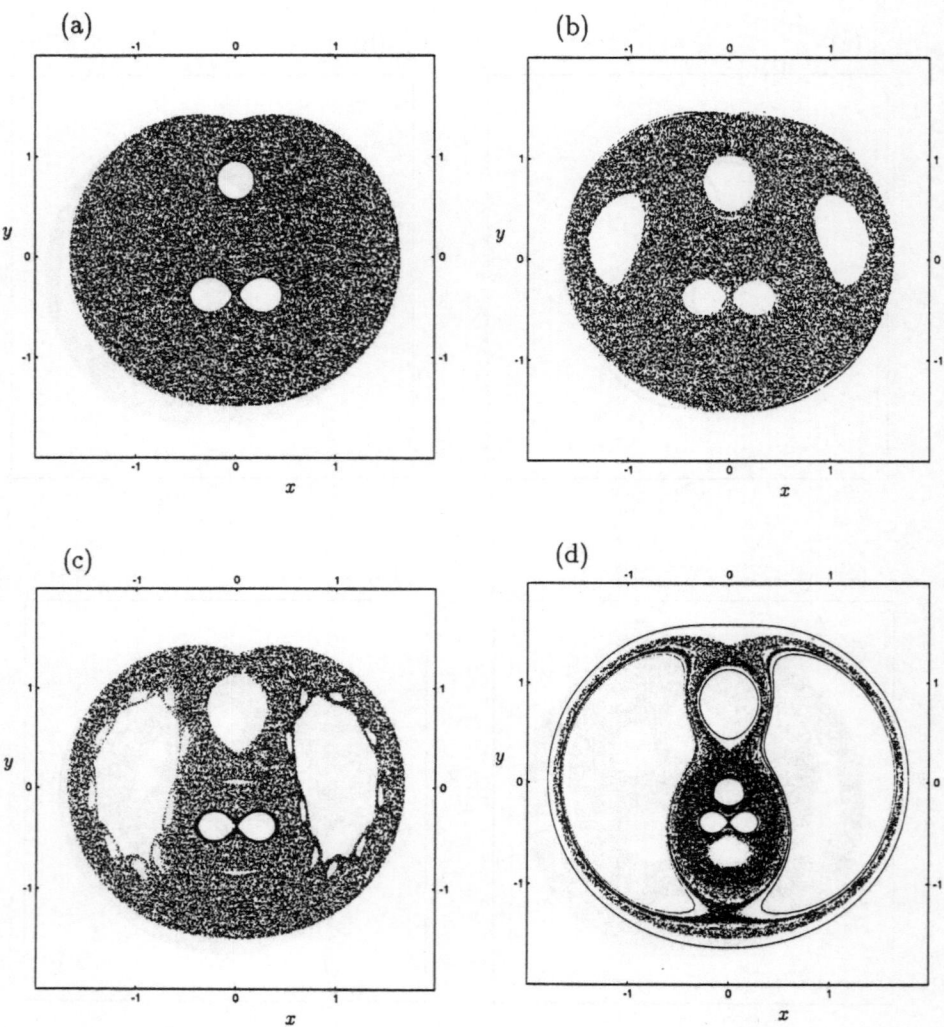

Figure 2: Poincare sections of tracer particle trajectories for different vortex geometries. Strong tracer chaotization for near-separatrix vortex flow: (a) $\Lambda = 0.707109 > \Lambda_c$, (b) $\Lambda = 0.70710668 < \Lambda_c$. Back to regular advection: (c) $\Lambda = 0.6052$, (d) $\Lambda = 0.4636$

(a)　　　　　　　　　　　　(b)

(c)　　　　　　　　　　　　(d)

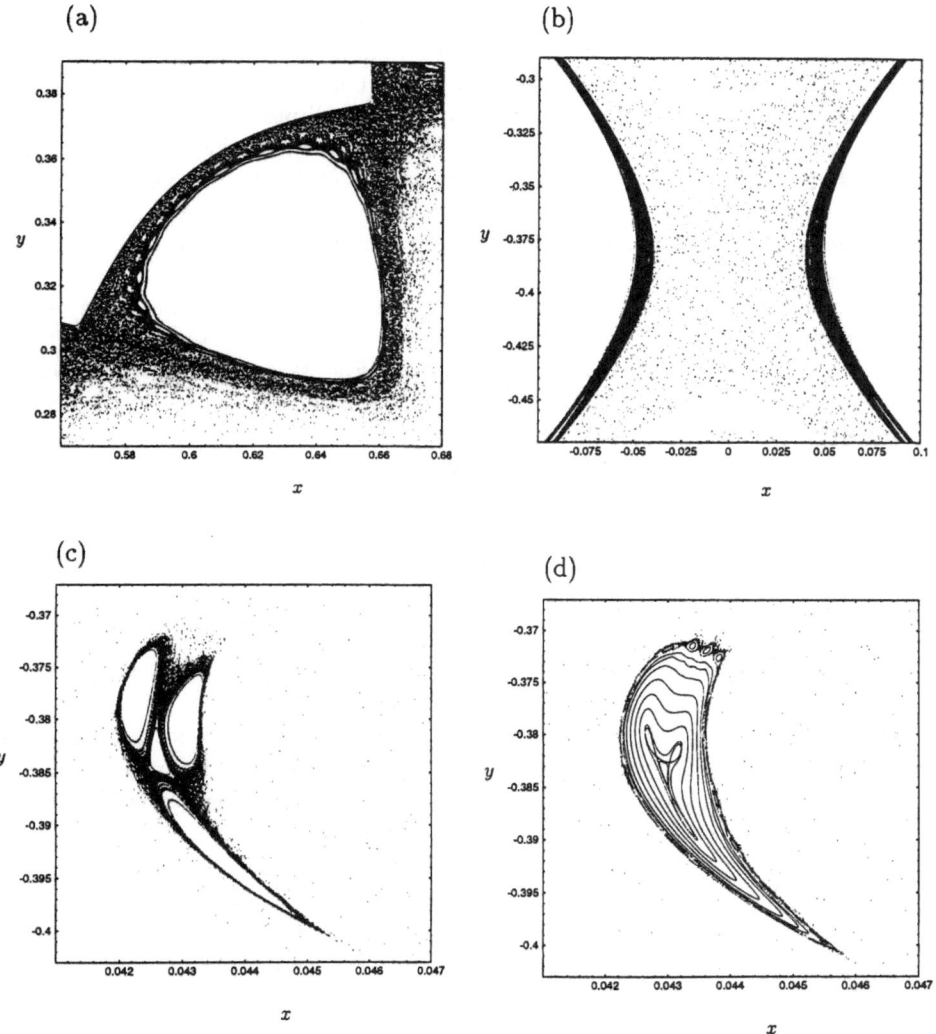

Figure 3: Structures in the mixing region. (a) Magnification of the piece of the island chain around elliptic island (Fig. 1c), $\Lambda = 0.94$. (b) Narrow sticky bands around vortex cores, zoom of central piece of Fig. 1d, $\Lambda = 0.71666$. Thin white areas inside the bands are regular island structures. Their zoom (stretched in x-direction) is shown for two close vortex geometries: (c) $\Lambda = 0.71682$, (d) $\Lambda = 0.716917$.

The strongest chaotization occurs when Λ is close to Λ_c. Elliptic islands vanish altogether, and the only structures remained in the stochastic sea are vortex cores (Fig 2a). Further decrease of Λ (below Λ_c) reverses the scenario (Fig. 2b, 2c, 2d) — stochastic sea gets smaller, elliptic islands reemerge and grow, and eventually chaos survives only in the narrow stochastic layers around the destructed separatrices of the limiting case $\Lambda \to 0$, when two of the vortices coalesce and advection becomes integrable. Note, that for $\Lambda > \Lambda_c$ all three cores are identical, while for $\Lambda < \Lambda_c$ one of the cores is considerably bigger (due to different symmetries (3.17),(3.19)).

Our simulations clearly indicate the existence of sticky singular zone on the border of the islands. Apart from the island chains around the elliptic islands (see Fig. 1c, 2c, and zoom of such an island in Fig. 3a), which resemble typical phase-space structures of Hamiltonian chaos, one can observe dark bands around the vortex cores (Fig. 1d, 2c), i.e. the boundary of the cores acts as a quasi-trap for tracers. Zoom of a piece of dark bands (Fig. 3b) shows, that they contain a multitude of small stretched subislands. These subislands, in turn, have a complex structure, very sensitive to the smallest variations of system parameters. Figures 3c, 3d show such an island (suitably magnified) for two close values of Λ. Sticky island-around-island chains can be responsible for particle trapping.

4 Strongly chaotic advection

Here we consider the most important for applications case $\Lambda \approx \Lambda_c$, when the chaotization of the tracer motion is the strongest. Numerical simulations (Fig. 2a, 2b) of the passive particle motion reveal a large chaotic component which fills completely approximately circular area of radius $R_{max} \sim 1.5\,L$, with the exception of almost circular patches of regular motion around the vortices (coherent vortex cores) of radius $\xi_{max} \sim 0.18\,L$ in case $\Lambda > \Lambda_c$. In case $\Lambda < \Lambda_c$ two of the cores have the same size, but the third core is bigger, with radius about $0.26\,L$. As Λ approaches to its critical value Λ_c, ξ_{max} practically does not change.

Existence of the robust vortex cores is a typical phenomenon in the advection in flows dominated by the regions of concentrated vorticity. Being one of the simplest situations in which the coherent vortex cores appear, 3-vortex flow allows to investigate analytically the problem of the size of the cores, and to point out relevant features of advection dynamics, determining the size of the cores in general case.

Let us consider the motion of vortices in the rotating frame where it is periodic (for $\Lambda \neq \Lambda_c$). Note, that the frequency of the overall rotation $\Omega(\Lambda)$ has no singularity at $\Lambda = \Lambda_c$, so we can put

$$\Omega(\Lambda) = \Omega(\Lambda_c) = 3/2\pi \qquad (4.1)$$

and concentrate our attention on the relative motion.

When Λ is close to Λ_c, vortices move in a particular way typical for near-separatrix dynamics. They perform relatively fast rearrangements (flips) between

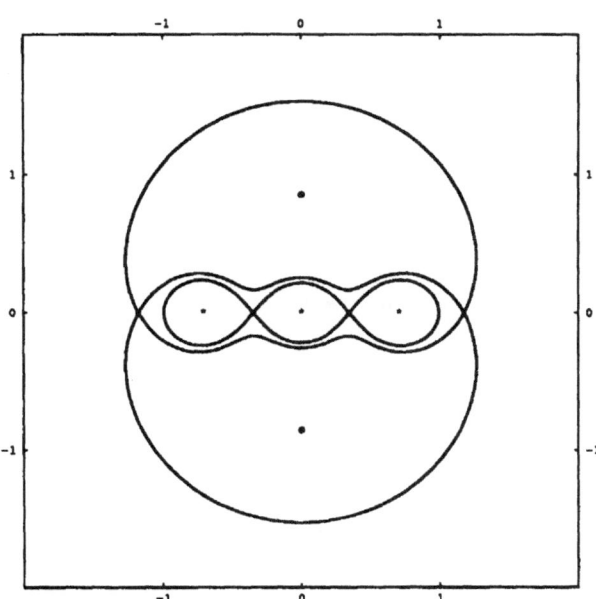

Figure 4: Unstable vortex equilibrium for $\Lambda = \Lambda_c$, separatrices of stream function (4.5). Vortex positions (4.2) are marked with asterisks. Solid circles — hyperbolic points (4.7), empty circles — elliptic points (4.8)

different unstable equilibria intervened with longer intervals spent in the neighborhoods of the equilibria, where they essentially stay at rest. These equilibria are close to the saddle points of 3-vortex system (3.16) at $\Lambda = \Lambda_c$, and up to an irrelevant phase factor they are given by (see Fig. 4, where equilibrium vortex positions are marked with asterisks):

$$\tilde{z}_i = \frac{1}{\sqrt{2}}, \quad \tilde{z}_j = 0, \quad \tilde{z}_k = -\frac{1}{\sqrt{2}}, \quad i \neq j \neq k; \quad i, j, k = 1, 2, 3 \qquad (4.2)$$

There are six ways to distribute three vortices in three locations, those with the same vortex in the center (at zero) can be obtained from each other by rotation, so finally we have three different equilibria which we will mark by the number of the central vortex (j). For every pair of these saddle points, there exists a family of separatrix soliton-like solutions $\tilde{z}_m^{(s)}(t - t_0)$ parameterized by t_0, the moment of the center of the soliton pulse. They describe the flips of the vortex configuration from one saddle point to another and are aperiodic (i.e. the flip takes an infinite time). All three of them are identical up to the renumbering of the vortices.

When $\Lambda > \Lambda_c$, vortices periodically rearrange themselves between the neighborhoods of the three equilibria in a way $(1) \rightarrow (2) \rightarrow (3) \rightarrow (1) \ldots$, or $(1) \rightarrow (3) \rightarrow (2) \rightarrow (1) \ldots$, depending on the initial orientation. For $\Lambda < \Lambda_c$ one of the vortices gets isolated, and since we agreed to mark this vortex with $m = 1$, the vortex configuration flips between the equilibria (2) and (3): $(2) \to (3) \to (2) \dots$. The characteristic time of the vortex configuration flip is given by the width of the separatrix pulse (B.28)

$$t_s \equiv \frac{\pi}{3\sqrt{3}} \tag{4.3}$$

where we put $L = 1$. The interval T between pulses is $T \equiv T_{rel}/3$ for $\Lambda > \Lambda_c$, and $T \equiv T_{rel}/2$ for $\Lambda < \Lambda_c$. Since $T_{rel} \sim \ln|\Lambda - \Lambda_c|^{-1}$ (see (B.27),(B.29)), when the value of Λ is sufficiently close to Λ_c, we have

$$T \gg t_s = \frac{\pi}{3\sqrt{3}} \tag{4.4}$$

The specific character of vortex motion in this case allows to find a mapping, describing the advection of the passive particle. As a first step, we look at the advection in the velocity field of the equilibrium vortex configuration (4.2). In the rotating frame vortices do not move, so the streamlines do not depend on time, and advection is regular in all plane. Substituting positions of vortices from (4.2) into (3.21) we obtain the stream function in the rotating frame:

$$\tilde{\Psi}_{sep}(\tilde{z}, \tilde{z}^*, t) = -\frac{1}{4\pi} \ln\left(|\tilde{z}|^2|\tilde{z}^2 - 1/2|^2\right) + \frac{3}{4\pi}|\tilde{z}|^2 \tag{4.5}$$

Apart from singularities at the vortex positions, it has six equilibrium points, which are the solutions of

$$\frac{\partial \tilde{\Psi}_{sep}}{\partial \tilde{z}} = 0 \tag{4.6}$$

They are two pairs of saddle points:

$$\tilde{z}_{hyp} = \pm\frac{1}{2}\sqrt{3 \pm \sqrt{19/3}} \tag{4.7}$$

and a pair of elliptic points:

$$\tilde{z}_{ell} = \pm\frac{i}{2}\sqrt{1 + \sqrt{11/3}} \tag{4.8}$$

Separatrices, connecting each pair of saddle points divide the plane into seven regions, see Fig. 4. In each of these regions action-angle variables (I, ψ) can be introduced, and the motion of advected particle can be described in a simple way:

$$I(t) = I_0, \qquad \psi(t) = \omega(I)t + \psi_0 \tag{4.9}$$

where

$$\omega(I) \equiv \frac{d\tilde{\Psi}_{sep}}{dI} \tag{4.10}$$

is the advection nonlinear frequency, I_0 and ψ_0 are constants.

When $\Lambda \approx \Lambda_c$, vortices spend a long time very close to equilibria (4.2), and equations (4.9) can be used to describe advection during this intervals. To complete the picture we need the relation between (I, ψ) before and after the vortex flip.

Let's denote by (I_n, ψ_n) the values of action and angle variables right before the beginning of the n-th flip. We are looking for a map

$$\hat{T} : (I_{n+1}, \psi_{n+1}) = \hat{T}(I_n, \psi_n) \qquad (4.11)$$

relating tracer coordinates between two consequent passings of the vortex system near the saddle points. Let us denote by $\Delta I(I_n, \psi_n)$ change of I during the n-th flip. Then the mapping (4.11) can be approximately written in the form:

$$\begin{cases} I_{n+1} = I_n + \Delta I(I_n, \psi_n) \\ \psi_{n+1} = \psi_n + \omega(I_{n+1})T \end{cases} \qquad (4.12)$$

where $\omega(I)$ is given by (4.10) (we have neglected phase detuning during the pulse).

At this moment we will restrict our consideration to the particular problem of the size of the coherent vortex cores, which was discussed in general in Section 2. That means, that we will be interested in the regions, surrounding the vortices. In those regions it is convenient to consider advection in the reference frame moving with the vortex which is located in the center of the region. We follow definition (3.22), and mark this particular vortex with $m = p$. (see formulae (3.22),(3.23),(3.25)).

To find $\Delta I(I_n, \psi_n)$, we note, that if the tracer is not too close to the border of the region (i.e. not too close to the inner separatrix in Figure 4), it spins rapidly around the vortex (due to logarithmic singularity in (3.25)). More specifically, with the exclusion of a narrow band around the separatrix, the advection non-linear frequency (4.10) is much larger than the inverse duration of the vortex flip (4.3), i.e.

$$\omega(I)t_s \gg 1 \qquad (4.13)$$

and we can treat the rearrangement of vortices as an adiabatic change of the parameters in (3.25). This property will be effectively used to find $\Delta I(I_n, \psi_n)$ in the first order of the adiabaticity parameter [45],[39].

Below we briefly outline necessary calculations, leaving the details for the Appendix C. First, we split stream function (3.25) in two parts:

$$\tilde{\Psi}(\zeta, \zeta^*, t) = H_0(\zeta, \zeta^*) + H_1(\zeta, \zeta^*, t) \qquad (4.14)$$

where H_0 is autonomous part:

$$H_0(\zeta, \zeta^*) = -\frac{k}{4\pi} \ln |\zeta|^2 + \frac{1}{2}\Omega|\zeta|^2 \qquad (4.15)$$

and H_1 depends on time through slowly varying parameters $\tilde{z}_{mp}(t)$, defined by (3.24):

$$H_1(\zeta, \zeta^*, t) = -\frac{k}{4\pi} \sum_{m \neq p}^{3} \left(\ln |\zeta - \tilde{z}_{mp}|^2 + \frac{\zeta}{\tilde{z}_{mp}} + \frac{\zeta^*}{\tilde{z}_{mp}^*} \right) \qquad (4.16)$$

In the region of interest $|\tilde{z}_{mp}|$ is always larger than $|\zeta|$, so the contribution of H_1 to the equation of motion (3.23) contains a small parameter $|\zeta/\tilde{z}_{mp}|$ (see Appendix C for details). The unperturbed solutions of (4.15) are circular rotations:

$$\zeta = \zeta_0 e^{i\omega t} \tag{4.17}$$

with the frequency

$$\omega(J) = \frac{1}{2\pi\zeta_0^2} - \Omega = \frac{1}{2\pi}\left(\frac{1}{|2J|} - 3\right) \tag{4.18}$$

where zeroth order "action" variable J is defined by:

$$J = \frac{i}{8\pi} \oint (\zeta^* d\zeta - \zeta d\zeta^*) = -\frac{1}{2}\zeta_0^2 \tag{4.19}$$

and angle variable is just a polar angle in the ζ-plane:

$$\theta = \arg\zeta \tag{4.20}$$

With the help of variables (J, θ) we can express Hamiltonian (4.14) as $\tilde{\Psi}(J, \theta; t)$ which has a slow (adiabatic) dependence on t through magnitudes $\tilde{z}_{mp}$, $\tilde{z}_{mp}^*$. Our next step will be introducing such variables $(\bar{J}, \bar{\theta})$ that dependence of $\tilde{\Psi}$ on $\bar{\theta}$ is killed in the first approximation.

In the lowest order of $|\zeta/\tilde{z}_{mp}|^2$ the required transformation is given by the generating function (see Appendix C)

$$S(\bar{J}, \theta) = \bar{J}\theta + \left[\frac{1}{8\pi i\omega(J)} \sum_{m \neq p} \frac{|2\bar{J}|e^{2i\theta}}{\tilde{z}_{mp}^2} + c.c.\right] \tag{4.21}$$

and new action $\bar{J}$ is related to the old one J by:

$$\bar{J} = J + \left[\frac{1}{4\pi\omega(J)} \sum_{m \neq p} \frac{|2J|e^{2i\theta}}{\tilde{z}_{mp}^2} + c.c.\right] \tag{4.22}$$

In fact $\tilde{z}_{mp}$ are slow functions of time, and so is $S(\bar{J}, \theta)$. New Hamiltonian function is given by:

$$\tilde{H}(\bar{J}, t) = \tilde{\Psi}(\bar{J}, t) + \frac{\partial S}{\partial t}(\bar{J}, \bar{\theta}, t) \tag{4.23}$$

and we get an equation for the evolution of the adiabatic invariant:

$$\dot{\bar{J}} = -\frac{\partial \tilde{H}}{\partial \bar{\theta}}(\bar{J}, t) = -\frac{\partial^2 S}{\partial \bar{\theta} \partial t}(\bar{J}, \bar{\theta}, t) = \frac{|2J|}{2\pi\omega(J)}\left[\sum_{m \neq p} e^{2i\theta}\frac{\dot{\tilde{z}}_{mp}}{\tilde{z}_{mp}^3} + c.c\right] \tag{4.24}$$

where we have replaced $\bar{\theta}$ by θ in the right-hand side with an accuracy of first approximation. The total change of $\bar{J}$ during the vortex flip is small, so it can be evaluated by integrating right side of (4.24) through the time of the flip:

$$\Delta \bar{J}(J, \theta^{(0)}) = \int_{\Delta t} \dot{\bar{J}} \, dt = \frac{|2J|}{2\pi\omega(J)} \left[e^{2i\theta^{(0)}} A(J) + c.c. \right] \tag{4.25}$$

where

$$A(J) \equiv \int_{\Delta t} e^{2i\omega(J)t} \sum_{m \neq p} \frac{\dot{\tilde{z}}_{mp}}{\tilde{z}_{mp}^3} \, dt \tag{4.26}$$

and Δt is time interval of the order T around the instant t_0 of the vortex flip with the phase $\theta^{(0)} = \theta(t_0)$. During the time of the pulse, vortex trajectories $\tilde{z}_{mp}(t; \Lambda)$ are practically indistinguishable from separatrix soliton-like trajectories (which can be obtained from (B.28)) corresponding to $\Lambda = \Lambda_c$, and can be replaced by the latter under the integral. That allows to extend the integration to $\pm$ infinity and to close the contour in the complex plane, so (4.26) will be given by the sum of contributions from singular points in the upper part of the complex plane. Due to the fast decay of the integrand up in the complex plane (ω is large, see (4.13)), it is enough to take only the singularity, which is closest to the real axis. It turns out to be at the point

$$t_0 = i \frac{\pi t_s}{2} = i \frac{\pi^2}{6\sqrt{3}} \tag{4.27}$$

and for the integral (4.26) we get:

$$A(J) = A_p \omega(J) e^{-\pi\omega(J)t_s} \tag{4.28}$$

where constants A_p, depending on the particular vortex and particular flip, were evaluated numerically:

$$|A_p| = \begin{cases} A_1 \approx 148. & \text{corner-center flip} \\ A_2 \approx 2.4 & \text{corner-corner flip} \end{cases} \tag{4.29}$$

here the first value A_1 should be taken when the vortex under consideration stays near the central equilibrium $\tilde{z}_p = 0$ before or after the flip, and the second value A_2 when it stays outside.

Now we return to the map (4.30). It is convenient to rewrite it in variables (J, θ):

$$\begin{cases} J_{n+1} = J_n + \Delta \bar{J}(J_n, \theta_n) \\ \theta_{n+1} = \theta_n + \omega(J_{n+1})T \end{cases} \tag{4.30}$$

since (4.12) requires specification in which region (I, θ) should be taken on each step, while (4.30) contains this information through the dynamics of the parameters $\tilde{z}_{mp}$. Eventually, (J, θ) are defined in a unique way for particles in the core.

The question to be answered is: given the initial value of J, is the motion regular or chaotic? Since the change in adiabatic invariant per pulse $\Delta \bar{J}(J_n, \theta_n)$ is small and the chaotization occurs mainly due to phase instability, we can use the stochasticity criterion in a form [10]

$$K \equiv \max_{\theta \in [0, 2\pi]} \left| \frac{\delta \theta_{n+1}}{\delta \theta_n} - 1 \right| > 1 \tag{4.31}$$

which for the map (4.30) gives:

$$\max_{\theta \in [0, 2\pi]} \left| \frac{d\omega(J)}{dJ} T \frac{d\Delta \bar{J}(J_n, \theta_n)}{d\theta_n} \right| > 1 \tag{4.32}$$

From (4.25) we obtain

$$\max_{\theta \in [0, 2\pi]} \left| \frac{d\Delta \bar{J}(J_n, \theta_n)}{d\theta_n} \right| = 2 \frac{|2J|}{\pi \omega(J)} |A(J)| \tag{4.33}$$

and (4.18), (4.19) gives

$$\frac{d\omega(J)}{dJ} = \frac{1}{4\pi J^2} \tag{4.34}$$

After substitution of (4.33), (4.34) into (4.32), we arrive at the following condition for the motion of the tracer to be chaotic:

$$\frac{T|A_p| \exp(-\pi \omega(J) t_s)}{\pi^2 |J|} > 1 \tag{4.35}$$

The value of action $J = J_c$ which makes left-hand side of (4.35) equal to one corresponds to the border between regular and chaotic motion and defines the radius of the core around the vortex:

$$\zeta_c \equiv \sqrt{2|J_c|} \tag{4.36}$$

Substituting $\omega(J)$ from (4.18) we get the following equation for ζ_c:

$$\zeta_c^2 \exp(t_s / 2\zeta_c^2) = \frac{2|A_p| \exp(\pi \Omega t_s)}{\pi^2} T(\Lambda) = \frac{2|A_p| \exp(\pi \sqrt{3}/6)}{\pi^2} T(\Lambda) \tag{4.37}$$

where we have used definitions (4.1),(4.3). Dependence of the right side of (4.37) on the background vortex flow enters only through the interval between the vortex flips $T(\Lambda)$, and is apparently quite strong, since T diverges logarithmically when $\Lambda \to \Lambda_c$, see (B.27)

$$T(\Lambda) = \frac{2\pi}{3\sqrt{3}} \ln |\delta \Lambda|^{-1} + C_1 + O(|\delta \Lambda| \ln |\delta \Lambda|) \tag{4.38}$$

with

$$C_1 = \frac{2\pi}{3\sqrt{3}} \ln \frac{14\sqrt{2}}{3} \tag{4.39}$$

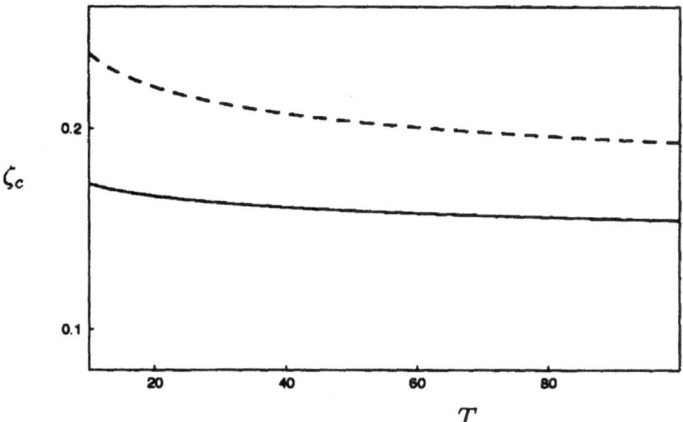

Figure 5: Radius of the vortex core vs. interval between vortex flips $T(\Lambda)$. Dashed line – outside vortex, $\Lambda < \Lambda_c$, $m = 1$ ($|A_p| = A_2$), solid line – flipping vortex, all other cases ($|A_p| = A_1$).

and we have denoted $\delta\Lambda \equiv \Lambda - \Lambda_c$. However, exponential dependence on the left-hand side of (4.37) makes core radius essentially independent from $T(\Lambda)$. It does depend on the type of the vortex rearrangement: if the vortex participates in corner-center flips ($|A_p| = A_1$) ζ_c turns out to be considerably smaller than for the "outside" vortex, when $|A_p| = A_2$. We plot solutions of (4.37) versus flip interval T in Figure 5 for both cases. When T changes by an order of magnitude, change of ζ_c is only about 10%. Comparison with Fig. 2a, 2b shows that values of core radii obtained from (4.37) are in good agreement with the results of numerical simulation.

When $\delta\Lambda$ is small and T is large, (4.37) gives

$$\frac{t_s}{2\zeta_c^2} = \ln C_2 T - \ln(t_s/2) \tag{4.40}$$

where

$$C_2 = 2|A_p| \exp(\pi\sqrt{3}/6)/\pi^2 \tag{4.41}$$

and we can write explicit formula for the core radius in terms of the geometrical parameter of vortex motion $\delta\Lambda$:

$$\zeta_c = \left[\frac{t_s}{2(\ln\ln|\delta\Lambda|^{-1} + \ln 4C_2)} \right]^{1/2} \tag{4.42}$$

Formulae (4.37) and (4.42) predict that the core radius should tend to zero as $\delta\Lambda \to 0$ and $T \to \infty$. On the other hand, direct simulation of advection equation does not show any variation in core size as initial positions of vortices are taken closer and closer to critical configuration (compare Fig. 1d, where $\delta\Lambda \approx 9.6 \cdot 10^{-3}$, $T \approx 6.5$, and Fig. 2a where $\delta\Lambda \approx 2 \cdot 10^{-6}$, $T \approx 16.6$). Asymptotic of ζ_c (4.42) provides an explanation for this effect — ζ_c decreases extremely slow as $\Lambda \to \Lambda_c$ (note the double logarithm in (4.42)), and the accuracy of the simulation is insufficient to notice any significant decrease in core radius. Even machine precision calculation ($\delta\Lambda \sim 10^{-16}$, $T \approx 45$) will result in less than 10% change of ζ_c. The number of digits, required to "visualize" the limit $\zeta_c \to 0$ as $\delta\Lambda \to 0$ is grotesque.

5 Case of the weak chaos

The onset of chaotic advection in three-point-vortex system can be studied analytically by Melnikov method, and in cases of weak chaotization ($\Lambda \approx 0$ and $\Lambda \approx 1$) the width of chaotic layers can be found using separatrix map [38],[10] (see also [46]). In this section we carry out the calculations for the case $\Lambda \approx 1$ and compare the results with numerical simulation.

Below we consider the deviation of Λ from unity

$$\epsilon = 1 - \Lambda \tag{5.1}$$

as a small parameter and look at the advection for the case $\epsilon \ll 1$. When $\epsilon = 0$, $\Lambda = 1$ vortices stay in the vertices of an equilateral triangle uniformly rotating with frequency

$$\Omega = \frac{3k}{2\pi L^2} = \frac{3}{2\pi} \tag{5.2}$$

In the rotating reference frame their positions are constant (see (3.14),(B.24)):

$$\tilde{z}_m(t) = \frac{L}{\sqrt{3}} e^{-\frac{4\pi i}{3}(m-1)} \quad m = 1, 2, 3 \tag{5.3}$$

and the stream function of the flow (3.21) does not depend on time:

$$\tilde{\Psi}_0(\tilde{z}, \tilde{z}^*) \equiv -\frac{k}{4\pi} \ln |\tilde{z}^3 - \rho^3|^2 + \frac{\Omega}{2}|\tilde{z}|^2 \tag{5.4}$$

Advection in this case is regular — tracers follow the level lines of $\tilde{\Psi}_0$. Figure 6 shows singular points and separatrices of this case.

In the lowest order in ϵ corrections to $\tilde{z}_m(t)$ can be found by consequitive expansion of the corresponding formulae of Appendix B (see Appendix D for details, compare to [12]):

$$\tilde{z}_m(t) = \frac{L}{\sqrt{3}} \left(e^{-\frac{4\pi i}{3}(m-1)} + \left(\frac{2\epsilon}{3}\right)^{1/2} e^{-\frac{2\pi i}{3}(m-1)} e^{-i\Omega t} \right) \quad m = 1, 2, 3 \tag{5.5}$$

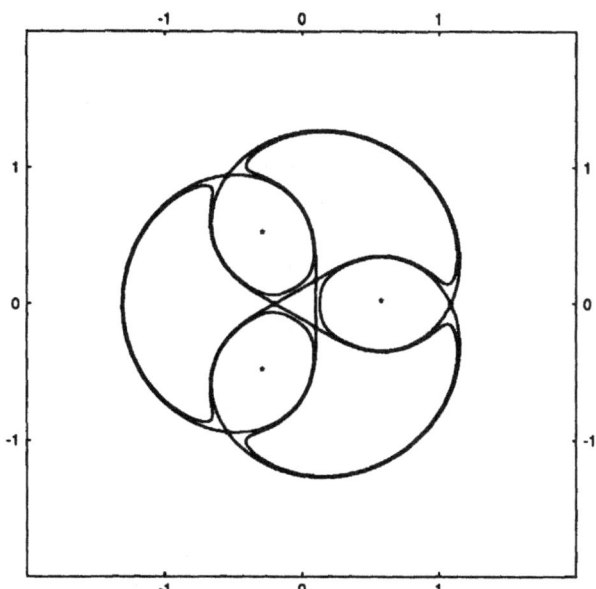

Figure 6: Separatrices of stream function (5.4), $\Lambda = 1$. Vortex positions (5.3) are marked with asterisks. Solid circles — hyperbolic points (5.11),(5.12).

Up to this order the stream function of the flow in the rotating frame is:

$$\tilde{\Psi}(\tilde{z}, \tilde{z}^*, t) = -\frac{k}{4\pi} \ln |\tilde{z}^3 - \rho^3 + (6\epsilon)^{1/2} e^{-i\Omega t} \tilde{z}\rho^2|^2 + \frac{\Omega}{2}|\tilde{z}|^2 + O(\epsilon) \qquad (5.6)$$

where we have introduced the distance from the vortex to the origin for stable ($\epsilon = 0$) configuration:

$$\rho \equiv L/\sqrt{3} = 3^{-1/2}. \qquad (5.7)$$

If we stay out of the immediate neighborhood of the vortices, i.e. $|\tilde{z} - \tilde{z}_m| \gg \sqrt{\epsilon}$, $m = 1, 2, 3$, we can rewrite (5.6) as:

$$\tilde{\Psi}(\tilde{z}, \tilde{z}^*, t) = \tilde{\Psi}_0(\tilde{z}, \tilde{z}^*) + \epsilon^{1/2} \tilde{\Psi}_1(\tilde{z}, \tilde{z}^*, t) + O(\epsilon) \qquad (5.8)$$

where the perturbation is given by:

$$\tilde{\Psi}_1(\tilde{z}, \tilde{z}^*, t) \equiv -\frac{k}{4\pi} 6^{1/2} \left(\frac{\tilde{z}\rho^2}{\tilde{z}^3 - \rho^3} e^{-i\Omega t} + c.c. \right) \qquad (5.9)$$

Equations of motion of a tracer follow from (5.8):

$$\dot{\tilde{z}}^* = [\tilde{z}^*, \tilde{\Psi}] = \frac{k}{2\pi i} \frac{3\tilde{z}^2}{\tilde{z}^3 - \rho^3} + i\Omega \tilde{z}^* - \frac{k}{2\pi i}(6\epsilon)^{1/2} \frac{\rho^2(\rho^3 + 2\tilde{z}^3)}{(\tilde{z}^3 - \rho^3)^2} e^{-i\Omega t} \qquad (5.10)$$

The separatrices of the unperturbed system split when $\epsilon \neq 0$ and are replaced by thin stochastic layers (see Fig 7). To proceed, we denote the two triples of hyperbolic points (Fig. 6) of $\tilde{\Psi}_0$ as

$$\tilde{z}_{s,1}(m) = \frac{2}{\sqrt{3}} \cos \frac{4\pi}{9} \exp(\pi i/3 + 2(m - 1/2)i\pi/3) \quad m = 1, 2, 3 \tag{5.11}$$

$$\tilde{z}_{s,2}(m) = \frac{2}{\sqrt{3}} \cos \frac{\pi}{9} \exp(2(m - 1)i\pi/3) \quad m = 1, 2, 3 \tag{5.12}$$

and corresponding values of $\tilde{\Psi}_0$ (separatrix energy) as:

$$\Psi_1^s \equiv -\frac{k}{4\pi} \ln |\tilde{z}_{s,1}^3 - \rho^3|^2 + \frac{\Omega}{2} |\tilde{z}_{s,1}|^2 \approx 0.2681 \tag{5.13}$$

$$\Psi_2^s \equiv -\frac{k}{4\pi} \ln |\tilde{z}_{s,2}^3 - \rho^3|^2 + \frac{\Omega}{2} |\tilde{z}_{s,2}|^2 \approx 0.2653 \tag{5.14}$$

For the separatrix map we define the pair (h_n, t_n) in a usual way: h_n is the deviation of the unperturbed stream function from its separatrix value

$$h_n \equiv \tilde{\Psi}_0 - \Psi_l^s \tag{5.15}$$

at the point, when the advected particle comes closest to the saddle point in its n-th passage through the saddle point neighborhood, and t_n is the moment of the center of the next velocity pulse. The mapping $\hat{T}_{sep} : (h_{n+1}, t_{n+1}) = \hat{T}_{sep}(h_n, t_n)$ has the form [38],[10]:

$$\begin{cases} h_{n+1} = h_n + \Delta h(t_n; \sigma_n) \\ t_{n+1} = t_n + \pi/\omega(h_{n+1}) \end{cases} \tag{5.16}$$

with the value of Melnikov integral given by

$$\Delta h(t_n; \sigma_n) \equiv \epsilon^{1/2} \int_{t_n-\infty}^{t_n+\infty} \{\tilde{\Psi}_0, \tilde{\Psi}_1\}(\tilde{z}_{\sigma_n}^{sep}(t - t_0), t)dt = \tag{5.17}$$

$$= 2i\epsilon^{1/2} \int_{t_n-\infty}^{t_n+\infty} \left(\frac{\partial \tilde{\Psi}_0}{\partial \tilde{z}} \frac{\partial \tilde{\Psi}_1}{\partial \tilde{z}^*} - c.c. \right) (\tilde{z}_{\sigma_n}^{sep}(t - t_0), t)dt \tag{5.18}$$

where the sign variable $\sigma_n = \pm 1$ indicates on which branch the separatrix solution $\tilde{z}_{\sigma_n}^{sep}(t - t_0)$ should be taken. In Appendix D the following expressions (D.16),(D.12) for $\Delta h(t_n; \sigma_n)$ and $\omega(h)$ (asymptotic for small h) are derived:

$$\Delta h(t_n; \sigma_n) = \frac{k}{\pi}(6\epsilon)^{1/2}\rho^2 \sin \Omega t_n M_{\sigma_n} \tag{5.19}$$

$$\omega(h) \approx \frac{2\pi\lambda}{\ln |h|^{-1}} \tag{5.20}$$

where constants λ, M_{σ_n} are listed in the Appendix D.

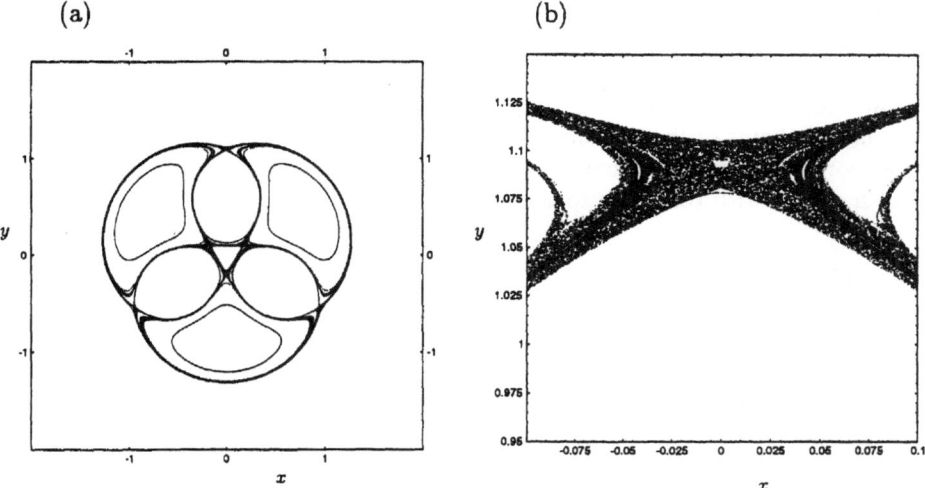

Figure 7: (a) Poincare section of tracer trajectories for case of weak chaos $1-\Lambda = \epsilon = 2.4 \cdot 10^{-7}$, $a = 0.4998$. Two thin stochastic layers around splitted separatrices did not merge yet, compare to Fig. 1a. (b) Zoom of the above picture near the saddle point.

The width of stochastic layer is found from the equation (see [38],[10])

$$\max \left| \frac{dt_{n+1}}{dt_n} - 1 \right| = \max \frac{\pi}{\omega^2} \left| \frac{d\omega}{dh} \frac{d\Delta h(t_n)}{dt_n} \right| \approx 1 \qquad (5.21)$$

Using (5.19),(5.20) we obtain the full width (both sides of separatrix) of the stochastic layer in stream function:

$$h_{sl} = \frac{\sqrt{6}(M_+ + M_-)}{4\pi^2 \lambda} \epsilon^{1/2} \qquad (5.22)$$

Evaluating the constants (see (5.20),(D.18),(D.19)) we find for the first (inner) triple of saddle points:

$$\Delta h_1 = 0.33\epsilon^{1/2} \qquad (5.23)$$

and for the second (outer):

$$\Delta h_2 = 0.29\epsilon^{1/2} \qquad (5.24)$$

The visible width of the layer, i.e. its space width on a plane, near the saddle points is given by

$$\Delta x = |\Delta h/\Psi''|^{1/2} \qquad (5.25)$$

where the derivative is taken along the direction transversal to the bisectrix of the angle between the separatrices. Separatrices of the system are very close to each other, see Fig. 6, 7 and (5.13),(5.14), and for fairly small $\epsilon = \epsilon_0 \sim 10^{-6}$ their stochastic layers merge together (see Fig. 1a), which puts a limit for the use of (5.22). We also have to keep in mind, that in reality the stochastic layer grows not smoothly, but step by step, absorbing thin stochastic layers around the island chains of the resonances just outside the main layer. Figure 7b shows the zoom of the outer stochastic layer near the saddle point for $\epsilon = 2.4 \; 10^{-7}$. Its width agrees with (5.25) up to the factor 1.5.

6 Conclusion

The main goal of the article was to provide a complete study of advected particle dynamics in the 3-point-vortex system. Applying different tools of the Hamiltonian chaos theory it was possible to estimate and derive different zones of stochastic and regular motion of a tracer particle. The most important situation is related to the generic case, when geometry of the vortices is on one hand not too close to the equilateral triangular, and on the other hand the distances between all three of them are still of the same order of magnitude, i.e. the system is not degenerated into perturbed two-point-vortex flow. In this case advection is strongly chaotic — the flow created by vortices has a large mixing region. However, there exist holes in the stochastic sea — non-mixing cores around vortices. External advected particle cannot approach any vortex closer than a critical distance ζ_c, while the particle originally inside such a hole, $\zeta < \zeta_c$, cannot escape from there. Moreover, in generic case ζ_c weakly depends on the initial geometry of 3-vortex system. This result explains different simulations where such holes were observed.

We also show that there exists stickiness of the advected particle to the boundary of the holes. This part of results is important for the problem of transport of particles, and it will be discussed in more detail elsewhere.

Acknowledgments

The authors would like to thank A. Provenzale, S. Boatto and T. Tél for providing the results of their work prior to publication. This work was supported by the US Department of Navy, Grant No. N00014-93-1-0218, and the US Department of Energy, Grant No. DE-FG02-92ER54184. One of us (LK) would like to acknowledge the support of Margaret and Herman Sokol Travel/Research Award.

Appendix

A   Equations for a point vortex trajectory

Here we derive equations for the trajectory of point vortex (2.10) from Helmholtz equation (2.8). To do so, we substitute in (2.8) expressions for vorticity in the form (2.4) and velocity in the form (2.7) and obtain:

$$k\delta'(x - x_1)\delta(y - y_1)\left(-\dot{x}_1 + \frac{k}{2\pi}\frac{y - y_1}{(x - x_1)^2 + (y - y_1)^2} + \frac{\partial \psi_R}{\partial y}\right) +$$

$$k\delta(x - x_1)\delta'(y - y_1)\left(-\dot{y}_1 - \frac{k}{2\pi}\frac{x - x_1}{(x - x_1)^2 + (y - y_1)^2} - \frac{\partial \psi_R}{\partial x}\right) +$$

$$\frac{\partial \omega_R}{\partial t} + (\mathbf{v}_R \nabla)\omega_R +$$

$$\left(\frac{k}{2\pi}\frac{y - y_1}{(x - x_1)^2 + (y - y_1)^2}\frac{\partial \omega_R}{\partial x} - \frac{k}{2\pi}\frac{x - x_1}{(x - x_1)^2 + (y - y_1)^2}\frac{\partial \omega_R}{\partial y}\right) = 0 \quad \text{(A.1)}$$

where

$$\omega_R \equiv -\frac{\partial^2 \psi_R}{\partial x^2} - \frac{\partial^2 \psi_R}{\partial y^2}; \qquad \mathbf{v}_R \equiv \left(\frac{\partial \psi_R}{\partial y}, -\frac{\partial \psi_R}{\partial x}\right) \quad \text{(A.2)}$$

are regular parts of vorticity and velocity at (x_1, y_1). Multiply this equation by $x - x_1$ and integrate over a circle of small radius ϵ centered at (x_1, y_1). Taking the limit $\epsilon \to 0$ we get

$$\dot{x}_1 = \frac{\partial \psi_R}{\partial y} \quad \text{(A.3)}$$

which is the first equation of (2.10). Multiplying (A.1) by $y - y_1$, integrating over the same circle and taking the limit $\epsilon \to 0$ we get the second equation of (2.10):

$$\dot{y}_1 = -\frac{\partial \psi_R}{\partial x} \quad \text{(A.4)}$$

B   Dynamics of three point vortices

In this section we obtain solution (3.14) of the equations of motion (3.2) of three point vortices of identical strength and polarization. We use the change of variables, introduced in [36], and repeat the derivation of the solution for the area variable $I(t)$ (see (B.18) below), presented there. Finally, we write down expressions for the "configuration angle" $\phi_1(t)$ and "rotation angle" $\phi_2(t)$ (see (B.30), (B.32)), so that the positions of vortices as functions of time (3.14), get completely specified.

Let us start from the equations (3.2):

$$\dot{z}_m^* = \frac{k}{2\pi i}\sum_{m \neq j}\frac{1}{(z_m - z_j)}, \qquad (j, m = 1, 2, 3) \quad \text{(B.1)}$$

As it was mentioned in the text, this is a Hamiltonian system with three independent integrals in involution. They are: Hamiltonian function itself

$$H = -\frac{1}{4\pi} \sum_{j \neq m} \ln |z_m - z_j|^2, \qquad (j, m = 1, 2, 3), \tag{B.2}$$

and two quadratic invariants:

$$Q^2 + P^2 = \sum_{i,j=1}^{3} z_i z_j^* \tag{B.3}$$

and

$$L^2 = \sum_{j=1}^{3} |z_j|^2. \tag{B.4}$$

The quadratic forms $Q^2 + P^2$ and L^2 can be simultaneously diagonalized by the following transformation to the new variables Q_n, P_n:

$$Q_n + i P_n \equiv \frac{L}{\sqrt{3}} \sum_{j=1}^{3} e^{i(2\pi n/3)(j-1)} z_j \qquad (n = 0, 1, 2) \tag{B.5}$$

which can be regarded as a discrete Fourier transform of an array of vortex positions. New variables have canonical Poisson brackets:

$$[Q_n, P_m] = \delta_{nm}, \quad [Q_n, Q_m] = 0, \quad [P_n, P_m] = 0, \quad (m, n = 0, 1, 2). \tag{B.6}$$

For $n = 0$ the transformation (B.5) shows that Q_0 and P_0 are proportional to the coordinates Q and P of the center of vorticity (3.6):

$$Q_0 = Q/\sqrt{3}, \qquad P_0 = P/\sqrt{3} \tag{B.7}$$

and are identically equal to zero, since we agreed to chose the origin of coordinates at the center of vorticity.

We have already reduced the number of variables to only two pairs (Q_n, P_n), $n = 1, 2$. It is convenient to use polar coordinates (J_n, θ_n) instead, defined by:

$$\sqrt{2 J_n} e^{i \theta_n} \equiv Q_n + i P_n, \qquad (n = 0, 1, 2) \tag{B.8}$$

because the invariant (B.4) is linear in J_1, J_2:

$$L^2 = 2(J_1 + J_2) \tag{B.9}$$

Vortex positions in terms of (J_n, θ_n) are given by an inverse Fourier transform of (B.5):

$$z_j = \frac{L}{\sqrt{3}} \sum_{n=1}^{2} \sqrt{2 J_n} e^{i \theta_n} e^{-2i\pi n(j-1)/3} \qquad (j = 1, 2, 3) \tag{B.10}$$

Substitution of (B.10) to (B.2) yields the Hamiltonian function in new variables:

$$H = -\frac{k}{4\pi}\ln[8(J_1^3 + J_2^3 - 2(J_1 J_2)^{3/2}\cos 3(\theta_2 - \theta_1))] \tag{B.11}$$

with Poisson brackets

$$[J_m, \theta_n] = \delta_{mn}, \quad [J_m, J_n] = 0, \quad [\theta_m, \theta_n] = 0, \quad (m, n = 1, 2) \tag{B.12}$$

Since (B.11) depends only on the angle difference $\phi_1 \equiv \theta_2 - \theta_1$, new canonical variables are useful:

$$\begin{array}{ll} I_1 = (J_2 - J_1)/2; & \phi_1 \equiv \theta_2 - \theta_1; \\ I_2 = (J_2 + J_1)/2; & \phi_2 \equiv \theta_2 + \theta_1; \end{array} \tag{B.13}$$

It follows from (B.9) that $I_2 = L^2/4$ is an integral of motion. Indeed, Hamiltonian, expressed in new variables

$$H = -\frac{k}{4\pi}\ln[16(I_2(I_2^2 + 3I_1^2) - (I_2^2 - I_1^2)^{3/2}\cos 3\phi_1)] \tag{B.14}$$

does not depend on ϕ_2. Variable I_1 has a geometrical interpretation, which is very helpful in the analysis of the types of vortex motions. From (B.5),(B.8),(B.13) it follows that

$$I_1 = \sigma A_{123}/\sqrt{3} \tag{B.15}$$

where A_{123} is the value of the area of the triangle with vertices in the current position of the point vortices (vortex triangle), and $\sigma = \pm 1$ is the orientation, which we chose to be $\sigma = +1$ for the counterclockwise arrangement of vortices with $m = 1, 2, 3$ along the perimeter of the vortex triangle, and $\sigma = -1$ otherwise.

The equation, governing the dynamics of I_1, follows from (B.14):

$$\dot{I}_1 = \frac{\partial H}{\partial \phi_1} = -\frac{12e^{4\pi H}}{\pi}(I_2^2 - I_1^2)^{3/2}\sin 3\phi_1 \tag{B.16}$$

Now we will use the geometrical parameter Λ defined in (3.11)

$$\Lambda = |z_1 - z_2||z_2 - z_3||z_3 - z_1|/L^3 = e^{-2\pi H} \tag{B.17}$$

instead of the value of the Hamiltonian H, and introduce new "area-variable" by

$$I \equiv (I_1/I_2)^2 = (16/3)A_{123}^2/L^4 \tag{B.18}$$

Using (B.14), we can exclude the angle ϕ_1 and get the equation for I only:

$$\left(\frac{dI}{d\tau}\right)^2 = -I[I^3 + 6I^2 + 3(3 - 8\Lambda^2)I + 8\Lambda^2(2\Lambda^2 - 1)] \equiv P_4(I; \Lambda) \tag{B.19}$$

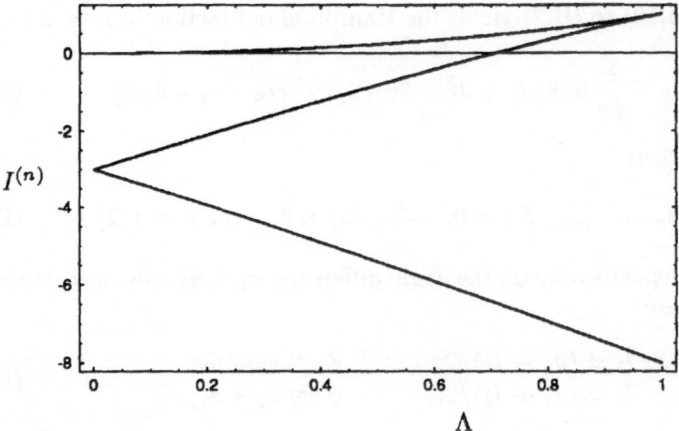

Figure 8: Roots of the cubic (B.21) vs. Λ.

where the rescaled time variable τ is:

$$\tau = \frac{3}{2\pi\Lambda^2 L^2} t \tag{B.20}$$

Dynamics of I_1 coincides with the motion of a particle of mass 2 in the potential $-P_4(I;\Lambda)$ with zero total energy, equation (B.19) being a law of energy conservation.

The potential $-P_4(I;\Lambda)$ has zeroes at $I = 0$ and at $I = I^{(n)}$, where $I^{(n)}$ are the roots of the cubic in (B.19):

$$I^{(n)}(\Lambda) = 2[(1 + 8\Lambda^2)^{1/2}\cos(1/3(2\pi n + \delta)) - 1], \quad n = 0, 1, 2 \tag{B.21}$$

with

$$\cos\delta(\Lambda) = (-8\Lambda^4 - 20\Lambda^2 + 1)/(1 + 8\Lambda^2)^{3/2} \tag{B.22}$$

They are always real and satisfy $I^{(1)} < I^{(2)} < I^{(0)}$. As Λ grows from 0 to 1, the largest root $I^{(0)}$ increases from 0 to 1, $I^{(2)}$ increases from -3 to 1, and $I^{(1)}$ decreases from -3 to -8. At $\Lambda = \Lambda_c = 1/\sqrt{2}$ root $I^{(2)}$ crosses zero, this value corresponds to the unstable collinear configuration $I = 0$ (saddle point), and aperiodic separatrix motion (see Fig. 8).

To write down the solution of (B.19), we arrange the zeros of potential in decreasing order, defining

$$J^{(0)} = I^{(0)}, \quad J^{(1)} = \max\{0, I^{(2)}\}, \quad J^{(2)} = \min\{0, I^{(2)}\}, \quad J^{(3)} = I^{(1)} \tag{B.23}$$

Now $I(t)$ can be expressed in terms of Jacobi elliptic functions:

$$I(\tau;\Lambda) = \frac{J^{(0)} - J^{(3)}\alpha^2 \mathrm{sn}^2(\gamma\tau)}{1 - \alpha^2 \mathrm{sn}^2(\gamma\tau)} \tag{B.24}$$

where

$$\alpha^2 = (J^{(0)} - J^{(1)})/(J^{(3)} - J^{(1)})$$
$$\gamma = (1/2)((J^{(0)} - J^{(2)})(J^{(1)} - J^{(3)}))^{1/2}$$
(B.25)

and modulus of the Jacobi elliptic function is:

$$\kappa = [(J^{(0)} - J^{(1)})(J^{(2)} - J^{(3)})]/[(J^{(0)} - J^{(2)})(J^{(1)} - J^{(3)})]$$
(B.26)

Solution (B.24) is periodic with period T equal to:

$$T(\Lambda) = \frac{4\pi}{3} \frac{\Lambda^2 L^2}{k} \frac{K(\kappa)}{\gamma}$$
(B.27)

where $K(\kappa)$ is the complete elliptic integral of the first kind. During the motion I stays between $J^{(1)}$ and $J^{(0)}$, it never reaches 0 for $\Lambda < \Lambda_c$, and does periodically for $\Lambda > \Lambda_c$.

Case $\Lambda = \Lambda_c$ is special, ($\kappa = 1$, period (B.27) has a logarithmic singularity), there is an unstable equilibrium solution $I = 0$, and a family of aperiodic soliton-like solutions given by:

$$I = \left[1 + \frac{2}{\sqrt{3}}\cosh(\sqrt{3}(\tau - \tau_0))\right]^{-1} = \left[1 + \frac{2}{\sqrt{3}}\cosh\left(\frac{3\sqrt{3}}{2\pi L^2}(t - t_0)\right)\right]^{-1}$$
(B.28)

where τ_0 (or t_0) corresponds to the center of the soliton and the scaling (B.20) was used with $\Lambda^2 = \Lambda_c^2 = 1/2$.

After one period T the vortex triangle repeats its shape, but the vortices get redistributed among the vertices of the triangle. It takes several periods T to return to the initial arrangement. This time defines the period of relative motion of the vortices, when the initial order of the vortices is restored, and is equal to:

$$T_{rel}(\Lambda) = \begin{cases} 2T & \text{if } 0 < \Lambda < \Lambda_c \\ 3T & \text{if } \Lambda_c < \Lambda < 1 \end{cases}$$
(B.29)

To specify the motion completely, we have to supply expressions for the angle variables ϕ_1 and ϕ_2. Conservation of H immediately yields for the "configuration angle" ϕ_1:

$$\cos 3\phi_1 = \frac{(1 + 3I) - 4\Lambda^2}{(1 - I)^{3/2}}$$
(B.30)

For "rotation angle" ϕ_2 we have to go back to Hamiltonian in the form (B.14), from which we get:

$$\dot{\phi_2} = -\frac{\partial H}{\partial I_2} = \frac{3}{4\pi\Lambda^2 L^2} \frac{4\Lambda^2 - I^2 - 3I}{1 - I}$$
(B.31)

Using (B.19) we can rewrite it as a quadrature:

$$\phi_2(t) = \int^{I(t)} \frac{4\Lambda^2 - I^2 - 3I}{2(1 - I)\sqrt{P_4(I)}} dI$$
(B.32)

where polynomial $P_4(I; \Lambda)$ is defined by (B.19).

During one period of relative motion, vortex triangle rotates by an angle $\Theta(\Lambda) = \phi_2(T_{rel}) - \phi_2(0)$, which can be found using (B.32). This defines the frequency of rotation of the vortex triangle:

$$\Omega \equiv \Theta(\Lambda)/T_{rel} = (\phi_2(T_{rel}) - \phi_2(0))/T_{rel} \qquad (B.33)$$

Finally we perform the inverse transformations, and write down an expression for vortex positions, similar to (B.10), in terms of constants of motion and functions $I(t)$, $\phi_1(t)$ and $\phi_2(t)$ defined by (B.24), (B.30) and (B.32):

$$z_m(t) = \frac{L}{\sqrt{6}} e^{i\phi_2(t)/2} \left[(1 - I^{1/2}(t))^{1/2} e^{-2\pi i(m-1)/3} e^{-i\phi_1(t)/2} + \right.$$

$$\left. + (1 + I^{1/2}(t))^{1/2} e^{-4\pi i(m-1)/3} e^{i\phi_1(t)/2} \right], \qquad (m = 1, 2, 3) \qquad (B.34)$$

C Derivation of adiabatic invariant

Here we derive the adiabatic invariant (4.22) and generating function (4.21) of the corresponding coordinate transformation.

Let us start from the stream function (4.14), expressed through the zero order action-angle variables (4.19),(4.20), considering slow time-dependence of $\tilde{z}_{mp}$ to be frozen:

$$\tilde{\Psi}(J, \theta; t) = H_0(J) + \epsilon H_1(J, \theta; t) \qquad (C.1)$$

with $H_0(J)$ obtained from (4.15),(4.19):

$$H_0(J) = -\frac{1}{4\pi} \ln |2J| - \Omega J \qquad (C.2)$$

and $H_1(J, \theta; t)$ from (4.16),(4.19),(4.20):

$$H_1(J, \theta) = -\frac{1}{4\pi} \sum_{m \neq p} \left[\ln |ie^{i\theta} \sqrt{|2J|} - \tilde{z}_{mp}|^2 + \frac{ie^{i\theta} \sqrt{|2J|}}{\tilde{z}_{mp}} - \frac{ie^{-i\theta} \sqrt{|2J|}}{\tilde{z}_{mp}^*} \right]$$
$$(C.3)$$

A parameter ϵ in (C.1) has been introduced to keep track of the order of different terms. It will be put $\epsilon = 1$ in the end of calculation.

Below we follow canonical perturbation theory (see [39] for the details of the method). Let us look for a transformation with generating function

$$S(\bar{J}, \theta) = \bar{J}\theta + \epsilon S_1(\bar{J}, \theta) + \dots \qquad (C.4)$$

such that, in new variables the Hamiltonian

$$\tilde{H}(\bar{J}; t) = \bar{H}_0(\bar{J}) + \epsilon \bar{H}_1(\bar{J}) + \dots \qquad (C.5)$$

does not depend on new angle up to the first order, and it depends adiabatically on time through $\tilde{z}_{mp}$. From (C.4) we get a coordinate transformation:

$$J = \bar{J} + \epsilon \frac{\partial S_1(\bar{J}, \bar{\theta})}{\partial \bar{\theta}} + \dots$$

$$\theta = \bar{\theta} - \epsilon \frac{\partial S_1(\bar{J}, \bar{\theta})}{\partial \bar{J}} + \dots \qquad (C.6)$$

so that Hamiltonian (C.5) is:

$$\tilde{H}(\bar{J}; t) = H(J(\bar{J}, \bar{\theta}), \theta(\bar{J}, \bar{\theta})) = H_0(\bar{J}) + \epsilon H_1(\bar{J}, \bar{\theta}) + \epsilon \omega(\bar{J}) \frac{\partial S_1}{\partial \theta} + \dots \qquad (C.7)$$

where we have used the frequency (4.18)

$$\omega(\bar{J}) = \frac{\partial H_0}{\partial \bar{J}} \qquad (C.8)$$

Now we expand $H_1(\bar{J}, \bar{\theta})$ (the second term in (C.7)), defined by (C.3) (we can put $\theta = \bar{\theta}$ in the first order), in Fourier series in θ, and introduce the averaged part:

$$< H_1 > \equiv \frac{1}{2\pi} \int_0^{2\pi} d\bar{\theta} H_1(\bar{J}, \bar{\theta}) = -\frac{1}{4\pi} \sum_{m \neq p} \ln |\tilde{z}_{mp}|^2 \qquad (C.9)$$

and the oscillating part:

$$\{H_1\} \equiv H_1 - < H_1 > = -\frac{1}{4\pi} \sum_{m \neq p} \left[\frac{|2J| e^{2i\theta}}{\tilde{z}_{mp}^2} + c.c. \right] + \dots \qquad (C.10)$$

where the dots in the last formula stay for the higher harmonics.

Since we require, that $\bar{H}_1(\bar{J})$ in (C.5) does not depend on θ, $\{H_1\}$ should get cancelled by the third term in (C.7), and we have the equation for the generating function S_1:

$$\omega(\bar{J}) \frac{\partial S_1}{\partial \theta} = -\{H_1\} \qquad (C.11)$$

which yields (4.21):

$$S(\bar{J}, \theta) = \bar{J}\theta + \left[\frac{1}{8\pi i \omega} \sum_{m \neq p} \frac{|2\bar{J}| e^{2i\theta}}{\tilde{z}_{mp}^2} + c.c. \right] \qquad (C.12)$$

Substituting it to (C.6), we obtain (4.22) for the first order adiabatic invariant $\bar{J}$.

D Stochastic layer width for the case $\Lambda \approx 1$

Solutions to the first order in $1-\Lambda$. As a first step to obtain the positions of the vortices up to the lowest order in ϵ (5.5), we expand the solution (B.24) of equation (B.19) in powers of ϵ. Using (B.21),(B.25),(B.26) we get:

$$\alpha^2 = -\frac{32\sqrt{2}}{81\sqrt{3}}\epsilon^{3/2} + O(\epsilon^2); \quad \gamma = 3/2 + O(\epsilon); \quad \kappa = \frac{256\sqrt{2}}{81\sqrt{3}}\epsilon^{3/2} + O(\epsilon^2) \quad \text{(D.1)}$$

and substituting to (B.24) obtain:

$$I(t) = 1 - 8/3\epsilon + O(\epsilon^{3/2}) \quad \text{(D.2)}$$

Substituting the above expression into (B.30) we find

$$\cos 3\phi_1 = \cos 3\tau \quad \text{(D.3)}$$

and taking into account that $\dot{\phi}_1 = -\frac{\partial H}{\partial I_1} > 0$ we get

$$\phi_1 = \tau \quad \text{(D.4)}$$

From (D.2) and the expression for ϕ_2 (B.32) it follows that up to the irrelevant constant phase

$$\phi_2 = \tau \quad \text{(D.5)}$$

Collecting everything (see (3.14)) we obtain (5.5):

$$\tilde{z}_m(t) = \frac{L}{\sqrt{3}}\left(e^{-\frac{4\pi i}{3}(m-1)} + \left(\frac{2\epsilon}{3}\right)^{1/2} e^{-i\tau - \frac{2\pi i}{3}(m-1)}\right) \quad \text{(D.6)}$$

where we have used definition (B.20) with $\Lambda = 1$.

Asymptotics of the frequency. The asymptotics of the frequency of near separatrix solution is defined by the motion in the neighborhood of the saddle point, in other words the period of motion is approximately equal to the time required to pass through this neighborhood. We introduce

$$\xi \equiv \tilde{z} - \tilde{z}_{s,l} \quad \text{(D.7)}$$

the distance from the saddle point, and expand stream function $\tilde{\Psi}_0$ (5.4) in its neighborhood (using definitions (5.2),(5.7)):

$$\tilde{\Psi}(\xi, \xi^*) = \Psi_l^s + \frac{1}{4\pi}\left[\frac{\tilde{z}_s^4 - 2\tilde{z}_s\rho^3}{\tilde{z}_s^3 - \rho^3}\frac{\xi^2}{2} + c.c.\right] + \frac{\Omega}{2}\xi\xi^* + O(\xi^3) \quad \text{(D.8)}$$

This expansion can be rewritten, using definition (5.15) as

$$h = a\xi^2 + 2b\xi\xi^* + a^*\xi^{*2} + O(\xi^3) \quad \text{(D.9)}$$

where

$$a \equiv \frac{1}{8\pi}\frac{\tilde{z}_{s,l}^4 - 2\tilde{z}_{s,l}\rho^3}{\tilde{z}_{s,l}^3 - \rho^3}; \quad b \equiv \Omega/4 \quad \text{(D.10)}$$

Hamiltonian (D.9) describes hyperbolic rotations with increment (decrement)

$$\lambda \equiv 4\sqrt{|a|^2 - b^2} \tag{D.11}$$

and for the time to pass the saddle point we have:

$$T^{(s)} \approx -\frac{\log|h|}{\lambda} \tag{D.12}$$

which is equivalent to (5.20).

Melnikov integral. To evaluate Melnikov integral, we use

$$\frac{\partial \tilde{\Psi}_0}{\partial \tilde{z}} = -\frac{k}{4\pi}\frac{3\tilde{z}^2}{\tilde{z}^3 - \rho^3} + \frac{\Omega}{2}\tilde{z}^* \tag{D.13}$$

and

$$\frac{\partial \tilde{\Psi}_1}{\partial \tilde{z}^*} = -\frac{k}{4\pi}6^{1/2}\rho^2\frac{\rho^3 + 2\tilde{z}^{*3}}{(\tilde{z}^{*3} - \rho^3)^2}e^{i\Omega t} \tag{D.14}$$

Substituting (D.13) and (D.14) into (5.18) and shifting integration variable we get:

$$\Delta h(t_n; \sigma_n) =$$
$$\frac{k}{\pi}(6\epsilon)^{1/2}\rho^2 \text{Im}\left[e^{-i\Omega t_n}\int_{-\infty}^{+\infty}e^{i\Omega t}\frac{\rho^3 + 2\tilde{z}^{*3}}{(\tilde{z}^{*3} - \rho^3)^2}\left(\frac{k}{4\pi}\frac{3\tilde{z}^2}{\tilde{z}^3 - \rho^3} + \frac{\Omega}{2}\tilde{z}^*\right)dt\right] \tag{D.15}$$

In this expression $\tilde{z}_{\sigma_n}^{sep}(t)$ is a solution of the unperturbed equation (5.10) on the separatrix, centered at $t = 0$. Particular branch of the separatrix is picked by $\sigma_n = \pm 1$. By an appropriate rotation the separatrix solution can be made symmetric with respect to the real axis, $\tilde{z}(-t) = \tilde{z}^*(t)$, so the value of the integral in (D.15) is real and we have

$$\Delta h(t_n) = \frac{k}{\pi}(6\epsilon)^{1/2}\rho^2 \sin \Omega t_n M_{\sigma_n} \tag{D.16}$$

where constants M_{σ_n} are defined by the integral:

$$M_{\sigma_n} \equiv 2\text{Re}\int_0^{+\infty}e^{i\Omega t}\frac{\rho^3 + 2\tilde{z}^{*3}}{(\tilde{z}^{*3} - \rho^3)^2}\left(\frac{k}{4\pi}\frac{3\tilde{z}^2}{\tilde{z}^3 - \rho^3} + \frac{\Omega}{2}\tilde{z}^*\right)dt \tag{D.17}$$

with $\tilde{z}_{\sigma_n}^{sep}(t)$ taken on the corresponding separatrix branch. Numerical evaluation of (D.17) gives for the inner separatrices:

$$M_+ = 1.70; \qquad M_- = 0.560 \tag{D.18}$$

and for the outer:

$$M_+ = 0.537; \qquad M_- = 0.790 \tag{D.19}$$

References

[1] H.Aref, Stirring by chaotic advection, J. Fluid Mech. **143**, 1 (1984)

[2] H.Aref, Chaotic advection of fluid particles, Phil. Trans. R. Soc. London **A 333**, 273 (1990)

[3] J. Ottino, *The kinematics of Mixing: Stretching, Chaos, and Transport* (Cambrige U. P., Cambrige, 1989)

[4] J. Ottino, Mixing, chaotic advection and turbulence, Ann. Rev. Fluid Mech. **22**, 207 (1990)

[5] V. Rom-Kedar, A. Leonard and S. Wiggins, An analytical study of transport mixing and chaos in an unsteady vortical flow, J. Fluid Mech. **214**, 347 (1990)

[6] S. Wiggins, *Chaotic Transport in Dynamical Systems* (Springer-Verlag, New York, 1992)

[7] A. Crisanti, M. Falcioni, G. Paladin and A. Vulpiani, Lagrangian Chaos: Transport, Mixing and Diffusion in Fluids, La Rivista del Nuovo Cimento, **14**, 1 (1991)

[8] A. Crisanti, M. Falcioni, A. Provenzale, P. Tanga and A. Vulpiani, Dynamics of passively advected impurities in simple two-dimensional flow models, Phys. Fluids A **4**, 1805 (1992)

[9] G. M. Zaslavsky, R. Z. Sagdeev and A. A. Chernikov, Stochastic nature of streamlines in steady-state flows, Sov. Phys. JETP **67**, 270 (1988)

[10] G. M. Zaslavsky, R. Z. Sagdeev, D. A. Usikov and A. A. Chernikov, *Weak chaos and quasiregular patterns* (Cambridge University Press, Cambridge, 1991)

[11] H.Aref and N. Pomphrey, Integrable and chaotic motion of four vortices, Phys. Lett. A **78**, 297 (1980)

[12] S. L. Ziglin, Nonintegrability of a problem on the motion of four point vortices, Sov. Math. Dokl. **21**, 296 (1980)

[13] Z. Neufeld and T. Tél, The vortex dynamics analogue of the restricted three-body problem: advection in the field of three identical point vortices, J. Phys. A: Math. Gen. **30**, 2263 (1997)

[14] S. Boatto and R. T. Pierrehumbert, Dynamics of a passive tracer in a velocity field of four identical vortices, unpublished

[15] T. H. Solomon and J. P. Gollub, Chaotic particle transport in time-dependent Rayleigh-Bénard convection, Phys. Rev. A **38**, 6280 (1988)

[16] T. H. Solomon and J. P. Gollub, Passive transport in steady Rayleigh-Bénard convection, Phys. Fluids **31**, 1372 (1988)

[17] V.V. Melezhko, M.Yu. Konstantinov, A.A. Gurzhi and T.P. Konovaljuk, Advection of a vortex pair atmosphere in a velocity field of point vortices, Phys. Fluids A **4**, 2779 (1992)

[18] L. Zanetti and P. Franzese, Advection by a point vortex in a closed domain, Eur. J. Mech., B/Fluids **12**, 43 (1993)

[19] G. Boffetta, A. Celani and P. Franzese, Trapping of passive tracers in a point vortex system, J. Phys. A: Math. Gen. **29**, 3749 (1996)

[20] Á. Péntek, T. Tél and Z. Toroczkai, Chaotic advection in the velocity field of leapfrogging vortex pair, J. Phys. A: Math. Gen. **28**, 2191 (1995)

[21] V.V. Meleshko, Nonstirring of an inviscid fluid by a point vortex in a rectangle, Phys. Fluids **6**, 6 (1994)

[22] G. M. Zaslavsky, M. Edelman, B. A. Niyazov, Self-similarity, renormalization, and phase space nonuniformity of Hamiltonian chaotic dynamics, Chaos **7**, 159 (1997)

[23] H.Aref, Integrable, chaotic and turbulent vortex motion in two-dimensional flows, Ann. Rev. Fluid Mech. **15**, 345 (1983)

[24] P.G. Saffman, *Vortex Dynamics* (Cambrige University Press, Cambrige, 1992)

[25] V.V. Melezhko, M.Yu. Konstantinov, *Dinamika vikhrevykh struktur*, (Naukova Dumka, Kiev, 1993) [in Russian]

[26] A. Babiano, G. Boffetta, A. Provenzale and A. Vulpiani, Chaotic advection in point vortex models and two-dimensional turbulence, Phys. Fluids **6**, 2465 (1994)

[27] R. Benzi, G. Paladin, S. Patarnello, P. Santangelo and A. Vulpiani, Intermittency and coherent structures in two-dimensional turbulence, J. Phys A **19**, 3771 (1986)

[28] R. Benzi, S. Patarnello and P. Santangelo, Self-similar coherent structures in two-dimensional decaying turbulence, J. Phys A **21**, 1221 (1988)

[29] J. B. Weiss, J.C. McWilliams, Temporal scaling behavior of decaying two-dimensional turbulence, Phys. Fluids A **5**, 608 (1992)

[30] J.C. McWilliams, The emergence of isolated coherent vortices in turbulent flow, J. Fluid Mech. **146**, 21 (1984)

[31] J.C. McWilliams, The vortices of two-dimensional turbulence, J. Fluid Mech. **219**, 361 (1990)

[32] D. Elhmaïdi, A. Provenzale and A. Babiano, Elementary topology of two-dimensional turbulence from a Lagrangian viewpoint and single particle dispersion, J. Fluid Mech. **257**, 533 (1993)

[33] G. F. Carnevale, J.C. McWilliams, Y. Pomeau, J. B. Weiss and W. R. Young, Evolution of Vortex Statistics in Two-Dimensional Turbulence, Phys. Rev. Lett. **66**, 2735 (1991)

[34] E. A. Novikov, Yu. B. Sedov, Stochastic properties of a four-vortex system, Sov. Phys. JETP **48**, 440 (1978)

[35] E. A. Novikov, Yu. B. Sedov, Stochastization of vortices, JETP Lett. **29**, 667 (1979)

[36] H.Aref and N. Pomphrey, Integrable and chaotic motion of four vortices. I. The case of identical vortices, Proc. R. Soc. Lond. a **380**, 359 (1982)

[37] V. K. Melnikov, On the stability of the center for time periodic perturbations, Trans. Moscow Math. Soc. **12**, 1 (1963)

[38] N. N. Filonenko, G. M. Zaslavsky, Stochastic instability of trapped particles and conditions of applicability of the quasi-linear approximation, Sov. Phys. JETP **25**, 851 (1968)

[39] A. J. Lichtenberg, M. A. Lieberman, *Regular and chaotic dynamics* (Springer-Verlag, New York, 1992)

[40] K.O. Friedrichs, *Special Topics in Fluid Dynamics* (Gordon and Breach, New York, 1966)

[41] E. A. Novikov, Dynamics and statistics of a system of vortices, Sov. Phys. JETP **41**, 937 (1975)

[42] H.Aref, Motion of three vortices, Phys. Fluids **22**, 393 (1979)

[43] J.L. Synge, On the motion of three vortices, Can. J. Math. **1**, 257 (1949)

[44] J. Tavantzis and L. Ting, The dynamics of three vortices revisited, Phys. Fluids **31**, 1392 (1988)

[45] L. D. Landau, E. M. Lifshits, *Mechanics* (Pergamon Press, New York, 1976)

[46] L. Kuznetsov and G.M. Zaslavsky, Hidden Renormalization Group for the Near-Separatrix Hamiltonian Dynamics, Phys. Reports **288**, 457 (1997)

Part 3:

Plasma and Turbulence

Tokamap: A Model of a Partially Stochastic Toroidal Magnetic Field

R. Balescu[1], M. Vlad[2] and F. Spineanu[2]

[1] Physique Statistique - Plasmas, Association Euratom - Etat Belge, Université Libre de Bruxelles, CP 231, Campus Plaine ULB, 1050 Bruxelles Belgium

[2] Permanent address: Institute of Atomic Physics, IFTAR, P.O. Box 7, Magurele, Bucharest Romania

1 Introduction

Magnetic confinement of a plasma (for the purpose of controlled thermonuclear fusion) is realized in tokamaks or stellarators by a magnetic field ideally structured by a set of nested toroidal *magnetic surfaces*. These are wound around a circular *magnetic axis*. The successive surfaces are labelled by the values of any *surface quantity* (i.e., a quantity that is constant on a magnetic surface) Ψ, playing the role of a radial coordinate. Each magnetic field line is tangent everywhere to a magnetic surface. Any point on such a surface is characterized by two angular coordinates: the poloidal angle θ (the short way around the torus) and the toroidal angle ζ (the long way around the torus): for convenience these angles are measured in radians divided by 2π.

To put the matter into quantitative form, we consider first this *ideal (unperturbed) situation*. We choose for the radial coordinate Ψ the *toroidal flux* $\widetilde{\psi}$, i.e., the magnetic flux through a surface perpendicular to the magnetic axis; for convenience, this quantity is made dimensionless by introducing $\psi = \widetilde{\psi}/B_0 a^2$, where B_0 is a characteristic magnetic field amplitude, and a is the minor radius of the tokamak. In the case of a circular torus, we have simply $\psi = r^2$, where r is the dimensionless radial coordinate (scaled with a). The magnetic axis thus corresponds to the value $\psi = 0$, and the edge of the torus to $\psi = 1$. The (stationary) magnetic field $\mathbf{B}(\mathbf{x})$ must satisfy the two constraints expressing its divergence-free nature, and its tangency to the magnetic surface $\Psi(\mathbf{x}) = const$:

$$\nabla \cdot \mathbf{B} = 0, \qquad \mathbf{B} \cdot \nabla \psi = 0. \tag{1}$$

The magnetic field satisfying these constraints is conveniently represented in the well-known *Clebsch form* [1] - [3]:

$$\mathbf{B} = \nabla \psi \times \nabla \theta - \nabla \alpha_0(\psi) \times \nabla \zeta, \tag{2}$$

where the surface quantity $\alpha_0(\psi)$ is the (dimensionless) *poloidal flux* [the magnetic field and the gradient operators are also made dimensionless by scaling

244

them with B_0 and a, respectively]. From this equation one finds the equations for the magnetic field lines expressed in the coordinates (ψ, θ, ζ) by using elementary geometrical formulae [3]. Using the toroidal angle as a running parameter, a field line is characterized by the two functions $\psi(\zeta)$, $\theta(\zeta)$ obeying the following differential equations:

$$\frac{d\psi}{d\zeta} = -\frac{\partial\alpha_0}{\partial\theta}, \quad \frac{d\theta}{d\zeta} = \frac{\partial\alpha_0}{\partial\psi} \tag{3}$$

As noted by many authors (e.g., [4] - [7]), the field line equations have *Hamiltonian structure:* α_0 plays the role of the Hamiltonian, ζ the role of "time", and ψ and θ appear as a pair of canonical variables [this property justifies the choice of ψ as a radial coordinate]. In the unperturbed case, when α_0 is a surface quantity, depending only on ψ, Eqs. (3) represent a 1-degree of freedom, hence integrable dynamical system:

$$\frac{d\psi}{d\zeta} = 0, \quad \frac{d\theta}{d\zeta} = W(\psi), \tag{4}$$

where the *winding number* (also called the *rotational transform*) is defined as folows:

$$W(\psi) = \frac{\partial\alpha_0(\psi)}{\partial\psi}. \tag{5}$$

[In the plasma physics literature, this quantity is often denoted by $\iota/2\pi$; its inverse $q = 1/W$ is called the *safety factor*]. Clearly, ψ is analogous to an action variable, a constant of the motion; the associated angle variable increases linearly in time.

The ideal structure described here is, however, strongly modified whenever some perturbation is present: the latter can be due to external features (such as imperfections in the coils producing the magnetic field) or to internal factors (i.e., instabilities or fluctuations). The topology of the magnetic field is then strongly modified: there appear *island chains*, together with undestroyed (but deformed) magnetic surfaces called *KAM barriers*, and in between these features there exist *chaotic orbits* filling a 3-dimensional region of space. This *incompletely chaotic structure* is generic for tokamaks; its understanding is a prerequisite for any realistic study of transport in such devices.

The *perturbed magnetic field* is conveniently represented in the Clebsch form (2), in which the unperturbed Hamiltonian is replaced by a function of all three coordinates:

$$\alpha_0 \to \alpha = \alpha_0(\psi) + K\,\delta\alpha(\psi, \theta, \zeta). \tag{6}$$

The perturbation Hamiltonian $K\,\delta\alpha$ is a 1-periodic function of the angles θ and ζ. The real, positive parameter K, is called the *stochasticity parameter*: it measures the strength of the perturbation. The corresponding equations of the field lines are now:

$$\frac{d\psi}{d\zeta} = -K\,\frac{\partial\,\delta\alpha(\psi,\theta,\zeta)}{\partial\theta},$$

$$\frac{d\theta}{d\zeta} = W(\psi) + K\,\frac{\partial\,\delta\alpha(\psi,\theta,\zeta)}{\partial\psi}. \tag{7}$$

These are the equations of motion of a $1\frac{1}{2}$ degrees of freedom dynamical system, which is, generically, non-integrable. This explains the appearance of the features described above.

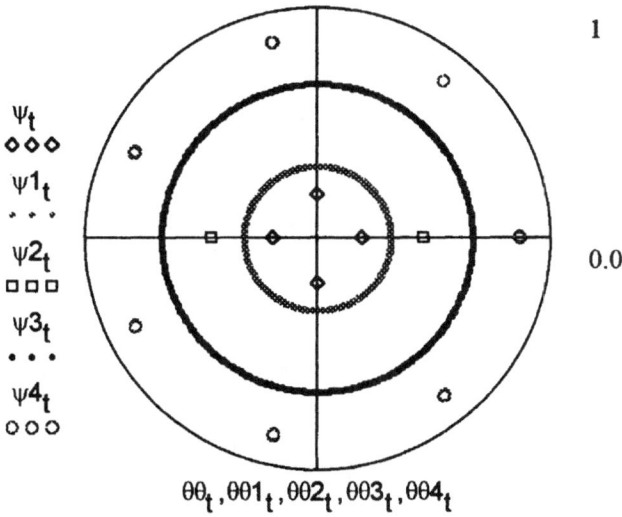

Fig. 1. Typical phase portrait of unperturbed system (polar representation).

The nature of the orbits is best studied by considering a Poincaré section on a plane perpendicular to the magnetic axis. For simplicity, we assume the cross-section of the torus to be circular. The Poincaré section of the *unperturbed* system consists of a set of concentric circles (coresponding to ergodic magnetic surfaces or KAM barriers) interspersed with discrete points (corresponding to rational values of the winding number) [Fig. 1]. An alternative graphical representation, which often provides a clearer picture, is obtained by making a cut starting from the center (magnetic axis), pulling the two lips apart, and expanding the point representing the magnetic axis into a line; after a mirror reflexion we obtain a square diagram; the radial coordinate ψ is represented on the vertical axis ($0 \le \psi \le 1$), and the poloidal angle is given, modulo 1, on the horizontal axis ($0 \le \theta \le 1$) [Fig. 2]. The singularity connected with the representation of the magnetic axis will be discussed below.

2 Construction of the tokamap

The solution of the field line equations (7) requires a very heavy numerical effort for achieving sufficient precision in the non-integrable case. It is therefore useful to construct simplified models based on discrete iterative *maps* rather than on differential equations in order to describe the Poincaré section of the magnetic field. With this mathematical tool, very long orbits are easily obtained even with a modest personal computer. It is generally not a simple matter to construct a map that is *exactly* equivalent to the starting differential equations (this would imply the solution of the latter, which is precisely what we want to avoid). One may wish to construct, instead, a model *ab initio*, and check its relevance *a posteriori*.

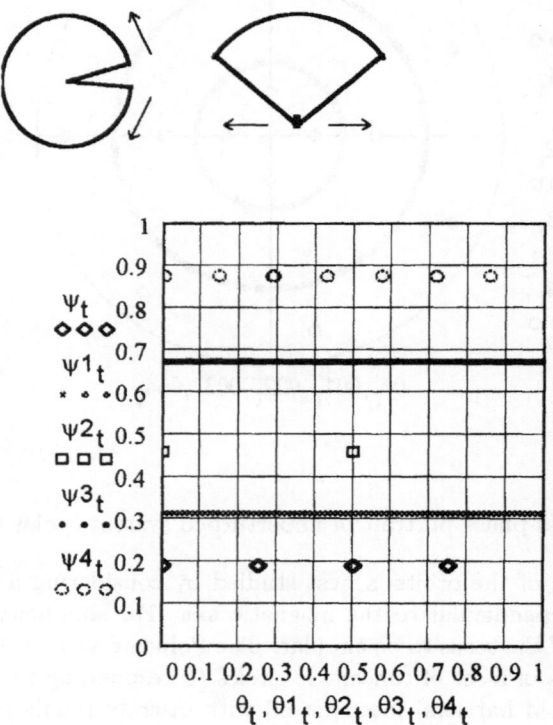

Fig. 2. Same as Fig. 1, "square representation".

Several authors have introduced maps representing various aspects of magnetic confinement devices. Punjabi, Boozer et al. [8] - [10] studied the X-point in a poloidal divertor geometry of a tokamap by means of very simple algebraic maps. Abdullaev and Zaslavsky studied the same divertor problem by means of a more sophisticated "separatrix map" [11], [12]. Abdullaev et al. recently constructed a map representing the effect of the dynamic ergodic divertor [13]. All these works are models of the edge region (scrape-off layer) of a tokamak. A global model of a specific stellarator (W VII-A) was introduced by Wobig in an important work [7] (see also [14]).

In the present work we construct a global model of the magnetic field in a tokamak by means of an *iterative two-dimensional* (ψ, θ) *map*. Such a map connects the values of the phase-space coordinates at "time" $\zeta = \nu + 1$ to their values at "time" $\zeta = \nu$, where ν is any non-negative integer: $\nu = 0, 1, 2, ...$:

$$\psi_{\nu+1} = P(\psi_\nu, \theta_\nu), \quad \theta_{\nu+1} = \Theta(\psi_\nu, \theta_\nu). \tag{8}$$

We hope that such a relatively simple model can be used as a starting point for the study of anomalous transport in an incompletely chaotic regime as prevails in a real tokamak.

The construction of such a map should satisfy a certain number of constraints. The first of these requires that the *Hamiltonian structure* of the differential equations (7) be reflected in the structure of the discrete model, which should be a *Hamiltonian* (or *symplectic*) map. The clearest way of constructing such a map starts from the fact [15] that in a Hamiltonian evolution the values of the canonical variables $\{\psi_{\nu+1}, \theta_{\nu+1}\}$ at time $(\nu + 1)$ are connected to their values $\{\psi_\nu, \theta_\nu\}$ at time ν by a *canonical transformation*. Such a transformation can be defined by means of a *generating function* of the new momentum and of the old angle [7]:

$$F(\psi_{\nu+1}, \theta_\nu) = \psi_{\nu+1}\theta_\nu + f(\psi_{\nu+1}, \theta_\nu). \tag{9}$$

The first term in the right hand side corresponds to the identity transformation. The transformation equations are:

$$
\begin{aligned}
\psi_\nu &= \frac{\partial F(\psi_{\nu+1}, \theta_\nu)}{\partial \theta_\nu} = \psi_{\nu+1} + \frac{\partial f(\psi_{\nu+1}, \theta_\nu)}{\partial \theta_\nu}, \\
\theta_{\nu+1} &= \frac{\partial F(\psi_{\nu+1}, \theta_\nu)}{\partial \psi_{\nu+1}} = \theta_\nu + \frac{\partial f(\psi_{\nu+1}, \theta_\nu)}{\partial \psi_{\nu+1}} \pmod 1
\end{aligned}
\tag{10}
$$

These equations define the map in a semi-implicit form (the explicit form is obtained by solving the first equation for $\psi_{\nu+1}$). The *unperturbed map* is obtained by taking: $f(\psi_{\nu+1}, \theta_\nu) = F_0(\psi_{\nu+1})$. We then obtain:

$$\psi_{\nu+1} = \psi_\nu, \quad \theta_{\nu+1} = \theta_\nu + W(\psi_{\nu+1}) \pmod 1. \tag{11}$$

This map represents the exact solution of the integrable Hamiltonian system (4). Indeed, ψ_ν remains constant and θ_ν increases by $W(\psi)$ upon each iteration (i.e., upon a toroidal turn of 2π). The winding number is simply related to the generating function:

$$W(\psi) = \frac{\partial F_0(\psi)}{\partial \psi}. \tag{12}$$

The phase portrait of this map is of the type of Figs. 1 or 2. For all the values of ψ such that $W(\psi) = \frac{n}{p}$ ($n, p \in \mathbb{Z}$), the p-th iterate coincides (modulo 1) with the starting point, i.e., we have a p-periodic orbit represented by a chain of p discrete points. For all irrational values of W, there is a KAM barrier, i.e. the orbit fills densely a horizontal segment in Fig. 2, or a circle around the origin in Fig. 1. The location of these features depends, of course, on the shape of the winding number function $W(\psi)$. Whenever this is a monotonous (growing or decreasing) function, Eq. (11) is called a *simple twist map*.

We now introduce a perturbation by considering a generating function of the form:

$$F(\psi_{\nu+1}, \theta_\nu) = \psi_{\nu+1}\theta_\nu + F_0(\psi_{\nu+1}) + K\,\delta F(\psi_{\nu+1}, \theta_\nu), \tag{13}$$

where K is the stochasticity parameter introduced in Eq. (6). The map (11) becomes:

$$\begin{aligned}
\psi_{\nu+1} &= \psi_\nu + K\,h(\psi_{\nu+1}, \theta_\nu), \\
\theta_{\nu+1} &= \theta_\nu + W(\psi_{\nu+1}) + K\,j(\psi_{\nu+1}, \theta_\nu).
\end{aligned} \tag{14}$$

From Eq. (10) we find the following definitions:

$$h(\psi_{\nu+1}, \theta_\nu) = -\frac{\partial\,\delta F(\psi_{\nu+1}, \theta_\nu)}{\partial\theta_\nu}, \quad j(\psi_{\nu+1}, \theta_\nu) = \frac{\partial\,\delta F(\psi_{\nu+1}, \theta_\nu)}{\partial\psi_{\nu+1}}. \tag{15}$$

It follows that:

$$\frac{\partial\,h(\psi_{\nu+1}, \theta_\nu)}{\partial\psi_{\nu+1}} + \frac{\partial\,j(\psi_{\nu+1}, \theta_\nu)}{\partial\theta_\nu} = 0. \tag{16}$$

Eqs. (14) with the functions h and j interrelated by Eq. (16) is the general form of a HAMILTONIAN MAP. It is easily checked that this map is area-preserving and possesses the symplectic property [16].

A simple realization of these constraints is obtained by taking $h = h(\theta_\nu)$, $j = j(\psi_{\nu+1})$: this corresponds to a *general twist map*. The fact that each function depends on a single variable greatly simplifies the analysis: the maps studied in most textbooks are of this type. An even more specific case is obtained by taking $h(\theta) = -(2\pi)^{-1}\sin 2\pi\theta$ and $j(\psi) \equiv 0$:

$$\psi_{\nu+1} = \psi_\nu - \frac{K}{2\pi}\sin 2\pi\theta_\nu, \qquad \theta_{\nu+1} = \theta_\nu + W(\psi_{\nu+1}). \tag{17}$$

Choosing also $W(\psi) = \psi$ we obtain the celebrated *Chirikov-Taylor standard map* studied by a great number of authors (e.g.: [17], [16], [18]).

The standard map is, however, *not a faithful model of a tokamak*, for several reasons. In the first place, the safety factor profile $q(r)$ is, in most tokamak experiments, a monotonously growing function of r, hence of $\psi = r^2$.[1] Typically, the safety factor on the magnetic axis is $q(0) = 1$, and at the edge: $q(1) = 4$. The value on axis may, however, be smaller $q(0) < 1$: this has an important effect on the properties of the discharge. It follows that in a tokamak *the winding number is a monotonously decreasing function of ψ*: this is just opposite of the standard map [2]. The map (17) with a monotonous W-profile will be referred to as the *standard twist map*.

A useful analytic form for this profile was derived by Misguich and Weyssow [19], by assuming that the density and electron temperature profiles in the tokamak are, respectively, $n(r) = n(0)\left[1 - r^2\right]$ and $T_e(r) = T_e(0)\left[1 - r^2\right]^2$; one then obtains, in the large aspect ratio limit:

$$W(\psi) = \frac{w}{4}\left(2 - \psi\right)\left(2 - 2\psi + \psi^2\right). \tag{18}$$

Here the positive constant $w = W(0)$ is the value of the winding number on the polar axis. By default it is taken equal to 1; however, the influence of its variation will be discussed below. It is easily checked that $W(\psi)$ is a monotonously decreasing function, reaching the value $w/4$ at the edge.

The standard twist map does not yet fulfill all the requirements for a faithful tokamak model. It follows from its geometrical meaning that the coordinate ψ *must be a definite positive number*; it may vary in the range $0 \leq \psi \leq \eta^2$, where $\eta = (R/a)$ is the aspect ratio of the torus, i.e. the ratio of the major radius to the minor radius. An indispensable condition is thus:

$$\text{If } \psi_0 > 0, \quad \text{then } \psi_\nu > 0, \quad \forall \nu. \tag{19}$$

On the other hand, the polar axis represented by $\psi = 0$ plays a special (singular) role in the toroidal (or cylindrical) geometry: As the radial coordinate can admit no negative values, the axis $\psi = 0$ represents (in the "square" representation, Fig. 2) a barrier that can never be crossed. This condition can be satisfied when the polar axis is globally invariant: an orbit starting on the axis remains forever on the axis:

$$\text{If } \psi_0 = 0, \quad \text{then } \psi_\nu = 0, \quad \forall \nu. \tag{20}$$

The no-crossing condition can, however, also be realized more weakly, by requiring that a point starting on the axis may either remain on the axis or move to a positive ψ (but never to a negative ψ), thus:

$$\text{If } \psi_0 = 0, \quad \text{then } \psi_\nu \geq 0, \quad \forall \nu. \tag{21}$$

[1] In recent experiments one produces a locally "reversed shear", i.e., a minimum of $q(r)$ near the magnetic axis; this has a beneficial effect on transport. In this case one has also $q(1) > q(0)$. This reversed shear configuration can easily be implemented in our map, but will not be discussed in the present article.

[2] Note, however that in a stellarator the winding number is an increasing function of ψ, though not as strongly as in the standard map. Wobig [7] models the W VII-A stellarator with $W(\psi) = \iota_0 + 0.01\psi$: a very small shear, indeed.

The standard twist map does not satisfy any of these conditions. It is well known that, for $K > 0$, an orbit starting at $\psi_0 = 0$ travels through both positive and negative values of ψ. Moreover, (in the globally chaotic regime), the orbits are 1-periodic in ψ (as well as in θ), hence $\psi = 0$ plays no special role: there is no unique polar axis in this map.

Wobig [7] took a first step towards solving the problem posed by the geometrical constraints by making the following ansatz for the generating function in Eq. (13): $\delta F(\psi_{\nu+1}, \theta_\nu) = -(2\pi)^{-2} \psi_{\nu+1} \cos 2\pi\theta_\nu$; using Eqs. (14), (15) we obtain: [3]

$$\psi_{\nu+1} = \psi_\nu - \frac{K}{2\pi} \psi_{\nu+1} \sin 2\pi\theta_\nu,$$

$$\theta_{\nu+1} = \theta_\nu + W(\psi_{\nu+1}) - \frac{K}{(2\pi)^2} \cos 2\pi\theta_\nu. \tag{22}$$

This *Wobig map* satisfies condition (20); it violates, however, condition (19) in a certain domain. Indeed, the explicit form of the first equation (22) is:

$$\psi_{\nu+1} = \frac{\psi_\nu}{1 + \frac{K}{2\pi} \sin 2\pi\theta_\nu}. \tag{23}$$

Clearly, whenever $K > 2\pi$, $\psi_{\nu+1}$ becomes *infinite* for a certain value of θ_ν and is *negative* for a range of θ_ν for all ψ_ν. Thus, Wobig's map cannot be accepted as a model valid over the whole parameter range [4].

We propose here an alternative ansatz for the generating function [5]:

$$\delta F(\psi_{\nu+1}, \theta_\nu) = -\frac{1}{(2\pi)^2} \frac{\psi_{\nu+1}}{1 + \psi_{\nu+1}} \cos 2\pi\theta_\nu, \tag{24}$$

which yields the following map:

$$\psi_{\nu+1} = \psi_\nu - \frac{K}{2\pi} \frac{\psi_{\nu+1}}{1 + \psi_{\nu+1}} \sin 2\pi\theta_\nu, \tag{25}$$

$$\theta_{\nu+1} = \theta_\nu + W(\psi_{\nu+1}) - \frac{K}{(2\pi)^2} \frac{1}{(1 + \psi_{\nu+1})^2} \cos 2\pi\theta_\nu. \tag{26}$$

The winding number profile considered here will be the one defined in Eq. (18), which depends on the parameter w. It should be clear, however, that the shape of $W(\psi)$ should be considered as being open to variation for studying

---

[3]Instead of $(2\pi)^2 \cos 2\pi\theta_\nu$, Wobig used a more general periodic function of θ_ν. This does not change our forthcoming argument.

[4]Wobig constructed his map as a specific model for the W VII-A stellarator; in that case, the empirical value of $K = 0.004$ is used. For this very small value, the map presents no problem and yields a quite satisfactory model of the device.

[5]A more general form of our map is obtained by replacing $\psi_{\nu+1} \to A\psi_{\nu+1}$ in Eq. (24) as well as in the argument of the winding number: $W(A\psi_{\nu+1})$; set also $K \to KA$ in Eq. (26). A is a positive parameter. This map displays very interesting symmetry properties, which will, however, not be discussed here. We thus set $A = 1$.

various regimes (just as in real experiments, the shape of the safety factor profile $q(r)$ can be "tailored"). The motivation of the construction of this map can be understood from the consideration of the "partially explicit" form:

$$\psi_{\nu+1} = \frac{\psi_\nu}{1 + \dfrac{K}{2\pi(1 + \psi_{\nu+1})} \sin 2\pi\theta_\nu}$$

Compared to the Wobig map (23) we note here a "self-healing" effect: whenever $\psi_{\nu+1}$ starts growing dangerously, the denominator $(1 + \psi_{\nu+1})$ decreases the factor of $\sin 2\pi\theta_\nu$ and the divergence is avoided.

In the form (25) the map is nonlinear: it possesses two solutions $\psi_{\nu+1}$ for given (ψ_ν, θ_ν). We make the following choice of the unique branch which provides us with the final definition of our map:

$$\psi_{\nu+1} = \frac{1}{2} \left\{ P(\psi_\nu, \theta_\nu) + \sqrt{[P(\psi_\nu, \theta_\nu)]^2 + 4\psi_\nu} \right\}, \tag{27}$$

where the function $P(\psi, \theta)$ is defined as:

$$P(\psi, \theta) = \psi - 1 - \frac{K}{2\pi} \sin 2\pi\theta. \tag{28}$$

Eqs. (26) - (28) define a one-valued iterative map that will be called the TOKAMAP. A simple analysis of Eqs. (25) - (28) shows that condition (19) is everywhere satisfied. It also shows that, for $K < 2\pi$, the polar axis is globally invariant, i.e., Eq. (20) is satisfied. For $K > 2\pi$, this invariance is lost (because $P(0, \theta) > 0$ in a certain range of θ), but Eq. (21) is always satisfied.

Summing up, *the tokamap is a Hamiltonian map, depending on two parameters (K and w), under which an initially positive radial coordinate ψ remains always positive, and the polar axis is a barrier that can never be crossed.*

3 Phase portraits of the tokamap

We show in Fig. 3 a typical phase portrait, corresponding to five orbits with $w = 1$ and $K = 2.55$ (400 iterations each). We see three island chains around periodic orbits corresponding to winding numbers $W = 1$, $1/2$ and $2/5$, as well as two KAM barriers. In the polar plot, the islands are represented as closed curves that do not encircle the origin, whereas the KAM barriers are curves enclosing the origin. The separatrix enclosing the $W = 1/2$ island chain is also visible, thus displaying the two hyperbolic (X-)points associated with the two elliptic (O-)points located at the center of the islands. The increased density of points near the X-points shows the beginning of a thin stochastic layer.

252

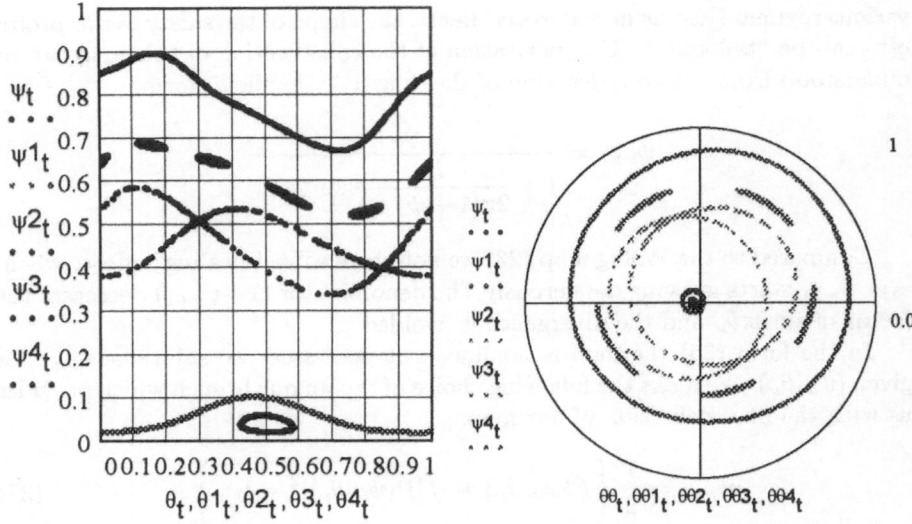

Fig. 3. Five tokamap orbits for $K = 2.55$.

In Fig. 4 we took a higher value of $K = 4.0$ and pictured four orbits (3000 iterations each). The remarkable feature here is that the outer part of the torus (near the edge) is now strongly chaotic around the islands $W = 1/2$, $1/3$; but the inner part of the phase space remains very robust and undestroyed. Pictures of this type might lead to a qualitative representation of a tokamap with an ergodic divertor.

We now note the following puzzling point. It is seen, especially in Fig. 3, that all island chains of period $p \geq 2$ have the "expected" structure predicted by the Poincaré-Birkhoff theorem [16], [18], which tells us that, whenever $W(\psi)$ is a monotonous function (i.e. for any twist map), a rational surface breaks under perturbation into an *even* number $2n$ of fixed points, n elliptic points alternating with n hyperbolic ones. This structure is not apparent for the period-1 fixed point appearing in Figs. 3 and 4: no corresponding X-point is visible. A careful analysis of the proof shows that the breakdown of the theorem is due to the special nature of the polar axis $\psi = 0$: the latter cannot be crossed by the map.

In order to study the problem quantitatively, we write down the equations determining the fixed points by using Eqs. (25 - 26) with $\psi_{\nu+1} = \psi_\nu = \psi$ and $\theta_{\nu+1} = \theta_\nu = \theta$:

$$\frac{\psi}{1+\psi} \sin 2\pi\theta = 0,$$

$$W(\psi) - \frac{K}{(2\pi)^2} \frac{1}{(1+\psi)^2} \cos 2\pi\theta = 0 \quad (\text{mod } 1) \tag{29}$$

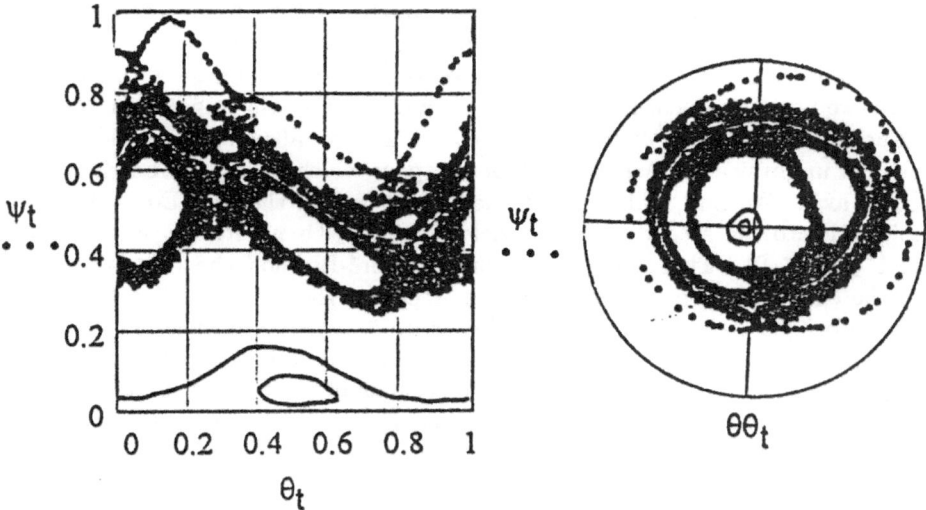

Fig. 4. Four tokamap orbits for $K = 4.0$.

The first equation is satisfied for $\theta = 0$ and $\theta = \frac{1}{2}$; upon substitution into the second equation, it is found numerically that the former yields no root for ψ in the physical domain $0 \leq \psi \leq 1$. The value $\theta = \frac{1}{2}$ yields, for every K, a physical root: it is precisely the fixed point seen in Figs. 3 and 4. A linear stability analysis (which we do not describe here) shows that this root corresponds, as expected, to an elliptic fixed point.

Next, we note that the first equation (29) is also satisfied by taking $\psi = 0$, in agreement with Eq. (20) expressing the invariance of the polar axis. At this point an important point should be made clear. The polar axis is represented as a single point in the polar representation (Fig.1), but shows up as a segment of length 1 in the "square" representation (Fig. 2). The axis should thus be thought of as being materialized by a very thin piece of wire of infinitesimally small radius, around which the various poloidal orientations are distinguishable. The invariance of the polar axis does *not* imply that each of its points is a fixed point: this is only true for $K = 0$. For finite K, setting $\psi_\nu = \psi_{\nu+1} = 0$ in Eq. (26) leaves us with a non-trivial one-dimensional map (which could be called "*aximap*") showing that, in general, any point of the axis moves under the map to a new poloidal position (in the polar language: moves around the axis by a certain angle):

$$\theta_{\nu+1} = \theta_\nu + w - \frac{K}{(2\pi)^2} \cos 2\pi\theta_\nu \quad (\text{mod } 1). \tag{30}$$

This map has fixed points [for which it reduces to the second equation (29)]:

for $w = 1$ there are two fixed points: $\theta = \frac{1}{4}$ and $\theta = \frac{3}{4}$, which are both physically acceptable. Stability analysis shows that both are X-points.

The inclusion of these points still does not save the Poincaré-Birkhoff theorem: we have now identified two X-points and one O-point in between: another O-point is "missing". The latter can be found by allowing a mathematical, though unphysical extension of the phase space which breaks the barrier of the polar axis. We find, indeed, that the root $\theta = 0$ of the first Eq. (29), substituted into the second equation yields a root with *negative* ψ in the range $(-1 \leq \psi \leq 0)$. Thus, for $w = 1$, the Poincaré-Birkhoff theorem is satisfied in the (unphysical) extension of the phase space: there are two fixed O-points and two fixed X-points in between (see Fig. 5, which is a blow-up of the region neighbouring the polar axis, for $K = 0.5$). It is important to note that this figure does *not* represent a chain of islands of period 2, but two *separate* islands of period 1. An orbit starting on the upper island remains there for ever (instead of wandering from one island to the other). The lower part ($\psi < 0$) of this picture is just a mathematical "ghost".

In a polar representation, the two X-points on the polar axis will not show up directly: they are squeezed into the single central point, which remains invariant although it does *not* represent a true fixed point. The role of the two X-points will appear, however, at higher values of K, even in the polar representation. The stochastic layer which develops around the period-1 island starts near the X-points, i.e., near the origin, in the $\theta = 1/4$ and $\theta = 3/4$ directions.

A very interesting phenomenon occurs when the parameter w is varied. When w increases beyond 1, the positive O-point moves slightly upwards, the left X-point moves to the left, the right X-point moves to the right and the negative O-point moves upwards. A *bifurcation value* $w = w_c(K)$ is reached when the formerly negative O-point and the two X-points merge into a single fixed point at the origin [6]. For $w > w_c$, the picture of Fig. 5 is completely changed. The "ghost" fixed points no longer exist; there remains only one positive O-point at $\theta = \frac{1}{2}$ and a positive X-point at $\theta = 0$ (Fig. 6). The Poincaré-Birkhoff theorem is satisfied in the positive-ψ sector.

It is instructive to compare the polar portraits (in the neighbourhood of the origin), for not too large K, below and above the bifurcation point. In the first case (Fig. 7), corresponding to the geometry of the upper part of Fig. 5, we see a set of closed curves centered (topologically) on the fixed O-point. There is no significant difference between cycles and KAM-curves in this representation; there is a smooth transition between curves that enclose the origin (KAM) and those that do not (cycles). Although the origin is invariant (in this representation), it is not a fixed point of the map: no islands develop around it. The fixed O-point thus represents the true *magnetic axis*, which has been shifted by the perturbation away from the geometrical polar axis.

---

[6] Keep in mind that $0 = 1 \pmod 1$!

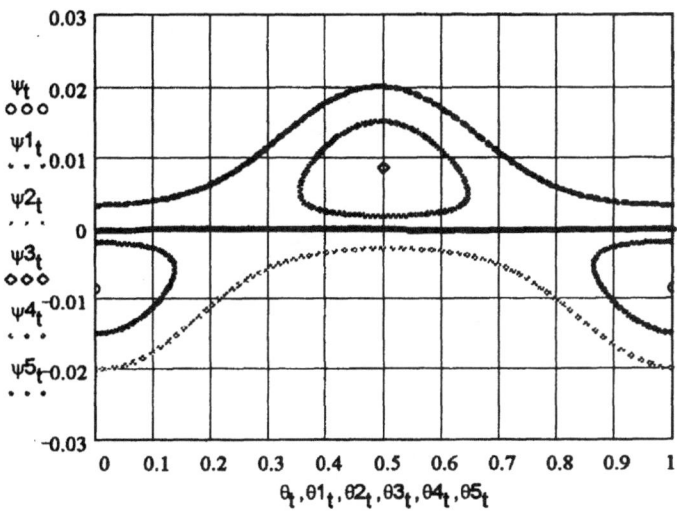

Fig. 5. Fixed points in the extended phase space for $K = 0.5$. $w = 1$.

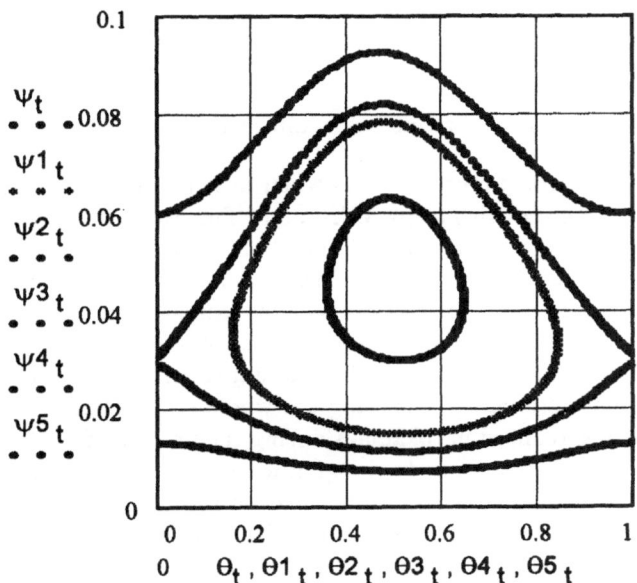

Fig. 6. Square phase portrait above bifurcation. $K = 0.5$, $w = 1.06$.

This is in agreement with the physics of tokamaks: it corresponds to an internal kink mode with $m = 1, n = 1$. It is known that this instability occurs whenever there is a $q = 1$ surface within the plasma, and that its effect is a radial displacement of the magnetic axis (see, e.g., [20], Sec. 6.4).

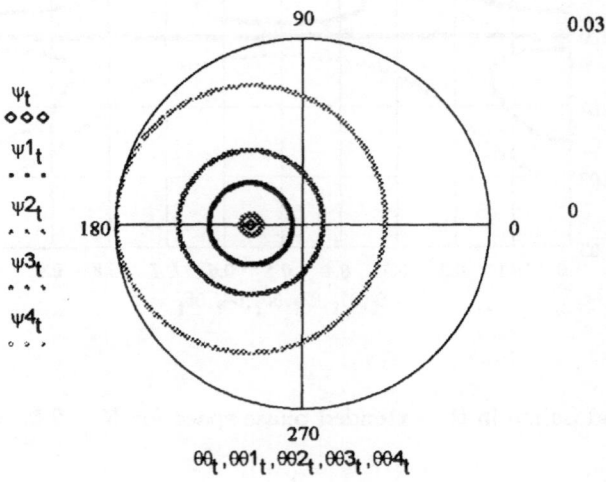

Fig. 7. Polar phase portrait below bifurcation. $K = 0.5$, $w = 1$.

An entirely different picture prevails when $w > w_c$ (Fig. 8). Here there is a well defined *separatrix*, with a single X-point. The cycles around the O-point in Fig. 6 become here crescent-shaped closed curves encircling the (displaced) magnetic axis. The lower KAM-curve which is outside the separatrix in Fig. 6, is however constrained by the barrier of the polar axis to remain above the latter. In the polar representation this yields a set of curves encircling the origin but remaining outside the separatrix. Finally, the upper KAM-curves of the square representation map into curves enclosing the whole separatrix. This seemingly complex topology is realized in tokamaks during the process of sawtooth oscillations. The "bubble" corresponds to a region with $q < 1$, which will be expelled in time by the development of an internal tearing instability. Fig. 8 is precisely identical with the field configuration predicted by the Kadomtsev model of the sawtooth instability (see [20], Sec. 7.6, [21], Sec. 7.2).

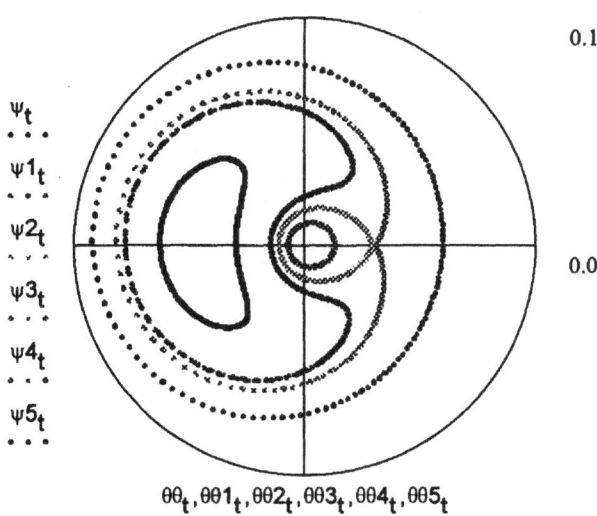

ψ_t
· · ·

$\psi1_t$
· · ·

$\psi2_t$
~ · ~

$\psi3_t$
· · ·

$\psi4_t$
· · ·

$\psi5_t$
· · ·

0.1

0.0

$\theta\theta_t, \theta\theta1_t, \theta\theta2_t, \theta\theta3_t, \theta\theta4_t, \theta\theta5_t$

Fig. 8. Polar phase portrait above bifurcation. $K = 0.5$, $w = 1.06$.

4 Tokamap and Continuous Time Random Walk

A quite different type of information about the tokamap dynamics is obtained
by a consideration of *time series*, in particular graphs of ψ_ν *vs.* ν. It is in
this representation that a very important aspect of the evolution is manifest:
stickiness. A chaotic orbit spends a long time near the boundaries of island
chains, KAM barriers and cantori. This property is due to the "braking" action
of the complex fractal structure of these boundaries (islands, around islands,
around islands,...). The fine structure of this process is discussed in other works
presented at this meeting. Here we take a rather different point of view in the
approach of this problem: we try to establish a "coarse grained" picture which
eventually leads to "macroscopic" equations of evolution.

This methodology was developed in a previous study of the standard map
[22]. It started from the observation that a graph of the successive iterations of
the radial coordinate presents rather regular oscillations in a certain "basin",
followed by a sudden jump to another mode of oscillation in a different basin,
which goes on for a certain time till another jump happens, etc. We just show
here that this type of behaviour is generic.

We consider, for $K = 3.7$, $w = 1$, a chaotic orbit starting at $\psi_0 = 0.20$,
$\theta_0 = 0.80$: this orbit remains confined in a region blocked by two KAM barriers
below and above a three-island chain (around $W = \frac{2}{3}$), as seen in the phase
portrait of Fig. 9 showing 3000 iterations.

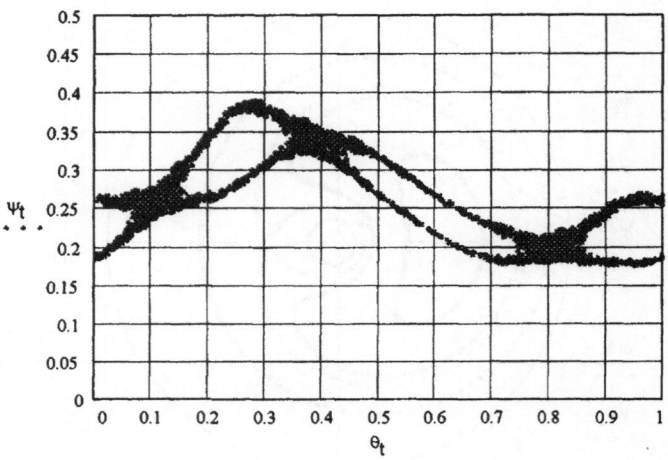

Fig. 9. Stochastic layer around 3-island chain; $K = 3.7$, $w = 1$.

In Fig. 10 we see a typical section of the time series (ψ_t vs. t) of this orbit, for $1700 \leq t \leq 2400$. The behaviour described above is manifest: three basins can be identified from three regimes of oscillation (mean position, amplitude, frequency). They correspond to motion encircling the island chain (basin O), motion above the islands (basin H) and motion below the islands (basin L).

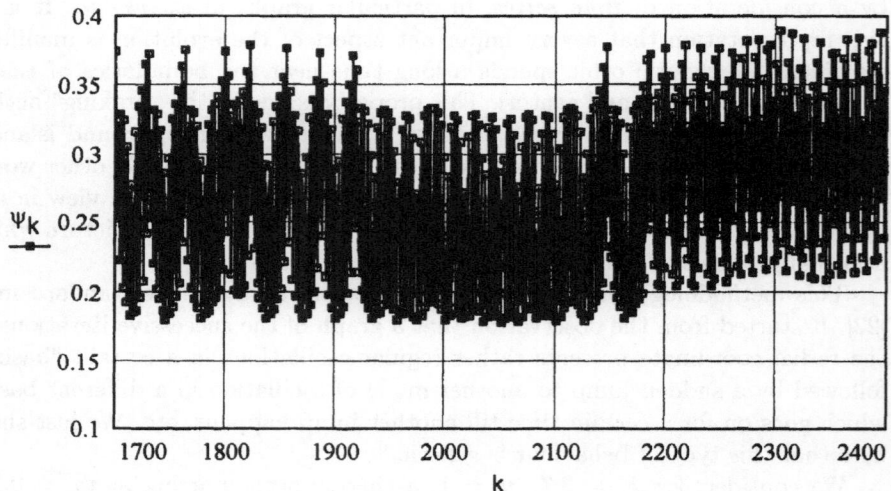

Fig. 10. Time series for motion in the stochastic layer of Fig. 9.

By plotting the phase space portrait for the various time ranges, this diagnostic is confirmed (Fig. 11).

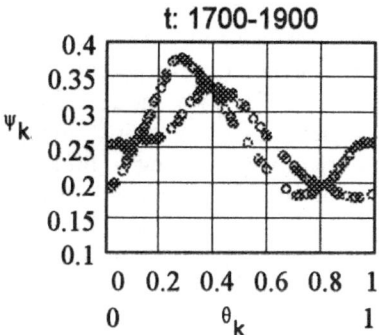

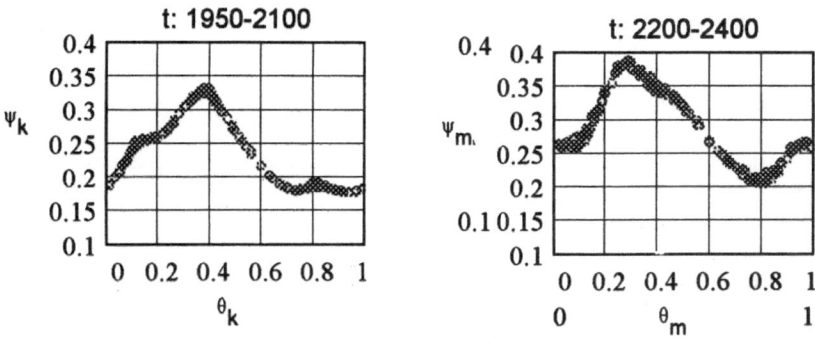

Fig. 11. Phase portraits of sections of the time series of Fig. 10.

This behaviour is exactly analogous to the one seen in the standard map [22]. It can be described by a *Continuous Time Random Walk (CTRW)*. The orbit is then globally described by a particle sojourning in a basin for a certain "time" ζ, making a transition to another basin, sojourning there, making another jump, etc. The process is completely defined by prescribing a *waiting time probability distribution* in basin m: $\psi_m(\zeta)$, and a transition probability from basin m to basin n: f_{nm}. Both quantities can be determined by an analysis of long time series.

Using then standard techniques of random walk theory, the probability density of finding the particle in basin m at time t, $n_m(\zeta)$ can be found exactly, and from there on quantities like the mean square displacement and the running diffusion coefficient can be calculated. We do not develop here these questions, as the method is exactly the same as in Ref. [22]; moreover, another paper on this subject has been presented at this meeting [23].

5 Conclusions

We have shown that a simple Hamiltonian map can be constructed, fulfilling the minimum requirements for a representation of a magnetic field in toroidal geometry. This tokamap describes a structure that is very robust in the central region, the stochasticity starting (for increasing K) in the edge region: the map could therefore prove useful as a model of a tokamak with an ergodic divertor. The central region has some quite interesting topological features, which can change dramatically (including a bifurcation) as the value of the safety factor on axis is varied. Typical configurations known from tokamak physics are qualitatively reproduced by the map.

Many more properties of the tokamap have been or will be studied in forthcoming works. These include questions such as the influence of the shape of the winding number, the dependence on of various physical properties, similarity and scaling properties. Last but not least, we intend to put charged particles in this magnetic field and study the transport properties in a partially chaotic tokamak configuration. This problem, which is very poorly understood, is of crucial importance for fusion physics.

6 Acknowledgment

Very fruitful discussions with R. White and J. Misguich are gratefully acknowledged.

References

[1] Hinton, F.L. and Hazeltine, R.D., Rev. Mod. Phys., **48**, 239 (1976)

[2] Boozer, A.H., Phys. Fluids, **27**, 2055 (1984)

[3] Balescu, R., *Transport Processes in Plasmas*, vol. 2, North Holland, Amsterdam, 1988

[4] Cary, J.R. and Littlejohn, R.G., Ann. Phys., **151**, 1 (1983)

[5] Salat, A., Z. Naturforsch., **40a**, 959 (1985)

[6] Elsässer, K., Plasma Phys., **28**, 1743 (1986)

[7] Wobig, H., Z. Naturforsch., **42a**, 1054 (1987)

[8] Punjabi, A., Verma, A. and Boozer, A., Phys. Rev. Lett., **69**, 3322 (1992)

[9] Punjabi, A., Verma, A. and Boozer, A., J. Plasma Phys., **52**, 91 (1994)

[10] Punjabi, A., Ali, H. and Boozer, A., Phys. Plasmas, **4**, 337 (1997)

[11] Abdullaev, S.S. and Zaslavsky, G.M., Phys. Plasmas, **2**, 4533 (1995)

[12] Abdullaev, S.S. and Zaslavsky, G.M., Phys. Plasmas, **3**, 516 (1996)

[13] Abdullaev, S.S., Finken, K.H., Kaleck, A. and Spatschek, K.H., *"Modelling of tokamak ergodic divertor by twist mapping"* (to be published) 1997

[14] Mendonça, J.T., Phys. Fluids, **B 3**, 87 (1991)

[15] Goldstein, H., *Classical Mechanics,* 2d. ed., Addison-Wesley, New York, 1980

[16] Ott, E., *Chaos in Dynamical Systems*, Cambridge Univ. Press, Cambridge, 1993

[17] Chirikov, B., Phys. Rep., **52**, 265 (1979)

[18] Lichtenberg, A.J. and Lieberman, M.A., *Regular and Stochastic Motion*, Springer, Berlin, 1983

[19] Misguich, J.H. and Weyssow, B., Euratom - CEA Internal Report NTΦ 8, Cadarache (1989)

[20] Wesson, J., *Tokamaks*, Clarendon Press, Oxford, 1987

[21] Kadomtsev, B.B., *Tokamak Plasma, a Complex Physical System*, Inst. of Phys. Publ., Bristol, 1992

[22] Balescu, R., Phys. Rev. **E 55**, 2465 (1997)

[23] Misguich, J., Reuss, J.D., Elskens, Y. and Balescu, R., (these proceedings)

Lagrangian Chaos and the Fast Kinematic Dynamo Problem

Edward Ott*

Institute for Plasma Research, University of Maryland, College Park, Maryland

Abstract. In this paper we review results on the fast kinematic dynamo problem, emphasizing the recent realization that Lagrangian chaos of the underlying flow is the key element for understanding of the problem. We also discuss the generic tendency for fractal magnetic field distributions with extreme cancellation properties. The relation of ergodic properties of the chaotic flow to properties of the dynamo (e.g., growth rate, fractal dimension) are also reviewed.

1 Background

One of the most basic observed facts of nature is the presence of magnetic fields wherever there is flowing electrically conducting matter. In particular, magnetic fields are observed to be present in planets with liquid cores, in the Sun and stars, and in the Galaxy. A natural question is why this is so. The most common approach to this question is to consider the kinematic dynamo problem: *Will a small seed magnetic field in an initially unmagnetized flowing electrically conducting fluid amplify exponentially in time?* If the answer is yes, then it is unnatural for magnetic fields not to be present. Note that the kinematic dynamo problem is essentially a problem of linear stability. Thus the structure of magnetic fields as they are currently observed is not directly addressed, since current fields presumably have evolved to a nonlinear saturated state.

The answer to the stability question posed by the kinematic dynamo problem depends on the flow field of the fluid and on the electrical conductivity of the fluid. For a given flow field one can, in principle, ask for the conductivity dependence of the exponential growth rate Γ of a magnetic field perturbation. Vainshtein and Zeldovich [1] suggest a classification of kinematic dynamos based on the electrical conductivity dependence of Γ. In particular, if Γ approaches a positive constant as the conductivity approaches infinity, then they call the dynamo a *fast* dynamo. Otherwise they call it a *slow* dynamo. This important distinction is illustrated schematically in Fig. 1. The horizontal axis in Fig. 1 is the magnetic Reynolds number, R_m, which can be regarded as the dimensionless electrical conductivity; $R_m = \mu_0 v_0 L_0 \sigma$, where μ_0 is the (mks) magnetic permittivity of vacuum, v_0 is a typical magnitude of the flow velocity, L_0 is a typical length scale for spatial variation of the flow, and σ is the electrical conductivity of the fluid. In the Sun, for example, $R_m > 10^8$. Thus only fast kinematic

* Also Department of Physics, Department of Electrical Engineering, and Institute for Systems Research

dynamos are of interest in such cases. In this paper we shall be concerned with fast dynamos.

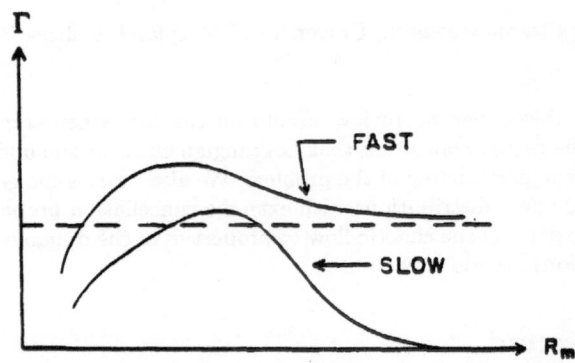

Fig. 1. Γ versus R_m for fast and slow kinematic dynamos.

We adopt the simplest MHD (magnetohydrodynamic) description. The basic equation (assuming $\nabla \cdot \mathbf{v} = 0$) is

$$\partial \mathbf{B}/\partial t + \mathbf{v} \cdot \nabla \mathbf{B} = \mathbf{B} \cdot \nabla \mathbf{v} + R_m^{-1} \nabla^2 \mathbf{B} \ , \qquad (1)$$

where t has been normalized to v_0/L_0, spatial scales have been normalized to L_0, and $\mathbf{v}$ has been normalized to v_0. Note that, for the kinematic dynamo problem, Eq. (1) is a linear equation in $\mathbf{B}$, because there is no linear response of the velocity, since the Lorenz force, $\mathbf{J} \times \mathbf{B} = \mu_0^{-1}(\nabla \times \mathbf{B}) \times \mathbf{B}$, is quadratic in $\mathbf{B}$. Thus we may regard $\mathbf{v}$ as an "equilibrium" field determined by factors (e.g., convection, stirring, rotation) not appearing in Eq. (1). Our main points are the following:

1. Lagrangian chaotic three dimensional flows typically yield fast kinematic dynamos.
2. In the limit $R_m \rightarrow \infty$ kinematic dynamo magnetic fields are expected to concentrate on a fractal set in space.
3. In the limit $R_m \rightarrow \infty$ the magnetic field has a singular tendency for flux cancellation.

[With respect to point 2 above, we shall be concerned with flows $\mathbf{v}(\mathbf{x}, t)$ that are "smooth" in that they possess no fractal properties of their own. In particular, they do not have power law wavenumber spectra. Thus we think of the spatial Fourier wavenumber spectrum of $\mathbf{v}$ as being peaked at some low wavenumber of order $1/L_0$ and decaying exponentially or faster with increasing wavenumber.]

2   The Relevance of Chaos

The equation describing the position $\mathbf{x}(t)$ of a fluid element is

$$dx(t)/dt = \mathbf{v}(\mathbf{x}(t), t) \ , \tag{2}$$

where $\mathbf{v}(\mathbf{x}, t)$ is the Eulerian velocity field of the flowing fluid. We say the flow is *Lagrangian chaotic* if the ordinary differential equations (1) exhibit chaos. In particular, consider two initial conditions, $\mathbf{x}(0)$ and $\mathbf{x}(0) + \delta\mathbf{x}(0)$, where $\delta\mathbf{x}(0)$ is a differential displacement. Let $\mathbf{x}(t)$ and $\mathbf{x}(t) + \delta\mathbf{x}(t)$ denote the orbits from these two initial conditions. Then for Lagrangian chaos $|\delta\mathbf{x}(t)|$ grows exponentially for an arbitrary choice for the orientation of $\delta\mathbf{x}(0)$,

$$|\delta\mathbf{x}(t)| \sim |\delta\mathbf{x}(0)| \exp(ht) \ , \tag{3}$$

where $h > 0$ is the (largest) Lyapunov exponent of the flow. The evolution of $\delta\mathbf{x}$ follows from taking a differential variation of Eq. (2)

$$d\delta\mathbf{x}/dt = \delta\mathbf{x} \cdot \nabla\mathbf{v}(\mathbf{x}(t), t) \ . \tag{4}$$

Now consider Eq. (1) in the "ideal limit" which corresponds to omitting the term $R_m^{-1}\nabla^2\mathbf{B}$,

$$d\tilde{\mathbf{B}}/dt \equiv \partial\tilde{\mathbf{B}}/\partial t + \mathbf{v} \cdot \nabla\tilde{\mathbf{B}} = \tilde{\mathbf{B}} \cdot \nabla\mathbf{v} \ , \tag{5}$$

where we use the symbol $\tilde{\mathbf{B}}$ for magnetic fields in the ideal limit. Comparing Eq. (4) for $\delta\mathbf{x}$ and Eq. (5) for $\tilde{\mathbf{B}}$ we see that the equations are the same. This is a consequence of the frozen in nature of the magnetic field at infinite conductivity, and means that the magnetic field grows in proportion to the stretching of magnetic field lines by the flow. The connection between fast dynamos and chaos is now clear: chaos implies exponential growth of $\delta\mathbf{x}$ in Eq. (4) and hence exponential field line stretching, and for a dynamo we need exponential growth of $\mathbf{B}$. There is a catch, however. In particular, the ideal equation (5) can never be fully justified even for very large R_M. What typically happens for chaotic flows is that as R_m becomes large $\mathbf{B}$ develops more fine scale structure, so that $R_m^{-1}\nabla^2\mathbf{B}$ in Eq. (1) remains of the same order as the other terms in (1). This implies that $\mathbf{B}$ varies on small spatial scales of order

$$\epsilon_* \sim R_m^{-1/2} \ . \tag{6}$$

(Recall that we use the normalizations introduced in (1) so that $\mathbf{v} \sim 0(1)$ and the typical scale for spatial variation of $\mathbf{v}$ is also $0(1)$.)

In spite of this the ideal treatment is still a powerful (and correct) indication that Lagrangian chaos is the key to fast dynamo action. This point was first explicitly made in the paper of Arnold, Zeldovich, Ruzmaikin and Sokoloff [2] who considered a chaotic flow in an abstract space of constant negative geodesic curvature (not the usual Euclidian space of classical physics), and the point was subsequently [3,4] made more physically relevant by considerations for flows in ordinary Euclidian space. By now this consideration is well-developed. An incomplete list of some representative papers is Refs. 5–24. An extensive summary is contained in the book by Childress and Gilbert [25].

Returning to the issue of how solutions of Eq. (1) for **B** at large R_m are related to those of the ideal equation (Eq. (5)) for $\tilde{\mathbf{B}}$, we note that Ref. 4 has pointed out that the dynamo growth rate for **B** obtained from (1) at large R_m should be related to solutions of the ideal equation (Eq. (5)). In particular, it is argued in Ref. 4 that the flux through finite open surfaces is robust under the limit $R_m \to \infty$. Specifically, let

$$\tilde{\phi}_S(t) = \int_S \tilde{\mathbf{B}} \cdot d\mathbf{A} \ , \tag{7}$$

where $\tilde{\phi}_S$ and $\tilde{\mathbf{B}}$ are the flux and magnetic field obtained via solution of the ideal equation with some arbitrarily chosen smooth initial condition $\tilde{\mathbf{B}}(\mathbf{x}, 0)$, and S a smooth fixed open surface. Then, according to Ref. 4, $\Gamma(R_m)$, the growth rate for the fastest growing unstable mode obtained from the solution of the linear instability problem for Eq. (1) at finite R_m, satisfies

$$\lim_{R_m \to \infty} \Gamma(R_m) \equiv \Gamma_* = \tilde{\Gamma} \ , \tag{8}$$

where $\tilde{\Gamma}$ is the *ideal* flux growth rate,

$$\tilde{\Gamma} = \limsup_{t \to \infty} [t^{-1} \ln \tilde{\phi}_S(t)] \ . \tag{9}$$

That is, for large finite R_m, the growth rate $\Gamma(R_m)$ is approximately $\tilde{\Gamma}$ which can be obtained from the ideal equation. Later in this paper we review tests of (8) from numerical solutions of (1) for smooth Lagrangian chaotic flows at high magnetic Reynolds number. To motivate (8) we note that, during a time interval Δt, the term $R_m^{-1} \nabla^2 \mathbf{B}$ diffuses the magnetic field over a distance of the order of $\sqrt{\Delta t / R_m}$. Setting $\Delta t = \tilde{\Gamma}^{-1}$ and recalling our normalizations, we have $\tilde{\Gamma} \sim 0(1)$ ($\tilde{\Gamma} \sim 0(v_0/L_0)$ in the unnormalized situation). Thus the diffusive rearrangement of the **B** field during a growth time occurs over the small scale $\epsilon_* \sim R_m^{-1/2}$ (Eq. (6)). Since for large enough R_m, the linear size of the fixed area S is large compared to ϵ_*, diffusive rearrangement of **B** over the scale ϵ_* during $\Delta t = \tilde{\Gamma}^{-1}$ has little effect on the flux growth through S, and (8) then follows.

The option of using the ideal equation and (8) to approximate $\Gamma(R_m)$ at large R_m is attractive because (8) is much easier to numerically solve than (1) at large R_m. Indeed the fractal and cancellation properties of the solution of (1) mentioned at the end of Sec. I have presented a severe limitation to the early pioneering attempts at such numerical solutions [26]. The solution of the ideal equation (Eq. (5)) to obtain the flux $\tilde{\phi}_S(t)$ through a surface S proceeds as follows: Choose a sufficiently fine grid of points on S, $\mathbf{x}_i$ for $i = 1, 2, \ldots, M$. For each of these M grid points evolve the position $\mathbf{x}_i$ *backwards* in time using the ordinary differential equation $d\mathbf{x}/dt = \mathbf{v}$ to obtain the location $\mathbf{x}_i^{(0)}$ that it started from at $t = 0$. Now starting from this point and using as initial condition $\tilde{\mathbf{B}} = \tilde{\mathbf{B}}(\mathbf{x}_i^{(0)}, 0)$, where $\tilde{\mathbf{B}}(\mathbf{x}, 0)$ is the initial magnetic field, integrate the ordinary differential equation $d\tilde{\mathbf{B}}/dt = \tilde{\mathbf{B}} \cdot \nabla \mathbf{v}$ (Eq. (5)) forward from $t = 0$ along the previously computed trajectory $\mathbf{x}(t)$ to obtain $\tilde{\mathbf{B}}(\mathbf{x}_i, t)$, the ideal magnetic field at the grid point $\mathbf{x}_i$ at time t. Do this for each of the M grid points, and use the results to obtain an approximation to $\tilde{\phi}_S(t)$,

$$\tilde{\phi}_S(t) \cong \frac{S}{M} \sum_{i=1}^{M} \mathbf{n}_i \cdot \tilde{\mathbf{B}}(\mathbf{x}_i, t) \ , \tag{10}$$

where $\mathbf{n}_i$ is the unit normal to S at $\mathbf{x}_i$. Note that only solutions of differential equations are involved in this procedure, and that the computations for different $\mathbf{x}_i$ can be done in parallel. Equation (1), in contrast, is partial differential and is typically solved on a three dimensional grid.

Another important property of kinematic dynamos is the result known as Cowling's antidynamo theorem which states that dynamo action is only possible if the magnetic field has three dimensional structure. While this result applies independent of R_m, it is instructive for us to illustrate it in the large R_m limit by using the ideal equation (Eq. (5)). In particular, even though Lagrangian chaos is possible in (time-dependent) two-dimensional flows we now show that fast kinematic dynamos are not possible in two dimensions. Consider a fluid confined to a rectangular region with rigid perfectly conducting walls as shown in Fig. 2. The line shown in Fig. 2(a) represents the initial configuration of a single field line. The dashed horizontal line segment S is crossed by the field line in the upward direction. After some time, during which the field line is stretched by the Lagrangian chaotic fluid flow (causing the field line length to increase exponentially with time), the configuration is as shown in Fig. 2(b). Although the number of crossings of S by the field line has increased, due to cancellation, the *net* upward flux through S in Fig. 2(b) is the same as in Fig. 2(a). The two dimensional topological constraint that the field line cannot cross itself prevents net exponential flux growth through S even though the flux line is exponentially stretched. (The situation is fundamentally different in three dimensions, and in what follows we only consider three dimensional situations.)

3 Dynamic Properties from Ergodic Characteristics of Lagrangian Chaotic Flows

In this section we report results on how one can obtain quantities such as the dynamo growth rate and fractal dimension of the fields from ergodic dynamical characteristics of the Lagrangian chaos of whatever specific velocity field $\mathbf{v}(\mathbf{x}, t)$ occurs in the dynamo problem under consideration. Note that, use of the Lagrangian chaos for this purpose implies that all calculations are done from the ordinary differential equations $d\mathbf{x}/dt = \mathbf{v}$ and $d\delta\mathbf{x}/dt = \delta\mathbf{x} \cdot \nabla \mathbf{v}$ (rather than the partial differential equation, Eq. (1)). In this section we shall only be reporting

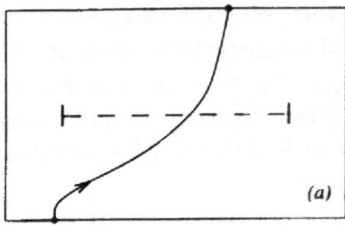 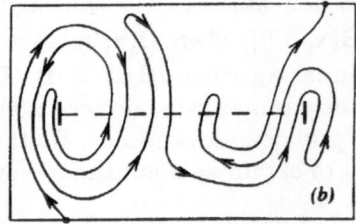

Fig. 2. Illustration of Cowling's antidynamo theorem for a two-dimensional Lagrangian chaotic flow at $R_m \to \infty$.

and illustrating results, not deriving them. For derivations see Refs. 8, 15, 16, and 24.

3.1 Cancellation Exponent

Before proceeding to discussions of the growth rate and fractal dimension it is necessary to consider the extreme cancellation properties of magnetic fields in the large R_m limit. In particular, from models [4,9,14] and numerical solutions [24] it is evident that at large time the magnetic fields can undergo rapid spatial variations in which their directions flip by 180°. This results in local alternating layers with opposing fields of thickness of order $\epsilon_* \sim R_m^{-1/2}$. Thus, as R_m increases this alternation becomes more rapid.

To quantitatively characterize situations with this kind of extreme tendency for cancellation, we use the cancellation exponent introduced in Ref. 14. We consider a magnetic field distribution **B** in three dimensional space, and we choose some planar surface S with unit normal **n**. We divide S into a grid of two dimensional ϵ by ϵ squares. Let ϕ_i be the flux through square i with respect to the unit normal **n** normalized to the total flux through S. If the quantity $\chi(\epsilon) = \sum_i |\phi_i|$ increases with decreasing ϵ as a power law in ϵ then we call that power the cancellation exponent and denote it κ,

$$\chi(\epsilon) = \sum_i |\phi_i| \sim 1/\epsilon^\kappa \ . \tag{11}$$

If there were no cancellation, $\phi_i \geq 0$ for all i, then $\sum_i |\phi_i| = \sum_i \phi_i \equiv 1$ and $\kappa = 0$. (It is assumed that κ is generically independent of the choice of S.) If the magnetic field $B_n(\mathbf{x})$ normal to S (where $\mathbf{x}$ is on S) is a smooth bounded function, then, for small enough ϵ,

$$\sum_i |\phi_i| \cong \left[\int_S |B_n| dA\right] / \left|\int_S B_n dA\right| < \infty \ ,$$

269

independent of ϵ, and κ is again zero. In practice, the magnetic field is always smooth on small enough scale; i.e., once we consider scales as small as $\epsilon_* \sim R_m^{-1}$, Eq. (6). Thus the practical meaning of (11) is that a plot of $ln\chi(\epsilon)$ versus $ln(1/\epsilon)$ shows a linear scaling range with slope κ for $\epsilon > \epsilon_*$.

As an example illustrating the cancellation index Fig. 3 shows a plot of $ln\chi(\epsilon)$ for a model with finite large R_m ($R_m = 10^{10}$) versus $ln(1/\epsilon)$. [We do not describe the model here. The interested reader should refer to Refs. 6, 14–16.] We see that the plot is well-fit by a straight line for $\epsilon > \epsilon_*$. The slope of this line is κ. For $\epsilon < \epsilon_*$ the curve flattens as expected.

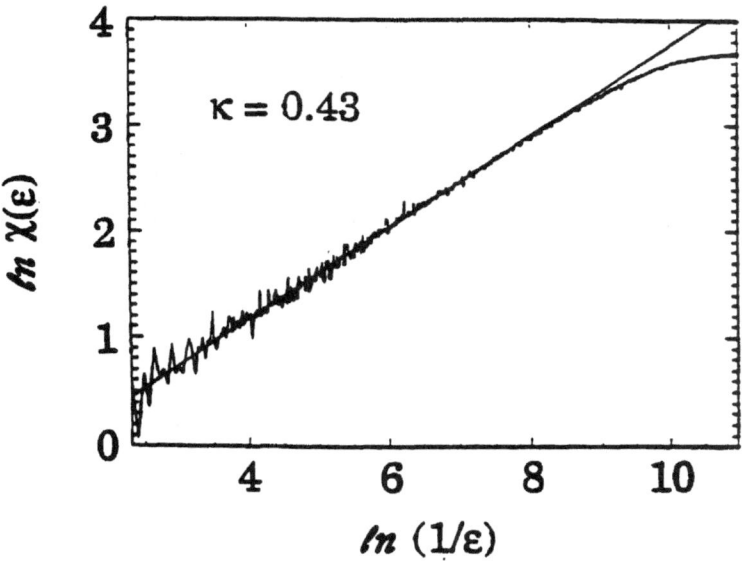

Fig. 3. $\ln\chi(\epsilon)$ versus $\ln(1/\epsilon)$ for the four strip model with cancellation at $R_m =: 10^{10}$.

3.2 Growth Rate Formula

Say we sprinkle around many initial conditions $x_{0j}(j = 1, 2, \ldots)$ in the chaotic region of a three dimensional dynamo flow. For each initial condition x_{0j} we consider a differential cube J of edge length δ (where δ is a differential) centered on the initial condition. We then use Eqs. (2) and (4) to evolve the cube forward in time by an amount t. This is illustrated in Fig. 4 which shows that the cube is deformed by the flow into a parallelipiped. We denote the three dimensions of the parallelipiped (the length, width and thickness) $L_{1j}\delta, L_{2j}, L_{3j}\delta$, where $L_{1j} \geq L_{2j} \geq L_{3j}$. For large t, we typically have that $L_{1j} \gg 1$ since the flow is chaotic. By volume conservation $L_{1j}L_{2j}L_{3j} = 1$, and so $L_{3j} \ll 1$. In what follows we assume that L_{2j} is typically greater than or equal to one. (In this case we expect and numerically observe that the magnetic field concentrates on a fractal set of sheet-like structures [15,16].) The quantities

$$h_i(\mathbf{x_{0j}}, t) = \frac{1}{t} \ln L_{ij} \; , \tag{12}$$

for $i = 1, 2, 3$, are the finite time Lyapunov exponents from point $\mathbf{x}_{0j}$.

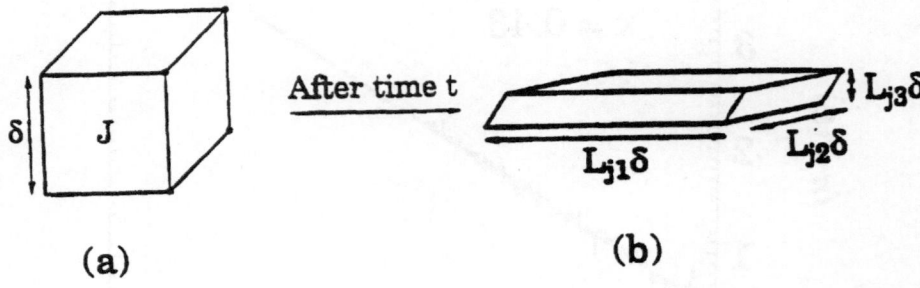

Fig. 4. Deformation of the differential cube J by the flow.

A formula using the L_{ij} and giving the dynamo growth rate in the large R_m limit (denoted Γ_*) has been obtain in Ref. 16,

$$\Gamma_* = \lim_{t \to \infty} \frac{1}{t} < L_1 L_3^\kappa > \; , \tag{13}$$

where the angle brackets denote an average over the initial conditions $\mathbf{x}_{0j}$ and the number of these initial conditions is taken to infinity. In application [16,24], one uses a large number of initial conditions and plots the quantity $\ln < L_1 L_3^\kappa >$ versus t. Fitting a straight line to such a plot and obtaining its slope then gives Γ_*. The cancellation exponent κ can be obtained from the ideal equation, Eq. (5) by calculating the magnetic field at a grid of points on some conveniently chosen surface S and then obtaining the slope of $\ln \sum |\phi_i|$ versus $\ln(1/\epsilon)$. As discussed in connection with Eq. (10), the ideal magnetic field on the surface S can be obtained purely from computations on ordinary differential equations.

Since $L_3 \ll 1$, Eq. (13) implies that cancellation always reduces the instability growth rate. Setting $\kappa = 0$, we have that if κ was positive, then

$$\Gamma_* < h_T \equiv \lim_{t \to \infty} \frac{1}{t} \ln < L_1 > \ . \tag{14}$$

Thus, h_T is an upper bound for the $R_m \to \infty$ growth rate Γ_*. The quantity h_T defined in (14) may, under certain circumstances, be identified with one of the fundamental quantities characterizing chaos in dynamical systems, namely the *topological entropy* (e.g., see Ref. 28, page 143). Thus, we have that the topological entropy is an upper bound on the growth rate Γ_* [4,23]. It should be noted that the limit on the right hand side of (14) is *not* the same as the largest Lyapunov exponent which is given by

$$h_1 = \lim_{t \to \infty} \frac{1}{t} < \ln L_1 > \ .$$

Since, the average of the log of a quantity is less than or equal to the log of the average of the quantity, $h_T \geq h_1$, and we should expect the inequality to typically apply.

3.3 Fractal Dimension

Let V be any three dimensional subregion of V_0, where V_0 denotes a finite volume in which the flow is confined. Then, we define a magnetic field based measure,

$$\mu(V) = \frac{\int_V |\mathbf{B}(\mathbf{x},t)| d^3\mathbf{x}}{\int_{V_0} |\mathbf{B}(\mathbf{x},t)| d^3\mathbf{x}} \ .$$

Now say we cover the volume V_0 by a grid of ϵ by ϵ cubes. Let μ_i denote the measure of the cube i. We then define a dimension spectrum depending on the continuous index q by [27,28] $I_q(\epsilon) \equiv (q-1)^{-1} \ell n (\sum_i \mu_i^q) \sim D_q \ell n (1/\epsilon)$.

For times t large enough so that the magnetic field has settled into a distribution with small scale variations at ϵ_* Du and Ott [15] present a formula giving the D_q dimension in terms of the L_i and the cancellation index. The result for $q = 1$ is

$$D_1 = 3 - \lim_{t \to \infty} \frac{< L_1 L_3^\kappa \ln(L_1 L_3^\kappa) > - < L_1 L_3^\kappa > \ln < L_1 L_3^\kappa >}{< L_1 L_3^\kappa \ln L_3^{-1} >} \ . \tag{15}$$

(Note that for $q \to 1$, we obtain from L'Hospital's rule, $I_1(\epsilon) = \sum \mu_i \ell n(1/\mu_i)$ and D_1 is called the information dimension [29].)

The case $q = 1$ is of most direct physical interest since it gives the dimension of the set on which the magnetic field concentrates. Thus, the prediction is that, if R_m is large and t is large enough so that the magnetic field varies on the limiting scale $\epsilon_* \sim R_m^{1/2}$ determined by finite resistivity, then a plot of $\sum \mu_i \ln(1/\mu_i)$ versus $\ln(1/\epsilon)$ has the slope D_1 given by (15) for ϵ small but larger than ϵ_*.

In the case where t is large but not large enough that ϵ_* has been reached, Antonsen and Ott [8] show that there is a scaling range in ϵ for which

$$D_1 = 3 - \lim_{t \to \infty} \frac{< L_1 \ln L_1 > - < L_1 \ln L_1 >}{< L_1 \ln L_3^{-1} >} \tag{16}$$

which is the same as (15) with κ set equal to zero. Thus, as time evolves, the large R_m fractal dimension of the magnetic field distribution crosses over from a value given by (16) to a value given by (15).

4 Numerical Computations of the Kinamatic Dynamo PDE at Large R_m

The general results discussed in the previous sections pertain to very large R_m. Although large R_m is relevant in nature, numerical computations at large R_m are difficult due to the necessity of resolving small scales, $\epsilon_* \sim R_m^{-1/2}$. Recently, Reyl et al. [24] have performed computations for a spatially smooth three dimensional flow at large enough $R_m (R_m = 10^5)$ that the realizations of the general properties we have discussed in this paper become feasible.

The flow considered in [24] is specifically chosen so as to facilitate large R_m computation. This flow is as follows,

$$\mathbf{v}(x, y, t) = \mathbf{x}_0 \tilde{v}_x(y) f(t) + \mathbf{y}_0 \tilde{v}_y(x) f(t - \frac{1}{3}T)$$
$$+ \mathbf{z}_0 \tilde{v}_z(x) f(t - \frac{2}{3}T),$$

where $f(t)$ is a periodic function with period T, $f(t) = 0$ for $T/3 + nT < t < (n+1)T$ with n integer, so that the flows in the x, y, and z directions are turned on sequentially. The Lagrangian chaotic dynamics generated by this flow can be analyzed by integrating $dx/dt = \mathbf{v}$ over one period T. This gives a three dimensional volume preserving map relating $\mathbf{x}$ at time $t = nT$ to $\mathbf{x}$ at time $t = (n+1)T$,

$$x_{n+1} = x_n + \tilde{v}_x(y_n),$$
$$y_{n+1} = y_n + \tilde{v}_y(x_{n+1}),$$
$$z_{n+1} = z_n + \tilde{v}_z(x_{n+1}),$$

where $f(t)$ is normalized so that $\int_0^T f(t)dt = \int_0^{T/3} f(t)dt = 1$. Because the partial differential equation, Eq. (1), is solved using a Fourier spectral representation, it is desired that $\tilde{v}_x$, $\tilde{v}_y$, and $\tilde{v}_z$ and their convolutions with the magnetic field have simple Fourier transformations. Thus, the $\tilde{v}$'s are chosen to be sinusoidal, $\tilde{v}_x = U_x \sin(K_y y + \theta_x)$, $\tilde{v}_y = U_y \sin(K_x x + \theta_y)$, $\tilde{v}_z = U_z \sin(K_x x + \theta_z)$. For discussion of the numerical techniques and other details see Ref. [24].

Using these computations various large R_m issues were addressed. These include the following: (i) the predicted equality of κ obtained from the ideal equation at large time and κ obtained from solution of (1) at times after ϵ_* has been reached; (ii) the predicted equality of the ideal flux growth rate (given by (9)), the theoretical growth rate result given by (13), and the growth rate from numerical solution of (1) at large R_m; and (iii) the predicted D_q from Ref. [15] for different q values (of which the $q = 1$ result is shown in Eq. (15)). In all cases, the predicted and computed results were consistent to within the limits of the numerical accuracy attained.

5 Conclusion

The fast kinematic dynamo problem displays a variety of interesting features connected with the singular small scale behavior of the magnetic field. The surprising result is that this behavior can be fully understood and quantitatively analized by use of concepts from chaotic dynamics.

This work was supported by the Office of Naval Research (Physics).

References

1 Vainshtein S. I., Zeldovich Ya. B. (1972): Sov. Phys. Usp. **15**, 159
2 Arnold V. I., Zeldovich Ya. B., Ruzmaikin A. A., Sokolov D. D. (1981): Sov. Phys. JETP **54**, 1083
3 Bayly B. J., Childress S. (1989): Geophys. Astrophys. Fluid Dyn. **44**, 211 (1988); ibid. **49**, 23
4 Finn J. M., Ott, E. (1988): Phys. Fluids **31**, 2992
5 Bayly B. J., Childress S. (1988): Geophys. Astrophys. Fluid Dyn. **44**, 211
6 Finn J., Hanson J., Kan I., Ott E (1989): Phys. Rev. Lett. **62**, 2965
7 Vishik M. M. (1989): Geophys. Astrophys. Fluid Dyn. **48**, 151
8 Ott E., Antonsen T. M. (1989): Phys. Rev. A **39**, 3660
9 Finn J., Ott E. (1990): Phys. Fluids B **2**, 916
10 Finn J., Hanson J., Kan I., Ott E. (1991): Phys. Fluids B **3**, 1250
11 Galloway D. J., Proctor M. R. E. (1992): Nature **356**, 691
12 Gilbert A. D., Bayly B. J. (1992): J. Fluid Mech. **241**, 199
13 Gilbert A. D. (1992): Philos. Trans. R. Soc. London Ser. A **339**, 627
14 Ott E., Du Y., Sreenivasan K. R., Juneja A., Suri A. K. (1992): Phys. Rev. Lett. **69**, 2654
15 Du Y., Ott E. (1993): Physica D **67**, 387
16 Du Y., Ott E. (1993): J. Fluid Mech. **257**, 265
17 Lau Y. -T., Finn J. (1993): Phys. Fluids B **5**, 365
18 Soward A. (1993): Geophys. Astrophys. Fluid Dyn. **73**, 179
19 Gilbert A. D., Otani N. F., Childress S. (1993): in *Theory of Solar and Planetary Dynamics*, edited by M. R. E. Proctor, P. C. Matthews, and A. M. Rucklidge (Cambridge University Press, New York,), 129–136
20 Otani N. F. (1993): J. Fluid Mech. **253**, 327
21 Ponty Y., Pouquet A., Sulem P. L. (1995): Geophys. Astrophys. Fluid Dyn. **79**, 239
22 Cattaneo F., Kim E., Proctor M., Tao L. (1995): Phys. Rev. Lett. **75**, 1522

23 Klapper I., Young, L. S. (1995): Comm. Math. Phys. **173**, 623
24 Reyl C., Ott E., Antonsen T. M. (1996): Phys. Plasmas **3**, 2564
25 Childress S., Gilbert A. D. (1995): *Stretch, Twist, Fold: The Fast Dynamo*, (Springer-Verlag, New York)
26 Galloway D., Frisch U. (1986): Geophys. Astrophys. Fluid Dyn. **36**, 53 (1986); V. I. Arnold and E. I. Korkiina, Vestn. Mosk. Univ. Mat. Mekh. **3**, 43 (in Russian)
27 Grassberger P., Procaccia I. (1983): Physica D **9**, 189
28 Ott E. (1994): *Chaos in Dynamical Systems* (Cambridge University Press), 78–85
29 Farmer J. D., Ott E., Yorke J. A. (1983): Physica D **7**, 153

Turbulence Scaling Laws in Fusion Plasmas

X. Garbet [1], R.E. Waltz [2]

[1] Association Euratom-CEA sur la Fusion, CE Cadarache, 13108 St Paul lez Durance, France
[2] General Atomics, P.O. Box 85608, San Diego, California 92186-9784

Abstract. This paper is a review of models for dimensionless scaling laws in fusion plasmas. The reasons why this subject is of particular interest are given. Models based on the effect of rotational shear flow and avalanches are described in details.

1 Introduction

Dimensionless scaling laws in thermonuclear fusion plasmas provide in principle a reliable way to extrapolate the performances of present day devices towards a reactor [1, 2]. The procedure can be understood as follows. For fixed geometry and profiles, and ignoring atomic physics processes, three main dimensionless parameters [3, 4, 5] describe a magnetized fusion plasma: the Larmor radius ρ_{s0} normalised to the plasma size a, called ρ^*, the ratio β of the kinetic pressure to the magnetic field pressure, and the Coulombian collision frequency normalised to a transit frequency (ion sound speed divided by the machine size), called ν^*. If n is the density, T the temperature, and B the magnetic field, these quantities scale as $\rho \equiv T^{1/2}/aB$, $\beta \equiv nT/B^2$, and $\nu^* \equiv na/T^2$. The values of the two latter parameters achieved in present day devices are already close to those relevant for a reactor. Thus, an extrapolation to a large size tokamak essentially involves the normalised Larmor radius ρ^*. The dependence of the confinement on this parameter will be our main concern in the following. It is easy to check that at fixed β and ν^*, the normalised power loss $Pa^{3/4}$ depends only on $\rho^* \equiv a^{-5/6}B^{-2/3}$. Let us now introduce a confinement scaling law. A common expression for the thermal diffusivity is $\chi \equiv T/eB \, [\rho^*]^\alpha$ (e is the proton charge). The well known Bohm scaling law corresponds to $\alpha = 0$. We will see later that the gyroBohm scaling law $\alpha = 1$ plays a central role. The confinement time, defined as the ratio of the kinetic energy content to the power loss P, scales as a^2/χ. One can then easily verify that the power loss scales as $[\rho^*]^{\alpha-5/2}$. This result proves that an extrapolation towards a large size reactor, i.e. towards small values of ρ^*, is more favorable for a gyroBohm scaling law $\alpha = 1$ than for a Bohm law $\alpha = 0$. For instance, the extrapolation of a discharge in the tokamak JET towards the ITER project predicts a power loss of 80MW when using a gyroBohm scaling law and 600MW for a Bohm law [2]. It turns out that a gyroBohm scaling law is predicted by conventional transport theories based on turbulent diffusion. This may be understood as follows. The correlation times τ_c are expected to scale as the inverse of the growth rates of localized eigenmodes, which depend only on the local gradients and scale as the ion sound speed divided by the plasma size c_{s0}/a. Furthermore, the correlation length are expected to scale as a Larmor radius. Using a random walk estimate, the local plasma diffusivity behaves as $T/eB\,[\rho^*]^\alpha$, with $\alpha = 1$, i.e. follows a gyroBohm scaling law. Thus, the common expectation is favorable for the extrapolation toward a reactor. Unfortunately, it disagrees with experimental results in some cases [1, 2]. More exactly, although the diffusivity for the electron heat channel is found to be gyroBohm, the scaling law for ions depends on the confinement regime. In the L-mode (Low confinement), ions are found to follow various laws, including the so-called Goldston scaling law $\alpha = -0.5$, whereas in the H- mode (High confinement), the scaling law is gyroBohm $\alpha = 1$. The latter regime is obviously more promising for a reactor. Understanding this unexpected behavior of ions is a difficult task. Several explanations have been proposed to explain the disagreement with experiment. The first explanation is based on the fact that the radial widths of linear eigenmodes (the so called global modes) scale as $\sqrt{\rho_{s0}a}$ [6, 7]. Using this linear scaling in a random walk estimate leads to a Bohm scaling. A second explanation relies on the stabilizing effect of a rotational shear flow (diamagnetic phase velocity plus diamagnetically induced ExB drift). Such a flow naturally exists in a tokamak plasma. An estimate of the poloidal flow is obtained by balancing the pressure gradient plus the electric force with the VxB Laplace force. Ones finds that the velocity scales as $\rho^* c_{s0}$ and its spatial derivative scales as $\rho^* c_{s0}/a$. Starting from a growth rate without shear flow γ_0, which scales as c_{s0}/a, simulations of ion turbulence indicate that the effective growth rate behaves as $\gamma_0 - |dV/dr|$, which scales as $c_{s0}/a\,(1 - \alpha^* \rho^*)$, where α^* depends on the details of the profiles. Assuming that the correlation lengths scale as ρ_{s0} and using again a random walk estimate with a correlation time inversely proportional to the growth rate, one finds that the heat diffusivity should scale as $T/eB\rho^*\,[1 - \alpha^* \rho^*]$ [8]. This prediction calls for two comments. First, it is not a monomial expression, i.e. the scaling law is not definite. This could explain why various scaling laws are found depending on the profiles. Second, since ρ^* is a small number in a tokamak (10^{-2} to 10^{-3}), this effect will play role if the growth rate without shear flow is small (α^* large), i.e. close to a turbulence threshold. A third explanation emphasizing the apparently non local plasma response was recently proposed by Diamond and Hahm [9]. This approach is related to the concept of Self-Organized Criticality (SOC) for which the paradigm is a sand pile automaton first introduced by Bak et al. [10]. In SOC models, avalanches play a central role [10, 11]. In a tokamak plasma, an avalanche corresponds to a fast radial propagation of a heat pulse. The mechanism can be understood as follows. A heating burst induces locally a transient steepening of the temperature profile. Once the temperature gradient exceeds the turbulence threshold, this steepening induces a burst of turbulence which expels the heat outward. The process is then renewed at the neighbouring radial position. A necessary condition for such a behavior is the existence of a turbulence threshold and a conservation law (here energy conservation). In sand pile numerical models, avalanches occur at all spatial and time scales. One consequence is a 1/f behavior of the frequency spectrum, which is commonly observed in these simulations [11]. Also, it is generally found that the time average slope of a sand pile automaton is sub-critical, i.e. it stays below the threshold value. Evidences of avalanches have been shown for simplified models of tokamak turbulence [12, 13, 14]. It was suggested by Diamond and Hahm that large scale avalanches break the gyroBohm expectation by introducing correlation length which scale as the machine size. This behavior may lead to a Bohm scaling. The purpose of this paper is to investigate these issues with a simplified 2D full radius code computing an Ion Temperature Gradient (ITG) turbulence in a tokamak. The code is simplified in the sense that the radial shape of each Fourier harmonic with respect to poloidal and toroidal directions is given and fixed. This allows a fast computation while keeping the main features of toroidal ITG turbulence. In particular the linear stability involves a turbulence threshold. This code has been first developed to work with a fixed temperature profile, essentially by filtering the low scale temperature fluctuations. This constraint forbids avalanches. The main results are the following [8]:

- well above the threshold, the scaling law is gyroBohm
- close to the threshold, the scaling law does not follow the gyroBohm prediction. It is shown that this is due to shear flow stabilization.
- the correlation lengths usually scale as a Larmor radius, whereas the correlation times depart from the gyroBohm prediction when the gradient is close to the threshold.

In a second step, this code has been extended to work at fixed thermal flux, allowing fluctuations of the temperature profile. The main results are the following [14]:

- the heat transport is intermittent and involve large scale avalanches.
- the profiles are super-critical, i.e. are above the marginal profile.

The question of scaling laws is still open in simulations at fixed flux. However, preliminar calculations show that the confinement scaling law is not gyroBohm at moderate fluxes. This result is true at fixed temperature gradient or at fixed flux, for the same temperature profile and plasma parameters. It suggests that the rotational shear stabilisation is the dominant effect. However, simulations at high fluxes, where a gyroBohm scaling is expected, are still to be done to confirm this result. The remainder of this paper is as follows. The simplified model which is used is described in the section 2. Results at fixed temperature gradient are shown in section 3 while those obtained at fixed flux are described in section 4. A conclusion follows in section 5.

2 A model for a tokamak ion turbulence

2.1 Fluid equations

We use a cylindrical equilibrium with a coordinate system (r, θ, φ) : r is the minor radius of a magnetic surface , θ (φ) the poloidal (toroidal) angle. The magnetic field is

$$\mathbf{B} = B_0 \left(1 - \frac{r}{R}cos(\theta)\right)\left(\mathbf{e}_\phi + \frac{r}{qR}\mathbf{e}_\theta\right)$$

where q(r) is the safety factor, R the major radius of the torus. Ion density, pressure and electric potential are written under the form

$$[n_i, p_i, \phi](r) = [n_{eq}, p_{eq}, \phi_{eq}](r) + \sum_{m,n} [n_i, p_i, \phi]_{m,n}(r,t)exp\left[i(m\theta + n\varphi)\right]$$

We start here with a set of fluid equations proposed by Nordman and Weiland [15] for toroidal ITG modes, which can be recast under the form [8, 14]

$$d_{\bar{t}}N = -i\omega_n^*.f.N + i\omega_D.(\hat{n}f.N + \tau P)$$

$$d_{\bar{t}}P = \frac{5}{3}i\omega_D.\left(\hat{p}f.N - \hat{T}^2\tau N + 2\hat{T}\tau P\right) + S \tag{1}$$

where the normalisations are

$$r \to \bar{r} = \frac{r}{\rho_{s0}} \; ; \; t \to \bar{t} = \frac{c_{s0}t}{a} \; ; \; \nabla \to \nabla = \rho_{s0}\nabla$$

$$\phi \to \Phi = \frac{a}{\rho_{s0}}\frac{e_i\phi}{T_{e0}} \; ; \; p_i \to P_i = \frac{a}{\rho_{s0}}\frac{p_i}{p_{e0}} \tag{2}$$

and n_{e0}, T_{e0}, p_{e0} are reference values (these are not the values on the magnetic axis), c_{s0} is the acoustic speed, $\rho_{s0} = m_i c_{s0}/e_i B_0$ is the corresponding ion Larmor radius, and 'a' is the minor radius. The ion temperature profile is supposed to be proportional to the electron temperature profile, τ being the ratio. We have assumed scale separation for the ion density but not for the temperature. The normalised profile $\hat{n}(r,t)$ is defined as

$$\hat{n}(r,t) = \int_0^{2\pi} \frac{d\theta d\varphi}{4\pi^2} \frac{n_i(r,\theta,\varphi,t)}{n_{e0}} \tag{3}$$

In practice, the normalised density profile $\hat{n}(r,t)$ is fixed (does not depend on time) and flat so that these simplications are not very important. Definitions similar to Eq.(3) hold for the normalised profiles $\hat{T}(r,t), \hat{p}(r,t)$. The other definitions are

$$\omega_D = -i\frac{2a}{R}\left(cos(\theta)\partial_{\bar{r}} + sin(\theta)\frac{1}{\bar{r}}\partial_\theta\right) \tag{4}$$

and

$$N = \hat{n}\left(\lambda\frac{\Phi}{\hat{T}} - \nabla_\perp^2\Phi\right) = f^{-1}.\Phi \tag{5}$$

where $\lambda = 0$ for m=0,n=0 radial modes and $\lambda = 1$ for helical m,n fluctuations. The time derivative advance is defined by

$$d_{\bar{t}} = \partial_{\bar{t}} + \mathbf{v_E}\cdot\nabla_{\bar{r}} = \partial_{\bar{r}} + \left[\partial_{\bar{r}}\Phi\frac{1}{\bar{r}}\partial_\theta - \frac{1}{\bar{r}}\partial_\theta\Phi\partial_{\bar{r}}\right] \tag{6}$$

and S is a source term. The linearised version of Eqs.(3) for the poloidal (m) and toroidal (n) components is

$$\partial_{\bar{t}} N_{mn} = -i\omega_n^*.f.N_{mn} + i\left[\omega_D.\left(\hat{n}f.N + \tau P\right)\right]_{mn}$$

$$\partial_{\bar{t}} P = -i\omega_p^*.f.N_{mn} + \frac{5}{3}i\left[\omega_D.\left(\hat{p}f.N - \hat{T}^2\tau N + 2\hat{T}\tau P\right)\right]_{mn} \qquad (7)$$

where

$$\left(\omega_n^* N\right)_{mn} = -a\frac{dn_{eq}}{n_{eq}dr}\frac{im}{\bar{r}} N_{mn} \qquad (8)$$

and vanishes for radial modes.

2.2 Quasi-ballooning representation

The equations above do not account for the parallel dynamics. The latter effects are known to localise the m,n Fourier components in the radial direction around the corresponding resonant surface $r = r_{m,n}$ such that $q\left(r_{m,n}\right) = -m/n$. To introduce this effect, each m,n mode is developed over a given set of functions

$$[N, P]_{mn}\left(\bar{r}, \bar{t}\right) = \sum_{m,n} [N, P]_{m,n,\ell}\left(\bar{t}\right) W_\ell\left(\frac{r - r_{mn}}{\Delta_{mn}}\right) \qquad (9)$$

Here, the set is limited to two functions, even ($\ell = 0$) and odd ($\ell = 1$). This is the minimum required to keep a non-vanishing geodesic curvature and account for shear flow generation. These functions are chosen to be gaussians, whose width $\Delta_{m,n}$ is proportional to the local ion Larmor radius $\rho_s\left(r_{m,n}, t\right) = \rho_{s0}\left[\hat{T}(r_{m,n}, t)\right]^{1/2}$. In the same way, the radial profiles are developed over a set of basis functions

$$[N, P]_{eq}\left(\bar{r}, \bar{t}\right) = \sum_k [N, P]_k\left(\bar{t}\right) W_k(r)$$

$$W_k(r) = \frac{J_0\left(\pi\alpha_k r/a\right)}{J_1\left(\pi\alpha_k\right)} \qquad (10)$$

where J_0, J_1 are Bessel functions and α_k is the k-th zero of $J_0(\pi r/a)$. The set (W_k) is orthonormal [16]. Note that this procedure imposes that all fields vanish at r=a. Eqs.(1) are then projected onto these m,n and radial shape functions. This reduces the dimensionality to 2. The details have already been given in the reference [8].

280

where $\chi_{gB} = \rho_{s0}^2 c_{s0}/a$ is the gyroBohm reference value. The link between the source amplitude and the total additional power P_{input} is given by the relation

$$P_{input} = 3n_0 T_0 V \frac{\rho_{s0} c_{s0}}{a^2} S_0 \qquad (17)$$

where $V = 2\pi R \pi a^2$ is the volume. In this model, the edge gradient adjusts itself such the time average of the heat outflux compensates the input flux, which is the radial integral of the source S.

3 Simulations at fixed temperature gradient

3.1 Eigenmodes and linear stability

Results in this section come essentially from the reference [8]. The system that was solved is close to Eqs.(1), with additional FLR effects vanishing in the limit of small $k_\perp \rho_{s0}$. To perform a ρ^* experiment in a controllable way, the following procedure was followed. Let us first consider the case of a slab geometry where the density profile is flat ($\nabla \hat{n} = 0$) and the temperature gradient length is constant. It is reminded here that for a flat density profile, the threshold is given by a critical value of the normalised gradient length L_{Ti}/R, which is thus the same everywhere. In this case, the eigenmodes are of the form

$$[N_{nm\ell}, P_{nm\ell}] = [N_{n\ell}, P_{n\ell}] \, exp\,[i(m - m_0)\theta] \qquad (18)$$

where θ_0 is the so-called ballooning angle.

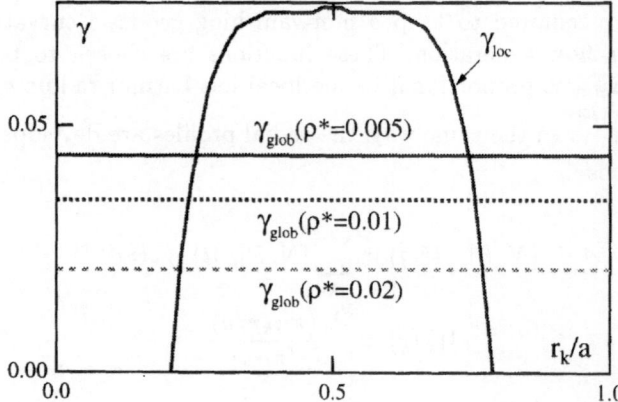

Fig.1: Global and local growth rates as a function of the resonant surface radius (proportional to m/n) for n=6. The global growth rates correspond to the most unstable mode and are calculated for 3 values of ρ^*. Increasing ρ^* is stabilizing, i.e. decreases the growth rates.

These modes extend all over the minor radius and the corresponding growth rates γ_{loc} do not depend on ρ^*. The critical temperature gradient length L_{Tc}/R is defined in this way. In fact, the temperature gradient length in an actual device is not constant and the eigenmodes are spatially localized by profile curvature. This localization is achieved by adding a dissipation in the edge and in the neighbourhood of the magnetic axis r=0.

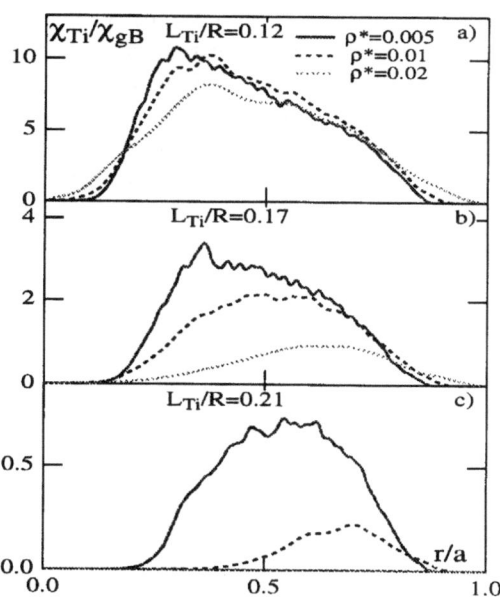

Fig.2: Ion heat diffusivity for three values of the temperature gradient length $L_{Ti}/R = 0.12$, 0.17, and 0.21 and three values of $\rho^* = 0.005$, 0.01, and 0.02. The threshold given by the lowest order of the ballooning calculation is $L_{Ti}/R = 0.29$.

This procedure leads to localised eigenmodes whose growth rates γ_{glob} depend on ρ^* as shown on Fig.1. It has to be stressed however that the generic case in a tokamak is a parabolic curvature. The present procedure tends to overestimate the effect of curvature in the ρ^* dependence of the growth rates. In a second step, a controlled amount of poloidal shear flow is added by introducing an ExB rotational shear $\gamma_E = dv_E/dr$ (the geometry is slab). It has been verified that the shear flow is stabilizing.

3.2   Scaling laws at fixed temperature gradient

For a given amount of shear flow, the thermal diffusivities are compared for three different values of ρ^* (ρ^*=0.005, 0.01 and 0.02) in 3 cases:

– well above the threshold, i.e. $L_{Ti}/R \ll L_{Tc}/R$

- close to the threshold $L_{Ti}/R \leq L_{Tc}/R$
- very close to the threshold $L_{Ti}/R \simeq L_{Tc}/R$

The results are shown in Fig.2. Well above the threshold, the scaling law is gyroBohm since the heat diffusivities normalised to the gyroBohm value are roughly the same for the values of ρ^*. Closer to the threshold, a departure from the gyroBohm prediction is observed. This breaking of the gyroBohm scaling law is obvious on the figure 2 when the gradient is very close to the threshold value. In the latter case, there is no turbulence for the largest value of $\rho^*=0.02$, whereas a gyroBohm law would predict the same diffusivity as in the case $\rho^*=0.005$. This observation is consistent with the picture along which the turbulence level is proportional to an effective growth rate $\gamma_{eff} = \gamma_0 - \gamma^* - \gamma_E$, where γ^* represents the effect of diamagnetic flow shear. The actual expression for the stabilizing term $\gamma^* + \gamma_E$ is still under discussion. A reasonable choice seems to take the radial derivative of the mode phase velocity in the laboratory frame. This picture is reinforced by an analysis of correlation functions. This analysis indicates that the correlation lengths scale as an ion Larmor radius, in agreement with the gyroBohm. However, when the scaling is not gyroBohm, the correlation times normalised to a/c_{s0} depends on ρ^*.

4 Simulations at fixed heat flux

4.1 Avalanches

The results in this section correspond to situations where the heat flux is fixed and the temperature profile is allowed to fluctuate. An example of temperature profile is shown in Fig.3 for the case where $\rho^*=0.01$ and $S_0=0.01$.
This profile calls for two remarks:

- the instantaneous profile exhibits a sequence of plateaux and steep slopes. The life time of these plateaux is of the order of a few time units a/c_{s0} (i.e. a few μs in a tokamak). They correspond to a transient quasilinear flattening.
- the time average of the temperature profile is smooth, as usually observed in a tokamak. The heat diffusivity is estimated from this time averaged profile.

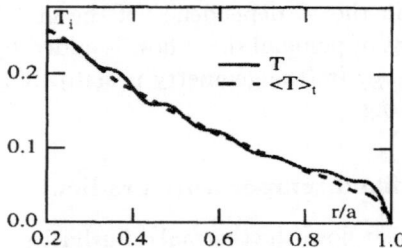

Fig.3: Instantaneous and time average temperature profiles for $S_0=0.01$, and $\rho^*=0.01$.

A remarkable behavior is that these plateaux are correlated all along the minor radius of the discharge, i.e. they correspond to avalanches. An avalanche can be understood as follows: a heating induces locally a steepening of the gradient, which exceeds the turbulence threshold and thus produces an increase of the turbulent heat flux, expelling the flux toward a larger radius. At this new position, this burst of heat will again increase the temperature and increase the turbulence. This process can be repeated many times, such as a domino effect.

Avalanches can be visualized by looking at the contour lines of the thermal flux and pressure (Fig.4). There exists avalanches at all spatial and time scales. This results in a 1/f frequency spectrum, which is also observed in sandpile automatons and a fluid model for an interchange turbulence in a cylinder. However, this is not a definite proof that this system is a SOC system. Another characterisric is the sub-criticality. It has been indeed observed in some realisations of SOC systems [17] that the slope lays below the critical value almost everywhere in the plasma. A stability analysis of the profiles show that this is not presently the case. The growth rate corresponding to a time average profile is indeed positive. Also, increasing the heating source leads to a profile which can exceed significantly the marginal profile. An important question is to know whether avalanches can break the gyroBohm scaling, as it has been suggested.

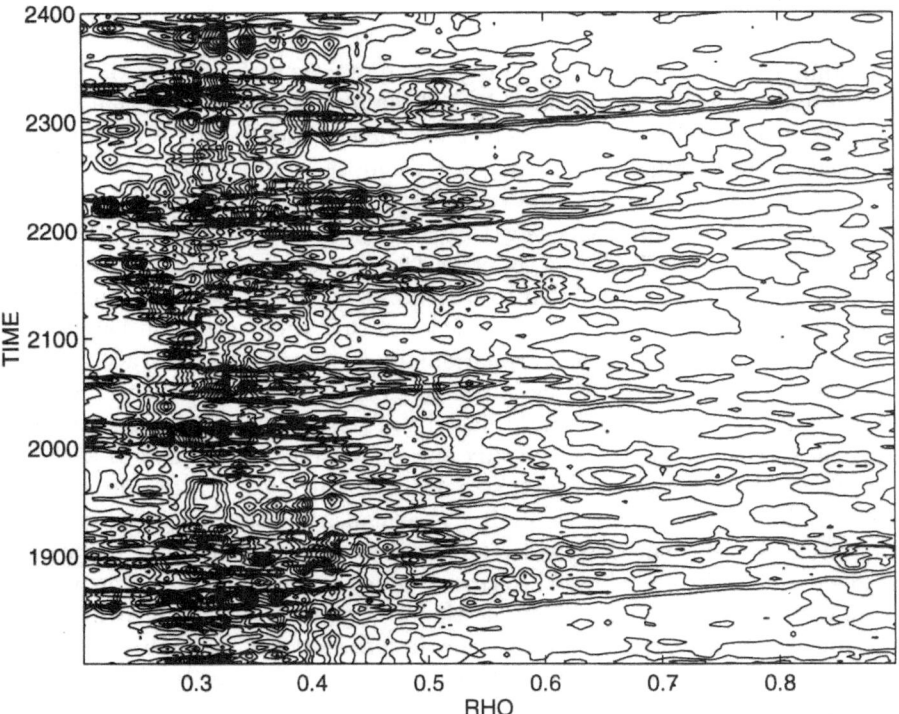

Fig.4: Contour plots of thermal flux as function of the radial position $\rho = r/a$ and time.

4.2 Scaling laws at fixed flux

A straightforward analysis of the heat equation Eq.(15) tells us that if the heat diffusivity normalised to the gyroBohm value scales as $[\rho^*]^{\alpha-1}$ ($\alpha = 1$ is gyroBohm, $\alpha = 0$ is Bohm), then the source amplitude S_0 has to scale as $[\rho^*]^{\alpha}$ to maintain the same temperature profile. To estimate a, the temperature profile corresponding to $\rho^*=0.01$ and $S_0=0.05$ is used as a starting point, and S_0 is adjusted to recover the same temperature profile for $\rho^*=0.02$. It is found that $S_0=0.0425$, corresponding to a negative value of $\alpha \simeq -0.2$, i.e. a scaling law worst than Bohm. This is confirmed by a local analysis of the heat diffusivities: the ratio of the normalised diffusivities is close to a Bohm scaling. The question arises whether this is due to the stabilising effect of rotational shear flow or to avalanches. The same simulations at fixed temperature profiles, where avalanches are forbidden, indicate that a Bohm scaling persists. This suggests that the rotational shear flow is responsible of this breaking. However, simulations at higher fluxes, where a gyroBohm scaling law is expected, remain to be done. A firm conclusion cannot be given here.

5 Conclusion

The confinement scaling laws of a magnetized plasma turbulence with respect to the normalized ion Larmor radius ρ^* has been investigated. Two situations have been studied depending on the control parameter which can be the temperature gradient or the thermal flux. In the case where the temperature gradient is the control parameter and is well above the critical value, it is found that the scaling law of the heat diffusivity is gyroBohm, i.e. the heat diffusivity behaves as $T/eB\,[\rho^*]^{\alpha}$, with $\alpha = 1$. This is no longer true closer to the threshold, where this rule is broken. This is attributed to the stabilizing effect of the rotational shear flow, which depends on ρ^*. A consequence is that the correlation length scales as a Larmor radius, but the correlation time normalised to an acoustic time a/c_{s0} depends on ρ^*. In this case, thereis no definite scaling since a parabolic behavior is expected for the heat diffusivity. The picture is rather different when the control parameter is the heat flux and the temperature profile is allowed to fluctuate. The main observation is that large scale events cross the discharge minor radius. These events are similar to avalanches which are observed in Self-Organised Systems such as sand pile automatons. However, the time averaged gradient is found to be above the critical value, in contrast with some SOC systems which are sub-critical. The question of scaling laws for a heat flux driven turbulence is still open. Preliminar calculations indicate that the scaling law is not gyroBohm in those systems. This effect persists when avalanches are forbidden, suggesting a dominant effect of rotational shear flow. More simulations are needed to give a firm conclusion.

References

[1] C.C. Petty, T.C. Luce, R.I. Pinsker, K.H. Burrell, S.C. Chiu, P. Gohil, R.A. James, D. Wroblewski, Phys. Rev. Lett. **74**, 1763 (1995).

[2] J.G. Cordey, B. Balet, D. Campbell, C.D. Challis, J.P. Christiansen, C. Gormezano, C. Gowers, D. Muir, E. Righi, G.R. Saibene, P.M. Stubberfield, and K. Thomsen, 23rd EPS Conference on Controlled Fusion and Plasma Physics, Kiev (1996), Plasma Physics and Controlled Fusion **38**, A67 (1996).

[3] B.B. Kadomtsev, Sov. J. Plasma Phys. **1**, 295 (1975).

[4] J.W. Connor, and J.B. Taylor, Nucl. Fusion **17**, 1047 (1977).

[5] R.E. Waltz, J.C. Deboo, and M.N. Rosenbluth, Phys. Rev. Lett. **65** (1990), 2390.

[6] J. W. Connor, J. B. Taylor, and H. R. Wilson, Phys. Rev. Lett. **70**, 1803 (1993).

[7] F. Romanelli and F. Zonca, Phys. Fluids B **5**, 4081 (1993).

[8] X. Garbet and R. Waltz, Phys. Plasmas **5**, 1907 (1996)

[9] P.H. Diamond, T.S. Hahm, Phys. Plasmas **2**, 3640 (1995)

[10] P. Bak, C. Tang, and K. Wiesenfeld, Phys. Rev. Lett. **59**, 381 (1987).

[11] T. Hwa and M. Kardar, Phys. Rev. A **45**, 7002 (1992).

[12] B.A. Carreras, D. Newman, V.E. Lynch, and P.H. Diamond, Phys. Plasmas **3**, 2903 (1996)

[13] Y. Sarazin, Ph. Ghendrih, and X. Garbet, "Intermittent transport due to particle flux drive of SOL turbulence", to appear in Controlled Fusion and Plasma Physics (Proc. 24th EPS conference, Berchtesgaden,1997).

[14] X. Garbet, R.E. Waltz, submitted to Phys. Plasmas

[15] H. Nordman , J. Weiland, Nucl. Fusion **29**, 253 (1989).

[16] P.M. Morse, and H. Feshbach, in Methods of Theoretical Physics, (McGraw Hill, 1953), vol. II, p. 1260.

[17] J.M. Carlson, J.T. Chayes, E.R. Grannan, and G.H. Swindle, Phys. Rev. Lett., **65**, 2547 (1990).

Bifurcation in First-Order Fermi Acceleration and the Origin of Cosmic Rays

M.A. Malkov

Max-Planck Institut für Kernphysik, D-69029, Heidelberg, Germany

Abstract. Strong astrophysical shocks are probably the most powerful accelerators in the universe. The main basis for the shock acceleration to perform so efficiently is believed to be created by accelerated particles themselves through their backreaction on the shock structure resulting in a substantial increase of the shock compression. Such an accelerating shock should thus be considered as a dynamical system with the pronounced self-organization. We review the current state of the theory of nonlinear shock acceleration. The main emphasis is on the bifurcation of the solutions for the acceleration efficiency in terms of the rate at which particles are drawn from thermal plasma (injection rate), their maximum energy (cut-off) and the Mach number of the shock. The bifurcation diagram shows that there exists a critical injection rate below which only a relatively inefficient acceleration is possible whereas above this quantity the acceleration may become extremely efficient. On the other hand, at least in a stationary regime in which all particles leave the system at or below the energy cut-off, the acceleration process can hardly be continued in a very efficient way to very high energies. A number of plasma processes inside the shock transition may drive the system to the inefficient acceleration regime. We speculate on a possible relevance of these results to the lack of evidence of high energy protons in supernova remnant shocks.

1 Introduction

Diffusive shock acceleration, also known as the first order Fermi process in shocks, has been formulated in its modern form in a number of papers about twenty years ago [1-4]. Initially, the interest in this process was motivated by earlier ideas that the cosmic rays (CRs), being so ubiquitous in the Galaxy must be born in strong shocks such as the supernova remnant (SNR) shocks. It was quickly realized, however, that the problem must be put in a more general physical context. Namely, if a strong shock propagates through the plasma, the question is which fraction of the shock energy can be channeled into accelerated particles (we will also use the term CRs for them). In a classical shock theory, a similar question of how the incident flow energy is distributed between the thermal motion of the gas behind the shock and its bulk motion is solved given the Mach number and the adiabatic index of the gas using solely the conservation of the fluxes of mass, momentum and energy across the shock transition (Rankine-Hugoniot relations). If, in addition, a certain fraction of charged particles can be accelerated to very high energies, the problem of distribution of the shock energy cannot be solved within the traditional gas dynamics since high energy particles can carry a sizable fraction of energy and momentum even if their mass density is negligible.

The above problem of energy distribution has basically two aspects. The first, relatively simple one, concerns primarily the age and the size of the shock. Indeed, particles gain energy upon recrossing the shock front to which they are bounded diffusively due to scattering off hydromagnetic disturbances which are assumed to be sufficiently strong in the shock vicinity. This takes time and at least the cut-off energy derives simply from the life time of the shock if losses do not come into play before. It is usually assumed that they begin to work when the Larmor radius of accelerated protons approaches a certain fraction of the shock size. Electrons start to loose their energy normally long before due to the synchrotron radiation or the inverse Compton scattering on a background photon field.

The second aspect finds its roots deeper in the fact that accelerated particles modify the flow which may substantially alter the acceleration process itself. In particular, in a strongly modified high Mach number shock the total compression ratio exceeds the conventional value of 'four' markedly, and as we shall see, the partial pressure of stationary accelerated particles becomes a nonintegrable function of momentum without an upper momentum cut-off. Therefore, in a quasi-stationary state, when the losses at the upper cut-off momentum p_1 are compensated through the injection around some slightly suprathermal momenta $p \sim p_0 \gtrsim p_{\text{th}}$, the acceleration efficiency will critically depend on the relation between these two. Moreover, the losses have the effect of increasing the total compression ratio, boosting CR production even further.

In the next section we briefly review the physical formulation of the problem and the simple test particle solution for diffusive shock acceleration.

2   Transport of Energetic Particles in Nonuniform Plasma Flows

A complete Maxwell-Vlasov analytic description of CRs in the turbulent shock environment is hardly possible. Since one needs to span typically about ten orders of magnitude in particle energy and since the diffusive spreading is proportional to particle energy and, thus, the spatial domain of interest is correspondingly large, the problem is inaccessible also for direct computer simulations. The most efficient reduction scheme of the Vlasov-Maxwell system consists of the following two steps. First, one derives a standard quasi-linear system under the assumption that the wave-particle interaction is due to the excitation of MHD waves via the cyclotron resonance with a slightly anisotropic energetic particle distribution emerging at the shock. This interaction leads to particle diffusion in pitch angle which is assumed to be the fastest process in the quasilinear kinetic equation. Since the hydromagnetic waves (scattering centers) propagate essentially at Alfvén velocity $v_A \ll U$ in the local plasma frame, where $U(x)$ is the bulk plasma speed, they are seen by energetic particles as frozen into the flow. Thus, the particle momentum distribution must be almost isotropic in the local plasma frame and an equation for the isotropic part $f(x, p, t)$ may be derived by averaging the quasi-linear equation over the pitch angle. The result is known as a diffusion-convection equation and may be written in the following form (see e.g., [5])

$$\frac{\partial f}{\partial t} + U \frac{\partial f}{\partial x} - \frac{\partial}{\partial x} \kappa \frac{\partial f}{\partial x} = \frac{1}{3} \frac{\partial U}{\partial x} p \frac{\partial f}{\partial p} \qquad (1)$$

Here the coordinate x is directed along the shock normal and $\kappa(p, x)$ is the spatial diffusion coefficient originating from the pitch angle scattering (wave-particle collisions). The smallest possible κ (the most rapid acceleration) corresponds to the Bohm regime in which the mean free path of the particles approaches the Larmor radius r_L so that $\kappa = r_L v/3$, where $v \simeq c$ is the particle velocity. The acceleration time-scale is then given by $\tau_{acc} \sim \kappa/U^2$.

3 Test Particle Solution

We begin with the simplest solution of equation (1) appropriate for an unmodified shock that is assumed to be running in the positive $x-$ direction, so that the flow profile in the shock frame is given by $U = -u_1$, $x > 0$ and $U = -u_2$, $x \leq 0$, where $u_1 > u_2$ are the constant upstream and downstream flow speeds, respectively. We also assume that there are no accelerated particles far upstream, at $x = \infty$, i.e., $f(x = \infty) = 0$. Then, the steady state solution of equation (1) in the upstream medium, $x > 0$, is

$$f = f_0(p) \exp\left(-u_1 x/\kappa\right) \qquad (2)$$

We assume here for simplicity that κ is independent of x. Downstream from the shock ($x \leq 0$) the only bounded solution is $f = f_0(p)$. We have used here the continuity of f at $x = 0$ since particles cross the shock balistically (the shock thickness is smaller than the mean free path). Integrating (1) across the shock transition gives the particle spectrum

$$f_0 = Q_{inj} p^{-q}, \quad \text{where} \quad q = \frac{3r}{r-1} \qquad (3)$$

and $r = u_1/u_2$ is the shock compression. The normalization constant Q_{inj} should be determined from matching the solution (3) with the thermal distribution downstream. Unfortunately, equation (1) is invalid for thermal plasma due to the strong anisotropy of particle distribution at the shock front for velocities $v \lesssim u_1 - u_2$ and the original quasi-linear equation should be used for this purpose [6]. We will consider parameter Q_{inj} as given.

4 Nonlinear Theory of Diffusive Shock Acceleration

So far we have considered a reactionless or test particle acceleration. If the parameter Q_{inj} is not vanishingly small, and the particle spectrum is extended to sufficiently high energies, the pressure exerted by these particles on the inflowing gas cannot be neglected while determining the shock structure. As a result, the latter acquires the form shown in Fig. 1 by solid line. The effect of the slowing down of the upstream flow may be very strong even for small Q_{inj} due to the following positive feedback: an increase of the CR pressure hardens the spectrum due to the higher compression which further increases the pressure. As a result the system jumps from a nearly test particle solution with small P_c to a solution in which the flow ram pressure is almost totally converted into the CR pressure. To describe this situation we must complement equation (1) with an equation for the flow profile $U(x)$. This can be obtained from the conservation of mass and momentum. Assuming the steady state, introducing for convenience $u(x) = -U(x)$ and $g(x,p) = p^3 f(x,p)$ the system of equations to determine both the particle distribution and the flow profile $u(x)$ takes the form

$$\frac{\partial}{\partial x}\left(ug + \kappa(p)\frac{\partial g}{\partial x}\right) = \frac{1}{3}\frac{du}{dx}p\frac{\partial g}{\partial p} \qquad (4)$$

$$\rho u = \rho_1 u_1, \qquad (5)$$

$$P_c + \rho u^2 = \rho_1 u_1^2 \qquad (6)$$

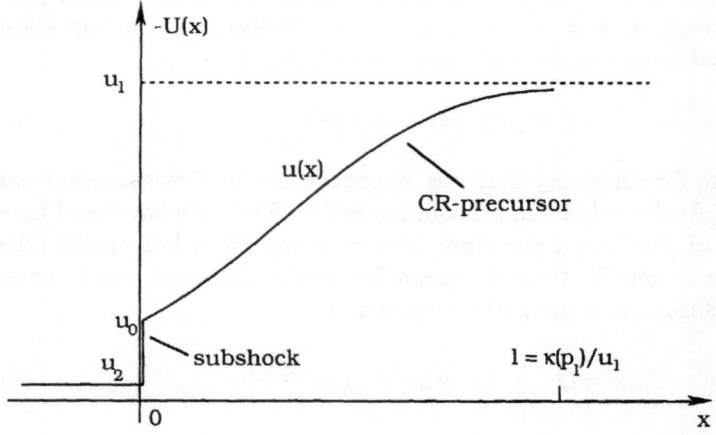

Fig. 1. The flow structure in a strongly modified CR shock.

Here $\rho(x)$ is the mass density, $\rho_1 = \rho(\infty)$, P_c is the CR pressure

$$P_c(x) = \frac{4\pi}{3} mc^2 \int_{p_0}^{p_1} \frac{pdp}{\sqrt{p^2+1}} g(p,x) \tag{7}$$

and $P_c(\infty) = 0$. The particle momentum p is normalized to mc. Equation (6) is written in the region $x > 0$ where we have neglected the contribution of the adiabatically compressed cold gas confining our consideration to sufficiently strong shocks with $M^2 \equiv \rho_1 u_1^2/\gamma P_{g1} \gg (u_1/u_0)^\gamma$, where γ is the specific heat ratio of the plasma (see [7] for a detailed discussion of this approximation). The subshock strength can be obtained from the familiar Rankine-Hugoniot condition for the gas

$$r_s \equiv \frac{u_0}{u_2} = \frac{\gamma+1}{\gamma-1+2M_0^{-2}} \tag{8}$$

where M_0 is the Mach number of the flow in front of the subshock. In the case of a purely adiabatic heating $M_0 = MR^{-(\gamma+1)/2}$ with $R = u_1/u_0$. Behind the shock ($x < 0$) we choose the same solution as for the test particle problem, i.e., $g = g_0(p)$, $u(x) = u_2 = const$.

4.1 Exact Solution for Arbitrary $\kappa(p)$ and $p_1 \to \infty$

Introducing the flow potential Ψ, such as $u = d\Psi/dx$ we seek the solution of (4) in the form

$$g = g_0(p) \exp\left\{ -\frac{1+\beta}{\kappa(p)} \Psi \right\}, \quad x > 0 \tag{9}$$

where $\beta(p) \equiv -(d\ln g_0/d\ln p)/3$. This is the key substitution in our analysis. By plugging it in (4) and separating the variables we obtain the following equations for $\Psi(x)$ and $\beta(p)$

$$\Psi \frac{d^2\Psi}{dx^2} - \lambda \left(\frac{d\Psi}{dx}\right)^2 = 0 \tag{10}$$

$$p\frac{d\beta}{dp} = (1+\beta)\left(\frac{d\ln\kappa}{d\ln p} - \frac{3}{\lambda}\beta\right) \tag{11}$$

where λ is a separation constant. Equation (10) may be readily integrated and yields the following scaling for the flow velocity in the CR precursor

$$u(x) - u_0 \propto x^{\lambda/(\lambda-1)} \tag{12}$$

The solution of equation (11) may also be found in a closed form

$$g_0(p) = C_1 p^3 \left[C_2 + \int \kappa p^{3/\lambda-1} dp \right]^{-\lambda} \tag{13}$$

The constants C_1 and C_2 should be determined from the solution of injection problem [6, 7]. For p not too close to the injection momentum p_0 the constant C_2 may be omitted. One sees then that if κ is a power-law then the particle spectrum g_0 must also be. A standard assumption $\kappa(p) = Kp^2(1 + p^2)^{-1/2}$, i.e., when the mean free path of a particle is proportional to its Larmor radius (here K is some reference diffusivity), leads to the following behaviour of the spectral slope β (this may be seen from (13)) $\beta = (2/3)\lambda$ for $p \ll 1$ and $\beta = (1/3)\lambda$ for $p \gg 1$ (see [8] for a complete solution).

What we obtained so far is a formally exact solution to equation (4) that requires a rather special flow profile $u(x)$. It is by no means guaranteed that this solution satisfies the Bernoulli's integral (6) (the continuity condition (5) can be easily satisfied since it does not depend on g). Indeed, the only unspecified quantity in this solution that can potentially be used to satisfy the functional relation (6) is the constant λ, which is, generally speaking, not enough. Fortunately this can be done nonetheless. Namely, substituting the solution (9) in (7) one obtains

$$P_c(\Psi) = \rho_1 u_1 (u_1 - u_0) - Q \left(\sqrt{\Psi(x)} - \sqrt{\Psi(0)} \right) \qquad (14)$$

where Q is a normalization constant proportional to the injection rate Q_{inj} introduced earlier. Using equation (12) one may see that the latter expression is indeed compatible with equation (6) if $\lambda = 1/2$ except for very small and very large x (see [7, 8] and the next subsection). For the particle spectral index downstream (in a standard normalization) $q = -d \ln f_0 / d \ln p$ this yields $q \equiv 3(\beta + 1) = 3\frac{1}{2}$ for relativistic particles ($p \gg 1$) and $q = 4$ for nonrelativistic ones. Remarkably, the $q = 3\frac{1}{2}$ index coincides with the plain test particle result for a strong shock in a purely thermal relativistic gas, whereas $q = 4$ coincides with that for a nonrelativistic gas. All these in spite of the facts that the flow profile is strongly modified and the total compression ratio may be much higher than 4 ($q = 4$) and even 7 ($q = 3\frac{1}{2}$) occurring in nonrelativistic and relativistic gases, respectively. Moreover, this spectral universality of strong shocks remains valid also for a broader class of possible dependencies $\kappa(p)$ [8].

It is worth while to comment on the asymptotic condition under which the above solution of the problem (4-7) is an exact one. The solution for $u(x) - u_0$ must be scale invariant for all $x > 0$ and, since the precursor length, where it is true, $l \sim \kappa(p_1)/u_1 \propto p_1$, then p_1 should tend to infinity. The power-law solution for g_0 is then also exact for all $p_0 < p < \infty$. For finite p_1 this solution is an asymptotic one, strictly valid for $x \ll l$ and $p \ll p_1$. As we mentioned the cut-off momentum p_1 may be very large, the values of $p_1/p_0 \sim 10^8$ are expected in SNR shocks. Therefore, the above solution must form a solid basis for an asymptotic theory, operating on $1 \ll p_1 < \infty$.

4.2 Asymptotic Theory for $p_1 < \infty$.
Reduction to Integral Equation

The above scale-invariant solution for $u - u_0$ and the corresponding power-law spectrum cannot be valid for all $0 < x, p < \infty$. The power-law cuts off at $p \sim p_1$ and, as a result, the scale-invariant behavior of $u(x)$ breaks at $x \sim l \sim \kappa(p_1)/u_1$ where u approaches u_1 to conform the asymptotics $u \to u_1$, $x \to \infty$, Fig. 1. Also at lower momenta $p \sim p_0$ the particle spectrum must conform to the subshock compression ratio. Accordingly, the flow profile deviates at the correspondingly small x from the self similar regime (see the next subsection and [7]). In treatment of this situation the so-called spectral function turns out to be very useful. It appears naturally in the derivation of equation for the particle spectrum g_0, exactly as we have already done to obtain equation (3). This time, however, we integrate equation (4) between $0-$ and $+\infty$ making use of (9). The result reads

$$-\frac{1}{3}\frac{\partial \ln g_0}{\partial \ln p} = \frac{1}{\bar{V}}\left(u_2 + \frac{1}{3}\frac{\partial \bar{V}}{\partial \ln p}\right) \tag{15}$$

Here the spectral function $\bar{V}$ is defined as follows

$$\bar{V}(p) = \int_{0-}^{\infty} e^{-s(p)\Psi}\, du(\Psi) \tag{16}$$

with

$$s(p) = \frac{1}{\kappa(p)\bar{V}(p)}\left[u_2 + \bar{V}(p) + \frac{1}{3}\frac{\partial \bar{V}}{\partial \ln p}\right] \tag{17}$$

The functional dependence of the variable s on $\bar{V}$ is not very critical here and (16) should be regarded logically as an integral transform $u(\Psi) \mapsto \bar{V}(p)$ rather than an equation for $\bar{V}(p)$ given $u(\Psi)$. The function $\bar{V}(p)$ reflects explicitly a degree of shock modification. In an unmodified shock $\bar{V}(p) \equiv \Delta u \equiv u_0 - u_2$, since then $du/dx = 0$ in the upstream region; the spectral index $q = 3\beta$ is just the conventional $q = 3u_2/(u_1 - u_2)$, (see equation (15)). In general $\Delta u \leq \bar{V}(p) \leq u_1 - u_2$ and $\bar{V}(p) \to u_1 - u_2$ as $p \to \infty$. Even if the shock is appreciably modified, one may show that at small $p \gtrsim p_0$ $\bar{V}(p) \simeq \Delta u$. The spectral index then corresponds simply to the subshock compression ratio, and at lower momenta we have

$$q \simeq q_0 = \frac{3u_2}{u_0 - u_2}, \quad g_0(p) = Q_{\text{inj}}\left(\frac{p}{p_0}\right)^{-q_0} \tag{18}$$

The injection solution [6, 9] produces essentially the same asymptotic result for $p \gtrsim p_0$, yielding thus the injection rate Q_{inj}. The solution $g_0(p)$ can be obtained then for all p using equation (15). To this end an independent equation for $\bar{V}(p)$ should be derived. Note, that the only equation available to derive an equation for $\bar{V}$ is the Bernoulli's integral (6). Using the solution (9) the latter can be rewritten as

$$\frac{d\Psi}{dx} + F(\Psi) = u_1 \tag{19}$$

where

$$F(\Psi) = \frac{4\pi}{3} \frac{mc^2}{\rho u_1} \int_{p_0}^{p_1} \frac{pg_0 dp}{\sqrt{p^2 + 1}} \exp\left[-\frac{1 + \beta(p)}{\kappa(p)}\Psi\right] \qquad (20)$$

The low energy asymptotics of g_0 is fixed by (18) determining the CR input from the thermal plasma. Substituting equation (15) into (19) the latter may be manipulated into the following integral equation for the normalized spectral function (see [7]) $J(p) = \bar{V}(p)/\bar{V}(p_0)$

$$J(\tau) = \frac{\zeta}{\varepsilon} \int_{\varepsilon}^{\varepsilon^{-1}} \frac{d\tau'}{\tau' + \tau} \frac{1}{\tau' J(\tau')} \exp\left[\frac{3}{\theta(r_s - 1)} \int_{\varepsilon}^{\tau'} \frac{d\tau''}{\tau'' J(\tau'')}\right] + 1. \qquad (21)$$

We have used the notations

$$\tau = \frac{\kappa_0 s}{\varepsilon}\left(1 - \frac{1}{r_s}\right), \quad \varepsilon^2 = \left(1 - \frac{1}{r_s}\right)\frac{p_0}{p_1}\theta \ll 1. \qquad (22)$$

The eigenvalue ζ is related to the injection rate ν through

$$\nu = \frac{\zeta}{R} \exp\left[\frac{3}{\theta(r_s - 1)} \int_{\varepsilon}^{1/\varepsilon} \frac{d\tau}{\tau J(\tau)}\right] \qquad (23)$$

The numerical factor $\theta \simeq 1.09$ and the injection rate

$$\nu \equiv \frac{4\pi}{3} \frac{mc^2}{\rho_1 u_1^2} p_0 Q_{\text{inj}} \simeq \frac{p_0 r_s}{r_s - 1} \frac{m n_c c^2}{\rho_1 u_1^2} \qquad (24)$$

where the number density of CRs is given by

$$n_c = 4\pi \int_{p_0}^{\infty} g_0(p) dp/p \qquad (25)$$

For simplicity, the diffusion coefficient κ is assumed to have a relativistic form $\kappa = \kappa_0 p/p_0 = \kappa_1 p/p_1$ for all p, since the relativistic particles are assumed to be dynamically much more important than the nonrelativistic ones. In order to close the system formed by equations (8) and (21) we need another equation to relate the three variables (ν, R, r_s) of which only one, say ν we consider as given. As an intermediate step we relate the precursor compression R and the spectral function J by inverting equation (16):

$$\frac{du}{d\Psi} = \frac{1}{2\pi i} \int_{-i\infty}^{i\infty} e^{s\Psi} \bar{V}(s) ds + \Delta u \delta(\Psi) \qquad (26)$$

where δ is a delta function corresponding to the jump of u at the subshock. Integrating then equation (26) over Ψ between $\Psi = 0+$ and ∞, using analytic properties of $J(\tau)$ that has two branch points at $\tau = -1/\varepsilon, -\varepsilon$, we get

$$u_1 - u_0 = \frac{\Delta u}{2\pi i} \int_{-1/\varepsilon}^{-\varepsilon} \frac{d\tau}{\tau} [J(\tau + i0) - J(\tau - i0)] \qquad (27)$$

We have put $\bar{V}(p_0) \approx \Delta u$ [7]. The integral around the cut $(-1/\varepsilon, -\varepsilon)$ may be evaluated with the help of equation (21) and the last equation rewrites

$$\frac{R-1}{1-r_s^{-1}} = \frac{\zeta}{\varepsilon} U \tag{28}$$

where

$$U = \int_\varepsilon^{1/\varepsilon} \frac{d\tau}{\tau^2 J(\tau)} e^{\Omega \phi(\tau)} \tag{29}$$

with

$$\Omega = \frac{3}{\theta(r_s - 1)} \quad \text{and} \quad \phi(\tau) = \int_\varepsilon^\tau \frac{d\tau'}{\tau' J(\tau')} \tag{30}$$

Equations (8), (21), and (28) form a closed system for describing nonlinear shock acceleration given the Mach number M, the cut-off momentum p_1 and the injection rate ν. It may have multiple solutions. On the other hand for sufficiently small injection rates ν there must always be a solution that corresponds to a test particle (linear) acceleration regime in which $J \to 1$ as $\nu \to 0$. This solution can be written down in terms of a Neuman series in ν or ζ as follows

$$J = 1 + \frac{\zeta}{\varepsilon^2} \ln \frac{\varepsilon^{-1} + \tau}{\varepsilon + \tau} + \mathcal{O}(\zeta^2) \tag{31}$$

where we have put $r_s = 4$ and $\theta = 1$ for simplicity. Solution (31) is essentially perturbative and cannot describe multiple solutions of equation (21) that appear beyond some $\nu > 0$. Namely at $\nu = \nu_1 > 0$ a pair of new solutions appears. They were studied in [7, 10]. As we shall see all the three solutions may be conveniently described by the single valued function $\nu = \nu(R)$ in the (R, ν) plane, Fig. 2 We term the solution with $R > R_1$ efficient, and that with $R_2 < R < R_1$ – intermediate. It merges with the inefficient solution at the point $\nu = \nu_2, R = R_2$. For a fixed $\nu \in (\nu_1, \nu_2)$ all the three solutions have different values of R and, hence, different subshock compression ratio r_s. It should be noted that the injection rate ν must be also calculated independently in terms of the subshock parameters yielding a function $\nu_s(R)$ which in combination with the bifurcation diagram should provide isolated solutions for R (shown by circles in Fig. 2).

4.3 Structure of the Shock Transition.
Overall Rankine-Hugoniot Relations

In a strongly nonlinear acceleration regime, the solution of equation (21) is dominated by the first term in the r.h.s. in contrast to the perturbative solution (31). Besides that the exponential factor in the r.h.s. may be replaced by unity since J turns out to be large enough. The resulting equation may be solved by using the smallness $\varepsilon \ll 1$. A uniformly valid (in τ) representation of this solution has the following form

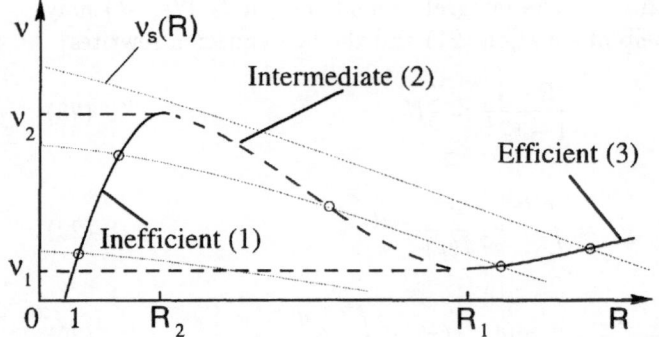

Fig. 2. *The nonlinear response R of an accelerating shock to the thermal injection of a rate ν represented in the form of a single valued function ν(R). Given ν ∈ (ν₁, ν₂) there are three substantially different acceleration regimes. Possible dependencies νₛ(R) (see text) are also drawn with the thin lines.*

$$J = \sqrt{\frac{\pi\zeta}{\varepsilon(\tau + \varepsilon)(1 + \varepsilon\tau)}} \tag{32}$$

Upon substitution of this solution in equations (26) we obtain the following equation for the flow potential

$$\frac{d\Psi}{dx} = u_0 + \sigma \int_0^\Psi e^{-a+\Psi'} I_0(a_-\Psi')d\Psi' \tag{33}$$

where I_0 is the modified Bessel function,

$$a_\pm = \frac{1}{2\kappa_0}\left(\frac{u_0}{\Delta u} \pm \theta\frac{p_0}{p_1}\right)$$

and

$$\sigma = \frac{u_0}{\kappa_0}\sqrt{\pi\zeta}.$$

A simple analysis shows that the most of the precursor can be divided into two different regions, described by the following approximate formulae:

$$u(x) - u_0 \simeq \begin{cases} u_1 x/l, & u_0/u_1 < x/l < 1 \\ (u_1 - u_0)\left[1 - \frac{1}{\pi}\sqrt{\frac{2l}{x}}\exp\left(-\frac{\pi x}{2l}\right)\right], & x > l \end{cases} \tag{34}$$

where the precursor scale height is $l = \pi\kappa_1/2\theta u_1$, $\theta \approx 1.09$. For smaller x the flow is dominated by the back reaction from low energy injection particles and may exhibit larger gradient with a quasi-singular behavior $u - u_0 \sim \sqrt{x}$ if the subshock is significantly reduced. This is, however, a small part of the shock transition in the case of efficient solution ($u_1 \gg u_0$).

From equation (28) one can easily determine the jump condition across the shock. The main result here is the following universal (Mach number independent) relation between the flow deceleration in the smooth and in the discontinuous parts of the shock transition in the efficient acceleration regime

$$\frac{u_1 - u_0}{u_0 - u_2} = \frac{\pi}{\theta}\eta p_1 \tag{35}$$

One sees that the flow deceleration upstream is, in fact, an amplified subshock jump. In other words, the modification effect disappears whenever does the subshock. The strength of the latter may be obtained from the conventional Rankine-Hugoniot relation (8).

4.4 Particle Spectrum

In the case of efficient solution the particle spectrum is given for all x by equation (9), where the downstream spectrum in a standard normalization $(q \to q + 3)$ being written as $f_0 \propto p^{-q}$ behaves as follows: $q = 3r_s/(r_s - 1)$ for $p \sim p_0$, and for smaller p joins smoothly the thermal distribution via the injection solution that should also yield a normalization constant, or the injection rate ν. For $p_0 \ll p \ll p_1$ the spectrum is completely universal, $q = 3\frac{1}{2}$. Taken together with the linear part of the flow profile from the last subsection this allows us to identify the efficient solution with the self-similar solution of the subsection 4.1. For p approaching p_1 the spectrum hardens to $q \approx 3.3$ but again, independently of any parameters.

5 Bifurcation of Acceleration Process

The method of integral equation developed in the preceding section allows one to describe the acceleration process on a universal basis in terms of the bifurcation analysis. In principle, our parameter space is two-dimensional and contains the Mach number M and the cut-off momentum p_1 (we will also use the parameter $p_1/p_0 \equiv \delta^{-1} \gg 1$). A convenient dependent variable is the flow compression R that obviously signifies the efficiency of acceleration. In the present study, however, we add to this parameter space also the injection rate ν, since this latter, even though being in principle calculable, may vary depending on the model of the subshock dissipation used [11]. Thus we perform our bifurcation analysis here in three-dimensional parameter space.

The character of bifurcation may be seen from the surface plots $\nu = \nu(R, M)$ at fixed p_1 and $\nu = \nu(R, p_1)$ at fixed M, respectively, Fig. 3. They are obtained on the basis of asymptotic solution of the system (8,21,23,28) for $\delta \ll 1$ and $M \gg 1$ [10]. The efficient part of the solution may be described by the following formula

$$\nu(R, M) = \frac{\zeta}{R} \exp\left(\frac{1}{\theta}\sqrt{\frac{2}{\zeta}\frac{1 + 3R^{8/3}M^{-2}}{1 - R^{8/3}M^{-2}}}\right) \tag{36}$$

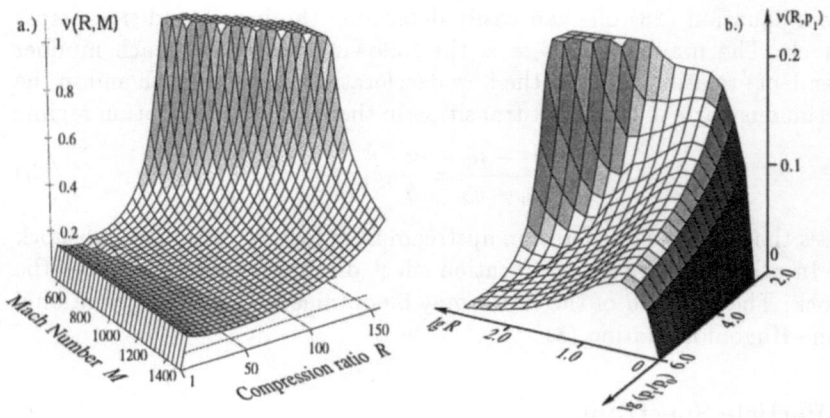

Fig. 3. *a.) The surface of stationary solutions $\nu(R, M)$ plotted for $\delta = 10^{-3}$. b.) The same as in Figure 3a but ν is given as a function of R and δ for $M = \infty$.*

with ζ to be taken from the relation

$$R - 1 = \frac{3\sqrt{2\zeta}}{4\varepsilon}\left(1 - \frac{R^{8/3}}{M^2}\right) \tag{37}$$

One sees that the multiplicity of the solution $R = R(\nu)$ is always present for sufficiently large values of M and p_1.

6 Discussion

The above analysis is formally limited to the case of a steady, plane shock acceleration with the parameterized injection rate and cut-off momentum. In realistic astrophysical situations these quantities may depend on time and/or they should be determined selfconsistently from current values of other shock parameters. However, our knowledge of the bifurcation diagrams of the solutions in the given parameter space allows us to draw some important conclusions about more realistic acceleration processes that are assumed a priori to be neither steady nor one-dimensional. A standard approach in analyses of complex dynamical systems is to consider the steady state manifolds like our $\nu = \nu(R, M, p_1)$ shown, e.g., in Fig. 3 or, at least certain parts of them, as attractors of the time dependent system. Then, as system parameters like

p_1 and ν, or dependent variables like the flow compression R evolve in time, the system should remain essentially on this manifold. Similarly, when the acceleration conditions vary along the shock front and the problem becomes three and two-dimensional, the acceleration parameters also change along the shock front but they do not leave this manifold. As in many other dynamical systems with "N" or "S" type characteristics (Fig. 2) the intermediate branch of the solution is likely to form an unstable manifold of this system. The stability of qualitatively similar hydrodynamic solutions for CR modified shocks have been studied recently by Mond & Drury [12]. They demonstrate that the intermediate part of the solution is indeed unstable, e.g., against deformations of the shock front (corrugational instability, e.g., [13]). The remaining part of the steady manifold also leaves room for ambiguity but, if $\nu < \nu_1(p_1, M)$, then the quasi-steady solution cannot be efficient whereas for $\nu > \nu_2$ the solution is by necessity efficient.

The main issue now is whether the actual injection rate can be maintained at a sufficiently high level while shock characteristics like the subshock Mach number and the compression ratio are driven to their extreme values by the backreaction of accelerated particles. As it is seen from equation (23) and Fig. 3 the constraints on injection become stronger when the system approaches the most efficient acceleration possible, i.e., when R is close to $M^{3/4}$ ($\nu(R, M) \to \infty$ as $R \to M^{3/4}$) and thus $r_{\rm s}$ is close to unity. On the other hand, the function $\nu(R)$ decreases with increasing p_1, for all R which should generally facilitate the efficient acceleration namely for larger cut-off momenta making it intrinsically unlimited. However, this is true only if no other than the adiabatic heating takes place in the CR precursor. Indeed, in the above consideration the Mach number at the subshock is $M_0 = M R^{-4/3}$. At the same time, as p_1 grows, the precursor length $l \sim \kappa(p_1)/u_1 \propto p_1$ grows as well and the role of resonant interaction of thermal particles with the MHD turbulence excited by the high energy particles must become more important. As a result, the formula for M_0 transforms to $M_0^{-2} = R^{8/3} M^{-2} + \eta(p_1, R)(v_{\rm A}/u)(R - 1)$, where η is the conversion factor of the turbulent energy into the plasma thermal energy. We have used equation (6) with the gas pressure added and a well known result that $v_{\rm A}/U$-fraction of the CR energy goes into MHD turbulence (see [3, 14]). Although the calculation of the function η is not straightforward, one may still assess the possible impact of this additional heating on the bifurcation considered in Sec. 5. Since M_0 enters the expression (23) for ν through the exponential factor, even a relatively slow increase of $\eta(p_1)$ will cause a quite rapid (probably exponential) increase of $\nu(p_1)$, starting from some p_1 where the exponential growth overcomes a power-law decay in p_1 dominating over the lower momentum range. Formally, a minimum of $\nu(p_1)$ appears already in expression (36), i.e., for $\eta \equiv 0$. However, the region of validity of this approximation shifts to higher R with growing p_1 (see also [10]) so that the function $\nu(p_1, R)$ decreases monotonically with p_1 for any fixed R, again if the heating of the precursor occurs only through the adiabatic compression. If η increases with p_1, the growth of $\nu(p_1)$ for large p_1 implies actually the existence of an intrinsic cut-off $p_{1\max}$ since the actual injection rate cannot be arbitrary large so that $\nu < \nu(p_{1\max})$. It should be born in mind that this is only valid for efficient acceleration, the system may still advance to higher p_1 in the inefficient acceleration regime, since the bifurcation curve $\nu(R)$ always passes through the point $\nu = 0$, $R = 1$. However, the estimates of acceleration efficiency required to maintain the galactic CR population at the observed level show that an averaged SNR shock must put 10–30% of its energy into CRs (see e.g., [15]). This suggests that in the sense of the above bifurcation analysis the shock must spend sufficiently long time in the "efficient" state in which, as argued above, the maximum energy may be seriously limited. On the other hand, observations of the CR background spectra suggest that a universal acceleration mechanism (presumably provided by SNR shocks) should be able to operate up to energies $\sim 10^2 - 10^3$ TeV. These data are, of course, rather indirect.

Direct observations of the SNRs in the TeV range are still rather inconclusive (see e.g., [16] and references therein). Moreover, some data seem to suggest that either there exists a spectral break in the 10–100 GeV-range or the spectrum cuts off there or at somewhat higher energies. This would definitely impose serious constraints on the SNR model of CR origin.

The above bifurcation analysis suggests that the acceleration process may indeed be terminated at relatively low energies if it operates in the efficient regime and the inflowing plasma is heated in the precursor noticeably faster than at the adiabatic rate. It should be noted, however, that the existence of different cut-off momenta in an ensemble of shocks that supposedly contribute to the observed CR background spectrum may still be reconciled with the smooth power-law character of this spectrum. Let us label the shocks in the ensemble by their cut-off momenta p_1 so that the spectrum at an individual shock is $f(p, p_1) = C(p_1)p^{-q_i(p_1)}H(p_1 - p)$, where H is the Heaviside function and C denotes the (unknown) weight function in the integral spectrum of a shock producing power-law spectra with the index q_i cut at $p = p_1$. In the simplest case $q_i(p_1) = const$, and $C \propto p_1^{-s}$, $s > 1$, averaging over p_1 yields for the integral spectrum $f(p) \sim p^{-q}$ with $q = q_i + s - 1$. If $q_i \simeq 3.5$, as it would be appropriate for efficient acceleration, one needs to have $s \simeq 1.7 - 1.9$ to obtain the spectrum of $q = 4.2 - 4.4$ inferred from observations. If this scenario is not totally wrong, one simple conclusion can be drawn immediately: it might be difficult to find the SNRs with sufficiently high values of the maximum momentum p_1. Indeed, since the distribution $C(p_1)$ must cover many orders of magnitude, say from $p_{\min}$ to $p_{\max}$, the probability to observe by chance a shock with the cut-off between $p_1 \gg p_{\min}$ and $p_{\max}$ is of the order of $(p_{\min}/p_1)^{s-1}$ or less if $p_1 \lesssim p_{\max}$. Even if the candidates for observations are carefully selected, this probability might still be insufficient to find them among all the SNRs suitable for observations.

I would like to thank Heinz Völk and Vladimir Ptuskin for interesting discussions. This work was done within the Sonderforschungsbereich 328, "Entwicklung von Galaxien" of the Deutsche Forschungsgemeinschaft (DFG).

References

1. Krymsky, G.F. (1977): Dokl. Akad. Nauk SSSR **234**, 1306 (Engl. Transl. Sov. Phys.-Dokl. 23, 327).
2. Axford, W. I., Leer, E., & Skadron, G. (1977): Proc. 15th ICRC (Plovdiv) **11**, 132
3. Bell, A. R. (1978): MNRAS , **182**, 147;443
4. Blandford, R. D., Ostriker, J. P. (1978): Astrophys. Journal , **221**, L29
5. Drury, L. O'C. (1983): Rep. Prog. Phys., **46**, 973
6. Malkov, M. A., Völk, H. J. (1995): Astron. Astrophys. , 300, 605
7. Malkov, M. A. (1997): Astrophys. Journal **485**, 638
8. Malkov, M. A. (1997): submitted to Phys. Rev. Lett.
9. Malkov, M. A., Völk, H. J. (1997): Proc. 25th ICRC, Durban, 4, 389
10. Malkov, M. A. (1997): Astrophys. Journal **491**, 584
11. Malkov, M. A. (1997): Proc. 25th ICRC, Durban, **4**, 393, submitted to Phys. Rev. E
12. Mond, M., Drury, L.O'C. (1998): Astron. Astrophys. in press
13. Landau, L.D., Lifshitz, E.M. (1988): Hydrodynamics, 4th Edition, Moscow, Nauka (in russian)
14. McKenzie, J.F., Völk, H. J. (1982): Astron. Astrophys. , **116**, 191
15. Drury, L.O'C., Markiewicz, W.J., Völk H.J. (1989): Astron. Astrophys. **225**, 179
16. Buckley, J.H. *et al.*, (1998): Astron. Astrophys. **329**, 639

Dynamical aspects of photon acceleration

J. T. Mendonça

GOLP/Centro de Física de Plasmas, Instituto Superior Técnico
1096 Lisboa Codex, Portugal

Abstract. We present here a review of recent work concerning the theory of photon acceleration in space and time varying plasmas. We will focus our atention on the dynamical aspects of the photon ray tracing equations and discuss the possible regimes of regular and stochastic photon acceleration. Photon trapping by an electron plasma wave and a model for photon Fermi acceleration will be presented. Our approach will be based on the Hamiltonian canonical equations for photons. A covariant Hamiltonian description will also be discussed.

1 Introduction

The concept of photon acceleration is quite recent in Plasma Physics [1-6]. It corresponds to an adiabatic frequency-shift of wavepackets propagating in a space and time-varying optical medium. This concept applies here to photons in their classical sense, as equivalent to elementary wavepackets. Using physical intuition, we could say that photons can be accelerated because they have an effective mass (except in vacuum). In an isotropic plasma, the photon mass is simply related to the electron plasma frequency. Moreover, photon acceleration is a non-resonant process which allows for the transfer of electromagnetic energy from one region of the spectrum to another. It contrasts with the well known resonant processes like the nonlinear wave mixing, where spectral transfer is controlled by well defined energy and momentum conservation laws. We can also say that photon acceleration is universal in the sense that it affects every photon in the medium, in contract with the well known Compton scattering which affects only a small fraction of the photons. In this paper we will restrict our discussions to the ray-tracing theory, which gives a very simple but accurate description of frequency shift of a single photon (a wavepacket) propagating in a non-stationary plasma [4]. The Hamiltonian version of the ray-tracing equations will be presented in Section 2. Using these equations we can extablish the conditions for photon trapping in the density modulations associated with an electron plasma wave. Of particular importance for the understanding of laser-plasma interaction are the relativistic electron plasma waves which can be excited by short laser pulses. Photon trapping by plasma waves is discussed in Section 3. If a given electron plasma wave is perturbed by a second plasma wave with a different phase velocity, or if it is modulated by a lower frequency wave of a different character (for instance, an ion acoustic wave), a fraction of the photon trajectories can become stochastic. This will lead to the transformation of an initial beam of nearly monocromatic radiation into white light. A similar effect is known in nonlinear optics as the generation of supercontinuum radiation[7]. The transition from regular to stochastic photon acceleration will be described in Section 4. Another exemple of a stochastic photon process is the ray-tracing model for the photon Fermi acceleration[7], which can occur when radiation is trapped inside an oscillating cavity, and will be described in Section 5. Finally, in Section 6 we will briefly discuss an alternative formulation of the ray-tracing equations which is based on a covariant Hamiltonian theory[9]. This covariant formulation is a direct consequence of the formal analogies between the ray-tracing equations and the equations of motion of a relativistic particle with a finite rest mass.

2 Canonical equations for photons

Let us assume an electromagnetic wave propagating in a space and time varying plasma. Its frequency ω and wavevector $\mathbf{k}$ are related by a local dispersion relation: $D(\omega, \mathbf{k}, \omega_p) = 0$ where ω_p is the electron plasma frequency, depending both on the position $\mathbf{r}$ and on time t. This expression stays valid as long as the space and time scales for plasma variation are much slower than $\mathbf{k}^{-1}$ and ω^{-1}. The wave electric field can be written in the form: $\mathbf{E}(\mathbf{r}, t) = \mathbf{E}_0 \exp i\phi(\mathbf{r}, t)$ The local frequency and wavevector appearing in the dispersion relation are defined by the space and time derivatives of the phase function ϕ:

$$\mathbf{k} = \frac{\partial}{\partial \mathbf{r}}\phi$$
$$\omega = -\frac{\partial}{\partial t}\phi \tag{1}$$

The local values of $\mathbf{k}$ and ω, for wavepackets propagating across the nonstationary medium will then change in space and time, following mean trajectories described by the ray tracing equations. It is well known that the ray tracing equations can be written in a canonical form:

$$\frac{d\mathbf{r}}{dt} = \frac{\partial \omega}{\partial \mathbf{k}}$$
$$\frac{d\mathbf{k}}{dt} = -\frac{\partial \omega}{\partial \mathbf{r}} \tag{2}$$

These ray tracing equations can also be seen as the photon trajectories. We notice that the canonical variables here are the photon position $\mathbf{r}$ and momentum or wavevector $\mathbf{k}$, while the energy or frequency ω plays the role of the Hamiltonian function. Here we will assume that the electron density perturbations associated with the nonstationary state of the plasma, move with a constant velocity $\mathbf{v}_f$. These perturbations can, for instance, be associated with an electron plasma wave with a relativistic phase velocity, as those produced by intense laser beams. These electron plasma waves are usually refered as the wakefields of the intense laser beam. Using the cold plasma dispersion relation, we can then write the photon Hamiltonian as:

$$\omega \equiv h(\mathbf{r}, \mathbf{k}, t) = \sqrt{k^2 c^2 + \omega_p^2(\mathbf{r} - \mathbf{v}_f t)} \tag{3}$$

This expression clearly shows that a photon propagating in a plasma behaves like a particle with a finite rest mass, proportional to the local plasma frequency.

What is interesting with the present Hamiltonian formulation of the photon motion is that there is no need for Lorentz transformations to the perturbation (or electron plasma wave) frame. The exact relativistic results are obtained just by using a canonical transformation from $(\mathbf{r}, \mathbf{k})$ to a new pair of canonical variables $(\boldsymbol{\eta}, \mathbf{q})$, such that:

$$\boldsymbol{\eta} = \mathbf{r} - \mathbf{v}_f t$$
$$\mathbf{q} = \mathbf{k} \tag{4}$$

The new Hamiltonian is now a constant of motion:

$$\omega' \equiv h'(\boldsymbol{\eta}, \mathbf{q}) = \sqrt{q^2 c^2 + \omega_p^2(\boldsymbol{\eta})} - \mathbf{v}_f \cdot \mathbf{q} \tag{5}$$

For each photon trajectory, the value of ω' can be determined by the initial values for its frequency and wavevector, ω_0 and $\mathbf{k}_0$, as $\omega' = \omega_0 - \mathbf{v}_v \cdot \mathbf{k}_0$ In the following we will apply these equations to study the photon motion in a wakefield.

3 Photon trapping

The plasma density perturbations associated with a wakefield can be described in a one-dimensional propagation model by:

$$\omega_p^2(\eta) = \omega_{p0}^2[1 + \epsilon \cos(k_p \eta)] \tag{6}$$

where ω_{p0} denotes the unperturbed plasma frequency, ϵ the plasma wave amplitude. and k_p is the plasma wave wavenumber. Looking at the photon trajectories we can easily recognize that this plasma wave or wakefield creates a nonlinear resonance in the photon phase space (η, q), with trapped orbits turning around an elliptic fixed point:

$$\eta_0 = \frac{\pi}{k_p}$$
$$q_0 = \frac{\omega_{p0}}{c} \beta \sqrt{\frac{1-\epsilon f_-}{1-\beta^2}} \tag{7}$$

where $\beta = v_f/c$. These trapped orbits are limited by a separatrix which connects the two hyperbolic fixed points determined by:

$$\eta_x = 0, \frac{2\pi}{k_p}$$

$$q_x = \frac{\omega_{p0}}{c}\beta\sqrt{\frac{1+\epsilon/\pm}{1-\beta^2}} \qquad (8)$$

The trapped photon orbits have initial frequencies in a range determined by:

$$\omega'_- < \omega' < \omega'_+ \qquad (9)$$

where $\omega'_\pm$ correspond to the extreme cases of motion at the elliptic fixed point and on the separatrix, respectively, and are defined by:

$$\omega'_\pm = \frac{\omega_{p0}}{\gamma}\sqrt{1 \pm \epsilon} \qquad (10)$$

Here γ is the relativistic gamma factor associated with the velocity v_f of the density perturbation. The minimum photon frequency shift occurs for trajectories near the elliptic fixed point and the maximum occurs for motion near the separatrix. The maximum frequency shift is determined by:

$$\delta\omega_{max} = 2\sqrt{2\epsilon}\omega_{p0}\gamma\beta \qquad (11)$$

4   Stochastic photon acceleration

Interesting new effects can occur when more than one wakefield is excited in the plasma. A photon propagating in such a plasma can eventually suffer a stochastic acceleration and two distinct photons, having nearly equal initial frequencies and positions, will follow exponentially divergent trajectories. This exponential divergence of a bunch of nearly monochromatic photons will then generate white light. In order to quantitavely determine the conditions under which white light generation can occur, let us assume that two distinct high phase velocity plasma waves propagate in the plasma. The space and time variation of the electron plasma frequency can be described by:

$$\omega_p^2(x,t) = \omega_{p0}^2[1 + \epsilon_1 \cos(k_1, \eta) + \epsilon_2 \cos(k_2(\eta - Vt))] \qquad (12)$$

where k_i and ϵ_i are the wavenumbers and amplitudes of the two distinct plasma waves $(i = 1, 2)$, the auxiliary variable η is now defined by $\eta = x - v_1 t$ and the difference between the two phase velocities is $V = v_1 - v_2$. In the photon phase space (q, η), these two waves build up two distinct nonlinear resonances, with hyperbolic fixed points located at q_1 and q_2, such that:

$$q_i = \frac{\omega_{p0}}{c} \beta_i \sqrt{\frac{1 - \epsilon_i}{1 - \beta_i^2}} \tag{13}$$

with $i = 1, 2$. On the other hand, it is easy to realise that the half-widths of these two resonances are determined by:

$$\Delta q_i = \frac{\omega_{p0}^2}{c^2} \sqrt{\frac{2\epsilon_i}{1 - \beta_i^2}} \tag{14}$$

It is well known that, in the phase space available around the two resonances, a transition from regular motion to large scale stochasticity can occur if the resonance widths are larger than a given threshold. A simple and quite accurate way to determine this threshold is to apply the overlaping criterion, which states that large scale stochasticity will occur if the sum of the two half-widths is larger than the distance between the two resonances:

$$|q_1 - q_2| \le \sum_{i=1,2} \Delta q_i \tag{15}$$

This threshold criterion for large scale stochastic motion in photon phase space can be rewritten in more explicit form as:

$$\frac{\gamma_1 \sqrt{2\epsilon_1} + \gamma_2 \sqrt{2\epsilon_2}}{|\beta_2 \gamma_2 \sqrt{1 - \epsilon_1}|} \le 1 \tag{16}$$

When this condition is satisfied we can say that white light can be generated from a pulse of nearly monochromatic light propagating in a plasma in the presence of two wakefields. This was confirmed by numerical calculations [4]. It is interesting to note that the above criterion is independent of the mean plasma frequency ω_{p0} and only depends on the amplitudes (ϵ_i) and on the phase velocities (β_i) of the two plasma waves.

5 Fermi acceleration of photons

We discuss now a different mechanism for photon acceleration which can occur inside an electromagnetic cavity with moving boundaries. This can be seen as the photon version of the well known mechanism for cosmic rays acceleration first proposed by Fermi, where charged particles can gain energy by bouncing forth and back between two magnetic clouds. In the case of photons, the clouds are replaced by mirrors or plasma walls with a sharp density gradient. We can assume t at one of the walls is fixed at $x = L_0$ and the other oscillates arround $x = 0$ with a frequency ω_m. The amplitude of this oscillation is assumed to be ϵL_0, with $\epsilon \ll 1$. It can be shown that the relative frequency $u = \omega/\omega_0$ of a photon trapped inside the oscillating cavity with an initial frequency ω_0 evolves according to the following discrete map:

$$u_{n+1} = u_n F(\phi_n) \tag{17}$$

$$\phi_{n+1} = \phi_n + G(u_{n+1}) \tag{18}$$

where the phase ϕ is defined by $\phi = \omega_m t$ and the quantities u and ϕ are determined after each successive reflection at the moving plasma wall. The functions $F(\phi)$ and $G(u)$ depend on the electron density profile of the reflecting plasma walls and, for a linear profile of the type $n_e(x) = 4\alpha x/L_0$, where α is the slope factor, they are given by:

$$F(\phi_n) = \frac{1 - \epsilon b \sin(\phi_n)}{1 + \epsilon b \sin(\phi_n)} \tag{19}$$

$$G(u_n) = 2b\left(1 + \frac{u_{n+1}^2}{a}\right) \tag{20}$$

where $b = L_0\omega_m/c$. A qualitative and numerical analysis of these mapping equations[8] show that the photon motion can become stochastic above a certain threshold value for the frequency, which mainly depends on the wall oscillating amplitude ϵ. This result contrasts with the usual Fermi acceleration of charged particles, which show that stochastic acceleration only takes place below a given threshold for the particle energy. This means that the above map looks like the inverse of the maps for charged particles. We conclude that a broad spectrum of radiation can be generated from nearly monochromatic light trapped in an oscillating cavity.

6 Covariant formulation

We have already noticed that a photon in a plasma behaves like a relativistic particle with an effective rest mass equal to $m_{eff} = \omega_p/c^2$. This means that the photon frequency can be determined by:

$$\omega = m_{eff}\gamma c^2 \tag{21}$$

where the photon gamma factor is:

$$\gamma = 1/\sqrt{1 - (\frac{\partial\omega}{\partial\mathbf{k}})^2} = \frac{\omega}{\omega_p} \tag{22}$$

This analogy with a relativistic massive particle suggests that a covariant description of the photon or ray-tracing trajectories can also be formulated in a covariant form [9]. Let us first define the 4-vectors photon position and momentum, such that:

$$\hat{r} = (ct, \mathbf{r})$$
$$\hat{k} = (\frac{\omega}{c}, \mathbf{k}) \tag{23}$$

With the aid of these two 4-vectors, we can construct a covariant Hamiltonian for the photon motion in a nonstationary plasma:

$$\hat{H}(\hat{r}, \hat{k}) = \frac{\hat{k}^2 c^2}{2\omega_p(\hat{r})} + \frac{1}{2}\omega_p(\hat{r}) \tag{24}$$

We should notice that the amplitude square of the 4-vector momentum is a negative quantity $\hat{k}^2 = -m_{eff}^2 c^2$ and that, by definition, we always verify $\hat{H} = 0$. The associated canonical equations in the 4-dimensional relativistic space will be given by:

$$\frac{dr^\alpha}{d\tau} = \frac{\partial \hat{H}}{\partial k_\alpha}$$
$$\frac{dk_\alpha}{d\tau} = -\frac{\partial \hat{H}}{\partial r^\alpha} \tag{25}$$

Here we have used the photon proper time $\tau = t/\gamma$. The plasma density perturbations associated with a wakefield can now be described by:

$$\omega_p^2(\hat{r}) = \omega_{p0}^2[1 + \epsilon \cos(\hat{q} \cdot \hat{r})] \tag{26}$$

where $\hat{q}$ is the 4-momentum associated with the wakefield. Let us write $\hat{q} \cdot \hat{r} = q^1 r^1 - q^0 r^0$. If we take $k^2 = k^3 = 0$ we can write the one-dimensional covariant Hamiltonian in adimensional form as $h = \hat{H}/\omega_{p0}$, such that:

$$h(x, v; y, u) = \frac{(v^2 - u^2)}{2\Omega_p(x, y)} + \frac{1}{2}\Omega_p(x, y) \tag{27}$$

where we have used the new variables $x = q^1 r^1$, $y = q^1 r^0$, $v = k_1 c/\omega_{p0}$ and $u = k_0 c/\omega_{p0}$. It can then be shown that the position of the nonlinear resonance built by the wakefield in the photon phase space is determined by $u_0 = -\gamma'$ and $v_0 = \beta\gamma'$, where β is the normalized phase velocity of the wakefield and $\gamma' = 1/\sqrt{1 - \beta^2}$ is the corresponding relativistic factor. It can also be easily recognized that the resonance half-width δu_0 is determined by:

$$\delta u_0 = \gamma'(\frac{\epsilon}{2})^2 \tag{28}$$

In this way, a covariant formulation equivalent to the description of traping photon motion arround a nonlinear resonance, as given in Section 2, can be established. The interaction of two such resonances resonances in the relativistic space will then lead to a stochastic photon acceleration.

7 Conclusions

In this work we have shown that photon propagation in nonstationary plasmas can be described in a rigorous but simple way by the Hamiltonian formulation of ray tracing theory. Explicit results for the photon frequency shift were obtained for photon motion in an electron plasma wave. Photon trapping, photon stochastic acceleration and a simple model for photon Fermi acceleration were discussed. As an alternative formulation, a covariant Hamiltonian version of the photon equations of motion was also described. It can be concluded that our Hamiltonian approach to photon acceleration provides a simple, rigorous and unified view of the physical processes leading to a frequency shift of electromagnetic wavepacket in a space and time varying plasma. However, it should be noted that a more exact approach, based on full wave calculations is also possible[10-12] which confirms the results of the above ray tracing description in its domain of validity but usually envolves more complicated calculations.

8 References

[1] S.C. Wilks, J.M. Dawson, W.B. Mori, T. Katsouleas and M.E. Jones, Phys. Rev. Lett., **62**, 2600 (1989).
[2]J.T. Mendonça, J. Plasma Phys. **22**, 15 (1979).
[3] W.B. Mori, Phys.Rev.Lett., **44**, 5118 (1991).
[4] J.T. Mendonça and L. Oliveira e Silva, Phys. Rev., **E 49**, 3520 (1994).
[5] J.M. Dias et al., Phys. Rev. Lett. **78**, 2271 (1997).
[6] R. Bingham, J.T. Mendonça and J.M. Dawson, Phys. Rev. Lett. **78**, 247 (1997).
[7] R.R. Alfano ed., "The Supercontinuum Laser Source", Springer-Verlag, N.Y. (1989).
[8] G. Figueira, J.T. Mendonça and L. Oliveira e Silva, *Transport,Chaos and Plasma Physics 2*, World Scientific, 237 (1996.
[9] J.T. Mendonça, K. Hizanidis, D.J. Frantzeskakis, L. Oliveira e Silva and J.L. Vomvoridis, J. Plasma Phys., **58**, 647 (1997).
[10] J.T. Mendonça and L. Oliveira e Silva, IEEE Trans. Plasma Sci. **24**, 147 (1996).
[11] L. Oliveira e Silva and J.T. Mendonça, IEEE Trans. Plasma Sci. **24**, 503 (1996).
[12] L. Oliveira e Silva, J.T. Mendonça and G. Figueira, Physica Scripta **T63**, 288 (1996).

Enhanced velocity diffusion
in slow-growing 1-D Langmuir turbulence

Isidoros Doxas and John R. Cary

Center for Integrated Plasma Studies
and
Department of Physics
University of Colorado
Boulder, CO 80309-0390

Abstract. Numerical simulations show an enhancement of the 1-D velocity diffusion coefficient over the quasilinear value in the regime where the autocorrelation time is much smaller than the linear growth time or resonance broadening time. The diffusion enhancement occurs when the resonance broadening time is small compared with the linear growth time. These simulations are self consistent, use a hybrid PIC/spectral symplectic integration method, and have enough modes to be in the continuous spectrum limit. That is, even at the initial amplitudes the intermode spacing is sufficiently small that the resonance overlap parameter is large. A possible mechanism for the enhanced diffusion (spontaneous spectrum discretization) is discussed.

1 Introduction

One-dimensional linear analysis shows that longitudinal plasma oscillations grow exponentially when the velocity distribution function is double-humped with a sufficiently deep valley. The resulting turbulent electrostatic oscillations lead to velocity diffusion, which causes the distribution function to evolve towards the stable, single-humped form. Quasilinear theory (Vedenov *et al*, 1962, Drummond and Pines, 1964) predicts the coefficient for this diffusion. Although quasilinear theory has been accepted for so long that it is in the textbooks (eg. Nicholson, 1983), theoretical work (Adam *et al*, 1979, Laval and Pesme, 1980, Laval and Pesme, 1983, Laval and Pesme, 1984) indicates that it can underestimate the diffusion coefficient, even in a regime where its validity is generally accepted. The simulations presented here show that this enhancement does occur, it has a dynamical basis, and we provide the enhancement factor as a function of the ratio of the linear growth time to the resonance broadening time.

These results have general impact on turbulence theory for several reasons. First, if the limits of validity of quasilinear theory, the simplest turbulence thory, are still not completely understood, then it is likely that we do not well understand the limits of validity of more complicated theories. Second, the present results, showing a dynamical basis for turbulent diffusion, indicate a direction for future investigations. Third, a large number of applications of turbulence theory rely on quasilinear theory in one form or another, either in direct calculations (Goldman *et al*, 1996) or as giving a form for the diffusion coefficient that is improved by adding fit parameters (Kotschenreuther *et al*, 1995). Finally, our results indicate the difficulty of obtaining valid numerical results for systems with small growth rates. Small growth rates can be problematic, as simulation noise can then be the dominant wave driver, and a greater number of computational cycles is needed. One cannot save computational cycles or reduce relatively the effect of noise by increasing the growth rate of instability for the present system, as this causes the needed inequality (of resonance broadening time shorter than growth time) to be reached only at large wave amplitudes - so close to saturation that determining the enhancement becomes problematic. To obtain unambiguous results one must have methods for carrying out low noise simulations. In the present simulations this is accomplished with a hybrid symplectic/PIC code, a PIC implementation of the symplectic integrator developed by Cary and Doxas (1993).

As noted by Laval and Pesme (1980), the reason that quasilinear theory may be wrong is that in the regime where the linear growth time is greater than the resonance broadening time, the nonlinear contributions to wave growth are comparable to the linear contributions. Because of energy conservation, enhanced wave growth rate implies enhanced diffusion. Subsequent heuristic arguments (Laval and Pesme, 1984) led them to suggest that in this regime the growth rate (diffusion coefficient) would be enhanced by a factor of two over the value given by linear (quasilinear) theory.

Since those early claims, there has been a seesaw of results in support or contradiction. Galeev *et al* (1980) ûstated that the nonlinear interaction of the harmonics does not change the structure of the equations, and that for quasilinear theory to apply, one need only have the autocorrelation time short compared with the diffusion time and the resonance broadening time. Early simulations (Theilhaber *et al*, 1987) may have seen some enhancement, but these results were inconclusive as at the point of measurement the time scale for diffusive change of the average distribution was of the order of the growth time for the modes. This loss of scale separation introduces large uncertainty in the measurement.

Cary *et al* (1990), noting that in the low-growth-rate regime the enhanced diffusion coefficient should be observable in test-particle simulations provided the correct ensemble is chosen, carried out a series of test-particle simulations for discrete spectra of randomly phased waves. They found that while the quasilinear value for the diffusion coefficient was recovered in the limit of large overlap, where the resonances widths greatly exceeded the phase-velocity spacing of the waves, in regions of intermediate overlap the diffusion coefficient could be enhanced over that of the quasilinear value by as much as a factor of 2.5.

The first experiments (Tsunoda *et al*, 1991) failed to find any overall enhancement, but they did see mode coupling. Further analytical (Liang and Diamond, 1993) and numerical work (Deeskow *et al*, 1991, Ishihara *et al*, 1992) failed to find any enhancement, but subsequent self consistent simulations (Cary *et al*, 1992) did find an overall enhancement of the growth rate for a spectrum initialized with an intermediate value of the overlap parameter. Enhancement has now also been seen by Berndtson *et al* (1994) for evolution in self-consistent fields at intermediate values of the overlap parameter and by Helander and Kjell-berg (1994) for test-particle simulations in prescribed fields. The most recent experimental results (Hartmann *et al*, 1995) show mode coupling but cannot operate in the regime of intermediate overlap needed to see the enhancement of the growth rate and diffusion coefficient.

While there is strong evidence now that mode coupling affects the growth rate of individual modes, and it has been verified by more than one group that the diffusion coefficient and the growth rate can be enhanced in a regime of intermediate overlap, it remains to be shown conclusively that enhancement can arise self-consistently for a continuous spectrum - one in which the waves are strongly overlapping. This is the purpose of the present work. Our simulations show that the enhancement of the diffusion coefficient can arise spontaneously in continuous-spectrum, self-consistent simulations. Our simulations are initialized with a sufficiently large number of modes such that the modes are strongly overlapping, and we measure the growth rate enhancement at a time both far from initialization and far from saturation. We find that that there is a modest enhancement ($\sim$ 30%) of the growth rate for values of resonance broadening frequency up to roughly 30 times the linear growth rate. That this is far below the 120% enhancement given by the turbulent trapping model (Laval and Pesme, 1984) shows the possible large uncertainties in such heuristic calculations. Our low value also shows that the 60% enhancement seen in Theilhaber *et al* (1987) was indeed spurious. However, it remains possible that the enhancement factor is greater at larger values of the resonance broadening frequency to growth rate ratio.

In the following section we review the classic system of the bump on tail instability including a discussion of the effects of modeling the system with a discrete spectrum. In the subsequent section we introduce our numerical model for this system. In Sec. 4 we illustrate that mode coupling sets in when the resonance broadening frequency exceeds the growth rate, as predicted by theory (Adam *et al*, 1979, Laval and Pesme, 1980, Laval and Pesme, 1983, Laval and Pesme, 1984) and seen in experiment (Tsunoda *et al*, 1991). In Sec. 5 we determine the conditions for carrying out definitive simulations to observe the growth enhancement. In Sec. 6 we show the results of the simulations: we find the enhancement factor as a function of the ratio of resonance broadening frequency to growth rate for values of resonance broadening frequency up to about 30 times the growth rate. Finally, we conclude and indicate new directions for research in Sec. 7.

2 The warm-beam instability

The bump-on-tail instability occurs when a weak, broad beam (taken to have density n_b) is present in a plasma. When the plasma is cold, its response can be assumed linear. The resulting evolution is given by the nonlinear interaction of the beam with the longitudinal waves supported by the background plasma (O'Neil *et al*, 1971, Mynick and Kaufman, 1978).

The evolution of the normalized beam distribution g in the electric field is given by the Vlasov equation,

$$\frac{\partial g}{\partial t} + v\frac{\partial g}{\partial x} + \frac{eE}{m}\frac{\partial g}{\partial v} = 0. \tag{1}$$

The electric field is given by Poisson's equation,

$$\frac{\partial E}{\partial x} = \rho_p - \rho_{p0} + \rho_{b0}\left[\int dvg(x,v,t) - 1\right] \tag{2}$$

where ρ_p and ρ_{p0} are the perturbed and average plasma charge densities, and ρ_{b0} is the average beam charge density. Standard linear analysis (Nicholson, 1983, Cary et al, 1992) of these or equivalent equations shows that for this system the modes for all wavenumbers oscillate at the plasma frequency ω_p, and that modes with phase velocity in the range,

$$v_a < v_\varphi \equiv \frac{\omega}{k} < v_b, \tag{3}$$

where the slope of the unperturbed, normalized distribution $g_0(v)$ is positive, are unstable, with a growth rate given by

$$\gamma_L = \frac{\pi}{2}\frac{\eta}{1+\eta}\omega_p v_\varphi^2 \frac{\partial g_0}{\partial v}(v_\varphi), \tag{4}$$

where

$$\eta \equiv n_b/n_p \tag{5}$$

is the ratio of beam density n_b to plasma density n_p. (This convention differs from that of Cary et al (1992), where g_0 was the normalized distribution for the plasma and the beam). Our goal is to analyze these equations nonlinearly.

The analysis rests on the approximation that the wave amplitude is slowly varying. With the electric field in the form,

$$E(x,t) = \frac{1}{2}\sum_{j=1}^{M}[E_j(t)\,e^{i(k_j x - \omega_p t)} + c.c.], \tag{6}$$

the complex amplitudes E_j can be assumed slowly varying, because the growth rate (4) is small, and the nonlinear effects are comparable. With this approximation, and that of the plasma response being linear, one can derive the equation

$$\frac{\partial\epsilon}{\partial\omega}(k_j,\omega_j)\dot{E}_j = -2\frac{\rho_{b0}}{k_j}\int\frac{dx}{L}\exp(-ik_j x + i\omega_j t)\int dvg(x,v,t) \tag{7}$$

for the evolution of the wave amplitudes. For a cold background plasma, the first factor is given by

$$\frac{\partial\epsilon}{\partial\omega}(k_j,\omega_j) = \frac{2}{\omega_p}. \tag{8}$$

Hence, the evolution of the electric field is given by

$$\dot{E}_j = -\frac{\omega_p \rho_{b0}}{k_j} \int \frac{dx}{L} \exp(-ik_j x + i\omega_j t) \int dv g(x, v, t). \tag{9}$$

In response to this field, standard theory states that beam particles will diffuse in velocity, with diffusion coefficient given by the quasilinear value

$$D_{QL}(v) = \frac{\pi}{2} \left(\frac{e}{m}\right)^2 \left\langle \frac{|E_j|^2}{k_j \delta v_{\varphi j}} \right\rangle \Bigg|_{v_{\varphi j} \approx v} \tag{10}$$

In this formula the brackets denote a local (in phase velocity) average of the energy density (per unit Doppler shifted frequency) for the modes with phase velocity near the velocity v at which the diffusion coefficient is being evaluated (Cary et al, 1992). The diffusion then reduces the slope of the distribution, stabilizing the waves.

As discussed early on by Dupree (1996), the theory giving the quasilinear diffusion coefficient relies on the autocorrelation time,

$$\tau_{ac} \equiv 1/(k_{typ} \Delta v_\varphi), \tag{11}$$

which is the time for decay of the force correlation along a constant-velocity (unperturbed) trajectory of a particle, being small compared with what came to be called the resonance broadening time,

$$\tau_{RB} \equiv 1/\nu_{RB} \equiv (k^2 D_{QL})^{-1/3}. \tag{12}$$

If the reverse is true, the constant-velocity trajectory approximation used to calculate the diffusion coefficient fails before the integral expression for the diffusion coefficient converges. Thus, the accepted criterion for validity of quasilinear theory has been

$$\tau_{ac} \ll \tau_{RB}. \tag{13}$$

This condition is equivalent to the criterion that the diffusion time be large compared with the autocorrelation time,

$$\tau_{ac} \ll \Delta v_\varphi^2 / D_{QL}. \tag{14}$$

Galeev et al (1980) states that the validity of quasilinear theory requires the validity of only conditions (13) and (14).

Since the work of Dupree, the question of quasilinear diffusion has been approached using the methods of nonlinear dynamics. As discussed by Chirikov (1979), the nature of the motion in the discrete spectrum (6) is determined by the overlap parameter,

$$A_{j,OL} \equiv \left(\frac{\pi}{2}\right)^2 \left(\frac{2\delta v_{rj}}{\delta v_{\varphi j}}\right)^2, \tag{15}$$

which is, up to a factor, the square of the ratio of the resonance width,

$$\delta v_{rj} \equiv 2\sqrt{e|E_j|/mk_j},\tag{16}$$

of mode j to the phase velocity separation,

$$\delta v_{\varphi j} \equiv \frac{\omega_p}{k_{j+1}} - \frac{\omega_p}{k_j}.\tag{17}$$

The result of Chirikov's work is that trajectories in the field (6) with fixed, constant phases and broad spectrum, become chaotic once the overlap parameter exceeds approximately unity, and diffuse with coefficient given by the quasilinear value (10). This work appeared to provide additional justification for quasilinear theory, as in the continuous spectrum limit (where the wave energy is divided among M waves with M becoming infinite), the overlap parameter becomes infinite (Cary *et al*, 1992, Rechester *et al*, 1979).

In terms of the present notation, the work of Adam, Laval, and Pesme (Adam *et al*, 1979) and Laval and Pesme (1980, 1983, 1984) states that the conditions (13) or, equivalently, (14) are not sufficient for validity of quasilinear theory. They state that one must also have

$$\nu_{RB} \ll \gamma_L.\tag{18}$$

Condition (18) ensures that the nonlinear contributions to the wave growth are small compared with the linear contributions. For ease of reference we define the parameter,

$$\mu \equiv \nu_{RB}/\gamma_L.\tag{19}$$

Then the regime of questioned validity of quasilinear theory is that of

$$\mu \gg 1.\tag{20}$$

3 Numerical model

Our nonlinear, dynamical system evolves according to Eq. (1) and Eq. (9). Our approach to intergrating the Vlasov equation is to represent the distribution as a collection of particles. This gives a discrete Hamiltonian system of N charged particles interacting with M modes. Having a Hamiltonian system allows us to use the methods of symplectic integration. In this section we present the equations of motion and discuss the conditions for validity of our simulation. We derive the conditions for the spectrum to be sufficiently finely discretized that it is in the continuous spectrum limit, and we determine the number of particles needed to well represent the distribution.

3.1 Discretization of the particle distribution

The Vlasov equation dictates that the distribution is constant along the characteristics. Hence, for the distribution,

$$g(x, v, t) = \alpha \sum_{i=1}^{N} \delta\left(x - x_i(t)\right) \delta\left(v - v_i(t)\right),$$ (21)

corresponding to a sum of particles, the positions and velocities of the particles obey the usual equations of motion,

$$\dot{x}_i = v_i$$

$$\dot{v}_i = \frac{e}{m} E = \frac{e}{m} \sum_{j=1}^{M} [E_{jR} \cos(k_j x_i - \omega_j t) - E_{jI} \sin(k_j x_i - \omega_j t)],$$ (22)

where E_{jR} and E_{jI} are the real and imaginary parts of the wave amplitudes. Normalization for a system of length L,

$$L = \int dx dv g(x, v, t) = N\alpha,$$ (23)

determines the coefficient

$$\alpha = L/N$$ (24)

in the representation (21).

The equation for the electric field for discrete particles can be put in the form

$$\dot{E}_{jR} = -\frac{\omega_p \rho_{b0}}{k_j N} \sum_{i=1}^{N} \cos(k_j x_i - \omega_j t)$$

$$\dot{E}_{jI} = \frac{\omega_p \rho_{b0}}{k_j N} \sum_{i=1}^{N} \sin(k_j x_i - \omega_j t)$$ (25)

To combine these equations with those for the evolution of the particles in a single Hamiltonian, we introduce the canonical coordinates,

$$Q_j \equiv \omega_p^{-3/2} \sqrt{\frac{N}{\eta}} \frac{e E_{jI}}{m}$$

$$P_j \equiv -\omega_p^{-3/2} \sqrt{\frac{N}{\eta}} \frac{e E_{jR}}{m}$$ (26)

With these definitions, our equations of evolution are Hamilton's equations,

$$\dot{x}_i = \frac{\partial H}{\partial v_i}$$

$$\dot{v}_i = -\frac{\partial H}{\partial x_i}$$

$$\dot{Q}_j = \frac{\partial H}{\partial P_j}$$

$$\dot{P}_j = -\frac{\partial H}{\partial Q_j} \tag{27}$$

with canonically conjugate x_i and v_i for the Hamiltonian,

$$H = \frac{1}{2}\sum_{i=1}^{N} v_i^2 + \sqrt{\frac{\eta}{N}} \sum_{i=1}^{N}\sum_{j=1}^{M} \frac{\omega_p^{3/2}}{k_j} \left[P_j \sin(k_j x_i - \omega_j t) - Q_j \cos(k_j x_i - \omega_j t)\right]. \tag{28}$$

Equivalently, if one uses the variables,

$$X_j \equiv Q_j \cos(\omega_j t) + P_j \sin(\omega_j t)$$

$$Y_j \equiv -Q_j \sin(\omega_j t) + P_j \cos(\omega_j t), \tag{29}$$

one obtains the Hamiltonian,

$$H = \frac{1}{2}\sum_{i=1}^{N} v_i^2 + \frac{\omega_p}{2}\sum_{j=1}^{M} \left[X_j^2 + Y_j^2\right]$$

$$+ \sqrt{\frac{\eta}{N}} \sum_{i=1}^{N}\sum_{j=1}^{M} \frac{\omega_p^{3/2}}{k_j} \left[Y_j \sin(k_j x_i) - X_j \cos(k_j x_i)\right]. \tag{30}$$

The Hamiltonian in (30) was derived by Mynick and Kaufman (1978) and Escande (1987) and used in some of our simulations.

3.2   Integration method

The Hamiltonian that we have derived is amenable to the symplectic integration method developed by Cary and Doxas (1993), which is based on the explicit symplectic schemes of Forest and Ruth (1990) and Candy and Rosmus (1991). We have augmented the previous scheme of Cary and Doxas (1993) by adapting particle-in-cell methods to it. The resulting integrator can achieve 5% accuracy in the value of the wave growth rate with a time step of the order of the inverse plasma frequency. We only briefly review this method here as it will be discussed in more detail elsewhere.

To have an explicit symplectic integrator one must be able to split the Hamiltonian into a sum of parts, each of which can be integrated exactly. The Hamiltonian (28) can be split into

$$H = H_1 + H_2, \tag{31}$$

where

$$H_1 = \frac{1}{2}\sum_{i=1}^{N} v_i^2, \tag{32}$$

and

$$H_2 = \sqrt{\frac{\eta}{N}} \sum_{i=1}^{N} \sum_{j=1}^{M} \frac{\omega_p^{3/2}}{k_j} \left[P_j \sin(k_j x_i - \omega_j t) - Q_j \cos(k_j x_i - \omega_j t) \right]. \tag{33}$$

Each of these Hamiltonians can be integrated exactly. For the alternate Hamiltonian (30), H_1 remains the same, while H_2 includes also the middle term of (30).

We define $T_1(\tau)$ to be the transformation that advances according to the Hamiltonian H_1 by a time step τ and $T_2(\tau)$ to the transformation that advances H_2 by the same amount. Then, the transformation that advances H by τ is given to second order by $T(\tau) = T_1(\tau/2)T_2(\tau)T_1(\tau/2) + O(\tau^4)$ (Candy and Rozmus, 1991, Forest and Ruth, 1990). By construction, this integrator is exactly symplectic and is therefore expected to better preserve the invariants of the Hamiltonian.

To carry out the present work, we implemented a particle-in-cell (PIC) version of this integrator. The PIC implementation of the integrator follows Cary and Doxas (1993) except for the following change. The Hamiltonian (33) is rewritten and then approximated as follows

$$\begin{aligned}
H_2 &= \sqrt{\frac{\eta}{N}} \sum_{i=1}^{N} \sum_{j=1}^{M} \frac{\omega_p^{3/2}}{k_j} \left\{ P_j \left[\sin(k_j x_i) \cos(\omega_j t) - \cos(k_j x_i) \sin(\omega_j t) \right] \right. \\
&\quad \left. -Q_j \left[\cos(k_j x_i) \cos(\omega_j t) + \sin(k_j x_i) \sin(\omega_j t) \right] \right\} \\
&\approx \sqrt{\frac{\eta}{N}} \sum_{i=1}^{N} \sum_{j=1}^{M} \frac{\omega_p^{3/2}}{k_j} \left\{ P_j \left[f_j(x_i) \cos(\omega_j t) - g_j(x_i) \sin(\omega_j t) \right] \right. \\
&\quad \left. -Q_j \left[g_j(x_i) \cos(\omega_j t) + f_j(x_i) \sin(\omega_j t) \right] \right\}
\end{aligned} \tag{34}$$

where $f_j(x_i)$ and $g_j(x_i)$ are interpolations of $\sin(k_j x_i)$ and $\cos(k_j x_i)$ from grid data. Use of interpolation reduces the computational effort from scaling as the number of particles times the number of waves to being proportional to just the number of particles. However, since the force equation

$$\begin{aligned}
\dot{v}_i = -\frac{\partial H}{\partial x_i} &= \sqrt{\frac{\eta}{N}} \sum_{j=1}^{M} \frac{\omega_p^{3/2}}{k_j} \left\{ P_j \left[f_j'(x_i) \cos(\omega_j t) - g_j'(x_i) \sin(\omega_j t) \right] \right. \\
&\quad \left. -Q_j \left[g_j'(x_i) \cos(\omega_j t) + f_j'(x_i) \sin(\omega_j t) \right] \right\}
\end{aligned} \tag{35}$$

involves the derivatives of the functions f and g, we have to use higher order interpolation. In order to use a second-order integrator, f' and g' must have continuous first derivatives, so we use cubic spline interpolation for f and g.

In general, if we have (regular) grid points $x_0, x_1, \ldots, x_{N_g}$, and we know the values of y on these points $y_0, y_1, \ldots, y_{N_g}$, then the value of the function $y(x)$ for $x_n < x < x_{n+1}$ is given by

$$y(x) = A y_n + B y_{n+1} + C y_n'' + D y_{n+1}'' \tag{36}$$

with

$$A = \frac{x_{n+1} - x}{\Delta x} \qquad B = 1 - A \tag{37}$$

The coefficients C and D are defined such that y has continuous first derivatives, and are given by

$$C = \frac{1}{6}(A^3 - A)(\Delta x)^2 \qquad D = \frac{1}{6}(B^3 - B)(\Delta x)^2 \tag{38}$$

where $\Delta x = x_{n+1} - x_n$ is the grid spacing.

For a function of the form $y = e^{ikx}$, the sequence y'' is given by $y'' = \alpha_k e^{ikx}$. Using Eqs. (36)–(38) and the continuity of y', we find

$$\alpha_k = \frac{3\cos(k\Delta x) - 1}{k^2(\Delta x)^2[2 + \cos(k\Delta x)]} \tag{39}$$

In particular, for the functions f and g in the hamiltonian, we get

$$f_{nj} = \sin(k_j x_n) \qquad f''_{nj} = \alpha_j \sin(k_j x_n)$$
$$g_{nj} = \cos(k_j x_n) \qquad g''_{nj} = \alpha_j \cos(k_j x_n)$$

for the values of the functions on the grid points.

Using the above for the force equation, we readily obtain

$$\dot{p}_i = -\sqrt{\frac{2\eta}{N}} \frac{1}{\Delta x} \Big[F_{n+1} - F_n + \frac{1}{6}(3B^2 - 1)(\Delta x)^2 F''_{n+1} - \frac{1}{6}(3A^2 - 1)(\Delta x)^2 F''_n +$$
$$G_{n+1} - G_n + \frac{1}{6}(3B^2 - 1)(\Delta x)^2 G''_{n+1} - \frac{1}{6}(3A^2 - 1)(\Delta x)^2 G''_n \Big] \tag{40}$$

where F and G are the fourier transforms of f and g respectively:

$$F_n = \sum_{j=1}^{M} \frac{\omega_p^{3/2}}{k_j} (P_j \cos(\omega_j t) - Q_j \sin(\omega_j t)) f_{nj}$$

$$G_n = \sum_{j=1}^{M} \frac{\omega_p^{3/2}}{k_j} (P_j \sin(\omega_j t) + Q_j \cos(\omega_j t)) g_{nj}$$

$$F''_n = \sum_{j=1}^{M} \frac{\omega_p^{3/2}}{k_j} (P_j \cos(\omega_j t) - Q_j \sin(\omega_j t)) f''_{nj}$$

$$G''_n = \sum_{j=1}^{M} \frac{\omega_p^{3/2}}{k_j} (P_j \sin(\omega_j t) + Q_j \cos(\omega_j t)) g''_{nj}$$

Advancing the particles therefore involves calculating two complex FFT's over the N_g grid points, plus N evaluations of Eq. (40). Since $N_g \ll N$, this is a number of operations that scales as N. We therefore see that the new method combines the advantages of the symplectic integrator of Cary and Doxas (1993) with the computational efficiency of PIC methods.

The code was benchmarked against the results of O'Neil *et al* (1971) for the cold beam case. The results, shown in Fig. 1, are that for small time steps ($\omega_p \Delta t = 0.1$) and small grid spacing ($k_{max} \Delta x = 0.1$) one can achieve high accuracy, of the order of one part in 10^4. Furthermore, the symplectic integration method does not suffer from the long-time spurious loss or gain of energy as does, for example, the Runge-Kutta method.

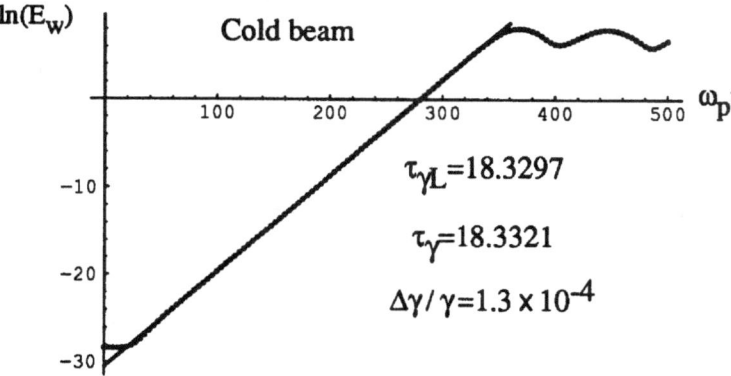

Fig. 1. The wave energy as a function of time for the cold beam instability with a beam to plasma density ratio of $\eta = 5 \times 10^{-4}$ modeled with 40000 particles. The timestep is $\omega_p \Delta t = 0.10$, and the grid spacing $k \Delta x = 0.10$. As indicated in the figure, the simulation gives the growth rate to a relative accuracy of 1.3×10^{-4}.

The code was also benchmarked against the warm beam-plasma instability, where a single wave interacts with a broad beam (Cary and Doxas, 1993). As shown in Fig. 2 we again found high accuracy for small time steps. Finally, we tested the code for larger time steps and found that for time steps of size $\omega_p \Delta t \approx 0.8$, and grid spacings of $k_{max} \Delta x \approx 0.8$, growth rates were accurate to within roughly 2%. These were the values that we chose to use in subsequent simulations.

3.3   Initial conditions

We chose our initial conditions for the simulations of Sec. 6 so as to minimize the ambiguity of our measurements. Our choice was to set up a system where the linear growth rate and the quasilinear diffusion coefficient were constant across the spectrum of waves and for particles of different velocities. Thus, our measurement of the average growth rate is not confused by such effects as the differential linear growth rate throughout the spectrum.

Eq. (4) plus the condition of constant linear growth rate gives a unique form

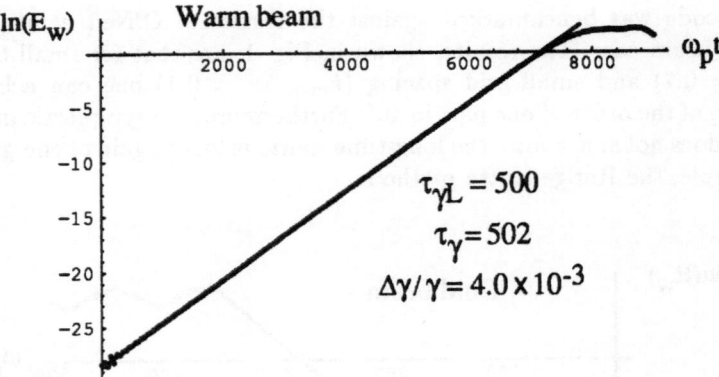

Fig. 2. The wave energy versus time for the warm beam instability with only a single wave excited, a beam to plasma density ratio of $\eta = 5 \times 10^{-4}$, and modeled by 512000 particles. The timestep is $\omega_p \Delta t = 0.25$ and the grid spacing $k\Delta x = 0.34$. As indicated in the figure, the simulation gives the growth rate to a relative accuracy of 4×10^{-3}.

for the beam distribution,

$$g_0(v) = C_1 + C_2 \left(1 - \frac{v_a}{v}\right)_{v_a < v < v_b} \tag{41}$$

where $g_0(v) = 0$ outside the range $v_a < v < v_b$. The coefficients C_1 and C_2 are determined by the normalization condition, $\int dv g_0(v) = 1$, and the value of the growth rate, Eq. 4. In terms of these parameters, the coefficients are

$$C_2 = \frac{2\gamma_L(1 + \eta)}{\pi \omega_p \eta v_a} \tag{42}$$

$$C_1 = \frac{1 + C_2[v_a \ln(v_b/v_a) - v_b + v_a]}{(v_b - v_a)} \tag{43}$$

In choosing the initial spectrum we demand that the quasilinear diffusion coefficient, which is determined by the local amplitudes of the waves, be constant across the spectrum. This is discussed in more detail in Cary *et al* (1990). Our choices are shown in Fig. 3. We chose $v_a = 0.25$ and $v_b = 0.38$ for all of our runs. The slope of the distribution is given by (41). The spectrum shown in Fig. 3b has 584 waves with wavenumbers given by

$$k_j = \frac{2\pi n_j}{L} \tag{44}$$

where L is the length of the system. The amplitude of the waves varies as $k^{-1/2}$, giving a quasilinear difussion coefficient that is constant in v (cf. Eq. 10).

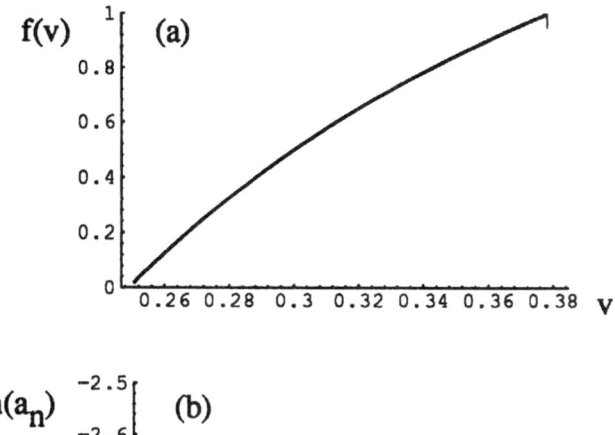

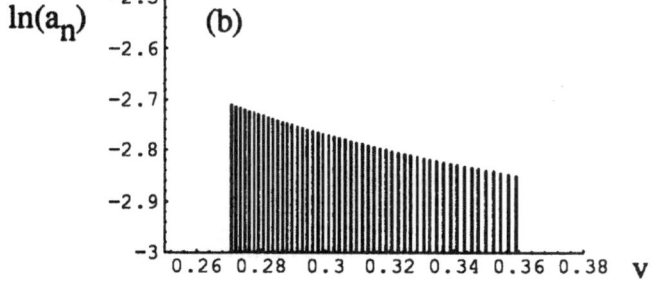

Fig. 3. The initial conditions for the particle distribution (a), and the wave amplitudes (b). For clarity, only one out of every five waves is shown.

4 Mode coupling

Turbulent trapping theory (Adam *et al*, 1979, Laval and Pesme, 1980, Laval and Pesme, 1983, Laval and Pesme, 1984) states that the enhancement of the diffusion coefficient over the quasilinear value is caused by modifications of the spectrum due to nonlinear mode-mode coupling. Part of this prediction, that significant mode coupling arises when $\mu > 1$, has been observed experimentally (Tsunoda *et al*, 1991, Hartmann *et al*, 1995), but the same experiments do not observe an enhancement of the average growth rate or diffusion coefficient. A possible explanation of this is that mode coupling sets in at moderate values of μ, while large values of μ are required for enhancement of the average growth rate (diffusion coefficient).

In this section we confirm that mode coupling sets in at $\mu \approx 1$. We performed a quiet start simulation in which the mode amplitudes initially vanish, and the beam particles are placed in a grid in phase space, such that initially the sums in Eqs. (25) vanish. As noted in Zekri (1993), this system is at an unstable equilibrium point. Roundoff error in the particles' initial positions is then sufficient to initiate exponential growth at the linear growth rate. This technique allows us to run simulations with very little noise.

Figure 4 shows the results of this simulation, which had 2×10^6 particles and 211 waves, and beam to plasma density ratio of $\eta = 2.7 \times 10^{-2}$. The spectrum grows for seven exponentiations through the point where the overlap parameter is unity maintaining its shape. This is the prediction of quasilinear theory: that the growth rates for individual modes are given by linear theory, which by construction gives a constant growth rate across the spectrum for this simulation (cf. Eq. 41).

However, once $\mu = 1$ is reached, the modes begin to grow differentially as can be seen in Fig. 4c. This is the effect of mode coupling. In addition, the amplitudes of three different modes are seen to diverge in Fig. 4d. This simulation demonstrates that mode coupling does onset at $\mu = 1$. Unfortunately, this particular simulation cannot be carried out further to see the average enhancement, as saturation begins to occur not much later. After the onset of saturation differential growth can no longer be unequivocally attributed to mode coupling, and the turbulent trapping regime $\mu \gg 1$ cannot be reached.

5 Simulation requirements

As noted in Fig. 4, it is difficult to achieve numerically the large μ regime needed to observe the enhanced growth rate. In essence, the problem arises because within one growth time after the simulation grows to the point $\mu = 1$, the particles become chaotic, as the Lyapunov time is of the order of the resonance broadening time (Stoltz and Cary, 1994). Thus, at this point spontaneous emission due to the random distribution of particles is present. To have an unequivocal measurement of the growth rate, the wave growth due to spontaneous emission must be small compared to the growth due to instability. Moreover, the waves must grow significantly past the point where spontaneous emission is relevant without reaching saturation, so that saturation dynamics do not confuse the measurement of the growth rate.

In addition, one must use enough modes that the initial overlap parameter is large. This ensures that the spectrum is essentially continuous. Hence, three conditions are needed for definitive simulations. Large initial overlap, large ratio of saturation energy to energy where spontaneous emission is significant, and large μ. In this section we determine the simulation parameters based on these conditions. We do so by estimating the spontaneous emission amplitude, the saturation amplitude, the linear growth rate, the resonance broadening frequency, and the overlap parameter for our numerical method. We then solve for the requisite number of particles, modes, and instability parameter.

5.1 Estimates of theoretical quantities

Growth rate The growth rate can be estimated from Eq. (4). For this estimate we note that the beam distribution (cf Fig. 3) is roughly triangular, centered at

325

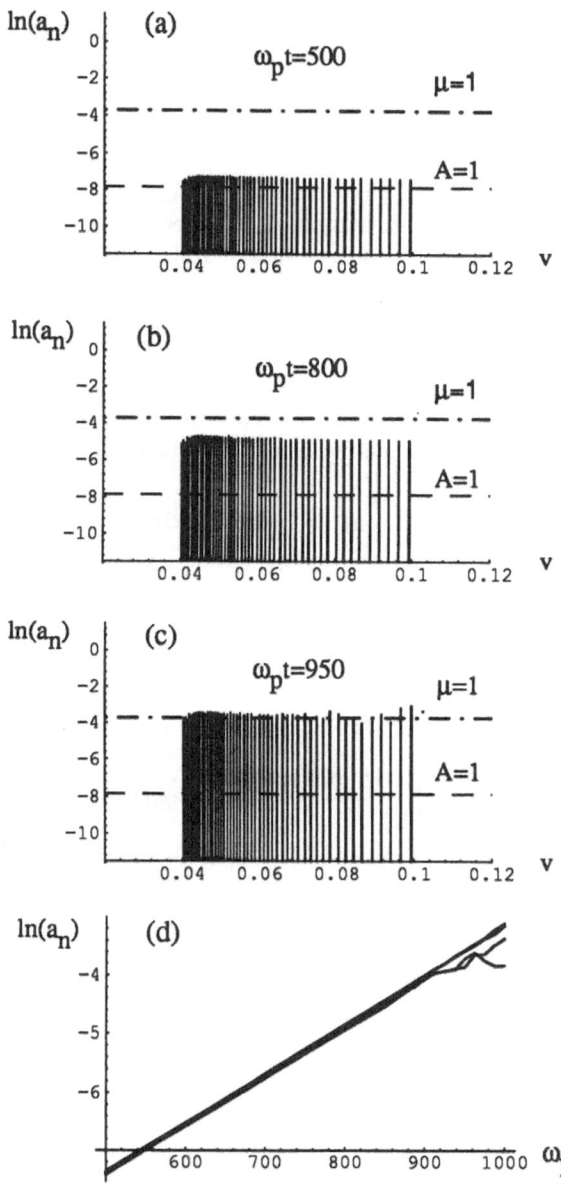

Fig. 4. The wave spectrum for a run initialized with a quiet start at successive times of $\omega_p t = 500$ (a), $\omega_p t = 800$ (b), and $\omega_p t = 950$ (c). The initial overlap parameter is $A = 0.03$. The dashed horizontal line corresponds to the amplitude at which $A = 1$, and the dash-dot line at the amplitude at which $\mu = 1$. All waves grow at the same rate until $\omega_p t \approx 840$, at which time $\mu \approx 0.6$ and after which differential growth is evident. The amplitudes of three different waves versus time are shown in (d).

some $\bar{v}$ and of width Δv. For such a distribution, the growth rate is estimated to be

$$\gamma_L/\omega_p \approx \pi\eta(\bar{v}/\Delta v)^2. \tag{45}$$

Saturation amplitude The saturation of the modes occurs when the distribution function has flattened so that there are no regions of positive slope where there are modes. Energy conservation then gives the amount of wave energy. This calculation is carried out in Sec. 10.5 of Ref. Krall and Trivelpiece (1986), for example. The result is

$$\frac{1}{2}\sum_j |E_{j,sat}|^2 = \frac{1}{6}mn_b\bar{v}\Delta v. \tag{46}$$

For purposes of estimation, we simply divide the total wave energy among the M modes of the simulation. Thus we find the saturation amplitudes to be given by

$$|E_{j,sat}|^2 \approx \frac{1}{3M}mn_b\bar{v}\Delta v. \tag{47}$$

Spontaneous emission amplitude Spontaneous emission can be calculated in the usual way (cf Ichimaru, 1973, Sec. 4.3C). One uses unperturbed particle trajectories in Eqs. (25) and calculates the change in the square of the amplitudes, averaged over the random initial positions of the particles. This calculation gives the rate of change of the electric field amplitude due to the spontaneous emission as

$$\left(\frac{d}{dt}\right)_{se}\frac{|E_j|^2}{2} = \frac{\pi m\omega_p^4}{k_j^3}\frac{n_b}{N}\eta g(v_{\varphi j}). \tag{48}$$

The spontaneous emission amplitude $|E_{jse}|$ is defined to be the amplitude at which the rate of change given by Eq. (48) equals that from the instability, $2\gamma_L|E_j|^2$. Thus,

$$|E_{j,se}|^2 \equiv \frac{\pi m\omega_p^4}{\gamma_L k^3}\frac{n_b}{N}\eta g(v_{\phi j}) \tag{49}$$

This varies across the distribution, but for the center of the distribution it is given by

$$|E_{j,se}|^2 \approx \frac{1}{N}n_b m\bar{v}\Delta v \tag{50}$$

where we used (45) for γ_L.

Number of growth times The number of growth times, N_γ, can now be estimated from the satuation amplitude (47) and spontaneous emission amplitude (50). We obtain

$$\Gamma \equiv e^{2N_\gamma} = |E_{j,sat}|^2/|E_{j,se}|^2 = N/3M. \tag{51}$$

Thus, we must have a large number of particles per mode to be able to see many efolds from the point where spontaneous emission and linear growth are competitive, to the point of saturation.

Quasilinear diffusion coefficient and resonance broadening width The measurements of the growth rate will occur at the point approximately halfway from $E_j = E_{j,se}$ to $E_j = E_{j,sat}$, where the field energy has grown by approximately the factor $\Gamma^{1/2}$. With the value of $E_{j,se}$ from Eq. (50), one can calculate the quasilinear diffusion coefficient using Eq. (10). The result is

$$\bar{D}_{QL} \approx \frac{\pi}{2} \frac{M}{N} \eta \Gamma^{1/2} \omega_p \bar{v}^2. \tag{52}$$

From this and Eq. (12) readily follows the typical resonance broadening width,

$$\bar{v}_{RB}/\omega_p = (\pi \eta M/2N)^{1/3} \Gamma^{1/6}. \tag{53}$$

Finally, from this and the growth rate (45), one can calculate the parameter,

$$\mu = \frac{\Gamma^{1/6}}{(2\pi^2 \beta^2)^{1/3}} \left(\frac{M}{N}\right)^{1/3} \tag{54}$$

that determines the degree to which the experiment is in the regime of enhanced growth. In the above equation

$$\beta = \eta \left(\frac{\bar{v}}{\Delta v}\right)^3 \tag{55}$$

Overlap parameter The overlap parameter (15) should initially be large. This assures that initially the system is in a state where the diffusion coefficient is given by the quasilinear value, and any subsequent enhancement over the quasilinear value is due to self-consistent evolution. From Eqs. (15-17) we obtain the typical value for the overlap parameter at the point where spontaneous emission equals linear growth as

$$\bar{A}_{OL} = (2\pi M)^2 \sqrt{\frac{\beta}{N}} \tag{56}$$

5.2 Calculation of simulation parameters

Equations (51), (54), and (56) can be solved to find the needed numbers of particles and modes and the parameter β in terms of the growth factor Γ, the turbulence strength parameter μ, and the initial overlap parameter $\bar{A}_{OL}$. We obtain

$$\beta = (2\pi^2 \mu^3)^{-1/2} (9\Gamma)^{-1/4} \tag{57}$$

$$M = \left(\frac{\bar{A}_{OL}}{4\pi^2}\right)^{2/3} \left(\frac{3\Gamma}{\beta}\right)^{1/3} \tag{58}$$

$$N = 3\Gamma M = \left(\frac{\bar{A}_{OL}}{4\pi^2}\right)^{2/3} \frac{(3\Gamma)^{4/3}}{\beta^{1/3}} \tag{59}$$

In addition, our simulations require a number of time steps proportional to the number of growth times and the number of steps per period. The number of timesteps needed is given by

$$N_t \approx \frac{N_\gamma}{\gamma_L \Delta t} \approx \frac{N_\gamma}{\pi \beta} \tag{60}$$

provided one can take large timesteps, $k \Delta v \Delta t = 1$, as possible for the Hamiltonian (28). For the Hamiltonian (30), which retains the high frequency in the amplitudes, the number of timesteps required is larger by $\bar{v}/\Delta v$.

The computational time for our simulations scales as the product of the number of time steps and the number of particles.

$$N N_t \approx 2 N_\gamma \bar{A}_{OL}^{2/3} \Gamma^{5/3} \mu^2 \tag{61}$$

This equation shows that increasing the parameter μ, which measures how far one is into the regime of enhanced growth, dramatically increases the computational time.

6 Numerical results for the growth enhancement

The calculations of the previous section allow us to choose the simulation parameters to have any value of turbulence strength parameter $\mu = \nu_{RB}/\gamma_L$. In practice, there is a limit to how large one can choose μ, as the numbers of modes (58) and particles (59) increase with μ, and the number of time steps (60) increases even more rapidly. As noted in the last section, the product of the number of time steps and the number of particles, increases rapidly with μ. Given our available computing resources, we were able to carry out simulations up to $\mu \approx 40$.

We arbitrarily chose $2.78 < k < 3.70$, $v_a = 0.25$ and $\bar{v} = 0.38$. This gives a linear autocorrelation time of $\omega_p \tau_{ac} \approx 2.5$, much shorter than any of the other times of our simulation. By varying the beam to plasma density ratio over the range, $4.1 \times 10^{-6} < \eta < 5.3 \times 10^{-4}$, we were able to vary μ by approximately a factor of 30. We chose to have 584 waves, and an initial overlap parameter of $A_{OL} \approx 200$. We measure the growth rate at a typical electric field value such that $|E_j|^2/|E_{j,se}|^2 \gg 1$, and $|E_j|^2/|E_{j,sat}|^2 \ll 1$, in order to avoid both spontaneous emission and saturation effects.

In this section we show results for the evolution of the turbulent system over this range and we obtain the summary plot showing the growth rate enhancement as a function of μ.

6.1 The quasilinear regime

Figure 5a shows the total wave energy plotted against time for the simulation having coupling constant of $\eta = 2.6 \times 10^{-4}$. For this particular run the timestep was $\omega_p \Delta t = 0.4$ and the grid spacing $k_{max} \Delta x = 0.25$. The linear growth rate for this simulation is $\gamma_L/\omega_p = 0.004$. We measured the growth rate by calculating the slope of the numerical curve (dotted line) between $\omega_p t = 240$ and $\omega_p t = 440$, that is after the initial transients have disappeared, but before the beam distribution function begins to distort due to the diffusion. Figure 5b shows the distortion of the distribution at the end of the growth rate measuring period. We estimate the uncertainty in the average growth rate due to using finite grid spacing and time step to be 2% on the basis of this being the accuracy found in the benchmarking simulations previously. At the center of the time of measurement of the growth rate, we had $\mu = \nu_{RB}/\gamma_L = 2.2$, close to the margin of the enhancement regime. We estimate our uncertainty in μ to be $\pm 20\%$, the amount of variation of μ over half a growth time. At that time, $|E_j|^2/|E_{j,se}|^2 \approx 25$ so spontaneous emission is of the order of 2%. The early faster growth can be attributed to spontaneous emission since at $\omega_p t = 100$ for example we have $|E_j|^2/|E_{j,se}|^2 \approx 5$, a 10% effect. The very fast early evolution ($\omega_p t < 20$) is due to the nonequilibrium nature of the initial conditions.

These results show that one can be significantly into what was expected to be the regime of enhanced growth rate and diffusion, yet see no enhancement. In Fig. 6 we show the mode evolution at the time of measurement of the growth rate, when $\mu \approx 2$. Differential growth of the modes is already evident, even though no enhancement of the average growth rate is observed.

6.2 The intermediate regime

Figure 7a shows the total wave energy plotted against time for the simulation having coupling constant $\eta = 6.6 \times 10^{-5}$. The linear growth rate for this simulation is $\gamma_L/\omega_p = 0.001$. For this case the growth rate was obtained by calculating the slope on the semilog plot between $\omega_p t = 720$ and $\omega_p t = 1440$, that is, again, after initial transients have disapeared, but before the beam distribution function begins to distort due to diffusion (cf Fig. 7b). At the time of measurement of the growth rate, $\mu = \nu_{RB}/\gamma_L = 5.5$. At this time the growth rate is observed to be enhanced over the linear value by approximately 14%. The ratio of wave energy to spontaneous emission is $|E_j|^2/|E_{j,se}|^2 \approx 21$ at this time, so spontaneous emission is of the order of 2%. The initial higher growth rate observed can be attributed to spontaneous emission, since at $\omega_p t = 200$ we have $|E_j|^2/|E_{j,se}|^2 \approx 3$, a 16% effect.

6.3 The enhancement regime

Figure 8a shows the total wave energy plotted against time for the simulation having coupling constant of $\eta = 4.1 \times 10^{-6}$. The linear growth rate for this simulation is $\gamma_L/\omega_p = 6.25 \times 10^{-5}$. For this simulation the growth rate was found by calculating the slope of the growth rate between $\omega_p t = 4800$ and $\omega_p t = 12800$. As shown in Fig. 8b, the measurement ends before the average velocity distribution has changed significantly. We see that in this regime the growth rate (and therefore the diffusion coefficient) is enhanced by approximately

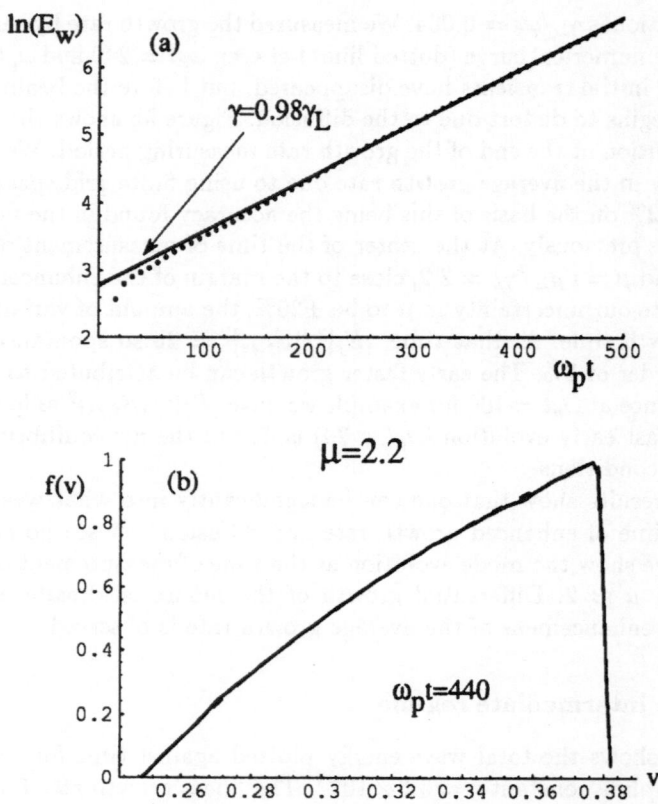

Fig. 5. (a) The total wave energy plotted against time for a run in the quasilinear regime ($\mu = 2.2$). The numerical growth rate is given by the slope of the dotted curve between $\omega_p t = 240$ and $\omega_p t = 440$. Numerical error is estimated to be $\sim 2\%$. (b) The initial beam function (dashed curve) and the distribution function at $\omega_p t = 440$ (solid curve). At the middle of our time of measurement, $|E_j|^2/|E_{j,se}|^2 \approx 25$ so spontaneous emission is a 2% effect.

30% over the quasilinear value. The ratio of wave energy to spontaneous emission at the time of measurement was $|E_j|^2/|E_{j,se}|^2 \approx 51$, so spontaneous emission is of the order of 1%.

Earlier in the simulation the growth rate is smaller than that observed between $\omega_p t = 4800$ and $\omega_p t = 12800$. This occurs because the quantity μ is smaller at these initial values, so that the growth rate is closer to that predicted by the combination of linear theory and spontaneous emission. This illustrates dynamically how the growth rate increases with μ.

The run in Fig. 8 used 6×10^6 particles. With $\gamma_L/\omega_p = 6.25 \times 10^{-5}$ and

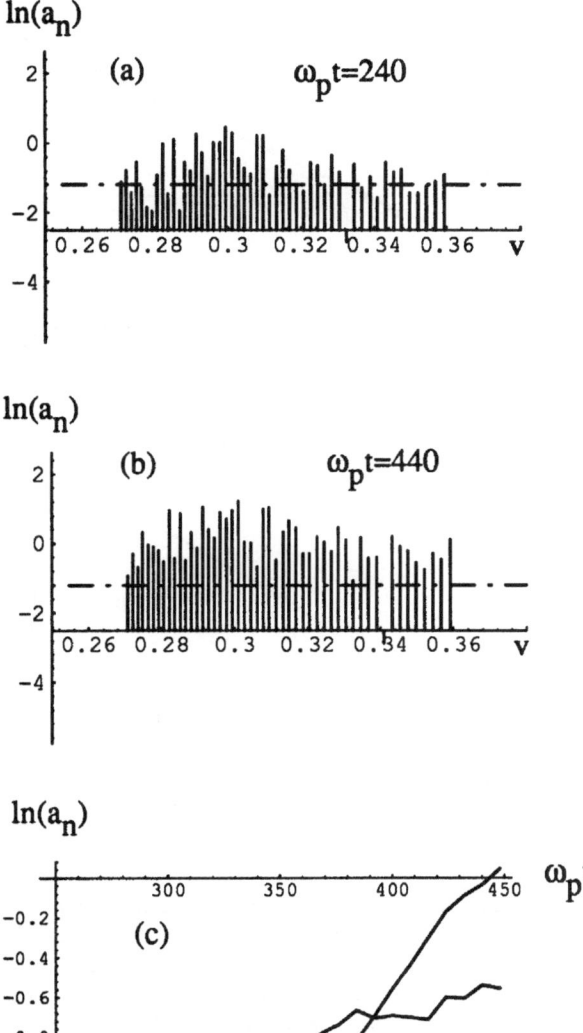

Fig. 6. Snapshots of the wave spectrum for the quasilinear-regime run of the previous figure at $\omega_p t = 240$ (a) and $\omega_p t = 440$ (b). Only one in every ten modes is shown, for clarity. Differential mode growth is seen in that the shape of the spectrum is not preserved by the evolution. (c) Plot of two mode amplitudes versus time, also showing differential mode growth.

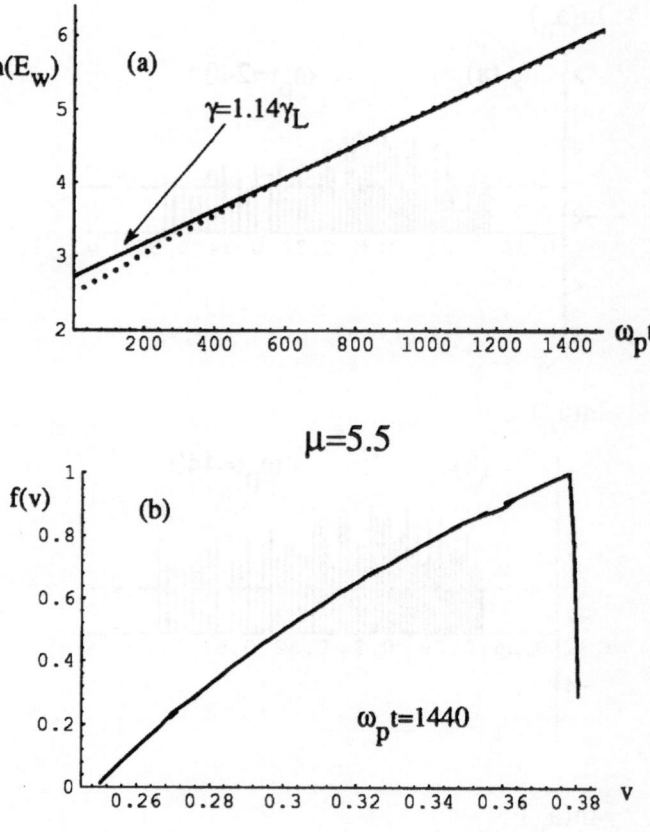

Fig. 7. (a) The total wave energy plotted against time for a run in the intermediate regime ($\mu = 5.5$). The numerical growth rate is given by the slope of the dotted curve between $\omega_p t = 720$ and $\omega_p t = 1440$. Numerical error is estimated to be $\sim 2\%$. (b) The initial beam function (dashed curve) and the distribution function at $\omega_p t = 1440$ (solid curve). At the middle of our time of measurement, $|E_j|^2/|E_{j,se}|^2 \approx 21$ so spontaneous emission is a 2% effect.

$\omega_p \Delta t = 0.4$, the integration for two growth times implies the calculation of 5×10^{11} particle pushes. For our cubic spline symplectic method, this required roughly two months of time on an IBM RS/6000 590.

6.4 Enhancement of growth rate

In all, we carried out a set of eight runs spanning a set of values of μ in the range $1.2 < \mu < 40$. For each of these runs we kept to the same methodology of ensuring

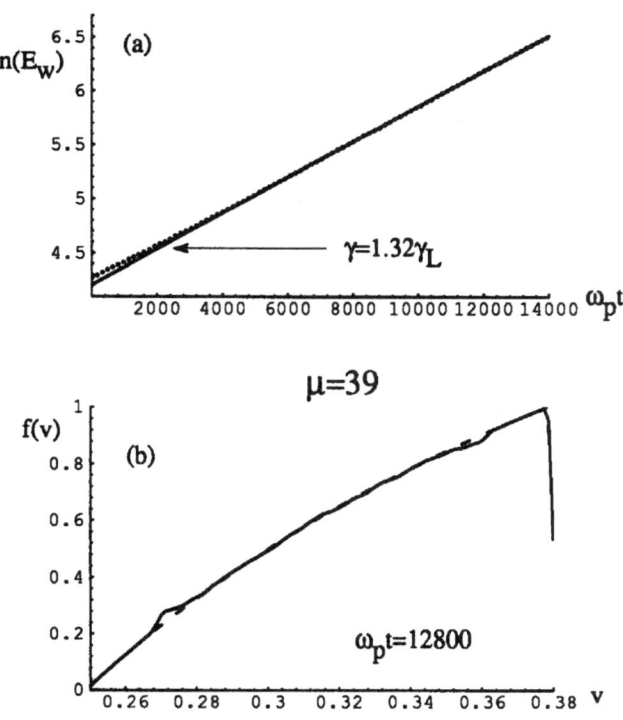

Fig. 8. (a) The total wave energy plotted against time for a run in the enhanced growth rate regime ($\mu = 39$). The numerical growth rate is given by the slope of the dotted curve between $\omega_p t = 4800$ and $\omega_p t = 12800$. Numerical error is estimated to be $\sim 2\%$. (b) The initial beam function (dashed curve) and the distribution function at $\omega_p t = 12800$ (solid curve). The ratio of wave energy to spontaneous emission at the time of measurement was $|E_j|^2/|E_{j,se}|^2 \approx 51$ so spontaneous emission is a 1% effect.

that the enhancement is measured at a time such that $|E_j|^2/|E_{j,se}|^2 \gg 1$ and $|E_j|^2/|E_{j,sat}|^2 \ll 1$, so that the system is at the same time far from saturation and far from the point where spontaneous emission dominates the growth. The summary result is shown in Fig. 9, which shows the enhancement factor as a function of μ. This plot shows that the enhancement does not begin until $\mu \approx$ 3 and from there it rises slowly. The vertical error bars (of 2%) in this plot may underestimate the error for the simulations, as they account for only the numerical integration error, not the errors due to spontaneous emission. Given the errors in our measurements, it is not yet clear whether the enhancement factor saturates at 1.3, the maximum value we observed, or whether it can be even larger for greater values of μ.

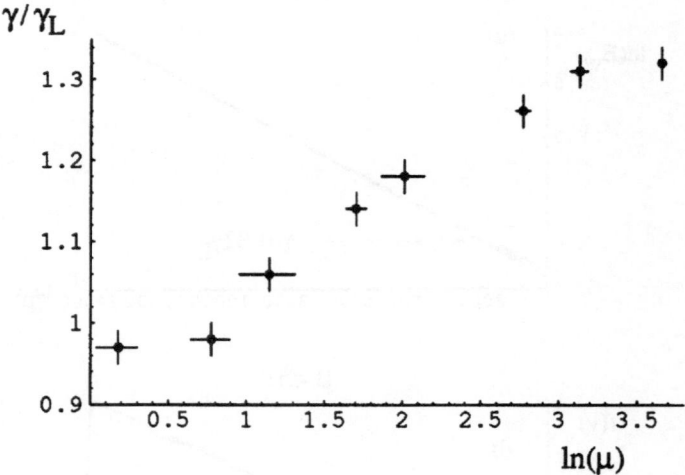

Fig. 9. The observed self-consistent enhancement factor, γ/γ_L, as a function of μ for a spectrum initialized with a high overlap parameter ($A > 200$). The horizontal error bars represent the change in ν_{RB} during the finite time required to measure γ. The vertical error bars show the size of a 2% error, corresponding to the numerical error of integration as determined in our benchmark runs (cf. Sec. III-b).

7 Summary and conclusions

We have observed enhancement of the spectral averaged growth in the regime of $\mu \equiv \nu_{RB}/\gamma_L \gg 1$. (Quasilinear results are recovered for $\mu \sim 1$.) This is in agreement qualitatively with earlier predictions (Adam *et al*, 1979, Laval and Pesme, 1980, Laval and Pesme, 1983, Laval and Pesme, 1984), but not quantitatively, as even for our largest values of μ (near 40), the observed enhancement is only slightly greater than 30%. In addition, we find that the requirement $\mu \gg 1$ is very strong, with large μ needed to see any enhancement; at $\mu = 1$ no enhancement is observed.

These results are consistent with the experiments carried out so far. The experiment of Tsunoda *et al* (1991) achieved a maximum of $\mu \approx 1.4$. In this experiment mode coupling was observed, but enhancement was not, consistent with the results shown in Figs. 5–6. A more recent experiment (Hartmann *et al*, 1995) was also not able to get into the regime $\mu \gg 1$. A new experiment (Guyomarc'h *et al*, 1996) is under construction for the purpose of getting farther into the $\mu \gg 1$ regime.

An important area of future research is to achieve a better understanding of why the enhancement occurs, and whether it can be larger. The answer to the second question awaits the development and use of more powerful computers and better algorithms as well as improved experiments. For the first we note that there does exist one conjecture (Cary *et al*, 1990), that mode coupling causes spontaneous discretization, so that the particles see a field marginally overlapping, where enhancement factors of $\sim 150\%$ have been observed in test particle simulations. To test this numerically, we need a better understanding of how this shows up statistically in a turbulent field, so that simulations can test for it. (The simulations shown here all contain many particle decorrelation lengths $\bar{v}/\nu_{RB}$, and so are the composition of many statistically independent systems.)

Another important area of future research is to determine whether such effects exist in more than one dimension. Such studies are at a very early stage, as the dynamics of particles in a field of many waves at conditions of marginal chaos are not well known. In addition, these dynamics have the complications of Arnold diffusion.

Finally, there are the questions of how this work applies in the many cases where quasilinear diffusion is used, either bare or as a starting point for parameterizing turbulent diffusion (Kotschenreuther *et al*, 1995). Is there a modification of the diffusion coefficient for these cases that cannot be seen except in very low-noise simulations that correctly model the nonlinear particle effects? Will turbulence saturated by wave decay mechanisms saturate at a greater amplitude?

References

A. A. Vedenov, E. P. Velikhov, and R. Z. Sagdeev, Nucl. Fusion Suppl. **2**, 465 (1962).

W. E. Drummond and D. Pines, Ann. Phys. (N.Y.) **28**, 478 (1964).

Nicholson, D.R., *Introduction to Plasma Theory* , (John Wiley & Sons, New York, 1983).

Adam, J.C., G. Laval, and D. Pesme, Phys. Rev. Lett. **43**, 1671 (1979).

Laval, G. and D. Pesme, Phys. Lett. A **80**, 266 (1980).

Laval, G. and D. Pesme, Phys. Fluids **26**, 52 (1983).

Laval, G. and D. Pesme, Phys. Rev. Lett. **53**, 270 (1984).

Goldman, M. V., D. L. Newman, J. G. Wang, and L. Muschietti, Physics Scripta **T63**, 28 (1996).

M. Kotschenreuther, W. Dorland, M. A. Beer, and G. W. Hamett, Phys. Plasmas **2**, 2381 (1995).

A. A. Galeev, R. Z. Sagdeev, V. D. Shapiro, and V. I. Shevchenko, Sov. Phys. JETP **52**, 1095 (1980).

Theilhaber, K., G. Laval and D. Pesme, Phys. Fluids **30**, 3129 (1987).

Cary, J.R., D. F. Escande, and A. Verga, Phys. Rev. Lett. **65**, 3132 (1990).

Tsunoda, S.I., F. Doveil and J.H. Malmberg, Phys. Fluids B **3**, 2747 (1991).

Liang, Y. M., and P. H. Diamond, Comments Plasma Phys. Controlled Fusion **15**, 139 (1993).

P. Deeskow, K. Elsasser, and F. Jestczemski, Phys. Fluids B **2**, 1551 (1991).

O. Ishihara, H. Xia, and A.Hirose, Phys. Fluids B **4**, 349 (1992).

Cary, J.R., I. Doxas, D.F. Escande and A.D. Verga, Phys. Fluids B **4**, 2062 (1992).

Berndtson, J.T., J.A. Heikkinen, S.J. Karttunen, T.J.H. Pättikangas, and R.R.E. Salomaa, Plasma Phys. Control. Fusion **36**, 57 (1994).

336

Helander, P., and L. Kjellberg, Phys. Plasmas **1** , 210 (1994).

D. A. Hartmann, C. F. Driscoll, T. M. O'Neil, and V. D. Shapiro, Phys. Plasmas **2**, 654 (1995).

T. M. O'Neil, J.H. Winfrey and J.H. Malmberg, Phys. Fluids **14**, 1204 (1971).

Mynick, H.E. and A.N. Kaufman, Phys. Fluids **21**, 653 (1978).

T. H. Dupree, Phys.Fluids **14**, 956 (1966).

B. Chirikov, Phys. Rep. **52**, 263 (1979).

A. B. Rechester, M. N. Rosenbluth, and R. B. White, Phys. Rev. Lett. **42**, 1247 (1979).

Escande, D.F., in *Hamiltonian dynamical systems*, R.S. MacKay and J.D. Meiss ed. (Adam Hilger, Bristol, 1987).

Cary, J.R., and I. Doxas, J. Comp. Phys. **107**, 98 (1993).

J. Candy and W. Rozmus, J. Comp. Phys. **92**, 230 (1991).

E. Forest and R. Ruth, Physica D **43**, 105 (1990).

Zekri, Stephane, Approche Hamiltonienne de la turbulence faible de Langmuir, Doctoral dissertation, Université de Provence, Marseille-Aix II (1993).

P. H. Stoltz and J. R. Cary, Phys. Plasmas 1, 1817 (1994).

N. A. Krall and A. W. Trivelpiece, *Principles of Plasma Physics*, (San Francisco Press, Inc. San Francisco, 1986).

S. Ichimaru, *Basic Principles of Plasma Physics, A Statistical Approach*, (W. A. Benjamin, Reading, MA, 1973).

Guyomarc'h, D., F. Doveil, and A. Guarino, Bull. Am. Phys. Soc. **41**, 1458 (1996).

Part 4:

Kinetics and Statistics

Statistical Mechanics of a Self Gravitating Gas

Y. Pomeau

LPS, laboratoire associé au CNRS, Ecole Normale Supérieure
24, rue Lhomond, 75231 Paris Cedex 05, FRANCE
and
Department of mathematics, University of Arizona,
Tucson, AZ 85721, USA

Abstract. Classical point particles interacting via a two-body Newtonian potential cannot be described by the Gibbs-Boltzmann statistics, because of fatal divergences at short distances of the partition function. The assumption of uniform filling of the phase space in the course of time must be replaced by the one of spreading in phase space going forever. What replaces the Gibbs-Boltzmann statistics then are asymptotic diffusion-like laws for this spreading process, where the time enters as a scaling parameter. Another possible description of systems of particles with long range interactions is the continuum Vlasov mean field equation. It is argued that solutions of these Vlasov-Newton equations have finite time singularities with spherical symmetry, and focusing of the energy with no mass, like focusing NLS in 3D.

More than three centuries ago, Isaac Newton solved the two-body problem in classical mechanics. The ingenuity and elegance of his geometrical proof still provokes the admiration. Over the centuries the three body problem has attracted the attention of distinguished minds as Lagrange, Poincaré and Kolmogoroff. The solution of the many body problem in the sense of an exact analytical solution is hopeless, since the three body problem is not integrable in general. It is tempting to have recourse then to statistical methods to study this many body problem, as did Boltzmann for dilute gases with short range forces. Even though this does not yield a solution of the many body problem in an explicit form, given initial conditions, it provides the statistical properties of the "mean solution", a concept bearing some hidden, and not so hidden subtleties, but that I shall take as granted. A core concept in statistical physics is the one of equilibrium probability distribution: one assumes, after Boltzmann, that, on average, the representation point is at random in phase space, with a probability measure given by the Liouville weight. The normalization constant for this measure is the so-called partition function, with various expressions depending on which "ensemble" one is considering (microcanonical, canonical, etc.). I shall deal with the microcanonical ensemble, the one of a system in a closed box, not exchanging energy or particles with the outside world. This microcanonical partition function reads:

$$Z_\mu(E) = \int \prod_i d\mathbf{q}_i dp_i \delta_D(E - \sum_i \frac{{p_i}^2}{2\mu} - \sum_{i>j} U(\mathbf{q}_i - \mathbf{q}_j)) \tag{1}$$

where E is the total energy, where I assumed a two body interaction potential $U(\mathbf{q}_i - \mathbf{q}_j)$, depending on the distance between the two interacting point particles of index i and j. Moreover $\frac{p_i^2}{2\mu}$ is the kinetic energy of the particle of index i, linear momentum $\mathbf{p}_i$ and mass μ. Finally, the boldface letters are for vectors in R^3 and δ_D is the Dirac function. The explicit calculation of this partition function is not possible in general, but for some cases rational expansion near well defined limits can be found. For instance for a dilute gas the contribution of the short range interaction in equation (1) can be neglected and the integral over the momenta can be carried out (Chavanis 1996). For long range forces, as the one I will consider here, the situation is more involved. One reason for this is that the very concept of long range force is not so clearcut. Actually, one has often in mind an interaction potential that decays slowly with the distance, but keeps the same dependance at short and large distance, like $1/r$ for the Newtonian and the electrostatic interactions. One immediate consequence is that this interaction potential diverges at short distance, like $1/r$. In situations relevant for plasma physics, the system is globally neutral, so that the long range Coulomb interaction is screened beyond the Debye length, and finally the Coulomb interaction is felt at short distances only. For realistic situations, quantum effects take care of any divergence due to the strong attraction at short distances between charges of opposite sign. The situation is completely different for Newtonian interactions: no charge cancellation at large distances, and for the problem of interest, that concerns *a priori* systems of macroscopic particles, like stars in a galaxy, no quantum effects at short distance, at least if one does not want to look at the inner structure of the interacting "point" masses. The consequence of this is that gravitational forces are really long range, contrary to what happens in plasmas with charge cancellation. Therefore the divergence of the partition function that will be just exhibited may be seen as a consequence of the "long range" character of the interaction, even though it is formally linked to the short distance behavior of the integrand in the coordinate space. Physically speaking, the divergence of the microcanonical partition function computed in a finite box is a result of the possibility of compensating any increase of the kinetic energy by lowering the potential energy when bringing two attracting masses at closer distance. The integral over the momenta transforms the partition function into:

$$Z_\mu(E) = S_{3N} \int \prod_i d\mathbf{q}_i (2\mu(E - \sum_{i>j} U(\mathbf{q}_i - \mathbf{q}_j))^{\frac{3N}{2}-1} \qquad (2)$$

where S_{3N} is the area of the sphere in 3N Euclidean dimensions. This integral diverges for N bigger than two, because of the short distance behavior of the integrand

$$(2\mu(E - \sum_{i>j} U(\mathbf{q}_i - \mathbf{q}_j))^{\frac{3N}{2}-1}$$

when a pair of particles get very close. Let r be this short distance. Since $U(\mathbf{q}_i - \mathbf{q}_j) = -\frac{G\mu^2}{|\mathbf{q}_i - \mathbf{q}_j|}$, G constant of gravitation, the integral will behave, for r very small like:

$$Z_\mu(E) = S_{3N}(\mu\Omega)^{N-2} \int dr r^2 (\frac{2\mu^3 G}{r})^{\frac{3N}{2}-1} \tag{3}$$

In this integral $(\mu\Omega)^{N-2}$ represents symbolically the (converging) result of the integration over the positions of the particles not belonging to the close pair. For r tending to zero, $Z_\mu(E)$ diverges if $N > 2$ as claimed. If N is large enough this divergence still exists for contributions arising from three particles close together, four particles, etc. But, generally, because of the $d\mathbf{r}$ volume element, this divergence becomes weaker and weaker when more particles are put into the collapsing swarm. Let ϵ be the shortest distance of approach of two particles, a short range cut-off, then the partition function is given by an integral that behaves like $\epsilon^{4-3N/2}$ as a function of ϵ when this length tends to zero, and if N is larger than two. This divergence takes into account configurations with only one pair of particles close together. If one considers situations when $M + 2$ particles get close together, the partition function diverges like $\epsilon^{4-3N/2-3M}$ because the volume element of the $M + 2$ particles is multiplied by ϵ^3 whenever a particle is added to the collapsing swarm. Therefore the strongest divergence comes from contributions where two particles are getting very close although the others are at finite distance of each other. This says almost nothing about the dynamics of a system with arbitrary initial conditions. This idea of arbitrary initial conditions with Newtonian interaction is already non trivial, because one cannot take as one of these conditions a configuration drawn at random in the microcanonical ensemble: we have just proved that, because of the divergence of the partition function, this statement is meaningless. Thus one can take for instance an arbitrary set of initial conditions, with a given total energy, but chosen with a well defined (=normalizable) probability distribution. This might be the uniform distribution on the energy surface, but constrained by the condition that no particle is closer to another one than a certain given distance. Furthermore, in order to avoid other divergences due to the large distance behavior, I shall assume that particles are constrained to live in a box, with perfectly reflecting boundaries. This puts the system quite far from astrophysical situations, where the masses seem to live in an almost infinite world. The next step consists in trying to use the information gained from the divergence of $Z_\mu(E)$ to predict the statistical properties of the average behavior of particles interacting with a Newtonian potential. The assumption I shall make is an attempt to extend the idea of Boltzmann, according to whom the phase space is filled "as uniformly as possible" in the course of time. Indeed no equipartition can occur in the present case, because of the divergence of the microcanonical partition function. To make a parallel, one may think to the diffusion of a Brownian tracer in a finite volume: after transients the distribution of the tracer in space will fill uniformly the finite volume. In the case of an infinite volume, instead, a tracer starting at finite distance will spread in the course of time. If this volume is the full 3D space, the long time distribution of concentration will be the familiar Gaussian, with a width increasing like the square root of time. This scaling law for the spreading of the distribution is a simple consequence of the scaling properties of the diffusion equation, and does not require any detailed knowledge of the solution of the diffusion equation. I used this approach to solve the problem of the long time behavior of a classical nonlinear field: there, because of the Jeans phenomenon, there are infinitely many degrees of freedom, and the long term dynamics is just a self similar spreading of energy in wavenumber space toward smaller and smaller scales, where the divergence of the volume of phase space is located. On the basis of an analysis of the equation of motion for the mode interaction, it turned out to be possible to predict the exponents for diffusion in wavenumber space, consistent with a selfsimilar behavior (Pomeau 1992). For the present problem, the situation is more complicated, because there is no dynamical theory, like the weak turbulence equations allowing to derive scaling laws. The only guidance is that the microcanonical partition function is diverging the more strongly when only one pair of particles are getting very close. Therefore I shall assume that for long times there is only one pair of masses in a close elliptic trajectory that is getting closer and closer as time goes on. The potential energy that is so generated is transferred to the kinetic energy of other particles in the field, by three body collisions (the close pair and another particle). During such a close encounter an exchange of particles might take place: the outgoing pair might eventually include the free incoming particle. The basic statement is that on average, the outgoing pair has a lower potential energy than the ingoing one, which allows to continuously transform potential energy into kinetic energy. An isolated pair of particles follows a Keplerian trajectory boosted by an arbitrary Galilean transformation. Because of the virial theorem, the kinetic energy of the relative motion in the pair is of the same order as the pair potential energy (but with a crucial sign difference, of course). I shall assume that the kinetic energy of the motion of the center of gravity is of the same order of magnitude too. This follows from a kind of Occam razor principle: there is no other large energy scale but the one given by the Newtonian interaction in the pair, and every other relevant energy is of the same order of magnitude. There is a weak point in this assumption: looking back at the divergence of the microcanonical partition function, one realizes that the divergence maintains the equipartition of kinetic energy: every particle contributes the same way to the sum of kinetic energies, clearly not compatible with the assumption just made that a close pair has as much kinetic energy as it has potential energy. If this were true, and if there were equipartition of the kinetic energies among all particles, this would lead to an obvious contradiction for a large number of particles in the box. If the potential energy of the close pair is of the same order as the kinetic energy of every other particle in the system, it cannot be that the sum of the kinetic energies of the many particles is of the order of the interaction energy of a single pair. This is not too dramatic however, because the reasoning nowhere requires equipartition of the kinetic energies. Said otherwise, not too much should be deduced from the partition function made formally convergent by a cut off in the interparticle distance: such a partition function is not a stationary distribution of the system, and does not have thus to be represented as a state of this system at any given time. Actually, the only information drawn from the divergence of the partition function is that close pairs are preferred configurations. If one had restricted the integral by assuming the kinetic energy of the close pair to be of the same order as its potential energy, one would have made the integral of the partition function slightly less diverging, but still diverging. Anyway a limited amount of information only can be drawn from the divergence of the partition function regarding the real dynamical process. Now, once the main assumption has been set up, one may find the scaling laws for the time evolution of the close pair. The parameters of this pair will change only through close encounters with a third mass. Because the pair is very close, one does not expect that the interaction with far away particles will have any important effect. Certainly too, their effect will be less and less important as time goes on and as the pair gets closer. Let $r(t)$ be the order of magnitude of the mutual distance inside the pair. The potential energy of the pair is $-\frac{G\mu^2}{r}$. Assuming only one energy scale, the velocity of the center of gravity of the pair is of order $(\frac{G\mu}{r})^{1/2}$ Given the number density n, the mean free collision time t_{mfp} for close encounters with a third mass , that is for encounters at a distance of order r is :

$$t_{mfp} = \frac{1}{nr^{3/2}(G\mu)^{1/2}}$$

I assume now that at each close encounter, there is an exchange of kinetic and potential energy with multiplicative constants of order unity. This is quite natural, because the target particle has a negligible speed, although the parameters of the close pair are all scaled with a single quantity, its large negative potential energy. Therefore, up to irrelevant constants and/or logarithms (counting the number of close encounters between time zero and t), one can substitute to t_{mfp} the actual time t elapsed since the system started. This gives the scaling law for the evolution of the size of the close pair:

$$r = (nt)^{-2/3}(G\mu)^{-1/3} \tag{4}$$

This picture of a dynamical statistical ensemble forbids to apply the virial theorem because its proof supposes a statistically time independent equilibrium state, perhaps one more explanation of the missing mass problem in the Universe, as this relies upon (among other assumptions) the application of the virial theorem to guess the Newtonian interaction energy from the distribution of velocities of stars. No simple relation between the kinetic and potential energy of the full system should exist then. Another consequence is that a classical system of points with gravitational interactions is never at equilibrium, something that might bear upon the fact that the Universe is manifestly not at equilibrium now, although the current theories of the Beginning, as I understand them, point to a very hot early Universe almost at thermal equilibrium. Finally, it is of interest to look at the consequence of the present considerations for the "fluid mechanics" of a large set of masses interacting gravitationally. Perhaps, this points to a two population dynamics, where most masses remain single, and where close pairs would keep heating the population of singles by transferring kinetic energy to them and lowering their potential energy meanwhile. In this respect it would be of interest to know if the observed population of twin stars is like a delta function in the statistics of distances from a star to its closest neighbour, meaning that there is a well defined population of pairs. For very large systems one should expect more than one pair of close particles: probably the number of close pairs decays slowly as a function of time through rare encounters of close pairs.

Another way of approaching the dynamics of particles interacting with Newton's inverse square force is the continuum description. Neglecting correlations, the density in phase space $f(r, p, t)$ is a solution of the Vlasov equation with a self-consistent gravitation field: this yields the Vlasov-Newton equation. The conditions of application of this approximation to a system of point masses are a bit unclear: contrary to the situation of plasmas, there is no well defined smallness parameter (the inverse number of particles in a Debye sphere in plasmas) allowing to neglect consistently the correlation in the Newtonian case. Hopefully some relevant information however can be extracted from the analysis of the Vlasov-Newton equation. I shall limit myself to the simple situation of concentric spherical shells, of negligible thickness , each with mass μ and radial position $r(t)$, in between 0 and ∞. The dynamical equation for this system are:

$$\frac{d^2 r_i}{dt^2} = -G\mu \sum_{j, r_j < r_i} \frac{1}{r_j^2} \qquad (5)$$

A shell at radius r_i is attracted by shells closer to the center than itself, and the force is as if every inside shell were at the center, a result due to Newton. Furthermore, I have neglected the self interaction of the shell, which is consistent with the Vlasov-Newton equation used later on. The energy of this system is:

$$H(p_i, r_i) = \sum_i \frac{p_i^2}{2\mu} + \sum_{i < j, r_j < r_i} \frac{G\mu^2}{r_j} \qquad (6)$$

Therefore, the microcanonical partition function reads:

$$Z_{micro}(E) = \int \prod_i dp_i dr_i \delta_D (E - H(p_i, r_i)) \qquad (7)$$

where the integral over the r_i's extends from 0 to $+\infty$, although the momenta p_i are integrated over the full real line. As for point particles, the momenta can be integrated out, giving:

$$Z_{micro}(E) = S_N \int \prod_i dr_i (2\mu(E - \sum_{i < j, r_j < r_i} \frac{G\mu^2}{r_j}))^{\frac{N}{2}-1} \Theta(E - \sum_{i < j, r_j < r_i} \frac{G\mu^2}{r_i}) \qquad (8)$$

where $\Theta(.)$ is the Heaviside step function imposing a positive kinetic energy. This partition function is diverging in two limits, at large and small r_i. The latter is not relevant for the present work, as I shall assume the system contained within a spherical enclosure of radius R. It remains to consider the divergence related to the short distance behavior of the integral giving the partition function. At this stage, the long time behavior of the solution might be analysed as before, by assuming that near the center two shells are getting closer and closer as time goes on, by giving their energy to other incoming shells reflected on the center. It is not very realistic though to continue this calculation, because of the many idealizations in this model. It seems more interesting to use the information gotten from the divergence of the partition function to make predictions for the continuous description, via the Vlasov-Newton equation. The main information is that the potential energy tends to be focused to a point, with spherical symmetry, but without focusing of the mass. This is reminiscent of the generic singularities of the focusing nonlinear Schrödinger equation (NLS) in dimension 3: the density of potential energy diverges at one point, with a finite amount of total energy at the singularity, but with a total mass tending to zero at the singularity (Mesurier and al. 1988). In NLS, this follows from the assumption of selfsimilar behavior of the solution of NLS near the singularity. In what follows, I shall use the same method (search of a selfsimilar solution) to study a possible blow-up in the evolution of the Vlasov-Newton equation. As it will appear, things are more complicated than for NLS, because Vlasov-Newton is a nonlinear integrodifferential equation not too easy to handle. This equation is a first order evolution equation for the probability distribution $f(r, v, t)$ of velocities v and positions r of a system of spherical shells with a Newton attraction law. This equation reads:

$$\frac{\partial f}{\partial t} + v \frac{\partial f}{\partial r} + E_s(r,t) \frac{\partial f}{\partial v} = 0 \tag{9}$$

The self consistent field $E_s(r, t)$ is related to $f(r, v, t)$ as

$$E_s(r,t) = -\frac{G\mu}{r^2} \int_0^r dr' \int_{-\infty}^{+\infty} dv' f(r, v', t)$$

This Vlasov-Newton equation has no formal general solution. To gain some understanding upon its possible solutions, I shall have to make use of the information coming from the divergence of the partition function. This function gives a diverging statistical weight to configurations where two shells are approaching $r = 0$ very closely. Therefore, I expect too a similar behavior of solutions of the Vlasov-Newton equation. A selfsimilar solution takes the functional form:

$$f(r, v, t) = (-t)^\alpha F(r(-t)^\beta, v(-t)^\gamma) \tag{10}$$

where the singularity happens at time zero, whence the $(-t)$ in the argument of the powers, since I am considering what happens before the singularity (negative times, t being small positive). The exponents α, β, γ are to be deduced from the Vlasov-Newton equation: by putting the self similar solution in the equation, one must get an equation in the stretched variables only, $r(-t)^\beta, v(-t)^\gamma$. By balancing the three terms one gets two relations: $\gamma - \beta = 1$, $\alpha + \beta = -1$. It remains to find another relation to determine the three exponents. This one can only be a consequence of the properties of conservation. If one assumes the mass conserved during the collapse, one obtains: $\alpha - \gamma - \beta = 0$, which yields $\alpha = -\frac{1}{3}, \beta = -\frac{2}{3}, \gamma = \frac{1}{3}$. Those exponents have the right sign, because they correspond to collapse to a single point (β negative), but they show the unphysical feature of a diverging energy in the collapse region. By itself this divergence is not too dramatic: the order of magnitude of each of the two contributions (potential and kinetic) to the energy diverges, although the total energy might have a different behavior, by exact compensation of the kinetic and potential energy. For a selfsimilar solution, that would imply that this total energy is zero. If such a blow up existed with a finite mass and zero total energy, it would be very different from what we expect from the divergence of the partition function, where the mass is the smallest possible, although the total energy is finite. That the total mass is zero is due to the continuum limit: therein the masses of the individual shells tends to zero, as the divergence of the partition function implies two shells only. This is the question I am going to consider: collapse with no mass and a finite total energy. Let E_T be this total energy. From the scaling relation already found and by using the energy E_T as supplementary scaling quantity, one gets the following scaling form for the solution of Vlasov-Newton:

$$f(r,v,t) = E_T^{\frac{1}{5}}(G\mu)^{-\frac{4}{5}}(-t)^{-\frac{1}{5}}F(R,V)$$

where

$$r = R(G\mu E_T)^{-\frac{1}{5}}(-t)^{\frac{4}{5}} \tag{11}$$
$$v = V(G\mu E_T)^{-\frac{1}{5}}(-t)^{-\frac{1}{5}} \tag{12}$$

define the stretched variables R and V, that are of order one. The total mass in the collapse domain behaves like:

$$M = (E_T)^{\frac{3}{5}}(G\mu)^{-\frac{2}{5}}(-t)^{\frac{2}{5}}$$

This tends to zero as the collapse time approaches (that happens when t tends to zero). The Vlasov-Newton equations become now a set of equations for the function $F(R,V)$, that reads:

$$\frac{1}{5}F + \frac{4}{5}R\frac{\partial F}{\partial R} - \frac{1}{5}V\frac{\partial F}{\partial V} + V\frac{\partial F}{\partial R} + E_s(R)\frac{\partial F}{\partial V} = 0 \tag{13}$$

where:

$$E_s(R) = -\frac{1}{R^2}\int_0^R dR'\int_{-\infty}^{+\infty}F(V,R') \tag{14}$$

To get some insight upon the the solution of these equations, I notice that by taking $E(R)$ as given, the equation for $F(R,V)$ can be solved by the methods of characteristics. The boundary conditions constrain already very strongly the solutions. To see that, I rewrite the equation for F as:

$$\frac{1}{5}F + A(R,V)\frac{\partial F}{\partial V} + B(R,V)\frac{\partial F}{\partial R} = 0 \tag{15}$$

where:

$$A(R,V) = -\frac{V}{5} + E_s(R)$$

and

$$B(R,V) = \frac{4R}{5} + V$$

The formal solution for F will be written in terms of the trajectories of the solutions of the two coupled ODE's:

$$\frac{dV}{d\Sigma} = A(R,V)$$

and

$$\frac{dR}{d\Sigma} = B(R,V)$$

where Σ is a dummy parameter("times" later on). This system of ODE's does not seem to be solvable by explicit quadrature, except in the simple case $E_s(R) = 0$. The solution for this simple case allows to get a fair understanding of the general solution. If $E_s(R) = 0$, $V^4(R+V)$ is a constant of the motion and the trajectories are like hyperbolea with the asymptotes $V = 0$ and $V+R = 0$. The phase portrait is drawn on figure (1). The only fixed point is $R = V = 0$. If $E_s(R)$ is not zero, this fixed point is shifted to $R = R^\star, V = V^\star$, where $R^\star$ and $V^\star$ are the root(s) of $A(R,V) = B(R,V) = 0$. $R^\star$ is the root of

$$R^\star = -\frac{5E_s(R^\star)}{4} \tag{16}$$

The field $E_s(R)$ is negative and I shall assume it to decrease from a finite or infinite value at $R = 0$ to zero at infinity. Therefore, one and only one root $R^\star$ exists. From the fixed point $R = R^\star, V = V^\star$ start two lines, the stable and unstable manifold. Although other possibilities exist, I shall assume that the perturbation brought by the attraction does not change the phase portrait too dramatically . This phase portrait is sketched in Figure (2). The maximum or minimum value of R on each trajectory is reached on the line where $\frac{dR}{d\Sigma} = B(R,V) = 0$, that is on the line $4R + 5V = 0$, which helped to draw the figure. The method of characteristics yields the values of F on each integral line of the velocity field. Let F_0 be value of F at some point R_0, V_0 of an integral line. By setting $\Sigma = 0$ for the value of Σ at R_0, V_0, the value of F at any point on the same trajectory is

$$F(\Sigma) = F(R_0, V_0)e^{-\frac{\Sigma}{5}} \tag{17}$$

where $F(\Sigma)$ is the value of F at the point R, V reached at Σ, given the starting point R_0, V_0 at $\Sigma = 0$. This excludes from the support of F any open trajectory running from $\Sigma = -\infty$ to $+\infty$, since the value of F for such a trajectory would become infinitely large at $\Sigma = -\infty$, and so make diverge any integral of F over V, like the one needed to get $E_s(R)$ out of F. From the phase portrait of the velocity field, this leaves two possibilities for trajectories as support of F: the unstable manifold of the fixed point running from $V = +\infty, R = 0$

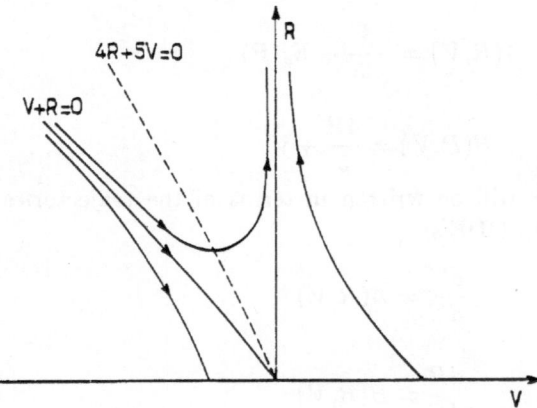

Figure 1: Phase portrait of the velocity field

$$\frac{dV}{d\Sigma} = A(R, V) \quad \text{and} \quad \frac{dR}{d\Sigma} = B(R, V) \quad \text{with } E(R) = 0.$$

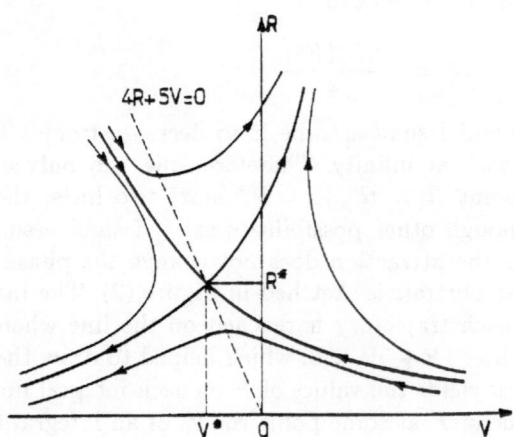

Figure 2: Phase portrait of the velocity field

$$\frac{dV}{d\Sigma} = A(R, V) \qquad \text{and} \qquad \frac{dR}{d\Sigma} = B(R, V)$$

with $E(R)$ negative of decreasing absolute value from $R = 0$ to $R = \infty$; The support of $F(R, V)$ is on the unstable manifold emanating from the fixed point and tending to $R = 0$ and $V = -\infty$. Other trajectories, not drawn, are also supporting $F(R, V)$, that are obtained by reflection of the unstable manifold on $R = 0$.

to the fixed point, plus eventually open flow lines underneath the stable and unstable manifold of the fixed point and above the $R = 0$ line. Actually the situation is quite involved: that trajectories take a finite "time" to travel from $R = 0, V = +\infty$ to $R = 0, V = -\infty$ depends on the behavior of $E_s(R)$ near $R = 0$, which is dependent itself upon the behavior of F near $R = 0$, making the whole thing self consistent. To keep the dicussion within a reasonable length, I will limit myself below to a single case, the one where the support of F is the unstable manifold of the fixed point. This makes appear two different problems,

1) For long negative "times" Σ, the trajectory reaches the fixed point. Therefore, there is a problem of integrability, in the sense of the possibility of calculating $E_s(R)$ via integration over V of F.

2) Assuming that the first problem can be solved, is the self consistent field found by integration of F over V the same as the field $E_s(R)$ used to compute F by the method of trajectories ?

The second problem is numerical, as there is no hope to get an explicit solution. I plan to study it elsewhere, as well as to discuss the otherl possibilities. The first problem concerns the behavior of F near the fixed point. Let δV and δR be the deviation of V and R away from the fixed point. The linearized dynamical equations for δV and δR read:

$$\frac{d\delta V}{d\Sigma} = -\frac{\delta V}{5} + \frac{dE_s(R)}{dR^\star}\delta R$$
$$\frac{d\delta R}{d\Sigma} = \frac{4\delta R}{5} + \delta V \tag{18}$$

The solution of this linearized system makes appear two eigenvalues, σ and this takes the general form: $\delta V, \delta R = e^{\sigma\Sigma}$. The eigenvalues σ are the two roots of the characteristic polynomial:

$$\sigma^2 - \frac{3}{5}\sigma + \frac{dE_s(R)}{dR^\star} - \frac{4}{25} = 0 \tag{19}$$

The sum of the roots is $\frac{3}{5}$, that is the divergence of the velocity field, and the roots are real, if $\frac{dE_s(R)}{dR^\star} < \frac{1}{4}$, as assumed. In this case, the solution for F near the fixed point and with its support on the unstable manifold takes the local form:

$$F(R, V) = \delta_D((\sigma_+ - \frac{4}{5})\delta R - \delta V)e^{\Sigma\sigma_+} \tag{20}$$

where σ_+ is the largest (positive) root of the characteristic equation and δ_D is the Dirac function. Along the unstable manifold, the relation between Σ and the local value of R or V follows from the integration of the linearized equations of motion. For instance, one has:

$$\delta V(\Sigma) = \delta V(0)e^{\Sigma\sigma_+} \tag{21}$$

One may now derive the sought behavior of F near the fixed point:

$$F(R, V) = \delta_D((\sigma_+ - \frac{4}{5})\delta R - \delta V)(\frac{\delta V(0)}{\delta V})^{\frac{1}{5\sigma_+}} \tag{22}$$

This is an integrable function if $\frac{1}{5\sigma_+}$ is less than 1, which is realized if if $\frac{dE_s(R)}{dR^*} <$ $\frac{6}{25}$. This condition is weaker than the one for the existence of a negative eigenvalue, that is $\frac{dE_s(R)}{dR^*} < \frac{4}{25}$. If both σ_+ and σ_- are positive, the singularity of F near the fixed point but along the less unstable eigenvalue is not integrable. Therefore, in both cases the only possible support for F is along the most unstable manifold. Of course, this cannot be considered as a proof that the whole schema works, since the inequalities depend on the numerical value of $\frac{dE_s(R)}{dR^*}$ that is itself determined self consistently. For positive Σ "times" the unstable manifold of the fixed point is asymptotic to the $R = 0, V = -\infty$ semi axis. The question of the boundary condition at $R = 0$ for F at $R = 0$ is relevant if F has not decayed to zero along its travel on the unstable manifold. F would decay to zero (because of the exponential $e^{-\frac{\Sigma}{5}}$ in the trajectory solution for F. Close to $R = 0$, $E_s(R)$ is expected to diverge and the equations of motion for $R(\Sigma)$ and $V(\Sigma)$ become close to the Hamiltonian system:

$$\frac{dV}{d\Sigma} = E_s(R)$$
$$\frac{dR}{d\Sigma} = V \tag{23}$$

From Liouville theorem, the density F in phase space becomes constant near $R = 0$, so that the mass density, obtained by integration of F over V tends to a constant at $R = 0$. Therefore, from its definition, $E_s(R)$ diverges like R^{-1} as R tends to zero, and finally by elementary scaling arguments it takes a finite Σ time to reach $R = 0$. Therefore, one must worry about the boundary condition at $R = 0$. The usual choice is the perfect reflection, which means there that the boundary value of F at $R = 0, V = -\infty$ is the same as the asymptotic value for F on the trajectory reflected of the unstable manifold. This reflected trajectory is defined by the condition of conservation of "energy" in the Hamitonian motion near $R = 0$. This reflected trajectory may either escape to infinity, because it is on the "left" side of the separatrices drawn by the stable and unstable manifold of the fixed point, or it may be thrown back underneath the stable manifold of the fixed point and later on hit again the $R = 0$ axis with an infinite negative velocity. Then the process continues. Notice that the energy of the near Hamiltonian motion decreases in the course of time, at least if the motion takes place close to the $R = 0$ axis. Let

$$H = \frac{V^2}{2} + U(R) \tag{24}$$

$\frac{dU}{dR} = -E_s(R)$ being the energy of the Hamiltonian part of the equations of motion. From these equations,

$$\frac{dH}{d\Sigma} = -\frac{V^2}{5} - E_s(R)\frac{4R}{5} \tag{25}$$

The positive term (recall that $R = 0$ is attracting) $-E_s(R)\frac{4R}{5}$ tends to a constant as R tends to zero, because $E_s(R)$ is like $1/R$ there, so that the dominant term in $\frac{dH}{d\Sigma}$ is the negative $-\frac{V^2}{5}$, which shows that near $R = 0$ the energy H decays as Σ grows. Accordingly an infinite set of trajectories is obtained by successive reflections on $R = 0$. Because of the decay of H, the limit set of these trajectories should have an infinitely negative energy, and so correspond to the point $R = 0$ in the R, V space. Because of the exponential decay of the amplitude of F as Σ increases, no divergence of the integrals of F over V yielding $E_s(R)$ seems to be associated to the infinitely many nested trajectories. This gives a consistent schema for the solution of the original equations, without divergence of the quantities involved. As mentionned already, the full solution of this problem cannot be completely given analytically because of the implicit and nonlinear relations between F and E_s.

Summary and perspectives:

The statistical mechanics of a self-gravitating gas shows that in the course of time close pairs should form and get closer and closer as time goes on. A somewhat similar situation is met when studying the mean field dynamics of this self-gravitating gas: there should be a collapse in a finite time, without mass , but with a finite amount of energy at the singularity. Indeed this has been studied for perfectly symmetric solutions, without angular momentum. Therefore, the connection between the two cases is not completely obvious, a question I plan to examine in the future.

References

Y. Pomeau (1992): Nonlinearity **5**, 707.

B.J. Le Mesurier, Papanicolaou G.C., Sulem C. and Sulem P.L. (1988): Physica **D31**, 78 and **D32**, 210.

P.H. Chavanis, PhD thesis, ENS-Lyon (France), December 1996; P.H. Chavanis, J. sommeria, and R. Robert, Astrophysical Journal **471**, 385.

The Arising and Evolution of the Passive Tracer Clusters in Compressible Random Media

A.I. Saichev and I.S. Zhukova

Radiophysics Department, University of Nizhny Novgorod 23 Gagarin ave. Nizhny Novgorod, 603600, Russia

1. Introduction

It is widespread opinion that the adequate description of temporal and spatial properties of random fields, in particular, hydrodynamic turbulence, have to based on spectral-correlation analysis. Nevertheless some important topographical information is maintained in probabilistic characteristics of random fields (see, for instance, [1-3]). Our report is devoted to description of some topographical peculiarities which one can extract from the probability distribution functions (pdf's) of random fields in compressible media. Such compressible fields give satisfactory mathematical models of broad class of physical phenomena, for instance, evolution of mass distribution in the Universe [4-7] or moving of floating tracer at the ocean surface [3]. Indeed, the divergence of the velocity field

$$(\nabla \cdot \mathbf{v}\,(\mathbf{x}, t)) = \frac{\partial v_1}{\partial x_1} + \frac{\partial v_2}{\partial x_2} + \frac{\partial v_3}{\partial x_3} \qquad (1.1)$$

in depth of the ocean is equal to zero. Let x_3 is the vertical and (x_1, x_2) are horizontal coordinates along the surface where the floating tracer is constrained. The divergence of the velocity horizontal component, which mostly determines the behavior of a floating tracer:

$$div_h \mathbf{v} = \frac{\partial v_1}{\partial x_1} + \frac{\partial v_2}{\partial x_2} = -\frac{\partial v_3}{\partial x_3} \neq 0 \qquad (1.2)$$

is not equal to zero, so floating passive tracer is subject to the laws of motion of a passive tracer in a 2D compressible flow.

The most distinctive peculiarity of tracer in the compressible media is the formation of stable cluster structure, i.e. the appearance of small regions of high density where almost all particles of passive tracer are putting together, surrounding by broad regions of low tracer density. The well known Large-Scale Structure of mass distribution in the Universe is the good example of such kind structure [4-7].

In this report we discuss some simple 2D and 1D models of compressible medium illustrating the arising and evolution of the cluster's structure.

2. Average areas and masses of clusters with raised density

The first example utilization of probabilistic properties for examination of the random fields topographical features is relation of some random field $\rho(\mathbf{x}, t)$

354

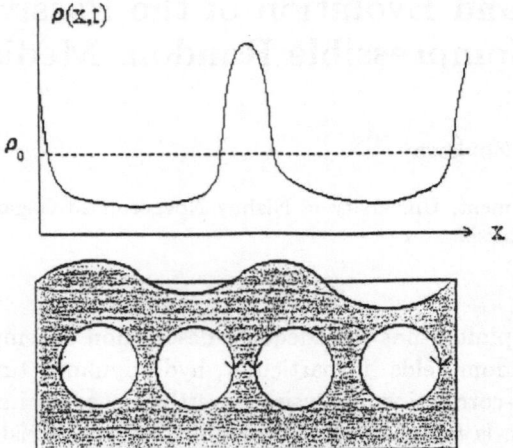

Fig. 1. Schematic illustration of floating tracer clusters on the surface of noncompressible liquid. Below - picture of moving liquid; above - plot of tracer's density.

pdf

$$w_E(\rho; \mathbf{x}, t) = \langle \delta\left(\rho\left(\mathbf{x}, t\right) - \rho\right)\rangle \tag{2.1}$$

with that area of space where $\rho(\mathbf{x}, t)$ exceeds the given level ρ:

$$\rho(\mathbf{x}, t) > \rho . \tag{2.2}$$

Angle brackets $\langle...\rangle$ mean the statistical averaging. Besides, to make this report more evident, we will illustrate general relations by the examples concerning random fields of some 1D or 2D compressible media, like in the case of floating tracer behavior at the surface of ocean.

Let us consider probability

$$P\left(\rho\left(\mathbf{x}, t\right) > \rho\right) = \langle \theta\left(\rho\left(\mathbf{x}, t\right) - \rho\right)\rangle = \int_{\rho}^{\infty} w_E(\rho'; \mathbf{x}, t)d\rho'$$

where $\theta(z)$ is the Heaviside function ($\theta(z) = 1$ for $z > 0$ and $\theta(z) = 0$ for $z < 0$).

Obviously, the integral of probability over some space domain Q with given area Q is equal to

$$\int_Q P\left(\rho\left(\mathbf{x}, t\right) > \rho\right) d\mathbf{x} = \langle S\left(Q, \rho, t\right)\rangle . \tag{2.3}$$

Here

$$S\left(Q, \rho, t\right) = \int_Q \theta\left(\rho\left(\mathbf{x}, t\right) - \rho\right) d\mathbf{x} \tag{2.4}$$

- area of regions inside of domain Q where inequality (2.2) is true.

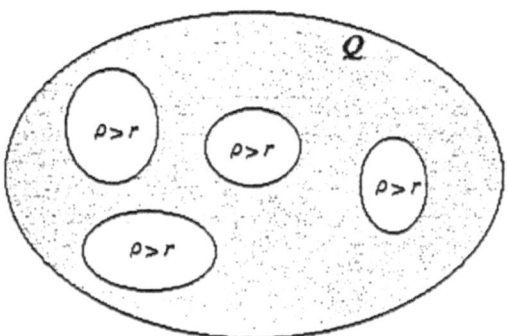

Fig. 2. Schematic picture of clusters: regions where $\rho(\mathbf{x}, t) > r > \rho^*(t)$.

If random field $\rho(\mathbf{x}, t)$ is statistically homogeneous, then limit

$$\lim_{Q \to \infty} \frac{1}{Q} \langle S(Q, \rho, t) \rangle = s(\rho, t) \tag{2.5}$$

is equal to the specific area over whole space where inequality (2.2) is true. On the other hand, this specific area just is equal to the integral of random field $\rho(\mathbf{x}, t)$ pdf:

$$s(\rho, t) = \int_{\rho}^{\infty} w_E(\rho'; t) d\rho' . \tag{2.6}$$

One can observe in addition that equality

$$\int_{\rho^*}^{\infty} w_E(\rho; t) d\rho = \frac{1}{2} \tag{2.7}$$

determines the threshold function $\rho^*(t)$ so that regions where

$$\rho(\mathbf{x}, t) > \rho > \rho^*(t) \tag{2.8}$$

form a system of "islands" - clusters, imbedded into the "ocean" where opposite inequality

$$\rho(\mathbf{x}, t) < \rho^*(t) \tag{2.9}$$

is valid. If $\rho(\mathbf{x}, t)$ is the density field of some passive tracer, then one can find the specific mass of tracer inside pointed out clusters with raised density $\rho(\mathbf{x}, t) > \rho$:

$$m(\rho; t) = \lim_{Q \to \infty} \frac{1}{Q} \langle M(Q, \rho, t) \rangle \tag{2.10}$$

where

$$M(Q, \rho, t) = \int_Q \rho(\mathbf{x}, t)\theta\left(\rho(\mathbf{x}, t) - \rho\right) d\mathbf{x}$$

is the mass of passive tracer contained at clusters. It is expressed via of pdf $w_E(\rho; t)$ of the statistically homogeneous density field by the next formula analogous to (2.6):

$$m(\rho; t) = \int_\rho^\infty \rho' w_E(\rho'; t) d\rho' \ . \tag{2.11}$$

3. Connections between of Lagrangian and Eulerian probability distributions of random fields

Above mentioned topographical characteristics of the passive tracer density fields were expressed through pdf (2.1) of density field $\rho(\mathbf{x}, t)$ at the Eulerian coordinate system. We will call such distributions of Eulerian fields as the *Eulerian probability distribution functions*. Many Eulerian statistical characteristics of the random fields are expressed via the pdf's of the same fields at the Lagrangian coordinate system i.e. via the *Lagrangian probability distributions*. Sometimes investigation of Lagrangian statistical characteristics of random fields is more convenient than investigation of Eulerian ones. So establishment of the relations between Eulerian and Lagrangian statistical properties of random fields promote to the comprehensive investigation of its various statistical and topographical peculiarities. To derive these relations let us consider the continuity equation for the density of some compressible passive tracer:

$$\frac{\partial \rho}{\partial t} + (\nabla \cdot \mathbf{v}(\mathbf{x}, t)\rho) = 0 \ . \tag{3.1}$$

Solution thereof has the form:

$$\rho(\mathbf{x}, t) = \int \rho_0(\mathbf{y})\delta\left(\mathbf{x} - \mathbf{X}(\mathbf{y}, t)\right) d\mathbf{y} \tag{3.2}$$

where $\rho_0(\mathbf{x})$ is the deterministic initial passive tracer density field, and

$$\mathbf{x} = \mathbf{X}(\mathbf{y}, t) \tag{3.3}$$

- random coordinates of some passive tracer's particle, initially (at time $t = 0$) situated at the point $\mathbf{y}$. Recall that $\mathbf{X}$ - Eulerian and $\mathbf{y}$ - Lagrangian coordinates of particle.

It is worthwhile note that averaging of equality (3.2) gives the first well known formula connecting of Eulerian and Lagrangian statistics. Indeed averaging yields:

$$\langle \rho(\mathbf{x}, t) \rangle = \int \rho_0(\mathbf{y}) w_L(\mathbf{x}; \mathbf{y}, t) d\mathbf{y} \ . \tag{3.4}$$

This relation, which makes up the *basic theorem of turbulent diffusion* statement [8], tied together the mean value of Eulerian density field $\langle \rho(\mathbf{x}, t) \rangle$ and the Lagrangian pdf of given particle coordinates:

$$w_L\left(\mathbf{x}; \mathbf{y}, t\right) = \langle \delta\left(\mathbf{x} - \mathbf{X}(\mathbf{y}, t)\right) \rangle \ .$$

As we saw earlier, to study of the random density topographical properties, we need in formulas expressing the Eulerian density pdf through Lagrangian one. To derive them let us consider the another useful density field representation which one can get from (3.2) using the Dirac-delta relation [9]:

$$\delta\left(\mathbf{x} - \mathbf{X}(\mathbf{y}, t)\right) = \frac{1}{J(\mathbf{y}, t)}\delta\left(\mathbf{y} - \mathbf{Y}(\mathbf{x}, t)\right) \ . \tag{3.6}$$

Here the vector function

$$\mathbf{y} = \mathbf{Y}(\mathbf{x}, t) \tag{3.7}$$

is the inverse to (3.2) and

$$J(\mathbf{y}, t) = \left| \frac{\partial \mathbf{X}(\mathbf{y}, t)}{\partial \mathbf{y}} \right| \tag{3.8}$$

is the Jacobian of the mapping of Lagrangian coordinates $\mathbf{y}$ to the Eulerian ones $\mathbf{x}$. Substituting (3.6) into (3.2) and utilizing of the Dirac-delta probing property one can obtain another useful expression for the passive tracer density:

$$\rho(\mathbf{x}, t) = \frac{\rho_0\left(\mathbf{Y}(\mathbf{x}, t)\right)}{J(\mathbf{Y}(\mathbf{x}, t), t)} = \frac{\rho_0\left(\mathbf{Y}(\mathbf{x}, t)\right)}{j(\mathbf{x}, t)} \ . \tag{3.9}$$

Here $J(\mathbf{y}, t)$ and $j(\mathbf{x}, t) = J(\mathbf{Y}(\mathbf{x}, t))$ are Jacobian in the Lagrangian and Eulerian coordinate systems respectively. Equality (3.9) means that the density field closely connected to the Jacobian which is measure of the medium dilation (if $J > 1$) or compression (if $J < 1$). So it is worthwhile to discuss at first the statistics of Jacobian field. Let us consider the joint Eulerian pdf

$$w_E(\mathbf{y}, j; \mathbf{x}, t) = \langle \delta\left(\mathbf{Y}(\mathbf{x}, t) - \mathbf{y}\right) \delta\left(j(\mathbf{x}, t) - j\right) \rangle \tag{3.10}$$

of Eulerian Jacobian field $j(\mathbf{x}, t)$ and Lagrangian coordinates $\mathbf{Y}(\mathbf{x}, t)$ of the particle getting into point $\mathbf{x}$ at some instant t. Having applied here the expression (3.6) one can obtain the promised relation

$$w_E(\mathbf{y}, j; \mathbf{x}, t) = j w_L(\mathbf{x}, j; \mathbf{y}, t) \tag{3.11}$$

connecting the Eulerian probability distribution with Lagrangian one

$$w_L(\mathbf{x}, j; \mathbf{y}, t) = \langle \delta\left(\mathbf{X}(\mathbf{y}, t) - \mathbf{x}\right) \delta\left(J(\mathbf{x}, t) - j\right) \rangle \ . \tag{3.12}$$

Factor j in (3.11), by which these two pdf's differ, takes into account the increase of dilated regions (where $j > 1$) contribution in the statistical ensemble

358

of Eulerian fields, as compared with the statistical ensemble of its Lagrangian counterparts.

For the physical applications much more interesting to establish the similar relations between of the Eulerian $\rho(\mathbf{x}, t)$ and Lagrangian

$$R(\mathbf{y}, t) = \rho(\mathbf{X}(\mathbf{y}, t)) = \frac{\rho_0(\mathbf{y})}{J(\mathbf{y}, t)} \tag{3.13}$$

density fields pdf's. It is easy to show from (3.10) — (3.13) that the desired relation has the form

$$\rho w_E(\mathbf{y}, \rho; \mathbf{x}, t) = \rho_0(\mathbf{y}) w_L(\mathbf{x}, \rho; \mathbf{y}, t) , \tag{3.14}$$

where

$$w_L(\mathbf{x}, \rho; \mathbf{y}, t) = \langle \delta\left(\mathbf{X}(\mathbf{y}, t) - \mathbf{x}\right) \delta\left(R(\mathbf{y}, t) - \rho\right) \rangle \tag{3.15}$$

is the joint Lagrangian pdf of the random fields $\mathbf{X}(\mathbf{y}, t)$, $R(\mathbf{y}, t)$, and

$$w_E(\mathbf{y}, \rho; \mathbf{x}, t) = \langle \delta\left(\mathbf{Y}(\mathbf{x}, t) - \mathbf{y}\right) \delta\left(\rho(\mathbf{x}, t) - \rho\right) \rangle \tag{3.16}$$

is the Eulerian joint pdf of the Lagrangian coordinate $\mathbf{Y}(\mathbf{x}, t)$ and the density field $\rho(\mathbf{x}, t)$. Integrating (3.14) over all $\mathbf{y}$, we arrive to the formula

$$\rho w_E(\rho; \mathbf{x}, t) = \int \rho_0(\mathbf{y}) w_L(\mathbf{x}, \rho; \mathbf{y}, t) d\mathbf{y}, \tag{3.17}$$

expressing the pdf of the Eulerian density field through the joint Lagrangian pdf of the random position of given particle, initially situated at point $\mathbf{y}$, and the tracer density $R(\mathbf{y}, t)$ in vicinity of that particle. Relation (3.17) generalizes the turbulent diffusion basic theorem (3.4). Indeed, one can get (3.4) just integrating of (3.17) over the all values ρ.

In the case of identical initial tracer density at the all space, that is if

$$\rho_0(\mathbf{x}) = \rho_0 = \text{const} \tag{3.18}$$

and in the case of statistically homogeneous random velocity field $\mathbf{v}(\mathbf{x}, t)$, the Lagrangian probability distribution at the right hand side of equality (3.17) depends only of difference of the Lagrangian and Eulerian coordinates:

$$w_L(\mathbf{x}, \rho; \mathbf{y}, t) = w_L(\mathbf{x} - \mathbf{y}, \rho; t) . \tag{3.19}$$

As a result expression (3.17) is turned into simple algebraic relation

$$\rho w_E(\rho; t) = \rho_0 w_L(\rho; t) \tag{3.20}$$

joining the Eulerian (2.1) and Lagrangian

$$w_L(\rho; \mathbf{y}, t) = \langle \delta\left(R(\mathbf{y}, t) - \rho\right) \rangle \tag{3.21}$$

density field pdf's. The corresponding connection between the Eulerian and Lagrangian pdf's of statistically homogeneous Jacobian field has the form

$$w_E(j;t) = jw_L(j;t) \ . \tag{3.22}$$

Since both pdf's in (3.22) have to satisfy the normalization condition, we automatically get from (3.22) two conservation laws

$$\langle J(\mathbf{y},t) \rangle = 1 \ , \quad \left\langle \frac{1}{j(\mathbf{x},t)} \right\rangle = 1 \ . \tag{3.23}$$

Both of them have the transparent physical sense. The first is the conservation law of the area occupied by the medium. The meaning of the second conservation law in (3.23) we also grasp in the case of identical initial density (3.18). Multiplying the second equality in (3.23) by ρ_0 and noticing that

$$\frac{\rho_0}{j(\mathbf{x},t)} = \rho(\mathbf{x},t) \tag{3.24}$$

is the Eulerian density field, we arrive to equality

$$\langle \rho(\mathbf{x},t) \rangle = \rho_0 \ . \tag{3.25}$$

The latter, in turn, is a consequence of the dynamic mass conservation law.

4. Lagrangian statistics of density and Jacobian

Let us discuss some statistical properties of the passive tracer density and Jacobian fields in Lagrangian representation. For this purpose let write out the corresponding equations in Lagrangian coordinate system. Namely the equation of particle trajectory

$$\frac{d\mathbf{X}}{dt} = \mathbf{v}(\mathbf{X},t) \ , \quad \mathbf{X}(\mathbf{y},t=0) = \mathbf{y} \tag{4.1}$$

and Lagrangian form of continuity equation (3.1)

$$\frac{dR}{dt} + u(\mathbf{X},t)R = 0 \ , \quad R(\mathbf{y},t=0) = \rho_0(\mathbf{y}) \ . \tag{4.2}$$

One can get the Jacobian evolution equation just by substitution of (3.13) into (4.2):

$$\frac{dJ}{dt} = u(\mathbf{X},t)R \ , \quad J(\mathbf{y},t=0) \equiv 1 \ . \tag{4.3}$$

Here occurs the new scalar field

$$u(\mathbf{x},t) = (\nabla \cdot \mathbf{v}(\mathbf{x},t)) \tag{4.4}$$

- divergence of the velocity field $\mathbf{v}(\mathbf{x},t)$. In compressible media under investigation it is a some random field, which we will suppose in what follows statistically homogeneous.

The solution of the Jacobian equation (4.3) has the form

$$J(\mathbf{y}, t) = \exp(L(\mathbf{y}, t)) \tag{4.5}$$

where

$$L(\mathbf{y}, t) = \int_0^t u\left(\mathbf{X}(y, t'), t'\right) dt' . \tag{4.6}$$

For the sufficiently large times t — much more than the time of random velocity field correlation τ_v:

$$t \gg \tau_v \tag{4.7}$$

it has sense, with agree of Central Limit Theorem, to suppose integral (4.6) is the Gaussian with the mean value

$$\langle L(\mathbf{y}, t) \rangle = \langle U \rangle t , \quad \langle U \rangle = \langle u(\mathbf{X}(\mathbf{y}, t)) \rangle \tag{4.8}$$

and variance

$$\sigma_L^2(t) = 2Bt , \quad B = \int_0^t \langle u\left(\mathbf{X}(\mathbf{y}, t'), t'\right) u\left(\mathbf{X}(\mathbf{y}, t), t\right) \rangle dt' . \tag{4.9}$$

Respectively the average of Jacobian (4.5) is equal to

$$\langle J(\mathbf{y}, t) \rangle = \exp\left[(\langle U \rangle + B) t\right] . \tag{4.10}$$

On the other hand according to the first conservation law (3.23) $\langle J \rangle = 1$. Thus we arrive to the useful connection

$$\langle U \rangle = -B . \tag{4.11}$$

Consequently the Lagrangian density field

$$R(\mathbf{y}, t) = \rho_0 \exp\left(-L(\mathbf{y}, t)\right) \tag{4.12}$$

has the logarithmically normal pdf

$$w_L(\rho; t) = \frac{1}{2\rho\sqrt{\pi\tau}} \exp\left[-\frac{\ln^2\left(\rho e^{-\tau}/\rho_0\right)}{4\tau}\right] \tag{4.13}$$

where it is used the dimensionless time $\tau = Bt$. In particular it follows from (4.13), that the median curve

$$\overline{\rho}_L(t) = \rho_0 e^\tau \tag{4.14}$$

corresponding to the equality

$$P\left(R(\mathbf{y}, t) < \overline{\rho}_L(t)\right) = \frac{1}{2} \tag{4.15}$$

grows exponentially with time. In other words, in the course of time, almost all particles of the tracer falls into the compressed regions (clusters) with an **exponentially growing density**.

5. Temporal behavior of the density field

More detailed information about the formation and evolution of the passive tracer clusters one can obtain by investigation of the Eulerian probabilistic properties of the random density field $\rho(\mathbf{x}, t)$. Using the Lagrangian and Eulerian pdf's connection (3.20) and already have specified earlier the Lagrangian density pdf (4.13), we turn up to the Eulerian pdf of $\rho(\mathbf{x}, t)$:

$$w_E(\rho; t) = \frac{1}{2\rho\sqrt{\pi\tau}} \exp\left[-\frac{\ln^2\left(\rho e^\tau/\rho_0\right)}{4\tau}\right] . \tag{5.1}$$

Note that the some statistical characteristics of density field $\rho(\mathbf{x}, t)$ are more

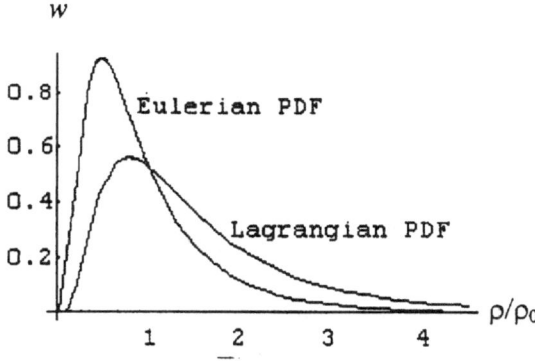

Fig. 3. Plots of the Eulerian and Lagrangian probability distributions of dimensionless density field ρ/ρ_0.

convenient to extract from Fokker-Planck equation

$$\frac{\partial w_E}{\partial t} = \frac{\partial^2}{\partial \rho^2}\left(\rho^2 w_E\right) , \qquad w_E\left(\rho; t = 0\right) = \delta\left(\rho - \rho_0\right) \tag{5.2}$$

to which pointed out Eulerian pdf (5.1) satisfies. For example, it is easy to get from (5.2) closed equations for the density field's statistical moments:

$$\frac{d}{d\tau}\langle\rho^n(\mathbf{x}, t)\rangle = n(n-1)\langle\rho^n(\mathbf{x}, t)\rangle , \qquad \langle\rho^n(\mathbf{x}, t = 0)\rangle = \rho_0^n . \tag{5.3}$$

The solutions thereof are

$$\langle\rho^n(\mathbf{x}, t)\rangle = \rho_0^n \exp(n(n-1)\tau) . \tag{5.4}$$

It should be emphasized that the density field possesses seemingly contradicting statistical properties inherent to the intermittency processes [10]. Indeed, as one can see from (5.4), the variance of Eulerian density field exponentially grows up with course of time. It seems to imply an appearance as t increases, of very high peaks of the density realizations. At the other hand it contradicts to the exponential decreasing of Eulerian pdf median curve

$$\rho^*(t) = \rho_0 e^{-\tau} . \tag{5.5}$$

The latter means that at large time $\tau \gg 1$ the Eulerian "probabilistic mass" is concentrated at the low density levels $\rho \ll \rho_0$.

Let us analyze the density temporal behavior in more detail. To this purpose recall that the Fokker-Planck equation (5.2) can be treated as equation for the transition pdf of an auxiliary Markov process $\rho(\tau)$, which satisfies to the stochastic equation

$$\frac{d\rho}{d\tau} + \rho = \xi(\tau)\rho , \quad \rho(\tau = 0) = \rho_0 . \tag{5.6}$$

Here $\xi(\tau)$ is a Gaussian white noise with covariance function

$$\langle \xi(\tau)\xi(s) \rangle = 2\delta(\tau - s) .$$

To our mind named resemblance between $\rho(\mathbf{x}, t)$ and $\rho(\tau)$ allows to assume that realizations of the density field $\rho(\mathbf{x}, t)$ at fixed point $\mathbf{x}$ and auxiliary process $\rho(\tau)$ realizations are similar. So one can assign the properties of auxiliary process $\rho(\tau)$ to the temporal behavior of density field $\rho(\mathbf{x}, t)$.

First of all notice that there exist exponentially decreasing majorant curves [11]

$$M(\tau) = A\rho_0 e^{-\mu\tau} \quad (0 < \mu < 1 , \quad A > 1) \tag{5.7}$$

such that with probability

$$P = 1 - A^{(\mu-1)} \tag{5.8}$$

realizations of process $\rho(\tau)$, for all t, are situated under the curve $M(\tau)$. In particular, one half of all realizations of $\rho(\tau)$ are located beneath the curve

$$M(\tau) = 4\rho_0 e^{-\tau/2} . \tag{5.9}$$

Moreover, the area below of $\rho(\tau)$ realizations

$$V = \int_0^\infty \rho(\tau)d\tau \tag{5.10}$$

bounded almost surely. To prove this statement let us consider the new process $V(\tau)$ satisfying to the stochastic equation

$$\frac{dV}{d\tau} + V = \xi(\tau)V + \rho_0 , \quad V(\tau = 0) = 0 . \tag{5.11}$$

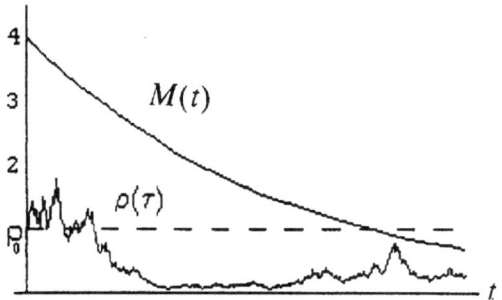

Fig. 4. Plots of monotonically decreasing majorant curve and typical realization of auxiliary process $\rho(\tau)$, modeling the temporal behavior of the passive tracer density field $\rho(\mathbf{x}, t)$.

The solution thereof has the form

$$V(\tau) = \rho_0 \int_0^\tau J(\tau, \tau') d\tau' \tag{5.12}$$

where $J(\tau, \tau')$ is the solution of the stochastic equation

$$\frac{dJ}{d\tau} + J = \xi(\tau)J , \quad J(\tau = \tau', \tau') = 1 . \tag{5.13}$$

Comparison of (5.10), (5.12) and (5.6), (5.13) lead to conclusion that random variables V (5.10) and $V(\tau = \infty)$ (5.12) are statistically equivalent, i.e. pdf's of V and $V(\infty)$ are coincide. The pdf $w(v; \tau)$ of Markov process $V(\tau)$ is submitted to the Fokker-Planck equation

$$\frac{\partial w}{\partial t} - \rho_0 \frac{\partial w}{\partial v} = \frac{\partial^2}{\partial v^2} \left(v^2 w \right) , \quad w(v; \tau) = \delta(v) . \tag{5.14}$$

As time increases, solution of this equation converges to the stationary pdf

$$\lim_{\tau \to \infty} w(v, \tau) = w_\infty(v) = \frac{\rho_0}{v^2} \exp\left(-\frac{\rho_0}{v}\right) , \quad (v > 0) . \tag{5.15}$$

It means that for any arbitrary probability $p < 1$ the area underneath of density field $\rho(\mathbf{x} = \text{const}, t)$ realizations less than $\rho_0/B \ln(1/p) < \infty$:

$$P\left(\int_0^\infty \rho(\mathbf{x}, t) dt < \rho_0/B \ln(1/p) < \infty\right) = p . \tag{5.16}$$

6. Evolution of clusters

Let us turn back to discussing of the topographical properties of density field. It follows from (2.6), (2.11) and (5.1) that the specific area and mass at the regions with raised density $\rho(\mathbf{x}, t) > \rho$ are equal to

$$s(t, \rho) = \Phi\left(\frac{\ln(\rho_0 e^{-\tau}/\rho)}{2\sqrt{\tau}}\right) , \qquad (6.1)$$

$$m(t, \rho) = \Phi\left(\frac{\ln(\rho_0 e^{\tau}/\rho)}{2\sqrt{\tau}}\right) , \qquad (6.2)$$

respectively. Here it is used the error function

$$\Phi(z) = \frac{1}{\sqrt{\pi}} \int_{-\infty}^{z} \exp\left(-y^2\right) dy . \qquad (6.3)$$

It is obvious from (6.1), (6.2), that for the large time, when $\tau \gg 1$, the specific area of such regions decreasing as

$$s(t, \rho) \approx \Phi\left(-\frac{\sqrt{\tau}}{2}\right) \sim \frac{1}{\sqrt{\pi\tau}} \exp\left(-\frac{\tau}{4}\right) \qquad (6.4)$$

independently from the ratio ρ/ρ_0. At the some time inside of these regions almost all tracer particles are putting together, so the specific mass of these regions tends to one:

$$m(t, \rho) \approx \Phi\left(\frac{\sqrt{\tau}}{2}\right) \sim 1 - \frac{1}{\sqrt{\pi\tau}} \exp\left(-\frac{\tau}{4}\right) . \qquad (6.5)$$

Correspondingly the threshold level $\rho^*(t)$, such that contours $\rho(\mathbf{x}, t) = \rho > \rho^*(t)$ separate the whole space onto cluster's regions of the raised density $\rho(\mathbf{x}, t) > \rho$ and remaining "ocean" of lower one, exponentially decreases (see (5.5)).

Basing on pointed out relations one can imagine the dynamics of the arising and evolution of the passive tracer clusters in chaotically moving compressible media.

Initial growth of specific area $s(\rho/\rho_0, \tau)$ is explained by symmetric density field fluctuations

$$\rho(\mathbf{x}, t) \approx \rho_0 - \int_0^t u(\mathbf{x}, t')dt' \qquad (6.6)$$

at the small time $\tau < 1$, when displacements of particles $\mathbf{X}(\mathbf{y}, t)$ from its initial coordinates $\mathbf{y}$ relatively small. Then, when make up the strong Jacobian fluctuations, clusters appear and specific area begins monotonically decrease. On the contrary, clusters specific mass monotonically increase and brings nearer to unity for big time ($\tau \gg 1$). It means that almost all passive tracer particles stick at the stable clusters.

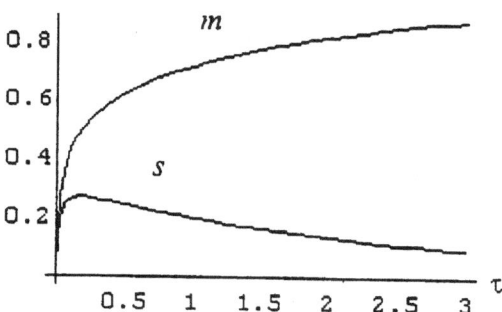

Fig. 5. Plots of the specific area $s(\rho/\rho_0, \tau)$ and specific mass $m(\rho/\rho_0, \tau)$ depending on dimensionless time τ for the ratio $\rho/\rho_0 = 1.2$.

7. Clusters in the presence of the molecular diffusion

Finally let us disclose some peculiarities of the passive tracer behavior in chaotically moving compressible media in presence of the molecular diffusion. Let us restrict ourselves by 1D case when density field $\rho(\mathbf{x}, t)$ depends only on the one spatial coordinate $\mathbf{x}$. Such model can describe the case of narrow channel, where the water has only a horizontal motion along of channel walls. The motion of a floating tracer at the surface of channel can then be described like motion of the passive tracer in a 1D compressible medium. In such a case the floating tracer density field obeys to the 1D continuity equation

$$\frac{\partial \rho}{\partial t} + \frac{\partial}{\partial x}\left(v(x,t)\rho\right) = \mu \frac{\partial^2 \rho}{\partial x^2} , \qquad \rho(x; t=0) = \rho_0(x) . \qquad (7.1)$$

It takes into account the molecular diffusion with diffusion coefficient μ. At the quiet surface when velocity identically equal to zero: $v(x, t) \equiv 0$, solution of this equation in the case of initially putting together particles $(\rho_0(x) = M\delta(x))$

$$\rho(x,t) = \frac{M}{2\sqrt{\pi\mu t}} \exp\left(-\frac{x^2}{2\mu t}\right) \qquad (7.2)$$

describes the monotonically spreading of passive tracer due to the mutually independent Brownian motion of the particles. It turns out, in presence of chaotic surface motion, when velocity $v(x, t)$ is the some stationary homogeneous random field, influence of chaotic compressions and dilations of floating tracer gives rise to the new seemingly unexpected physical effect - localization of clusters due to the competition of hydrodynamic chaotic motion and molecular diffusion. Let us describe this effect in more detail.

At first let us looking for the asymptotic solution of continuity equation (7.1) for the small molecular diffusion coefficient μ. To this end represent $\rho(x, t)$ in the form

$$\rho(x,t) = \overline{\widetilde{\rho}(x,t)}$$

where $\widetilde{\rho}$ is satisfied to the auxiliary first order equation

$$\frac{\partial \widetilde{\rho}}{\partial t} + \frac{\partial}{\partial x}\left(v(x,t)\widetilde{\rho}\right) = \eta(t)\frac{\partial \widetilde{\rho}}{\partial x} \qquad (7.3)$$

$\eta(t)$ — Gaussian white noise with covariance function

$$\overline{\eta(t)\eta(t+\tau)} = 2\mu\delta(\tau)$$

where the upper bar means averaging over $\eta(t)$ ensemble.
Solution of the equation (7.3) has the form

$$\widetilde{\rho}(x,t) = \int \rho_0(y)\delta(\widetilde{X}(y,t)-x)dy . \qquad (7.4)$$

Here $\widetilde{X}(y,t)$ — particle trajectory taking into account the molecular diffusion and satisfying to the characteristic equation

$$\frac{\partial \widetilde{X}}{\partial t} = v(\widetilde{X},t) + \eta(t) , \qquad \widetilde{X}(y,t=0) = y . \qquad (7.5)$$

Let us split $\widetilde{X}(y,t)$ onto two parts:

$$\widetilde{X}(y,t) = X(y,t) + z(y,t) \qquad (7.6)$$

where $X(y,t)$ describes purely hydrodynamic motion of the floating tracer and obeys to the equation

$$\frac{\partial X}{\partial t} = v(X,t) , \qquad X(y,t=0) = y . \qquad (7.7)$$

At the same time $z(y,t)$ takes into account deviation of particle from $X(y,t)$ due to the Brownian motion, and submitted to the equation:

$$\frac{\partial z}{\partial t} = v(X+z,t) - v(X,t) + \eta(t) , \qquad z(y,t=0) = 0 .$$

Let l_v is the typical scale of spatial variability of the random velocity field $v(x,t)$. In what follows we will suppose that the next inequality is true:

$$z \ll l_v . \qquad (7.8)$$

If it is the case, then one can replace the equation for $z(y,t)$ by the linear equation

$$\frac{\partial z}{\partial t} = u(X,t)z + \eta(t) , \qquad z(y,t=0) = 0 \qquad (7.9)$$

where

$$u(x,t) = \frac{\partial}{\partial x}v(x,t) .$$

In frame of pointed out linear approximation $z(y, t)$ is the Gaussian process with zero mean and variance, satisfying to the equation:

$$\frac{\partial \sigma^2}{\partial t} = 2u(X,t)\sigma^2 + 2\mu , \quad \sigma^2(y, t = 0) = 0 . \tag{7.10}$$

Substituting (7.6) into (7.4) and averaging obtained expression over Gaussian ensemble of process $z(y, t)$, we arrive to the promised asymptotic formula for 1D floating tracer density field taking into account both particles hydrodynamic motion and molecular diffusion:

$$\rho(x,t) = \int \rho_0(y) \frac{1}{\sqrt{2\pi}\sigma(y,t)} \exp\left[-\frac{(x - X(y,t))^2}{2\sigma^2(y,t)}\right] dy . \tag{7.11}$$

Let at the initial time $t = 0$ all particles are concentrated in physically infinitesimal vicinity of the origin. Then one should take as initial density the $\rho_0 = M\delta(x)$, where M - whole floating tracer mass. As a result the density field acquires the form

$$\rho(x,t) = R(x - X(t),t) , \tag{7.12}$$

where $X(t) = X(0, t)$ and

$$R(x,t) = \frac{M}{\sqrt{2\pi}\sigma(t)} \exp\left(-\frac{x^2}{2\sigma^2(t)}\right) \tag{7.13}$$

- the "shape" of cluster. Here $\sigma(t) = \sigma(0, t)$ - its effective width.

In what follows for the sake of concreteness we will suppose that the velocity field $v(x, t)$ is Gaussian and delta-correlated in time with covariance function

$$\langle v(x,t)v(x + s, t + \tau)\rangle = a(s)\delta(\tau) , \quad a(s) = 2D - Bs^2 + ... \tag{7.14}$$

Recall that everywhere angle brackets indicate the statistical averaging over ensemble of random velocity field $v(x, t)$.

¿From stochastic equations (7.7), (7.10) and from (7.14) it follows that collection $\{X(t), \sigma(t)\}$ forms the two-dimensional Markov process whose pdf

$$f(x, \sigma; t) = \langle \delta(X(t) - x)\delta(\sigma(t) - \sigma)\rangle$$

satisfies to the next Fokker-Planck equation [2]

$$\frac{\partial f}{\partial t} + \mu \frac{\partial}{\partial \sigma}\left(\frac{f}{\sigma}\right) = D\frac{\partial^2 f}{\partial x^2} + B\frac{\partial^2}{\partial \sigma^2}\left(\sigma^2 f\right) , \quad f(x, \sigma; t = 0) = \delta(x)\delta(\sigma) . \tag{7.16}$$

Statistical and dynamical interpretation of this equation consequences makes possible to understand the clusters' behavior under competition of the molecular diffusion and compressible medium chaotic hydrodynamic motion. First of all notice that the solution of equation (7.16) falls into the product of two pdf's:

$$f(x, \sigma; t) = \varphi(x;t)g(\sigma; t) \tag{7.17}$$

where the first

$$\varphi(x;t) = \frac{1}{\sqrt{2\pi Dt}} \exp\left(-\frac{x^2}{4Dt}\right)$$

is responsible for hydrodynamic cluster's diffusion. The second one obeys to the equation

$$\frac{\partial g}{\partial t} + \mu \frac{\partial}{\partial \sigma}\left(\frac{g}{\sigma}\right) = B \frac{\partial^2}{\partial \sigma^2}\left(\sigma^2 g\right) \ , \quad g(\sigma;t=0) = \delta(\sigma) \qquad (7.18)$$

and describes the fluctuations of the cluster's effective width (see Fig.6).

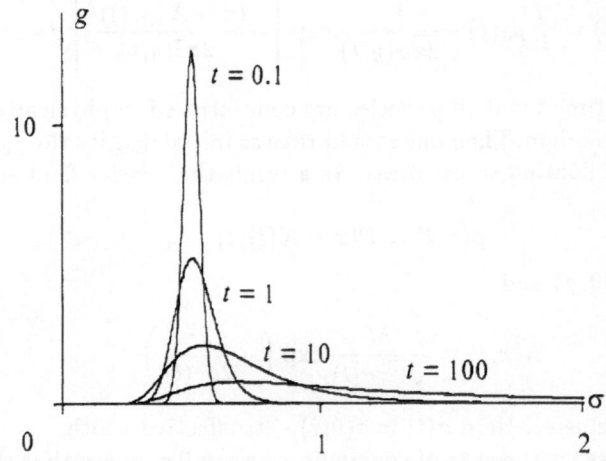

Fig. 6. The temporal behavior of the cluster's width pdf for $l = 1$.

In particular it follows from (7.18) that variance of the cluster's width submits to the equation

$$\frac{\partial \langle \sigma^2 \rangle}{\partial t} = 2\mu + B \langle \sigma^2 \rangle \ , \quad \langle \sigma^2(t=0) \rangle = 0 \ ,$$

solution thereof

$$\langle \sigma^2(t) \rangle = \frac{1}{M} \int x^2 \langle R(x,t) \rangle \, dx = \frac{2\mu}{B}\left(e^{Bt} - 1\right) \qquad (7.19)$$

has evident physical sense: namely, as long as $Bt \ll 1$, the molecular diffusion dominates and cluster's width variance grows with agree of classical diffusion law

$$\langle \sigma^2(t) \rangle \cong 2\mu t \ . \qquad (7.20)$$

Later the medium dilations and compressions begin dominate. As a result $\langle \sigma^2(t) \rangle$ grows up much more faster and when $Bt \gg 1$ the linear law (7.20) is replaced by exponential:

$$\langle \sigma^2(t) \rangle \cong \frac{2\mu}{B} e^{Bt} \ .$$

It should be emphasized however that conclusions about cluster's behavior based on mean variance (7.19) analysis don't reflect properly the actual cluster's behavior with course of time. More accurate analysis have to be rested on the probabilistic properties of cluster's width. So let us investigate some probabilistic consequences of the equation (7.18). It is easy to show that this equation has a stationary solution

$$g_\infty(\sigma) = \lim_{t\to\infty} g(\sigma; t) = \sqrt{\frac{2}{\pi}} \frac{l}{\sigma^2} \exp\left(-\frac{l^2}{2\sigma^2}\right), \quad l = \sqrt{\frac{\mu}{B}}, \qquad (7.21)$$

describing the statistical properties of cluster's width for case when $Bt \gg 1$.

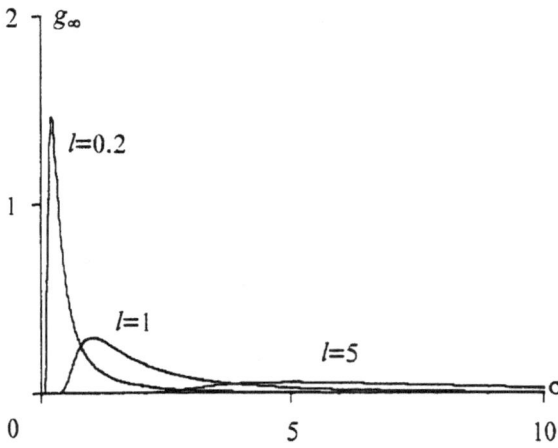

Fig. 7. Plot of cluster's width stationary probability distribution function $g_\infty(\sigma)$ depending on dimensionless cluster's width σ/l.

Existing of the stationary distribution (7.21) means that medium chaotically compressions prevent from molecular diffusion action and leads to the cluster's localization. Indeed, averaging of the cluster's shape (7.13) with help of pdf (7.21) one can obtain that mean cluster's density profile

$$\langle R_\infty(x) \rangle = \lim_{t\to\infty} \langle R(x, t) \rangle = \frac{Ml}{\pi} \frac{1}{x^2 + l^2} \qquad (7.22)$$

relatively well localized along the axis x. This result don't contradicts to tending to infinity of mean variance for $t \to \infty$ because the corresponding integral

$$\langle \sigma^2(t = \infty) \rangle = \int \langle R_\infty(x) \rangle \, dx = \infty$$

is equal to infinity with agree of (7.19).

Described localization effect means that despite of immovable media, where the maximal density of cluster, according to (7.2), monotonically decreases to zero:

$$R(t) = \max \rho(x,t) = \rho(0,t) = \frac{M}{2\sqrt{\pi \mu t}} \; ,$$

the maximal density of cluster in chaotically moving 1D medium

$$R(t) = \max \rho(x,t) = R(x = 0, t) = \frac{M}{\sqrt{2\pi}\sigma(t)}$$

has stationary pdf

$$w_\infty(R) = \frac{2}{\gamma\sqrt{\pi}} \exp\left(-\frac{R^2}{\gamma^2}\right) \; , \qquad \gamma = M\sqrt{\frac{B}{\pi\mu}}$$

with non-zero mean and bounded variance

$$\langle R \rangle = \frac{\gamma}{\sqrt{\pi}} \; , \quad \sigma_R^2 = \gamma^2 \left(\frac{1}{2} - \frac{1}{\pi}\right) \; .$$

This report was supported in parts by Grants 97-02-16521 and 95-IN-RU-723 of Russian Foundation of Fundamental Researches.

REFERENCES

1. Isichenko M.B. Percolation, statistical topography, and transport in random media. Rev. Modern. Phys. **64**, No.4, pp. 961-1043 (1992).
2. Saichev A.I., Woyczynski W.A. Probability distributions of passive tracers in randomly moving media. In: Stochastic Models in Geosystems, S.A. Molchanov and W.A. Woyczynski, eds., IMA Volumes, Springer-Verlag, Berlin, New York, **85**, pp. 359-399 (1997).
3. Klyatskin V.I., Saichev A.I. Statistical theory of the diffusion of a passive tracer in a random velocity field. JETP, **84**, No.4, pp. 716-724 (1997).
4. Gurbatov S.N., Malakhov A.N., Saichev A.I. Nonlinear Random Waves and Turbulence in Nondispersive Media: Waves, Rays and Particles, Manchester University Press, Manchester, 1991.
5. Saichev A.I., Woyczynski W.A. Evolution of Burgers' turbulence in the presence of external forces. J. Fluid. Mech. **331**, pp. 313-343 (1997).
6. Saichev A.I., Woyczynski W.A. Density fields in Burgers and KdV-Burgers turbulence. SIAM J. Appl. Math. **56**, No. 4, pp. 1008-1038 (1996).
7. Saichev A.I., Woyczynski W.A. Advection of passive and reactive tracers in multi-dimensional Burgers' velocity field. Physica D, **100**, pp. 119-141 (1997).
8. Csanady G.T. Turbulent Diffusion in the Environment. D. Reidel, Dordrecht, The Netherlands, 1974.

9. Saichev A.I., Woyczynski W.A. Distributions in the Physical and Engineering Sciences. Volume 1, Birkhauser, Boston, 1997.

10. Zeldovich Ya.B., Molchanov S.A., Ruzmaykin A.A. Intermittency in random medium. Sov. Phys. Usp. **152**, pp. 3-32 (1987).

11. Klyatskin V.I., Saichev A.I. Stochastical and dynamic localization of plane waves in randomly layered media. Sov. Phys. Usp. **35**, No.3, pp. 231-247 (1992).

Anomalous Diffusion in the Strong Scattering Limit: A Lévy Walk Approach

E. Barkai[1], and J. Klafter[2]

[1] School of Physics and Astronomy, Tel Aviv University, Tel Aviv 69978, Israel.
[2] School of Chemistry, Tel Aviv University, Tel Aviv 69978, Israel.

Abstract. The continuous time random walk (CTRW) is a powerful stochastic theory developed and used to analyze regular and anomalous diffusion. In particular this framework has been applied to sublinear, dispersive, transport and to enhanced Lévy walks. In its earlier version the CTRW does not include the velocities of the walker explicitly, and therefore it is not suited to analyze situations with randomly distributed velocities. Experiments and theory have recently considered systems which exhibit anomalous diffusion and are characterized by an inherent distribution of velocities. Here we develop a modified CTRW formalism, based on a velocity picture in the strong scattering limit, with emphasis on the Lévy walk limit. We consider a particle which randomly collides with unspecified objects changing randomly its velocity. In the time intervals between collision events the particle moves freely. Two probability density functions (PDF) describe such a process: (a) $q(\tau)$, the PDF of times between collision events, and (b) $F(\mathbf{v})$, the PDF of velocities of the particle. In this renewal process both the velocity of the random walker and the time intervals between collision events are independent, identically distributed, random variables. When either $q(\tau)$ or $F(\mathbf{v})$ are long-tailed the diffusion may become non-Gaussian. The probability density to find the random walker at $\mathbf{r}$ at time t, $\rho(\mathbf{r}, t)$, is found in Fourier-Laplace space. We discuss the role of initial conditions especially on the way $P(\mathbf{v}, t)$, the probabilty density that the particle has a velocity $\mathbf{v}$ at time t, decays to its equilibrium. The phase diagram of the regimes of enhanced, sublinear and normal types of diffusion is presented. We discuss the differences and similarities between the Lévy walk collision process considered here and the CTRW for jump processes.

1 Introduction

Anomalous diffusion [1-5] is a well established phenomenon, found in a broad range of fields. It is characterized by

$$\langle r^2 \rangle \sim t^\beta \tag{1}$$

with $\beta \neq 1$. Various mechanisms are known which lead to enhanced diffusion ($\beta > 1$), or to subdiffusion, also called dispersive or slow diffusion ($\beta < 1$). For these cases the goal is to find the probability density $\rho(\mathbf{r}, t)$ to be at $\mathbf{r}$ at time t, from which transport moments can be obtained. Stochastic frameworks which describe phenomenologically such behaviors include fractal Brownian motion [6], fractional calculus [7-12], generalized diffusion equation [13-17] and a generalized Langevin equation approach [18, 19]. Two other approaches, relevant especially here, are: (a) the continuos time random walk (CTRW), [20, 21] and (b) a velocity approach whose details are given below [2,22-28]. Our main aim in this work is to generalize the velocity approach and show its relation to the CTRW.

All these stochastic theories are used to describe processes for which the conditions that ensure the validity of the central limit theorem (CLT) are not satisfied. Lévy and Khintchine [29] introduced a generalization of the CLT to the case where x_i in the sum $X_N = \sum_{i=1}^{N} x_i$ are independent, identically distributed, random variables with a long-tailed distribution so that the existence of the first two moments is not necessarily assumed. The corresponding random walk, X_N, is called a Lévy flight. For such random walks the mean squared displacement $\langle X_N^2 \rangle$ diverges for all $N > 0$ and, hence, a Lévy flight cannot represent an anomalous diffusion of the type in Eq. (1).

A way to overcome the divergence in the mean squared displacement is to introduce a velocity into the random walk scheme, the result being that the divergence of the mean square displacement found for Lévy flights is replaced by enhanced diffusion (see e.g. [25]). Such random walks are called Lévy walks [23] and can be defined within the context of the CTRW. Within the CTRW framework the pausing times between successive steps, as well as the lengths of the steps, are random variables. Schemes in which the pausing times and step lengths are either decoupled or coupled [1, 30] have been investigated thoroughly [20, 21]. For both cases the jumping events are instantaneous so that the random walker pauses at some location for a finite (random) time and then performs a jump, with a vanishing jumping time, to another location. Within the CTRW the stochastic evolution is described in terms of $\psi(\mathbf{r}, t)$, the probability density of making a jump of length $\mathbf{r}$ in the time interval between t and $t + dt$. Klafter, Blumen and Shlesinger [30] introduced a coupled space-time distribution

$$\psi_{ct}(\mathbf{r}, t) = C r^{-\mu_{ct}} \delta\left(r - t^{\nu_{ct}}\right) \qquad (2)$$

where, through the Dirac δ function, r and t are coupled. The idea behind this coupling is that the longer is the step, the more time it takes to be performed. For this coupling one can find normal, enhanced or sublinear types of diffusion, depending on the choice of exponents μ_{ct} and ν_{ct} (the subscript ct stands for CTRW). Recently, in [31] such a coupled CTRW was used to analyze statistical properties of chaotic trajectories generated by a deterministic non-linear map.

Although the jump process described by the CTRW with space-time coupling, Eq. (2), is a powerful tool describing anomalous diffusion, it considers explicitly the positions of the random walkers and not their velocities. Therefore it is less suited to describe systems where the velocity of the random walkers is an important stochastic ingredient of the transport process. However, recently much attention has been drawn to systems which exhibit anomalous type of diffusion, and for which the velocities of the walker are randomly distributed according to a PDF $F(\mathbf{v})$ [32-36].

Approaches which include a constant velocity were developed [2,22-25], where a particle moves with a constant speed for some time interval and then a new direction of motion is chosen (the kinetic energy of the particle is a constant of motion). In one-dimension such a model is a two state model with the velocity having only the values $\pm v$. When the times of free motion with a constant velocity are independent random variables whose distribution decays algebraically, then, under certain conditions, the diffusion is enhanced and displays a Lévy walk.

Here we consider the more general case for which the magnitude of the velocity is not necessarily constant, but may rather change due to collision events. We investigate a d-dimensional stochastic strong collision model, according to which a particle moves freely between turning points (collision events), and at each turning point the velocity of the particle (both direction and magnitude) is randomized. The process is renewed after each collision. The time intervals between collision events are described by a probability density function (PDF) $q(\tau)$ and the velocities of the particle are described by a PDF $F(\mathbf{v})$. Both the times and the velocities are assumed to be independent, identically distributed, random variables. We call such a process a Lévy walk collision process (LWCP).

An example for such a process is the anomalous Knudsen diffusion proposed by Levitz [37], where the velocities of the gas particles are Maxwell distributed and scatterers are randomly distributed on a fractal structure in such a way that the PDF of the times between collision events decays algebraically. Other examples can be found in non-linear chaotic dynamics and in turbulent systems [32, 35, 36] and also in the statistics of single ion trajectories in optical lattices which exhibit Lévy walks [27, 34].

In figures (1) and (2) two different one-dimensional realizations of the LWCP are shown for the case where the mean time between collisions diverges,

$$q(\tau) \sim \tau^{-3/2}. \tag{3}$$

As can be seen the velocity is not a continuous function of time due to the strong collision events. Figure (1) shows a realization of the LWCP for the case where $F(\mathbf{v})$ is a Gaussian and this should be compared with the two state velocity model {i.e.$F(\mathbf{v}) = 0.5[\delta(v - 1) + \delta(v + 1)]$} shown in figure (2) and considered previously in [25]. Characteristic of the stochastic process shown in figures (1) and (2) are long time intervals in which no collision events take place. In appendix A we provide a short algorithm which generates time intervals whose PDF decays algebraically with time, Eq. (3) being an example.

When the collision times are Poisson distributed and the velocity PDF decays fast enough the LWCP reduces to the well known strong collision model (the Drude model) widely used in the context of plasma and condensed matter physics [38, 39].

If the time between collision events is a constant τ_0, the number of collision events in the interval $(0, t)$ is a non random variable N. For this case the LWCP reduces to a Lévy flight with a diverging mean squared displacement, provided that the first or the second moments of the velocity PDF diverge. This case has been investigated recently by Zanette and Alemany [40] (and see also [41]).

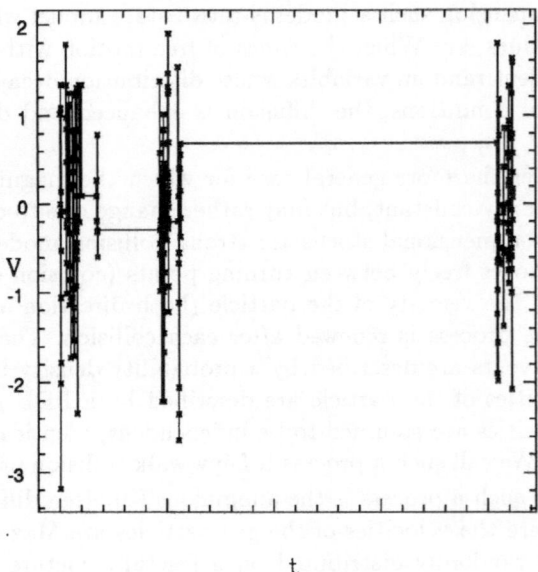

Fig. 1. Velocity vs time (dimensionless units) for a one-dimensional Lévy walk collision process. We generate such a process by generating two random numbers per collision. The first is the velocity of the particle, described by the PDF $F(\mathbf{v})$, and the second is the time of free motion described with the PDF $q(\tau)$ Eq. (3). Here the particle has encountered 200 collisions and the velocity PDF $F(\mathbf{v})$ is a Gaussian with a variance that equals unity. Notice the long time intervals in which no collision takes place, this is an important characteristic feature of the PDF in Eq. (3).

In figure (3) we show a realization of a one-dimensional LWCP for which both the velocity PDF

$$F(|v|) \sim |v|^{-3/2} \tag{4}$$

and the collision time PDF $q(\tau)$, Eq. (3), decay slowly and are characterized by heavy tails. Here we observe, in addition to the long time intervals, in which no collision takes place, also rare events in which the velocity of the particle becomes very large. The longer we observe the process the longer are the free time intervals and higher are the velocities that are observed. Generally we expect a cutoff in the velocity spectrum and then the algebraic decay is not valid for very high velocities.

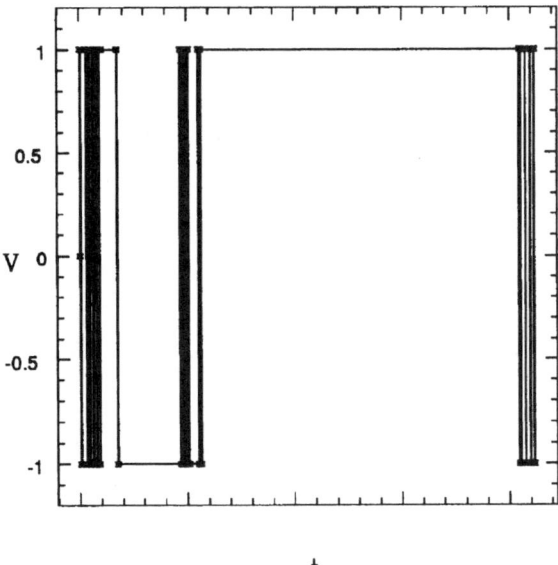

Fig. 2. A two state model where the velocity has the values ±1. The collision times are the same as found in figure (1).

The LWCP and the CTRW process, are intimately related as can be seen by considering realizations of the two processes. A CTRW process is defined through pausing times τ_i^p $(i = 1, 2, \cdots)$ and jump lengths $\mathbf{x}_i$, and a sequence of these pausing time and jump lengths is characterized by

$$\{[\tau_1^p, \mathbf{x}_1], \cdots [\tau_n^p, \mathbf{x}_n], \cdots [\tau_{N-1}^p, \mathbf{x}_{N-1}], [\tau_N^p]\}.$$

The pausing times satisfy $\sum_{n=1}^{N} \tau_n^p = t$, with t being the observation time and the location of the random walker is $\mathbf{x}(t) = \sum_{n=1}^{N-1} \mathbf{x}_n$. Notice that the N-th pausing time τ_N^p is not related to a displacement $\mathbf{x}_N$. The CTRW is a jump process and therefore at the time of observation t the particle is trapped motionless somewhere in the system. This is not the case for the LWCP where the time of free motion, τ_i^f, is related to a displacement $\mathbf{x}_i$. In this case a sequence is characterized by

$$\{[\tau_1^f, \mathbf{x}_1], \cdots [\tau_n^f, \mathbf{x}_n], \cdots [\tau_N^f, \mathbf{x}_N]\}.$$

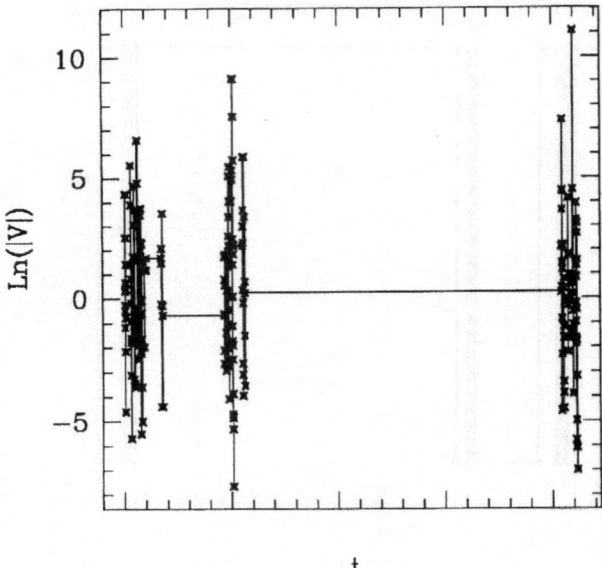

Fig. 3. A LWCP for which both the times between collisions and the velocities are described by PDFs with long tails. The LWCP is characterized by long time intervals in which no collision events take place, as well by events in which the particle gains very high velocities. Notice that we present $\ln(|v|)$ vs time. As discussed in the text, for such a process the mean squared displacement diverges.

The total displacement is $\mathbf{x}(t) = \sum_{n=1}^{N} \mathbf{x}_n$, and, unlike the CTRW, the summation includes the N-th term. One might expect that for a proper choice of coupling and PDFs the same asymptotic behavior will be found for both processes. Indeed, as expected, we find that when diffusion is normal, $\beta = 1$ in Eq. (1), the two pictures converge for long times. However, for systems exhibiting anomalous diffusion we find differences between the two approaches thus emphasizing the importance of investigating the LWCP.

The paper is organized as follows. In section (2) a solution of the model, in Fourier-Laplace space, is derived. Then, the role of initial conditions is considered in section (3). The influence of the first waiting time PDF and of the initial distribution of the velocities on the process are investigated. In section (4) the asymptotic behavior of our model is investigated. Depending on the parameters of the model we find sublinear, enhanced and diverging mean squared displacements. A comparison between the new results, the coupled CTRW, Eq. (2), and other velocity approaches is given in section (5).

2 The Lévy Walk Collision Process (LWCP)

Let $q\left(\tau\right)$ be the PDF of the independent time intervals between strong collision events. The survival probability

$$W(t) = 1 - \int_0^t q\left(\tau\right) d\tau, \tag{5}$$

is the probability that no collision event has taken place in the time interval $(0,t)$. Let $F\left(\mathbf{v}\right)$ be the PDF of the velocity $\mathbf{v}$ of the particle. Between collision events the particle moves freely according to the law [see Eq. (2) and [28, 30]]

$$\mathbf{r}(t) = \mathbf{r}(0) + \mathbf{v}t^\nu \tag{6}$$

with $\nu \geq 0$. When $\nu = 0$ the generalized velocity $\mathbf{v}$ is a displacement. Notice that only when $\nu = 1$, $\mathbf{v}$ has the dimensions of [length/time]. After each free motion event the particle's velocity is resampled from the PDF $F(\mathbf{v})$ and the LWCP is renewed.

We label the collision events in the interval $(0,t)$ according to $\{1, 2, ...s, ...\}$, and define $\eta_s\left(\mathbf{r},t\right) d\mathbf{r}dt$ as the probability that the s collision event takes place in the interval $(\mathbf{r}, \mathbf{r} + d\mathbf{r})$ during the time interval $(t, t + dt)$. The PDF $\eta_s\left(\mathbf{r},t\right)$ is normalized

$$\int_0^\infty dt \int d\mathbf{r} \eta_s\left(\mathbf{r},t\right) = 1, \tag{7}$$

where the spatial integration is over the whole space. The PDF $\eta_s\left(\mathbf{r},t\right)$ satisfies the recursion relation

$$\eta_s\left(\mathbf{r},t\right) =$$

$$\int d\mathbf{v} \int_0^t d\tau \eta_{s-1}\left(\mathbf{r} - \mathbf{v}\tau^\nu, t - \tau\right) F\left(\mathbf{v}\right) q\left(\tau\right) \tag{8}$$

for $s \geq 1$. The initial condition of starting from $\mathbf{r} = 0$ at $t = 0$ is incorporated by the condition

$$\eta_0\left(\mathbf{r},t\right) = \delta\left(\mathbf{r}\right) \delta\left(t\right). \tag{9}$$

The PDF $\eta_s\left(\mathbf{r},t\right)$ is related to the PDF $\rho\left(\mathbf{r},t\right)$ according to

$$\rho\left(\mathbf{r},t\right) =$$

$$\sum_{s=0}^\infty \int d\mathbf{v} \int_0^t d\tau \eta_s\left(\mathbf{r} - \mathbf{v}\tau^\nu, t - \tau\right) F\left(\mathbf{v}\right) W\left(\tau\right). \tag{10}$$

We now introduce for the Fourier-Laplace transforms the covention that the arguments of a function indicate in which space the function is defined, e.g.

$\rho(\mathbf{k}, u)$ is the Fourier- Laplace of $\rho(\mathbf{r}, t)$. Using the convolution theorem, Eq. (8) can be written as:

$$\eta_s(\mathbf{k}, u) = \overline{q}(\mathbf{k}, u) \eta_{s-1}(\mathbf{k}, u), \qquad (11)$$

for $s \geq 1$, and $\eta_0(\mathbf{k}, u) = 1$. In Eq. (11) we have used the definition

$$\overline{q}(\mathbf{k}, u) \equiv L\left[\tilde{F}(\mathbf{k}\tau^{\nu}) q(\tau)\right], \qquad (12)$$

where the operator L is the Laplace transformation, and

$$\tilde{F}(\mathbf{k}\tau^{\nu}) \equiv \int d\mathbf{v} e^{i\mathbf{k}\cdot\mathbf{v}\tau^{\nu}} F(\mathbf{v}), \qquad (13)$$

is the $\mathbf{v} \to \mathbf{k}\tau^{\nu}$ Fourier transform of $F(\mathbf{v})$. Reverting Eq. (10) to the Fourier-Laplace space we find

$$\rho(\mathbf{k}, u) = \sum_{s=0}^{\infty} \overline{W}(\mathbf{k}, u) \eta_s(\mathbf{k}, u), \qquad (14)$$

where $\overline{W}(\mathbf{k}, u)$ is defined by Eq. (12), where $q(\tau)$ is replaced by $W(\tau)$. Using Eq. (11), which implies $\eta_s(\mathbf{k}, u) = \overline{q}^s(\mathbf{k}, u)$, we find:

$$\rho(\mathbf{k}, u) = \frac{\overline{W}(\mathbf{k}, u)}{1 - \overline{q}(\mathbf{k}, u)}. \qquad (15)$$

It is easy to show that $\rho(\mathbf{k} = 0, u) = 1/u$, as it should from the normalization condition. Let as also note that Eq. (15) can be derived from a generalized master equation (in analogy to [30, 42])

$$\frac{\partial \rho(\mathbf{r}, t)}{\partial t} = \int d\mathbf{r}' \int_0^t d\tau K(\mathbf{r} - \mathbf{r}', t - \tau) \rho(\mathbf{r}', \tau') \qquad (16)$$

with the memory kernel

$$K(\mathbf{k}, u) = \frac{u\overline{W}(\mathbf{k}, u) - 1 + \overline{q}(\mathbf{k}, u)}{\overline{W}(\mathbf{k}, u)}. \qquad (17)$$

To derive Eq. (17) one should Fourier-Laplace transform Eq. (16) and compare the result with the solution of our model, Eq. (15).

3   The Role of Initial Conditions

3.1   Other Renewal Processes

The choice of the initial condition can play a significant role in determining the nature of the anomalous transport [43]. In some cases it is important to distinguish between the PDF $h(t_1)$ of the time which elapses between the start of observation $t = 0$ and the first collision event at t_1, and the PDF $q(\tau)$. In our derivation of Eq. (15) we have assumed $h(t_1) = q(t_1)$. However, one may encounter situations where $h(t_1) \neq q(t_1)$. An example is the equilibrium renewal process [43] where

$$h_{eq}(t_1) = \frac{1 - \int_0^{t_1} q(\tau)\, d\tau}{\langle \tau \rangle}, \tag{18}$$

and the mean time between collisions $\langle \tau \rangle = \int_0^\infty \tau q(\tau)\, d\tau$ is assumed to be finite.

Another issue we consider here is the initial distribution of velocities. We denote by $P_0(\mathbf{v}, 0)$ the PDF of the velocity of the particles at time $t = 0$; generally $P_0(\mathbf{v}, 0) \neq F(\mathbf{v})$. While for a normal process initial conditions usually decay exponentially with time, for some LWCPs exhibiting anomalous diffusion the initial condition decays as a power law. As for normal transport systems it is of interest to find the relaxation patterns of the velocity PDF, and relate them to the fluctuations characterized, for example, by $\langle r^2(t) \rangle$ derived in a following section.

To calculate $\rho(\mathbf{k}, u)$ we first define the survival probability

$$Z(t) = 1 - \int_0^t h(\tau)\, d\tau \tag{19}$$

which is the probability that no collision event has taken place from the start of the observation, at $t = 0$, to time t. Then, similarly to Eq. (15) we find that

$$\rho(\mathbf{k}, u) = \overline{Z}_0(\mathbf{k}, u) + \frac{\overline{h}_0(\mathbf{k}, u)\, \overline{W}(\mathbf{k}, u)}{1 - \overline{q}(\mathbf{k}, u)}. \tag{20}$$

where the functions $\overline{Z}_0(\mathbf{k}, u)$ and $\overline{h}_0(\mathbf{k}, u)$ are defined according to:

$$\overline{f}_0(\mathbf{k}, u) \equiv L\left[f(\tau)\, \tilde{P}_0(k\tau^\nu, 0) \right]. \tag{21}$$

When $P_0(\mathbf{v}, 0) = F(\mathbf{v})$ and $h(\tau) = q(\tau)$, Eq. (20) reduces to Eq. (15).

3.2   The Time Dependence of the Velocity PDF

We define $P(\mathbf{v}, t)$ to be the PDF to find the particle at time t with a velocity $\mathbf{v}$, assuming an initial condition $P_0(\mathbf{v}, 0)$. Since $Z(t)$, Eq. (19), is the probability that no collision event has taken place in $(0, t)$, and since a single collision event is needed to relax $P(\mathbf{v}, t)$ to the equilibrium PDF $F(\mathbf{v})$, we obtain

$$P(\mathbf{v}, t) = Z(t)\, P_0(\mathbf{v}, 0) + [1 - Z(t)]\, F(\mathbf{v}). \tag{22}$$

The PDF $P(\mathbf{v}, t)$ satisfies the differential equation:

$$\frac{\partial P(\mathbf{v}, t)}{\partial t} = [P(\mathbf{v}, t) - F(\mathbf{v})] \frac{\partial}{\partial t} \ln[Z(t)]. \tag{23}$$

Assuming that the process is described by a rate α, so that $h(t_1) = \alpha e^{-\alpha t_1}$ we have

$$P(\mathbf{v}, t) = \exp(-\alpha t) P_0(\mathbf{v}, 0) + [1 - \exp(-\alpha t)] F(\mathbf{v}). \tag{24}$$

For this case the process is described by

$$\frac{\partial P(\mathbf{v}, t)}{\partial t} = -\alpha [P(\mathbf{v}, t) - F(\mathbf{v})]. \tag{25}$$

This Poissonian process is identical to the strong collision model [38, 39]. Comparing Eq. (23) and Eq. (25) we realize that our strong collision model, Eq. (23) is similar to the standard strong collision model, Eq. (25), with a time dependent rate, determined by the transformation

$$\alpha \to -\frac{\partial}{\partial t} \ln[Z(t)]. \tag{26}$$

Characterizing anomalous transport processes by time dependent transport coefficients is well known but should be used with care as pointed out in [44]. Here we have been able to justify this approach for the evolution of $P(\mathbf{v}, t)$.

It is also possible to describe the process using an integro-differential equation

$$\frac{\partial P(\mathbf{v}, t)}{\partial t} = \int d\mathbf{v}' \int_0^t d\tau K(\mathbf{v} - \mathbf{v}', t - \tau) P(\mathbf{v}, \tau). \tag{27}$$

Assuming an initial condition $P(\mathbf{v}, t = 0) = \delta(\mathbf{v} - \mathbf{v}_0)$, the memory kernel in Fourier-Laplace space $(\mathbf{v}, t) \to (\mathbf{l}, u)$ is found by an approach similar to that used to derive Eq. (17),

$$K(\mathbf{l}, u) = \frac{u h(u) \left[F(\mathbf{l}) - e^{i\mathbf{l}\cdot\mathbf{v}_0} \right]}{e^{i\mathbf{l}\cdot\mathbf{v}_0} + h(u) \left[F(\mathbf{l}) - e^{i\mathbf{l}\cdot\mathbf{v}_0} \right]}. \tag{28}$$

The solution of the nonlocal integro-differential, Eq. (27), and the differential equation Eq. (23), which is local in time, is Eq. (22). Thus, we see that the two approaches can be used to describe the anomalous process.

4 The Asymptotic Behavior of $\rho\,(\mathbf{k}, u)$

We investigate the small $(\mathbf{k}, u)$ behavior of $\rho(\mathbf{k}, u)$, Eq. (15), for different $F(\mathbf{v})$ and $q(\tau)$. We consider functions which satisfy $F(\mathbf{v}) = F(v)$ and so $\tilde{F}(\mathbf{k}\tau^{\nu}) = \tilde{F}(k\tau^{\nu})$, which yields $\rho(\mathbf{k}, u) = \rho(k, u)$. Such a choice means that we have translational symmetry (i.e. $\langle \mathbf{v} \rangle = 0$) and so there is no net drift. We choose the PDFs which behave according to:

$$\tilde{F}(k\tau^{\nu}) \sim 1 - C_1 k^{\delta} \tau^{\delta\nu} \qquad 0 < \delta \leq 2 \tag{29}$$

where $(k\tau^{\nu})$ is small. For $\delta < 2$ the variance of $F(v)$ diverges. We also choose:

$$q(\tau) \sim \tau^{-(1+\gamma)}, \tag{30}$$

valid for large τ. For $0 < \gamma < 1$ the mean time between collisions diverges, while when $1 < \gamma < 2$, $\langle \tau \rangle$ is finite, all other integer moments of $q(\tau)$ diverge. The results which follow below will be compared with the coupled CTRW of Ref. [30].

The small (k, u) behavior of $\rho(k, u)$ is determined in the following way. We first expand the denominator and nominator of Eq. (15) in the small parameter k,

$$\rho(k, u) \sim \frac{1}{u}\left[\frac{1 - C_1 k^{\delta} f_1(u)}{1 + C_1 k^{\delta} f_2(u)}\right] \tag{31}$$

with

$$f_1(u) = u\frac{L\left[\tau^{\delta\nu}W(\tau)\right]}{1 - q(u)} \tag{32}$$

and

$$f_2(u) = \frac{L\left[\tau^{\delta\nu}q(\tau)\right]}{1 - q(u)}. \tag{33}$$

We then analyze the small u behavior of $f_1(u)$ and $f_2(u)$. Two important parameters $\delta^* \equiv \delta\nu$ and γ control the asymptotic behavior of these functions. Notice also that for $\delta < 2$ Eq. (31) implies that for $t > 0$, the mean squared displacement diverges, this is expected since the variance of the velocity PDF $F(v)$ diverges.

We consider now the case when $1 < \gamma < 2$. Expanding Eqs. (32) and (33) to the lowest order of approximation in u, we find

$$f_1(u) \sim \begin{cases} u^{\gamma - \delta^* - 1}B_1^{\gamma\delta^*}/C_1 & \gamma - 1 < \delta^* \\ \\ B_1^{\gamma\delta^*}/C_1 & \delta^* < \gamma - 1, \end{cases} \tag{34}$$

and

$$f_2(u) \sim \begin{cases} u^{\gamma - \delta^* - 1}B_2^{\gamma\delta^*}/C_1 & \gamma < \delta^* \\ \\ u^{-1}B_2^{\gamma\delta^*}/C_1 & \gamma > \delta^*. \end{cases} \tag{35}$$

Here $B_i^{\gamma\delta^*}$, with $i = 1, 2$, are coefficients determined from Eqs. (32) and (33) for a given $q(\tau)$. Notice that for $\gamma < \delta^*$ the power law behavior of $f_1(u)$ and $f_2(u)$ is given by the same exponent $\gamma - \delta^* + 1$. This means that for this case both functions are needed in order to obtain the correct asymptotic behavior of $\rho(k, u)$. Using Eqs. (31), (34) and (35) we find

$$\rho(k, u) \sim \frac{u^{\delta^* - \gamma} - B_1^{\gamma\delta^*} k^\delta / u}{u^{\delta^* - \gamma + 1} + B_2^{\gamma\delta^*} k^\delta} \qquad \gamma < \delta^*$$

$$\rho(k, u) \sim \frac{1 - B_1^{\gamma\delta^*} k^\delta u^{-\delta^* + \gamma - 1}}{u + B_2^{\gamma\delta^*} k^\delta} \qquad \delta^* < \gamma < \delta^* + 1$$

$$\rho(k, u) \sim \frac{1 - B_1^{\gamma\delta^*} k^\delta}{u + B_2^{\gamma\delta^*} k^\delta} \qquad \delta^* + 1 < \gamma \quad . \tag{36}$$

When $\delta^* + 1 < \gamma$ one can approximate

$$\rho(k, u) \sim \frac{1}{u + \left(B_1^{\gamma\delta^*} + B_2^{\gamma\delta^*}\right) k^\delta} \tag{37}$$

which, when inversed Laplace transformed, yields the familiar Lévy PDF

$$\rho(k, t) \sim \exp\left[-t \left(B_1^{\gamma\delta^*} + B_2^{\gamma\delta^*}\right) k^\delta\right]. \tag{38}$$

When $\delta = 2$, the mean squared displacement is finite. ¿From Eq. (31) it is easy to show that

$$\langle r^2(u) \rangle = \frac{2C_1}{u} [f_1(u) + f_2(u)]. \tag{39}$$

Using this equation, the large t behavior of the mean squared displacement is then found to be

$$\langle r^2(t) \rangle \sim \begin{cases} 2\dfrac{B_1^{\gamma 2\nu} + B_2^{\gamma 2\nu}}{\Gamma(2\nu - \gamma + 2)} t^{2\nu - \gamma + 1} & \gamma < 2\nu \\[2ex] 2B_2^{\gamma 2\nu} t & \gamma > 2\nu, \end{cases} \tag{40}$$

for $1 < \gamma < 2$. When $\gamma < 2\nu$ the diffusion is enhanced and when $2\nu < \gamma$ the diffusion is normal.

We now find the small (k, u) behavior when $0 < \gamma < 1$. Expanding Eq. (32)

$$f_1(u) \sim \begin{cases} u^{1-\gamma} B_1^{\gamma\delta^*} / C_1 & \delta^* < \gamma - 1 \\[2ex] u^{-\delta^*} B_1^{\gamma\delta^*} / C_1 & \gamma - 1 < \delta^*, \end{cases} \tag{41}$$

and from Eq. (33)

$$f_2(u) \sim \begin{cases} u^{-\delta^*} B_2^{\gamma\delta^*} / C_1 & \gamma < \delta^* \\[2ex] u^{-\gamma} B_2^{\gamma\delta^*} / C_1 & \gamma > \delta^*. \end{cases} \tag{42}$$

Using Eq. (31) we find

$$\rho(k, u) \sim \frac{u^{\delta^*-1} - B_1^{\gamma\delta^*} k^\delta / u}{u^{\delta^*} + B_2^{\gamma\delta^*} k^\delta} \qquad \gamma < \delta^*$$

$$\rho(k, u) \sim \frac{u^{\gamma-1} - B_1^{\gamma\delta^*} k^\delta u^{-\delta^*+\gamma-1}}{u^\gamma + B_2^{\gamma\delta^*} k^\delta} \qquad \delta^* < \gamma < \delta^* + 1$$

$$\rho(k, u) \sim \frac{u^{\gamma-1} - B_1^{\gamma\delta^*} k^\delta}{u^\gamma + B_2^{\gamma\delta^*} k^\delta} \qquad \delta^* + 1 < \gamma \qquad (43)$$

For $\gamma < \delta^*$ the exponents in Eq. (43) are independent of γ and for $\delta = 2$

$$\langle r^2(t) \rangle \sim \begin{cases} 2\frac{B_1^{\gamma 2\nu} + B_2^{\gamma 2\nu}}{\Gamma(2\nu+1)} t^{2\nu} & \gamma < 2\nu \\ \\ 2B_2^{\gamma 2\nu} t^\gamma & \gamma > 2\nu. \end{cases} \qquad (44)$$

We see that when $\gamma < 2\nu$ the diffusion is enhanced for $\nu > 1/2$ and slow for $\nu < 1/2$. The regime $\gamma > 2\nu$ exhibits a subdiffusive behavior.

All these different types of behaviors of the mean squared displacement are summarized in figure (4). Such a phase diagram can be derived also from the coupled CTRW, Eq. (2). As far as we know this diagram was presented first in [30], where slightly different notations were used, and later by [28]. Other, related, though more complex, phase diagrams were found by Weeks et al [45] in the context of tracer diffusion in rotating flows. They have taken into account also the possibility of sticking events in their random walk scheme.

4.1 An Example

As an example consider the case $\nu = 1$ with Gaussian velocities,

$$\tilde{F}(k\tau) = \exp\left(-k^2\tau^2 a^2/2\right). \qquad (45)$$

For $q(\tau)$ we choose

$$q(\tau) = \frac{\gamma}{(1+\tau)^{\gamma+1}}, \qquad (46)$$

which yields

$$q(u) \sim \begin{cases} 1 - \Gamma(1-\gamma)u^\gamma & 0 < \gamma < 1 \\ \\ 1 - \langle\tau\rangle u - \Gamma(1-\gamma)u^\gamma & 1 < \gamma < 2 \end{cases} \qquad (47)$$

with $\langle\tau\rangle = 1/(\gamma - 1)$. A simple calculation, using Eqs. (31)-(33), shows

$$\rho(k, u) \sim \frac{u - 0.5a^2(1-\gamma)(2-\gamma)k^2 u^{-2}}{u^2 + 0.5a^2(1-\gamma)\gamma k^2} \qquad (48)$$

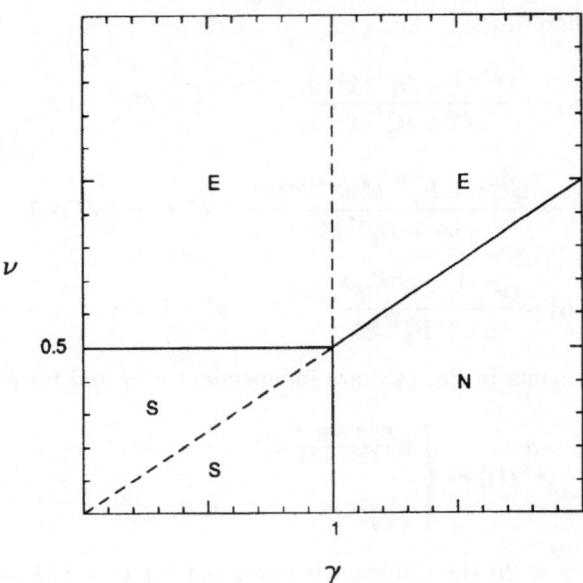

Fig. 4. The phase diagram, for $\delta = 2$, showing the different types of behaviors: (a) N for normal diffusion (b) E for enhanced diffusion and (c) S for sublinear, dispersive, diffusion. The different types of behaviors are specified in Eqs. (40) and (44).

for $0 < \gamma < 1$, and

$$\rho(k, u) \sim \frac{u^{2-\gamma} - 0.5a^2(\gamma - 1)\Gamma(3 - \gamma)k^2/u}{u^{3-\gamma} + 0.5a^2\gamma(\gamma - 1)\Gamma(2 - \gamma)\gamma k^2} \tag{49}$$

for $1 < \gamma < 2$. Using Eqs. (48)-(49), the mean squared displacement is:

$$\langle r^2(t) \rangle \sim \begin{cases} a^2(1 - \gamma)t^2 & 0 < \gamma < 1 \\ 2a^2 \frac{(\gamma-1)}{(2-\gamma)(3-\gamma)}t^{3-\gamma} & 1 < \gamma < 2. \end{cases} \tag{50}$$

The regime $0 < \gamma < 1$ is called the ballistic regime. The behavior $\langle r^2 \rangle \sim t^{3-\gamma}$ in Eq. (50), has been recently reported in a large number of systems [4]. When $\gamma > 2$ diffusion is normal and $\langle r^2 \rangle \sim t$.

5 CTRW vs the Lévy Walk Collision Process

A comparison is made between the results obtained here and those obtained using the framework of the coupled and uncoupled jump CTRW [21, 20]. First consider $\nu = 0$, which according to Eq. (6) implies that the generalized velocity $\mathbf{v}_{\nu=0}$ is in fact a displacement. Since the displacement at each step is statistically independent of the time interval between steps, it is not surprising that for $\nu = 0$ our Lévy walk collision model is similar to the decoupled version of the CTRW. To see this, note that from Eq. (12)

$$\bar{q}_{\nu=0}\left(\mathbf{k}, u\right) = \tilde{F}\left(\mathbf{k}\right) q\left(u\right) \tag{51}$$

and $\overline{W}_{\nu=0} = \tilde{F}\left(\mathbf{k}\right) W\left(u\right)$. ¿From Eq. (15) we find

$$\rho_{\nu=0}\left(\mathbf{k}, u\right) = \frac{1 - q\left(u\right)}{u} \frac{\tilde{F}\left(\mathbf{k}\right)}{1 - \tilde{F}\left(\mathbf{k}\right) q\left(u\right)}. \tag{52}$$

For the decoupled version of the CTRW where $\psi\left(\mathbf{r}, t\right) = F\left(\mathbf{r}\right) q\left(t\right)$ one finds [20, 21, 30]

$$\rho_{ct}\left(\mathbf{k}, u\right) = \frac{1 - q\left(u\right)}{u} \frac{1}{1 - \tilde{F}\left(\mathbf{k}\right) q\left(u\right)}. \tag{53}$$

Eqs. (52) and (53) clearly differ. The factor $\tilde{F}\left(\mathbf{k}\right)$ which appears in our collision model results from the free evolution in the time interval between the last collision event in the sequence and the observation time t. For the CTRW the jumping random walker, is fixed in a "deep trap" during this time interval. This seemingly small difference in the models can become important as we demonstrate.

We compare now the transformed probability density $\rho\left(\mathbf{k}, u\right)$, Eq. (15), and the one obtained within the framework of the coupled CTRW. According to Eq. (21) in Ref. [30] the CTRW result for the coupled kernel, Eq. (2), is

$$\rho_{ct}\left(\mathbf{k}, u\right) = \frac{1 - \psi\left(u\right)}{u} \frac{1}{1 - \psi\left(\mathbf{k}, u\right)} \tag{54}$$

and $\psi\left(u\right) = \psi\left(\mathbf{k} = 0, u\right)$. We rewrite Eq. (15) as

$$\rho\left(\mathbf{k}, u\right) = \frac{1 - q\left(u\right)}{u} \frac{1}{1 - \bar{q}\left(\mathbf{k}, u\right)} + \frac{\overline{W}\left(\mathbf{k}, u\right) - \overline{W}\left(0, u\right)}{1 - \bar{q}\left(\mathbf{k}, u\right)} \tag{55}$$

where

$$\overline{W}\left(0, u\right) = \frac{1 - q\left(u\right)}{u}. \tag{56}$$

The two models can be compared to each other if we identify

$$\bar{q}\left(\mathbf{k}, u\right) = \psi\left(\mathbf{k}, u\right) \tag{57}$$

and hence $\overline{q}(\mathbf{k} = 0, u) = q(u) = \psi(u)$. To quantify the differnce between the two models we define:

$$\Delta \equiv \rho(\mathbf{k}, u) - \rho_{ct}(\mathbf{k}, u) = \frac{\overline{W}(\mathbf{k}, u) - \overline{W}(0, u)}{1 - \overline{q}(\mathbf{k}, u)}. \tag{58}$$

For $\mathbf{k} = 0$ we have $\Delta = 0$, as expected from the normalization condition.

Let us examine now the difference $\Delta(k, u)$ using the small k expansion Eq. (29),

$$\Delta(k, u) \sim -C_1 \frac{k^\delta}{u} f_1(u), \tag{59}$$

where $f_1(u)$ has been defined already in Eq. (32). The CTRW result can be written in our notation using Eqs. (31), (58) and (59)

$$\rho_{ct}(k, u) \sim \frac{1}{u} \left[\frac{1}{1 + C_1 k^\delta f_2(u)} \right]. \tag{60}$$

Hence, all our results reduce to the CTRW results when we assign $f_1(u) = 0$ in Eqs. (31) and (39), which yields $B_1^{\gamma\delta^*} = 0$ in Eqs. (36) and (43). The question remains whether one can approximate $\rho_{ct}(k, u)$ by $\rho(k, u)$. Comparing Eq. (31) and Eq. (60) one reaches the conclusion that only if

$$f_1(u) << f_2(u) \tag{61}$$

such an approximation is justified. When all moments of $q(\tau)$ exist [e.g. when $q(\tau) = \alpha \exp(-\alpha\tau)$] then Eq. (61) is satisfied when u is small, meaning that $\rho(\mathbf{r}, t) \simeq \rho_{ct}(\mathbf{r}, t)$ for large t. However, comparing Eq. (34) with (35) and Eq. (41) with (42) we see that the condition in Eq. (61) is not always satisfied. This means that the contribution from $f_1(u)$ cannot be neglected and both $f_1(u)$ and $f_2(u)$ determine the long time behavior of $\rho(\mathbf{r}, t)$. For these cases one cannot approximate the CTRW PDF $\rho_{ct}(\mathbf{r}, t)$ with the LWCP PDF $\rho(\mathbf{r}, t)$ even for long times.

The difference between the two results can be easily understood when $\delta = 2$. Then $\langle r^2 \rangle$ is non diverging and

$$\langle r^2(t) \rangle - \langle r^2(t) \rangle_{ct} = L^{-1} \left[-\frac{\partial^2 \Delta(\mathbf{k}, u)}{\partial \mathbf{k}^2} |_{\mathbf{k}=0} \right], \tag{62}$$

where L^{-1} is the inverse Laplace operator. We find, using Eq. (40),

$$\langle r^2 \rangle - \langle r^2 \rangle_{ct} \sim \begin{cases} 2 \frac{B_2^{\gamma 2\nu}}{\Gamma(2\nu-\gamma+2)} t^{2\nu-\gamma+1} & \gamma < 2\nu \\ 0 & \gamma > 2\nu, \end{cases} \tag{63}$$

when $1 < \gamma < 2$, and from Eq. (44)

$$\langle r^2 \rangle - \langle r^2 \rangle_{ct} \sim \begin{cases} 2 \frac{B_2^{\gamma 2\nu}}{\Gamma(2\nu+1)} t^{2\nu} & \gamma < 2\nu \\ 0 & \gamma > 2\nu, \end{cases} \tag{64}$$

when $0 < \gamma < 1$. We see that the LWCP is more "efficient" then the CTRW process (i.e. $\langle r^2 \rangle - \langle r^2 \rangle_{ct} \geq 0$), due to a considerable difference in the prefactors of the two processes. Such a behavior, is very different from ordinary diffusion processes where random walks provide a good approximation to simple collision models.

Even though the deviations between the two processes exist we emphasize that the exponents appearing in the long time limit of the mean squared displacement for the CTRW and the LWCP are identical once the correspondence Eq. (57) is made. In other words, the non vanishing exponents appearing in Eqs. (63) and (64) are identical to those in Eqs. (40) and (44). Thus the difference between the two processes is characterized by different prefactors which are the cause for our finding $\langle r^2 \rangle - \langle r^2 \rangle_{ct} \geq 0$ for long times.

The deviations between the LWCP and the coupled jump version of the CTRW can be understood based upon the following argument. For the CTRW at the observation time t the random walker is trapped in a lattice point. The random walker has occupied this trap for a time $t - t_l^{ct}$, where t_l^{ct} is the location on the time axis at which the last jump in the sequence of jumps was executed. On the other hand, for the LWCP the particle is always moving according to the law in Eq. (6). This evolution includes the time interval $t - t_l$ and here t_l is the location on the time axis at which the last collision in the sequence has occurred. To see this better, assume that for the LWCP during the last time interval in the sequence the particle does not evolve but rather stays fixed at the location of the last collision event in the sequence occurred. Then we have to replace Eq. (10) with

$$\rho(\mathbf{r}, t) =$$

$$\sum_{s=0}^{\infty} \int_0^t d\tau \eta_s (\mathbf{r}, t - \tau) W(\tau), \tag{65}$$

leaving Eqs. (5)-(9) unchanged. Then, following the same procedure we have followed to derive Eq. (15), we derive the result:

$$\rho(\mathbf{k}, u) = \frac{W(u)}{1 - \bar{q}(\mathbf{k}, u)} \tag{66}$$

which is the CTRW result, Eq. (53), provided that the condition in Eq. (57) is satisfied. For normal systems the additional evolution in the last interval in the sequence does not contribute significantly to $\rho(\mathbf{r}, t)$ when t is large, and so the collision process and the CTRW give practically the same results. However, if the last interval in the sequence is very long, in an averaged sense, then the difference between the coupled CTRW process and the LWCP may become large.

As mentioned in the Introduction, our generalized approach maps onto the previous approach when the magnitude of the velocity is a constant $|\mathbf{v}| = 1$, and the collisions change only the directions of motion. We summarize now similarities and differences between the results obtained within these two frameworks. First, when the variance of $F(\mathbf{v})$ diverges there is no place for comparison, since for the previous theories $\langle r^2 \rangle$ is finite while our approach yields a diverging $\langle r^2 \rangle$. When the first two moments of $F(\mathbf{v})$ exist, the exponents controlling the diffusion in Eqs. (40) and (44) are identical to those obtained previously [25, 28] while the prefactors are different.

Another difference between the LWCP and the other constant velocity approaches concerns the wings of the PDF $\rho(\mathbf{r}, t)$. For the case $|\mathbf{v}| = 1$ one finds that

$$\rho(\mathbf{r}, t) = 0 \quad \text{when} \quad |\mathbf{r}| > t. \tag{67}$$

This result is obvious, since if the particle has a maximal speed there is probability zero to find it beyond $|\mathbf{v}|t$. What is found for instance in Ref. [46], is that delta peaks appear in the solution for $\rho(\mathbf{r}, t)$, at $|\mathbf{r}| = t$. Now, if we choose $F(\mathbf{v})$ to be a Gaussian, Eq. (67) is not valid since the probability of finding the particle with a velocity $\mathbf{v} < \infty$ is nonzero. The delta peaks are not expected for this $F(\mathbf{v})$. To see this we define

$$H(\mathbf{r}, t) \equiv$$

$$\int d\mathbf{v} \int_0^t d\tau \, \delta(\mathbf{r} - \mathbf{v}\tau^\nu) \, \delta(t - \tau) \, F(\mathbf{v}) \, W(\tau). \tag{68}$$

The function $H(\mathbf{r}, t)$, is the $s = 0$ term in the sum that appears on the right hand side of Eq. (10). It describes the contribution to $\rho(\mathbf{r}, t)$ from trajectories for which the particle did not encounter collisions. We consider the one-dimensional case with $\nu = 1$ and a Gaussian $F(\mathbf{v})$. Then

$$H(x, t) = \frac{W(t)}{\sqrt{2\pi}t} \exp\left[-\left(x^2/2t^2\right)\right]. \tag{69}$$

While for the two state model, $F(v) = 0.5[\delta(v + v_0) + \delta(v - v_0)]$, then

$$H(x, t) = \frac{W(t)}{t} 0.5[\delta(x/t + v_0) + \delta(x/t - v_0)]. \tag{70}$$

We see that $H(x, t)$ behaves differently for the two choices of the PDF $F(\mathbf{v})$. For the Gaussian process it is centered around $x = 0$, while for the constant velocity approach, delta peaks appear at $x = \pm|v|t$. Thus, the choice of $F(\mathbf{v})$ has an influence on the asymptotic shape of $\rho(\mathbf{r}, t)$.

6 Summary

In this work we have investigated a strong collision model which we have called the Lévy walk collision process (LWCP). The LWCP scheme can be viewed as a generalization of the CTRW for the case when the velocities of the random walkers are randomly distributed. An extension of the normal Brownian motion to anomalous diffusion with sublinear, enhanced or diverging diffusion has been given. The CTRW framework with coupled kernels also results in such diffusional patterns. However the CTRW, describing a jump process, considers positions of the random walkers rather then their velocities and therefore it is not suited to describe the situation we are interested in. Furthermore, we have shown that differences exist between the CTRW and our results. Thus, even though the exponents and the dynamical phase diagram of the two models are the same, the two approaches are nonidentical. The LWCP is found to be more "efficient" then the CTRW.

Unlike the previous velocity models, our work is not restricted to the condition that $|\mathbf{v}|$ is a constant, rather velocities are random variables described by the PDF $F(\mathbf{v})$. Our model allows to consider the case when $F(\mathbf{v})$ is long-tailed. Even for the case when the variance of $F(\mathbf{v})$ is finite the asymptotic behavior of $\rho(\mathbf{r}, t)$ behaves differently for the two approaches.

Here we have considered in some detail the case where $F(\mathbf{v})$ is symmetric with a zero mean. In general one can include the case where the mean velocity is finite and then an anomalous drift is expected. We shall discuss this drift in the velocity field $F(\mathbf{v})$ in a future publication.

The LWCP assumes strong collisions, which means that there are no correlations between the velocity of a particle just before and just after a collision event. A Gaussian one-dimensional model with either weak or strong collisions has been investigated recently by one of the authors in Refs. [47, 48].

Since our model considers the random distribution of velocities, one may follow the evolution the velocity PDF $P(\mathbf{v}, t)$ to the equilibrium $F(\mathbf{v})$. We have shown that $P(\mathbf{v}, t)$ can be derived from a differential equation with a time dependent relaxation coefficient $\alpha(t)$. This equation generalizes the strong collision model which assumes an exponential process and which has been used frequently in different fields. We believe our approach will find its applications for non-Gaussian diffusion processes.

Acknowledgement: The authors acknowledge fruitful discussions with V. Fleurov, M. Shlesinger and G. Zumofen and the support of a grant from GIF.

Appendix A

Ways to generate random variables described by different types of PDFs (e.g. the exponential, the Lorentzian and the Gaussian PDFs) can be found in Ref. [49]. Here we show how to generate random variables whose PDF follows the rule

$$q(\tau) \sim \tau^{-(1+z)} \tag{71}$$

for $\tau \to \infty$, $0 < \tau < \infty$. We have in mind cases where the exact behavior of the PDF for small τ is irrelevant. Two main methods to generate random variables are usually used [49] (a) an accept-reject method, which is not efficient and (b) a transformation method. Here we give a simple transformation rule which generates random variable described by a longed tailed PDF.

We use a random number generator which generates a random variable u which is distributed uniformly in the interval $0 < u < 1$. Then we define the transformation (for $\xi > 0$)

$$\tau = \left[\tan\left(\frac{u\pi}{2}\right)\right]^\xi \tag{72}$$

and hence

$$q\left(\tau\right) = p(u)\left|\frac{du}{d\tau}\right| = \left(\frac{2}{\pi\xi}\right)\frac{\tau^{(1-\xi)/\xi}}{(1+\tau^{2/\xi})} \tag{73}$$

and $p(u)$ is the uniform PDF. For long times we have

$$q\left(\tau\right) \sim \left(\frac{2}{\pi\xi}\right)\tau^{-(1+1/\xi)} \tag{74}$$

and hence if we identify $1/\xi = z$ our goal is accomplished.

References

1. M. F. Shlesinger J. Klafter and Y. M. Wong *J. of Stat. Phys.* **27** (1982) 499.
2. J. P. Bouchaud and A. Georges, *Physics Report* **195** (1990) 127.
3. M.F. Shlesinger, G. M. Zaslavsky and U. Frisch ed. Lévy Flights and Related Topics in Physics (Springer-Verlag Berlin 1994).
4. J. Klafter, M. F. Shlesinger and G. Zumofen, *Physics Today* **49** (1996) 33.
5. R. Balescu Statistical Dynamics Matter Out of Equilibrium (Imperial College Press London 1997).
6. B. Mandelbrot The Fractal Geometry of Nature, (Freeman, San Francisco, 1982).
7. W. R. Schneider and W. Wyss, *J. Math. Phys* **30** (1998) 134.
8. R. Metzler, W. G. Glöcke and T. F. Nonnenmacher *Physica A* **211** (1994) 13.
9. H C. Fogedby, *Phys. Rev. E* **50** (1994) 1657.
10. R. Hifler and L. Anton, *Phys. Rev. E* **51** (1995) R848.
11. A. Compte, *Phys. Rev. E* **53** (1996) 4191.
12. G. M. Zaslavsky, M. Edelman and B. A. Niyazov, *Chaos* **7** (1) (1997) 159.
13. L. F. Richardson, Proc. R. Soc. London, Ser. A **110**, (1926) 709.
14. G. K. Batchelor, Proc. Cambridge Philos. Soc. **48** (1952) 345.
15. A. Okubo, *J. Oceanol. Soc. Jpn.* **20** (1962) 286.
16. A. S. Monin and A. M. Yaglom, Statistical Fluid Mechanics, (MIT, Cambridge, MA, 1971), Vol 1; (1975) Vol. 2.
17. H. G. E. Hentschel and I. Procaccia, *Phys. Rev. A* **29** (1984) 1461.
18. R. Muralinder, D. Ramkrishna, H. Nakanishi and D. Jacobes, *Physica A* **167** (1990) 539.
19. K. G. Wang, L. K. Dong, X. F. Wu, F. W. Zhu and T. Ko, *Physica A* **203** (1994) 53.
20. E.W. Montroll and M.F. Shlesinger in: Nonequilibrium Phenomena II, From Stochastics To Hydrodynamics ed. J.L. Lebowitz and E.W. Montroll (North Holland Amsterdam 1984).
21. G. H. Weiss Aspects and Applications of the Random Walk (North Holland, Amsterdam, 1994).
22. T. Geisel, J. Nierwetberg and A. Zachrel, *Phys. Rev. Let.* **54** (1985) 616.
23. M. F. Shlesinger, B. West and J. Klafter, *Phys. Rev. Let.*, **58** (1987) 1100.
24. J. Masoliver, K. Lindenberg and G. H. Weiss, *Physica A* **157** (1989) 891.
25. G. Zumofen and J. Klafter, *Phys. Rev. E* **47** (1993) 851.
26. G. Trefán, E. Floriani, B. J. West and P. Grigolini, *Phys. Rev. E* **50** (1994) 2564.
27. S. Marksteiner, K. Ellinger, and P. Zoller, *Phys. Rev. A* **53** 5 (1996) 3409.

28. Hermann Schulz-Baldes, *Phys. Rev. Lett.* **78** (1997) 2176.
29. B. V. Gnedenko and A. N. Kolmogorov, *Limit Distributions for Sums of Independent Random Variables* (Addison - Wesley, Reading, MA, 1968).
30. J. Klafter, A. Blumen and M. F. Shlesinger, *Phys. Rev. A* **35** (1987) 3081.
31. E. Barkai and J. Klafter, *Phys. Rev. Let.* **79**, (1997) 2245.
32. T. H. Solomon, E. R. Weeks and H. L. Swinney, *Phys. Rev. Let.* **71**, (1995) 23.
33. A. E. Hansen, E. Schröder, P. Alstrom, J. S. Andersen and M. T. Levinsen, *Phys. Rev. Let* **79** 10 (1997) 1845.
34. H. Katori, S. Schlipf and H. Walther, *Phys. Rev. Let.* **79** 12 (1997) 2221.
35. O. V. Tel'kovskaya and K. V. Chukbar, *JETP* **85** 1 (1997) 87.
36. A. E. Hansen, D. Marteau and P. Tabeling, *Phys. Rev. E* (1997) submitted.
37. P. Levitz *Europhysics Letters*, **39** 6 (1997) 593.
38. R. Kubo, M. Toda and N. Hashitsume, Statistical Physics 2 (Springer-Verlag, Berlin) 1991.
39. Barkai and V. Fleurov, *Phys. Rev. E* **52** (1995) 1558.
40. D.H. Zanette and P. A. Alemany, *Phys. Rev. Let* **75** (1995) 366.
41. C. Tsallis, S. V. F. Levy, A. M. C. Souza and R. Maynard *Phys. Rev. Let.*, **75** (1995) 3589.
42. J. Klafter and R. Silbey, *Phys. Rev. Lett.* **44** (1980) 55.
43. J. W. Haus and K. W. Kehr, *Physics Report* **150** (1987) 263.
44. B. Berkowitz and H. Scher, *Water Resources Research* **31** (1995) 1461.
45. R. Weeks, J. S. Urbach and H. L. Swinney, *Physica D* **97**, (1996) 291.
46. J. Klafter and G. Zumofen *Physica A* **196** (1993) 102.
47. E. Barkai and V. Fleurov, *Chemical Physics* **212** (1996) 69.
48. E. Barkai and V. Fleurov, *Phys. Rev. E* **56** (1997) 6355.
49. W.H. Press, S.A. Teukolsky, W.T. Vetterling and B. P. Flannery *Numerical Recipes in Fortran* Cambridge University Press (New York) 1992.

On the Equilibrium Distribution of Like-Signed Vortices in Two Dimensions

Igor Mezić[1] and Inki Min[2]

[1]Department of Mechanical Engineering, University of California, Santa Barbara, CA 93106-5070, USA   and [2]The Aerospace Corporation, El Segundo, CA, USA

Abstract. We study the equilibrium statistics for a system of point vortices of different circulations in two-dimensions. We use the methods of Lundgren and Pointin [2] who have analyzed this problem in the past for vortices of the same strength. This necessitates the development of the probability density functions for *each* of the vortices. We find a power-law relationship between the probability distributions of vortices with different circulations. These distributions are verified numerically.

1 Introduction

Recent visualizations of turbulent flows (experimental and numerical) have shown the presence of distinct vortex elements as being the key driver of the flow [3], [1]. Further, Saffman [10] proposes " that turbulence should be modelled or described as the creation, evolution, interaction and decay of these [discrete vortical] structures". Many studies of two-dimensional turbulent flows have shown that concentrated vortices are an important feature of such flows. In this context, it is of obvious interest to find the statistical properties of the collection of discrete vortex elements, singular or with a core, and compare the results obtained to numerical simulations and experiments. The concentrated vortices that occur in the previously mentioned studies are of different circulations. In this paper we study the statistical mechanics of a system of singular vortices in which each vortex can have a different positive circulation. Previous studies in this context [2], [6] treated vortices of the same circulation (while possibly of different signs). In this case vortices can have different position probability distribution, depending on their circulation, as opposed to the case with the same circulation. We uncover a power law relationship between probability densities for vortices of different circulations. Our theory is supported by numerical calculations.

2 Derivation

Consider a system of N point vortices in a plane. The circulation of vortex i is $\Gamma_i > 0$. The isolating integrals (constants of motion) are

$$\mathcal{H} = -\frac{1}{2\pi} \sum_{i<j}^{N} \Gamma_i \Gamma_j \ln r_{ij}/l_0 \ , \quad r_{ij} = |\mathbf{r}_i - \mathbf{r}_j| \tag{1}$$

$$\mathbf{R} = \sum_{i=1}^{N} \frac{\Gamma_i}{\bar{\Gamma}} \mathbf{r}_i \qquad (2)$$

$$L^2 = \sum_{i=1}^{N} \frac{\Gamma_i}{\bar{\Gamma}} (\mathbf{r}_i - \mathbf{R})^2 \ , \qquad (3)$$

where $\mathcal{H}$ is the energy, $\mathbf{R}$ the center of vorticity, and L^2 a constant associated with the angular momentum. We let

$$\bar{\Gamma} = \frac{\sum_{i=1}^{N} \Gamma_i}{N} \ .$$

We denote by $P_N(\mathbf{r}_1, \mathbf{r}_2, \dots, \mathbf{r}_N)$ the probability density of the system of vortices such that $P_N d\mathbf{r}_1 d\mathbf{r}_2 \cdots d\mathbf{r}_N$ is the probability that $\mathbf{r}_1$ is in $d\mathbf{r}_1$, $\mathbf{r}_2$ is in $d\mathbf{r}_2$ etc. In equilibrium, P_N is a function of the integrals of motion $(\mathcal{H}, \mathbf{R}, L^2)$. We use the microcanonical ensemble

$$P_N = \delta(\rho\mathcal{H} - E)\delta\left(\sum_{i=1}^{N} \frac{\Gamma_i}{\bar{\Gamma}}(\mathbf{r}_i - \mathbf{R})^2 - NL^2\right) \delta\left(\sum_{i=1}^{N} \frac{\Gamma_i}{\bar{\Gamma}}\mathbf{r}_i - N\mathbf{R}\right) /Q(E, L^2) \qquad (4)$$

which is based on the ergodic hypothesis. The normalizing factor Q must satisfy

$$Q(E, L^2) = \int \delta(\rho\mathcal{H} - E)\delta\left(\sum_{i=1}^{N} \frac{\Gamma_i}{\bar{\Gamma}}(\mathbf{r}_i - \mathbf{R})^2 - NL^2\right)$$
$$\delta\left(\sum_{i=1}^{N} \frac{\Gamma_i}{\bar{\Gamma}}\mathbf{r}_i - N\mathbf{R}\right) d\mathbf{r}_1 \cdots d\mathbf{r}_N. \qquad (5)$$

In the case of vortices with different circulations, it is not necessarily true that the reduced probability densities

$$P_{1_i}(\mathbf{r}_i) = \int P_N \, d\mathbf{r}_1 \cdots d\mathbf{r}_k \cdots d\mathbf{r}_N \ , \ k \neq i \qquad (6)$$

are independent of i. Similarly,

$$P_{2_{ij}}(\mathbf{r}_i, \mathbf{r}_j) = \int P_N \, d\mathbf{r}_1 \cdots d\mathbf{r}_k \cdots d\mathbf{r}_N \ , \ k \neq i \ , k \neq j \ . \qquad (7)$$

Using Eq. (6),

$$P_{1_i} = Q(E, L^2)^{-1} \int \delta(\rho \mathcal{H} - E) \delta \left(\sum_{j=1}^{N} \frac{\Gamma_j}{\bar{\Gamma}} (\mathbf{r}_j - \mathbf{R})^2 - NL^2 \right)$$

$$\delta \left(\sum_{j=1}^{N} \frac{\Gamma_j}{\bar{\Gamma}} \mathbf{r}_j - N\mathbf{R} \right) d\mathbf{r}_1 \cdots d\mathbf{r}_k \cdots d\mathbf{r}_N \quad k \neq i .$$

$$(8)$$

Differentiating with respect to $\mathbf{r}_i$,

$$\frac{\partial P_{1_i}(\mathbf{r}_i)}{\partial \mathbf{r}_i} = \Gamma_i \left[\frac{\rho}{2\pi} \sum_{\substack{j=1 \\ j \neq i}}^{N} \Gamma_j \int (\frac{\partial}{\partial \mathbf{r}_i} \ln r_{ij}) \frac{1}{Q} \frac{\partial}{\partial E} Q P_{2_{ij}}(\mathbf{r}_i, \mathbf{r}_j) d\mathbf{r}_j \right.$$

$$\left. -2(\mathbf{r}_i - \mathbf{R}) \frac{1}{N\bar{\Gamma}} \frac{1}{Q} \frac{\partial}{\partial L^2} Q P_{1_i}(\mathbf{r}_i) - \frac{1}{N\bar{\Gamma}} \frac{1}{Q} \frac{\partial}{\partial \mathbf{R}} Q P_{1_i}(\mathbf{r}_i) \right] . \quad (9)$$

Using

$$\lambda = \frac{\rho}{4\pi NkT} \sum_{i<j} \Gamma_i \Gamma_j , \quad (10)$$

the closure assumption

$$P_{2_{ij}}(\mathbf{r}_i, \mathbf{r}_j) = P_{1_i}(\mathbf{r}_i) P_{1_j}(\mathbf{r}_j) , \quad (11)$$

and thermodynamic relationships similar to those in [2], we obtain

$$\frac{\partial P_{1_i}(\mathbf{r}_i)}{\partial \mathbf{r}_i} = \Gamma_i \sum_{\substack{j=1 \\ j \neq i}}^{N} \frac{2\lambda \Gamma_j}{\frac{1}{N} \sum_{k<l} \Gamma_k \Gamma_l} \left\{ \int (\frac{\partial}{\partial \mathbf{r}_i} \ln r_{ij}) P_{1_i}(\mathbf{r}_i) P_{1_j}(\mathbf{r}_j) d\mathbf{r}_j \right\}$$

$$-2 \frac{1}{\bar{\Gamma}} \frac{(\mathbf{r}_i - \mathbf{R})}{L^2} (1 + \lambda) P_{1_i}(\mathbf{r}_i) . \quad (12)$$

Using the change of variables

$$p_{1_i}(\eta_i) = L^2 P_{1_i}(\mathbf{r}_i) , \quad \eta_i = (\mathbf{r}_i - \mathbf{R})/L , \quad (13)$$

we have

$$\frac{\partial}{\partial \eta_i} \ln p_{1_i} = \frac{\Gamma_i}{\bar{\Gamma}} \left\{ \sum_{\substack{j=1 \\ j \neq i}}^{N} \left(\frac{2\lambda N\bar{\Gamma}\Gamma_j}{\sum_{k<l} \Gamma_k \Gamma_l} \int (\frac{\partial}{\partial \eta_i} \ln \eta_{ij}) p_{1_j}(\eta_j) d\eta_j \right) - 2(1 + \lambda)\eta_i \right\} .$$

$$(14)$$

Using the fact that each of the probability distributions is rotationaly symmetric under the assumption of ergodicity [5] we can consider only the magnitude of η_i

$$p_{1_i}(\eta_i) = C_i \exp 2\frac{\Gamma_i}{\bar{\Gamma}} \left\{ \sum_{\substack{j=1 \\ j \neq i}}^{N} \frac{2\lambda N \bar{\Gamma} \Gamma_j}{\sum_{k<l} \Gamma_k \Gamma_l} \int \ln \eta_{ij} p_{1_j}(\eta_j) d\eta_j - (1+\lambda)\eta_i^2 \right\} \quad (15)$$

where C_i is obtained from $\int_{\eta_i} p_{1_i}(\eta_i) d\eta_i = 1$. Clearly then, for N large enough (i.e. when the fact that one of the terms in the sum for each vortex is different is irrelevant)

$$\left(\frac{p_{1_i}(\eta)}{C_i}\right)^{1/\Gamma_i} = \left(\frac{p_{1_j}(\eta)}{C_j}\right)^{1/\Gamma_j} . \quad (16)$$

i.e

$$p_{1_i}(\eta) = \left(\frac{p_{1_j}(\eta)}{C_j C_i^{1/\Gamma_i}}\right)^{\Gamma_i/\Gamma_j} = C p_{1_j}(\eta)^{\Gamma_i/\Gamma_j} . \quad (17)$$

This is the main result of this paper. The last equation implies that larger vortices will on the average spend more time closer to the center of vorticity than smaller vortices.

3 Numerical Simulations

Numerical simulations of N-vortex dynamics were performed to complement the derivations of the previous section. The methodology is similar to that employed in previous studies [8], [9], except as required to accomodate the multi-Γ_i configurations. We briefly summarize the relevant details. $L = 1, R = 0, H$ varied.

In figure 1, a numerical check of Eq. (16) is shown for the case of $\Gamma_1 = -.3, N_1 = 75, \Gamma_2 = -1.7, N_2 = 75$, and $\lambda = 0$. A very good match is seen. In figure 2 we see that for the case of same configuration as figure 1, there is very good match to the predicted PDF of

$$p_{1_i}(r) = \frac{\Gamma_i}{\pi \bar{\Gamma}} \exp\left(-\frac{\Gamma_i}{\bar{\Gamma}} r^2\right) .$$

Figure 3 shows a snapshot of the distribution of the vortices in physical space.

399

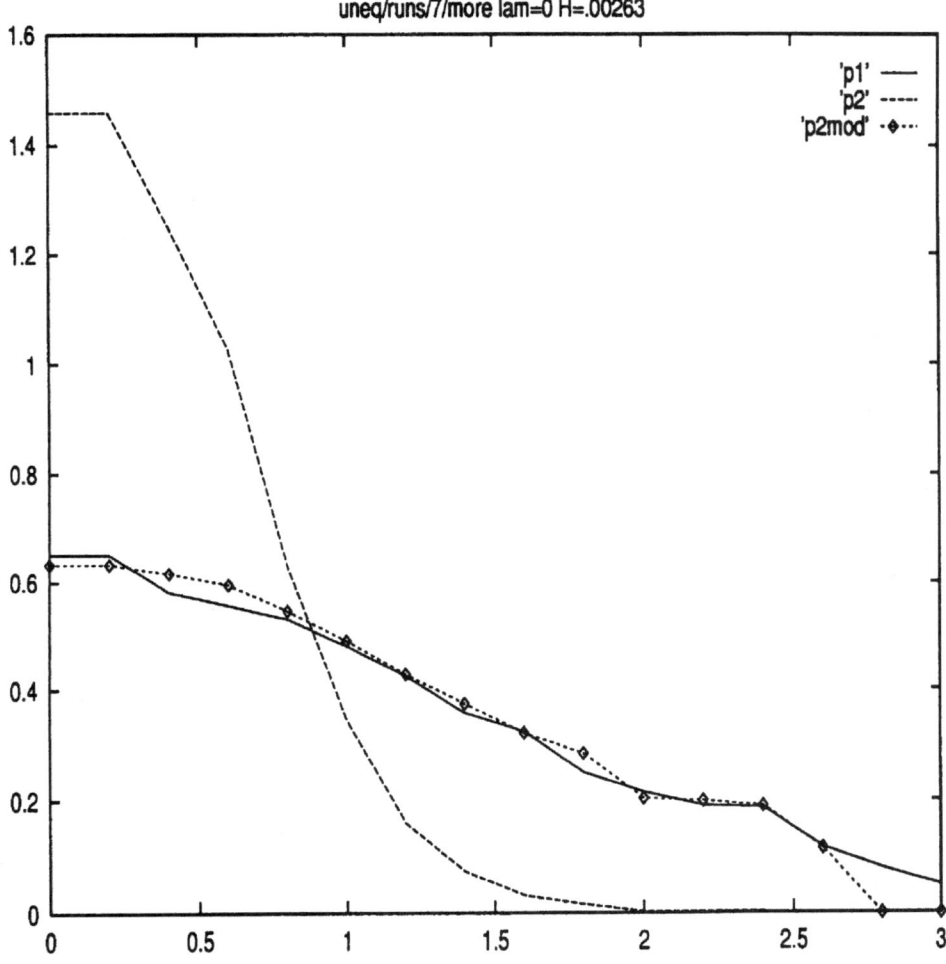

Fig. 1. comparison of PDFs p_{1_1} and $p1_{1_2}^{\frac{r_2}{r_1}}$ for $\lambda = 0$

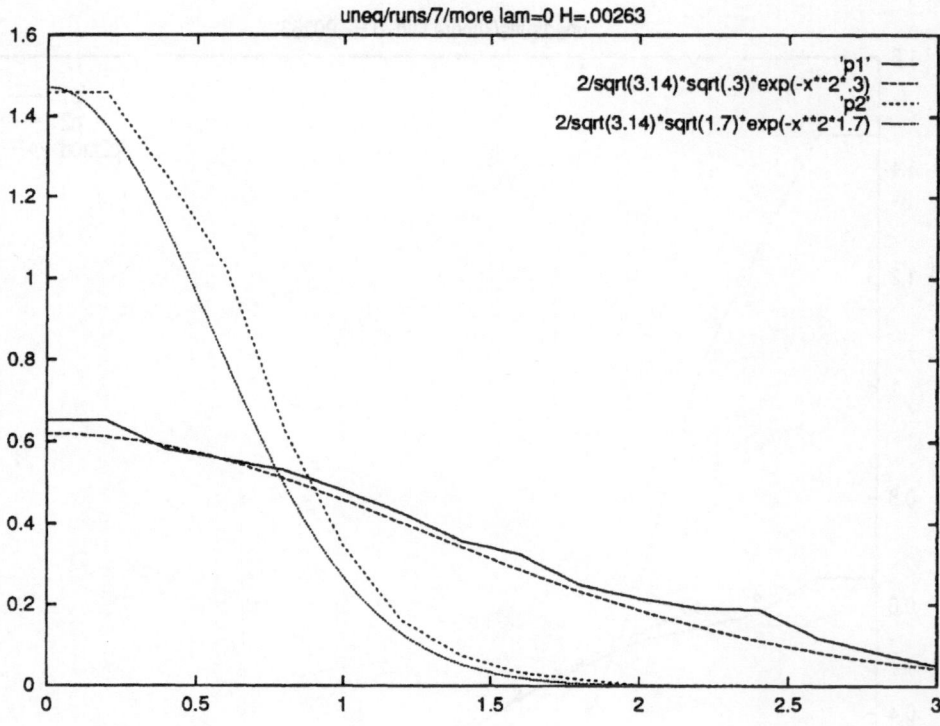

Fig. 2. theoretical versus numerical PDFs p_{1_1} and p_{1_2} for $\lambda = 0$

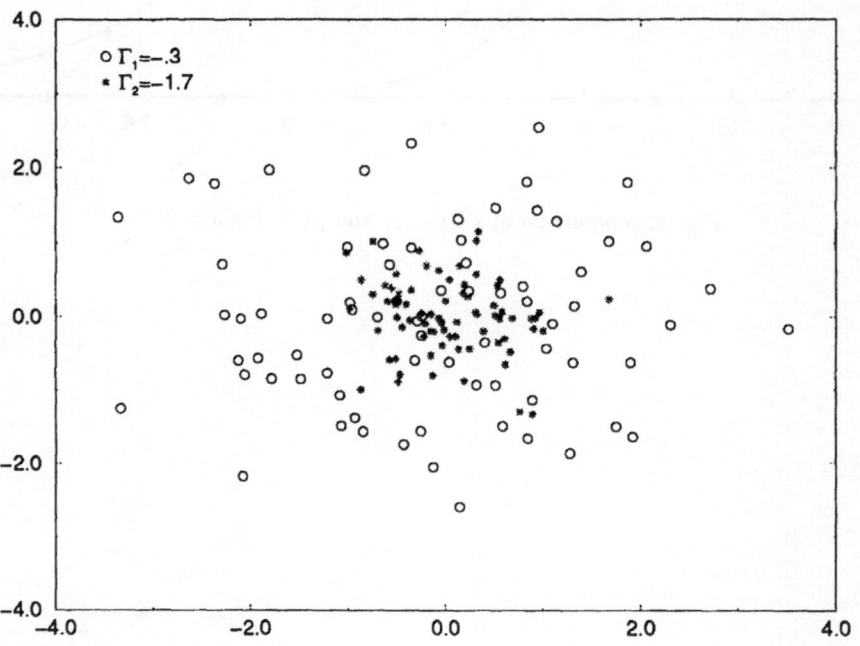

4 Conclusions

In this paper, we have derived statistical mechanics for a system of vortices with positive but possibly different circulations. We have uncovered a power-law relationship between probability density for different vortices. This relationship implies that larger vortices will on the average spend more time near the center of vorticity than smaller vortices.

A number of interesting questions can be pursued as an extension to this work. For example, what are the probability density functions for velocity and velocity derivatives induced by the considered system of point vortices? We have answered this question for same-circulation vortices in our previous work [9].

Acknowledgements

This work was supported by the ONR grant N00014-98-1-0056. We are grateful to Professors T. Lundgren and A. Leonard for useful comments.

References

[1] W. J. A. Dahm, K. B. Southerland, K. A. Buch , (1991): Phys. Fluids A3(5), 1115.

[2] T. S. Lundgren and Y. B. Pointin, "Statistical mechanics of two-dimensional vortices" , (1977): J. Stat. Physics, 17(5) 323.

[3] A. Vincent and M. Meneguzzi, "The spatial structure and statistical properties of homogeneous turbulence" (1991): J. Fluid Mech. 225, 1.

[4] J. B. Weiss and J. C. McWilliams, "Nonergodicity of point vortices" (1991): Phys Fluids A 3, 835.

[5] I. Mezić, "On the statistical properties of motion of point vortices" (1995): UCSB preprint.

[6] D. Montgomery and G. Joyce, "Statistical mechanics of 'negative temperature' states" , (1993): Phys. Fluids 17, 1139, (1974).

[7] P. D. Koumoutsakos, Ph.D. thesis, California Institute of Technology.

[8] I. A. Min, "Transport, stirring and mixing in two-dimensional vortex flows" (1994): Ph.D. Thesis, California Institute of Technology.

[9] I. A. Min, I. Mezić, A. Leonard, "Lévy stable distributions for velocity and velocity difference in systems of vortex elements" , (1996): Phys. Fluids, 8(5), 1169-1180.

[10] P. G. Saffman, "Vortex Interaction and Coherent Structures in Turbulence" (1981): *Transition and turbulence* Proceedings of a symposium conducted by the Mathematics Research Center, the University of Wisconsin –Madison, October 13-15, Richard E. Meyer, ed..

Nonuniversality of Transport for the Standard Map

S. Kassibrakis[1], S. Benkadda[1], R. B. White[2] and G. M. Zaslavsky[3,4]

[1] UMR 6633, CNRS-Université de Provence. Centre de St. Jérôme, Case 321. F-13397 Marseille Cedex 20, France
[2] Plasma Physics Laboratory, Princeton University, P.O.Box 451,Princeton, New Jersey 08543
[3] Courant Institute of Mathematical Sciences, New York University, 251 Mercer Street, New York, New York 10012, and
[4] Physics Department, New York University, 4 Washington Place, New York, New York 10003

Abstract. Chaotic transport is investigated for the standard map for different values of stochasticity parameter K. It is found that the form of the transport coefficient is a strong function of K, and that this variation is associated with the formation and disappearance of complex multi-island structures. For some values of the parameter K superdiffusion is associated with the existence of exact chains of self-similar islands attached to accelerator islands, giving rise to long time stickiness of orbits. Close to threshold other types of multi-island structures cause stickiness. In both cases the phase space of these traps, and the exponents of the characteristic long time tails associated with them are determined. Computational procedures for the anomalous exponents and intermediate asymptotics are discussed in many details.

PACS numbers: 05.45.+b, 47.52.+j, 05.60.+w, 47.53.+n

1 Introduction

Non diffusive, or anomalous transport is a well established phenomenon. It occurs in many different physical systems (see for example Weeks *et al.* 1996, Venkataramani *et al.* 1997, Benkadda *et al.* 1997). Transport can be characterised by the time evolution of the mean-squared-displacement $< r^2(t) >$. In the case of diffusive processes, the mean-squared-displacement increases linearly with time $< r^2(t) > \sim t$, whereas in nondiffusive processes (anomalous), its time dependence is $< r^2(t) > \sim t^\mu$ with $\mu > 1$ (superdiffusive) or $\mu < 1$ (subdiffusive). A good illustration of such phenomena is shown in the experimental study of a quasi two dimensional flow in a rotating vessel (Solomon *et al.* 1993), where the azimuthal displacement of tracers exhibits a superdiffusive behaviour. Numerical evidence of anomalous transport for Hamiltonian systems with few degrees of freedom has also been found (Zaslavsky *et al.* 1993, Benkadda *et al.* 1997, Afanas'ev *et al.* 1991).

One of the fundamental problems arising in understanding chaotic transport such as mixing and diffusion by chaotic advection is that how the Hamilton's equations lead to a specific normal or anomalous kinetics. The phase space topology of Hamiltonian systems is fairly rich and complex. It consists in different sets of islands which can appear, disappear, bifurcate etc. depending on the value of a control parameter. In order to investigate the transport properties of Hamiltonian systems, it is crucial to understand the geometrical structures of phase space which certainly rule the mixing of wandering orbits.

Recently, considerable effort has been made in the attempt to understand these processes, named 'strange kinetics' in (Shlesinger *et al.* 1993), but a global theory is still missing. Strategies used involve new notions, such as Lévy flights, fractional brownian motion and recently fractional kinetics (Zaslavsky *et al.* 1993, Zaslavsky 1994, Zaslavsky *et al.* 1997).

In this paper, we study the transport properties and their links to the phase space topology in a two-dimensional area preserving map introduced by Chirikov and Taylor (Chirikov 1979) and called the standard map. In spite of its simple form, this map captures much of the complexity and canonical behavior of more complicated systems. In particular it models many problems of plasma physics as well as fluid systems exhibiting Rayleigh-Bénard convection with large aspect ratio (Solomon and Gollub 1988). The standard map has the form

$$p_{n+1} = p_n + K sin(x_n) \tag{1}$$
$$x_{n+1} = x_n + p_{n+1} \tag{2}$$

For any orbit initiated in the stochastic sea, and for $K > K_c$ ($K_c = 0.971635406$ (Chirikov 1979)) the phase space motion is unbounded in p, making statistical study for these trajectories possible, and in particular the study of time dependence of the mean-squared-displacement $< p^2 >$. It was found that for almost all K values ($K > K_c$), the process describing the statistics of the action p is diffusive, providing a diffusion coefficient D with $D = \lim_{t\to\infty} \frac{<P^2>}{t}$. The theory giving D as a function of K has been derived in (Rechester *et al.* 1981) and confirmed by numerical experiments. Nevertheless, for particular values of K, D is not well defined ($D \to \infty$) indicating that for these values $< p^2 >$ grows faster than t.

An important feature of chaotic orbits in area preserving maps is the existence of long time correlations of orbits, with an inverse power law distribution $P(t) \sim t^{-\gamma}$ of the sticking times. The chaotic orbits spend long times in hierarchical structures due to the stickiness of island chains (Karney 1983, Ishizaki *et al.* 1990, Klafter and Zumofen 1994, Zaslavsky 1994, Benkadda *et al.* 1997). Many descriptions of these phenomena have been proposed (Karney 1983, Beloshapkin and Zaslavsky 1983, Chirikov and Shepeliansky 1984, Ishizaki *et al.* 1990, Zumofen and Klafter 1994, Klafter *et al.* 1995, Zaslavsky 1994). Recently, the idea that 'strange kinetics' could be linked to the topological structure of the islands has been suggested (Zaslavsky *et al.* 1997), offering a possibility of a predictive theory.

Recently we examined the case of anomalous transport in the standard map due to accelerator modes (Benkadda *et al.* 1997). Let us note that the existence of accelerator islands is a necessary condition for anomalous transport in p, but it is not sufficient; not all values of K for which the accelerator islands exist produce anomalous transport. For the particular case of self-similar islands around islands, a theory based on the fractional generalization of the Fokker-Planck-Kolmogorov equation and on the renormalization group transport has been derived, giving $\mu = \frac{ln(\lambda_s)}{ln(\lambda_t)}$ where λ_s is the scaling parameter for the islands area and λ_t is the scaling parameter for a characteristic time (Zaslavsky 1994). In this particular case, long correlations are due to orbits wandering inside self-similar structures in time and space, providing stickiness around the accelerator mode.

The more general case for which the islands around islands structure is not exactly self similar would need generalization of the theory. The goal is to characterize the nature of the island structures leading to trapping, and to relate the time and space scales of the structures to the anomalous diffusion.

To observe long time trapping, it is simplest to look for the associated flight. We are interested in finding values of K which produce a significant modification of transport, so we scan values of K and measure the averaged mean squared displacement for an ensemble of initial conditions. Near an accelerator island, trapping produces flights in p. In Fig. 1 is shown the result of such a scan, showing the value of the transport coefficient μ as a function of the map parameter K in the vicinity of the accelerator mode. The exact value of μ depends somewhat on the number of orbits used and the time of the simulation, but the gross features, i.e. domains of K in which μ is large, persist.

Near threshold for $K \geq K_c$ trapping near an island structure leads to long flights along x with $-\pi < p < \pi$. In this case a systematic search for the primary trapping structures reveals a well identified topological structure responsible for long correlations (White *et al.* submitted) consisting of a multi-layer island structure surrounding a mother island. In this structure self similarity holds for the time scales of nested islands but not for the space scale.

In order to improve our insight concerning the underlying processes leading to 'strange kinetics', it is also useful to study different time distributions such as Poincaré recurrence time, sticking time and escape time distributions.

In this paper we will review transport properties and their link to the topological structure of phase space for different values of stochasticity parameter K with a particular emphasis on the computational methods used and their accuracy. Section two deals with these numerical procedures used to characterize chaotic transport. These tools are applied to analyse transport near accelerator modes with self-similar chains of islands. In section three we discuss other topological structures responsible for the anomalous transport in the standard map for various values of K. In section four we present results that show a link between the transport and the probability $P(x,t)dx$ for a particle to be located between x and $x + dx$ at time t . Finally the last section is devoted to discussion and conclusion.

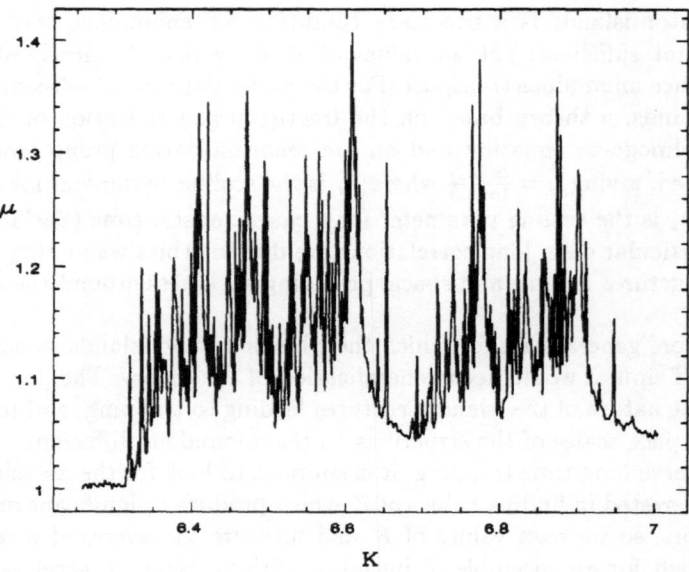

Fig. 1. Exponent μ for anomalous transport in p versus K near accelerator mode.

2   Self-similarity and chaotic transport

The usual way to study transport is to analyze the temporal behaviour of different moments. However recently other transport quantities associated with characteristic times of the system have been successfully used for characterizing chaotic transport (Easton W. *et al.* 1993, Meiss 1997, Benkadda *et al.* 1994, Zaslavsky and Tippet 1991, Zaslavsky and Niyazov 1997, Zaslavsky *et al.* 1997). In what follows, we will discuss different numerical procedures for computing these transport quantities and we will comment on their accuracy.

2.1   Extraction and characterization of self similar chains

Due to periodicity of the standard map in x and p, the phase space can be studied on the torus $[0, 2\pi] \times [0, 2\pi]$. In this representation, the phase space is made of islands embedded in the stochastic sea as shown in Fig. 2. Because of the Poincaré-Birkhoff theorem, in the neighborhood of a generic periodic orbit there are satellite elliptic orbits of smaller sizes, each of which in turn has satellite elliptic orbits, and so forth. These elliptic orbits are ordered forming chains of islands around islands. For particular values of K the island chains can assume a self similar character. In order to make explicit this self-similarity of islands we found values of the perturbation parameter K for which there is a self-similar hierarchy of subislands in the boundary layer.

The existence of this fine structure and its properties are fundamental in the study of anomalous transport. Indeed these islands and their boundaries alter the pattern of diffusion dynamics.

The number of islands surrounding an island is a function of K, and it has been conjectured (Zaslavsky 1994) that there exist values of K producing self similar structure. The search for K values is made by dichotomy. By changing the K value one constrains successively each generation to contain the desired number of islands. The required precision of K increases with the generation order.

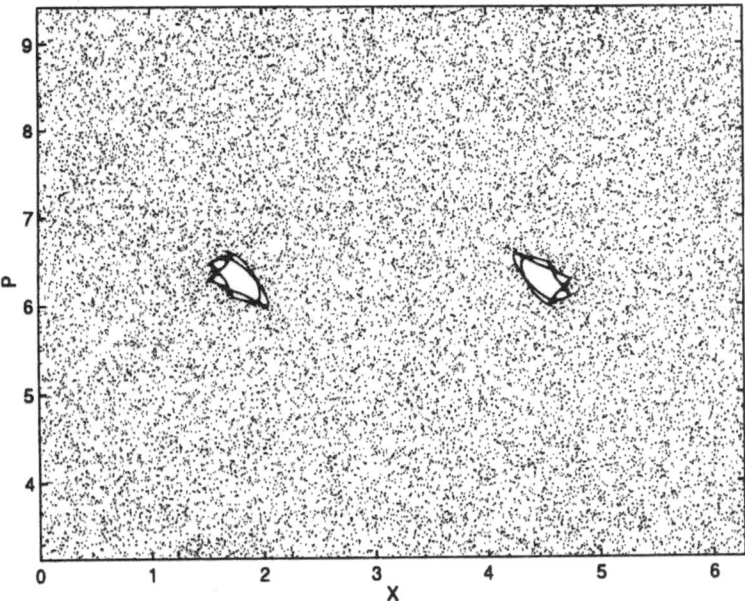

Fig. 2. Poincaré section for $K = 6.476939$. Two unconnected islands are embedded in the stochastic sea. Each island is urrounded by islands chain and so forth.

Self similar islands chains

In Fig. 3 are shown four generations of island chains related to the period 5 accelerator mode for $K_c = 6.476939$, where a significant peak is observed in Fig. 1. The generation 0 island is the 5 accelerator mode top right and subsequent generations are seen by proceeding counterclockwise. To obtain this example we searched in the vicinity of the critical perturbation parameter value K_c related to this accelerator mode, until we found 3 generations of island chains of 11 islands each for $K_c = 6.476939$. Let us emphasize that we typically need six digits precision to produce this 5-11-11-11 structure.

For the visualization of each Poincaré section magnification, it is convenient
to initiate a large number of initial conditions near one of the hyperbolic fixed
points of the studied chain in order to get enough detail. The island chain structure naturally appears as being the dark part of Poincaré plots because of the
stickiness of trajectories in its vicinity .

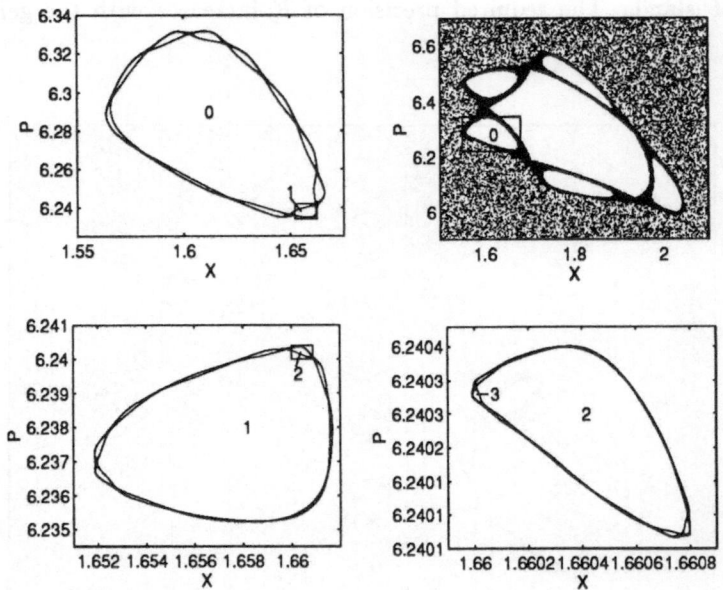

Fig. 3. Four generations of self-similar islands related to the accelerator mode,
$K = 6.476939$.

For generation 3 of the given example, one needs to perform a magnification
of the phase space area contained in a square of typical size $\Delta x.\Delta p$ with $\Delta x \sim
\Delta p \sim 10^{-4}$. At these scales each plot is extremely sensitive to the value of K
and reveals very complicated structures of 'parasite' islands.

At each step one has to concentrate on the first island chain occuring outside the island of the previous generation. Because of the extreme sensitivity
of Poincaré sections to the K value at small scales, it can happen that during
the dichotomy procedure, an island chain crosses the stability border and enters
the island of the previous generation. To guard against this the number of iterations used to build a plot must be large enough to allow trajectories to leave
the vicinity of the islands chain and to reach the stochastic sea. Of course after
each step, one has to check that the number of islands of previous generations
has not been modified.

The range of time scales needed to reach the third generation of this self-
similar islands hierarchy is already of the order of 10^6. As the statistics for

transport exponents involve the same range of steps (as it will be shown in the next section devoted to transport) there is no need to explore the hierarchy further than the third generation.

Other examples of self-similar island chain hierarchies have been reported in a previous work (Zaslavsky *et al.* 1997) and will be presented in this section.

We are now interested in the quantitative characterization of the topological properties of these islands around islands structures. Besides the number of islands for each generation, two scaling parameters can characterize this self-similarity. One is related to the area of an island and the other one is related to an appropriate time scale attached to the island (Zaslavsky 1994, Zaslavsky *et al.* 1997, Zaslavsky and Niyazov 1997).

Renormalization of areas

In order to investigate the spatio-temporal properties related to self-similarity we need to determine accurately the phase space domains embedding the accelerator islands and susceptible to contain self-similar chains of islands. Long time trapping is only caused by such domains, and for numerical reasons the domain of initial conditions used should correspond to them. To make rapid numerical calculations of trapping time or exit time for a given set, we divide that part of phase space of interest into $N \times N$ domains with a rectangular grid with boundaries $x_1, x_1 + \Delta x$ and $p_1, p_1 + \Delta p$. The grid provides an immediate coarse graining of phase space x, p into i, j with

$$j = (x - x_1) * N/\Delta x + 1 \tag{3}$$
$$i = (p - p_1) * N/\Delta p + 1 \tag{4}$$

Each cell represents an elementary domain of the phase space of area $\Delta x \times \Delta p$. The procedure consists in initiating initial conditions (10^3) in the stochastic sea, applying the map a large number of times (typically 10^5), allocating the value 1 to all nonvisited cells and 0 to the visited ones. To compute the area of an island it is sufficient to count the number of cells composing it. The typical grid used was $N = 1024$ which provides enough accuracy, but errors can arise in the matrix construction. A more convenient way to build the matrix is to initiate x, P near one of the hyperbolic points belonging to the next generation of islands chains. Then one has only to check whether the computed area is not over-estimated by comparing the border of the island identified in the matrix and a stable orbit as near as possible to the separatrix.

The scaling parameter λ_s related to the area of islands, characterizing one toplogical property of the structure, is obtained by dividing the total area occupied by islands of generation $(k - 1)$ by the total area occupied by islands of generation k.

$$\Delta S_{k-1} = \lambda_s \Delta S_k, \quad \lambda_s > 1 \tag{5}$$

Another way to compute the area scaling parameter is to use the area of polygons built with elliptic fixed points of an island chain. This procedure is more convenient because elliptic fixed points can be very accurately located using the standard Newton-Raphson method and because of its systematic character. Values reported in Table I (see below), show that the parameter λ_s is independent of generation if each island chain contains the same number of islands.

Renormalization of characteristic times

Let us consider now the determination of the scaling parameter λ_t related to characteristic time of an island chain. Unlike the determination of λ_s, it is possible to use many a priori unrelated definitions for the characteristic time. For the theory to be consistent, λ_t must be of course independent of the definition used. This has been conjectured in (Zaslavsky 1994) and already checked for the web map (Zaslavsky et al. 1997, Zaslavsky and Niyazov 1997). The independence of λ_t with respect to the definition of the characteristic time is the first step to test the theory.

Winding number

The first characteristic time used to display the temporal self-similarity of the islands, is the winding number. It is the rotation frequency of a stable orbit $\omega = 1/\Delta t$ with Δt as the time interval required to completely circle the island O point. The winding number ω is thus a function of the location inside an island. It can be computed for the O point itself, becoming a well defined characteristic time attached to an island chain. Using again a Newton-Raphson method, one can locate with high accuracy the elliptic fixed point of an island of generation k and also calculate the tangent mapping M. The winding number is then given by (Green 1979):

$$\omega_k(O) = \frac{2\pi \times \arccos(\frac{1}{2}.tr(\prod_{i=1}^{q} M_i))}{q} \tag{6}$$

where q is the periodicity of the fixed point. As for λ_s, the scaling parameter λ_t, is defined as :

$$\omega_{k-1} = \lambda_t \omega_k, \quad \lambda_t > 1 \tag{7}$$

It is also independent of the generation if each chain contains the same number of islands. This characteristic time is related to properties of stable trajectories, and is thus a priori uncorrelated with properties of unstable trajectories used in the study of transport.

TABLE 1

k	n_k	$\omega_k(0)$	$\lambda_T(\omega)$	σ_k	$\lambda_T(\sigma)$	ΔS_k	λ_S
0	5	2.3433×10^{-2}	–	1.189×10^{-1}	– –	2.7803×10^{-2}	–
1	5×11	2.4257×10^{-3}	9.791	1.429×10^{-2}	8.318	1.8938×10^{-3}	14.646
2	$5 \times 11 \times 11$	2.2632×10^{-4}	10.58	1.316×10^{-3}	10.86	6.5110×10^{-5}	29.156
3	$5 \times 11 \times 11 \times 11$	2.1877×10^{-5}	10.36	1.219×10^{-4}	10.80	2.2131×10^{-6}	29.421

Lyapunov exponent

We also used the largest Lyapunov exponent σ as a characteristic time for the island. It gives the mean exponential rate of divergence of two initially arbitrarily close trajectories. The Lyapunov exponent is associated with a given trajectory and written as :

$$\sigma = \lim_{t \to \infty} \lim_{d(0) \to 0} \left(\frac{1}{t}\right) \ln \frac{d(x_0, t)}{d(x_0, 0)} \tag{8}$$

By definition a random initial condition chosen in the vicinity of a given island chain will reach the stochastic sea after a certain time. As a consequence, the general definition can not be used as a characteristic time attached to a given generation. To avoid this difficulty one can use two different approaches. The first one consists in using trajectories initiated from hyperbolic fixed points. Their localization can be accurate enough to get a reliable characteristic time by using the associated tangent mapping, and its eigenvalues. The Lyapunov exponent is then given by :

$$\sigma = \frac{\ln \lambda}{q} \tag{9}$$

where λ is the largest eigenvalue of $\prod_{i=1}^{q} M_i$, and q is the periodicity of the hyperbolic fixed point.

Values of λ_t are given in Table I which shows a very good agreement with values of λ_t computed with the previous definition. Like the winding number, the Lyapunov exponent is not directly linked to properties of trajectories which will be involved in the transport study, because hyperbolic fixed points are not reachable from the stochastic sea, from where are initiated orbits for the transport study.

The second approach consists in using a modified definition for the Lyapunov exponent. We compute a mean exponential rate of divergence for only that part of trajectories belonging to the vicinity of a given generation.

$$\sigma = \frac{1}{N} \sum_{i=1}^{N} \sum_{t=1}^{t=T_i} \lim_{d(0) \to 0} \left(\frac{1}{t}\right) \ln \frac{d(x_i, t)}{d(x_i, 0)} \tag{10}$$

Where the first sum represents the average over initial conditions initiated in the generation k , T_i holds for the time spent by the ith trajectory in the vicinity of generation k and only k.

In order to characterize generation 1, we use two matrices, already defined. One is attached to generation 1 and the other to generation 2. It is necessary to define two matrices because the boundary domain of generation 2 is embedded in that of generation 1. A convenient way to build the boundary domain of an island chain is to initiate particles near a hyperbolic fixed point, letting them evolve under the mapping sufficiently long to generate the domain but not so long as to let them escape from the domain. A domain can be enlarged by adding to it, mesh points within some range $\Delta i, \Delta j$. For both matrices we use the typical size $N = 1024$. In Fig. 4 it is a shown magnification of phase space area represented by the two matrices. Shown dotted is the generation 1 boundary domain enlarged by $\Delta i \sim \Delta j = 4$, in order to avoid possible missing cells, due

412

to a lack of initial conditions used for the domain construction. In solid line, is shown the boundary domain of generation 2.

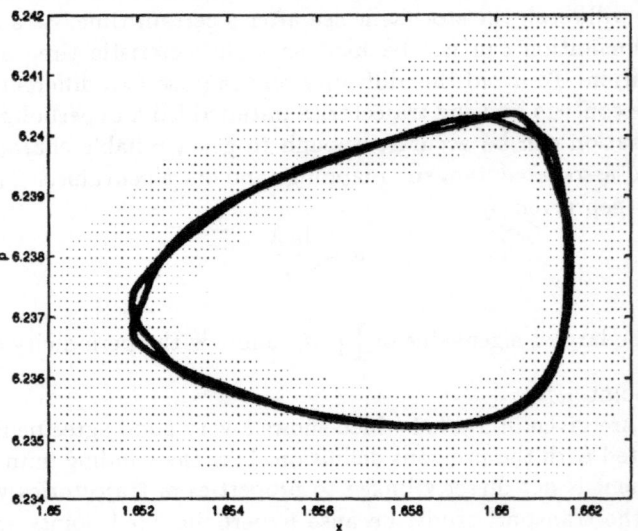

Fig. 4. Domains used for Lyapunov exponent computation. Dots represent generation 1 and solid line, generation 2.

The local Lyapunov exponent is computed using standard methods (Benettin *et al.* 1976), averaging 10^4 trajectories initiated near the hyperbolic fixed point $(1.65, 6.236)$. The divergence rate of nearby trajectories is computed at each step using the tangent mapping, only while the trajectories remain in generation 1.

In order to detect when a trajectory leaves the boundary of generation 1, we introduce two different observation periods. One is the periodicity of the accelerator mode 5, and the other one is the periodicity of islands of generation 1: 5×11. Each 5 steps we check whether the trajectory has left the generation 1 boundary for the stochastic sea by looking at the corresponding cells in matrix 1. In the same manner, each 55 steps we check whether the trajectory left the generation 1 boundary for the generation 2 boundary by looking at the corresponding cell in matrix 2. So the time spent in generation 1 is determined up to a precision of 5 steps when the trajectory leaves the islands chain 1 to reach the stochastic sea and up to a precision of 55 steps when the trajectory leaves the island chain of generation 1 to reach the next generation. This approximation is not relevant because of the large time scale associated with time spent in each boundary island chain.

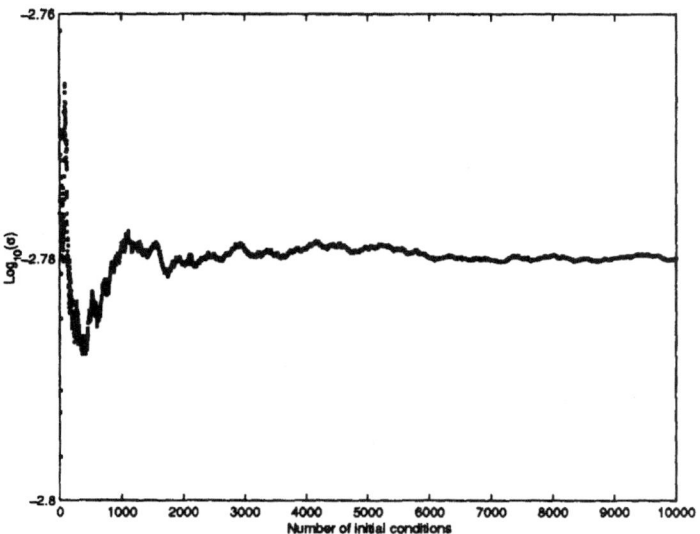

Fig. 5. Convergence of Lyapunov exponent associated with generation 1 as a function of initial conditions used for its computation.

In Fig. 5 is shown σ for generation 1 as a function of the number of initial conditions used in the average. We see a good convergence towards the value 1.66×10^{-3}. This quantity measures the e-fold time for orbit separation within the first generation of islands. Fig. 6 shows the same convergence for σ associated to the second generation around the value 1.72×10^{-4}. From these values we find $\lambda_t = 9.65$, in good agreement with values reported in Table 1. The difference comes from two aspects of the computation. First, the procedure used for the construction of matrices does not give rigourously the same representation of domains for each generation. Second, the random presence of 'parasite' islands in each boundary island chain modifies the σ computation because of correlations which they introduce. To visualize this effect, we show in Fig. 7 histograms for σ attached to generations 1 and 2. We see that they both present an asymmetry due to correlations previously mentioned.

In spite of computational drawbacks, this characteristic time gives a value of λ_t in good agreement with the one derived from the winding number . Other self-similar accelerator island chains have been investigated (Zaslavsky *et al.* 1997) and the relation between the scaling parameters and the transport exponent $\mu = \frac{ln(\lambda_s)}{ln(\lambda_t)}$ has been checked. Let us emphasize that the latter relation is well verified in the case of self-similar chains of islands.

414

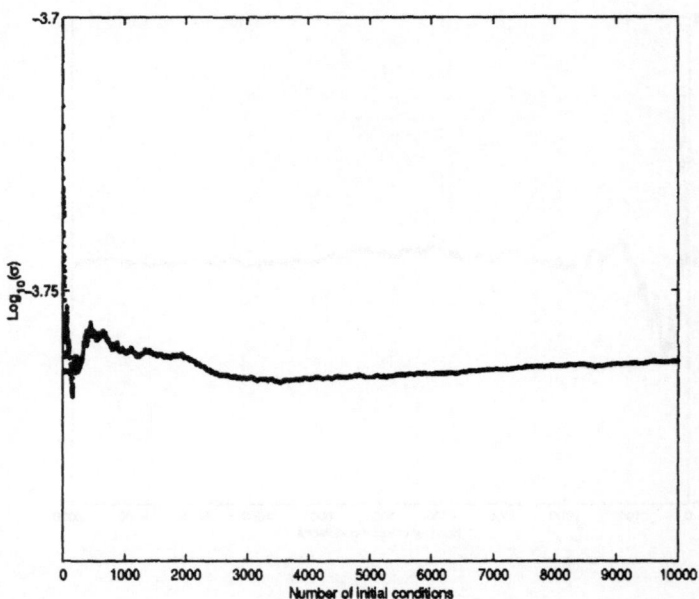

Fig. 6. Convergence of Lyapunov exponent associated with generation 2 as a function of initial conditions used for its computation.

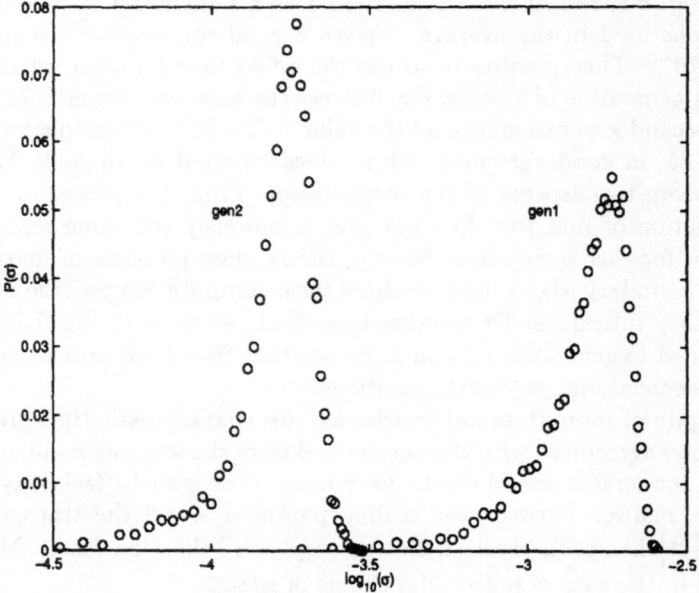

Fig. 7. Lyapunov probability associated with generation 1 and 2. Both generations present the same shape shifted. This shift is of order of $\lambda_t = 10.2$.

2.2 Distribution of characteristic times

In the previous section we showed how to characterize topological properties of the phase space domain involved in anomalous transport. In what follows, we will present computational results of distributions of characteristic times relevant for analyzing transport properties..

Sticking time distribution

Let us consider a domain $\Delta\Gamma$ in phase space, accessible from the stochastic sea. During its motion, a wandering orbit enters the domain at times $\{t_i^{in}\}$ and leaves it at times $\{t_i^{out}\}$ providing a set $\{\tau_i\} = \{t_{i+1}^{out} - t_i^{in}\}$ of times spent in $\Delta\Gamma$. These times are functions of $\Delta\Gamma$, and we are interested in their probability distribution function $\Psi_{stick}(\tau;\Delta\Gamma)$, with the normalization condition

$$\int_0^\infty \Psi_{stick}(\Delta\Gamma;\tau)d\tau = 1 \tag{11}$$

In order to characterize the two accelerator main islands we use a domain defined by the two squares surrounding them as shown in Fig. 8. We construct the corresponding matrix $b(i,j)$ so that, $b(i,j) = 1$ within the domain and zero otherwise. At each step t we evalute $b(t) = b(i,j)$, then $b(t) - b(t-1) = 1$ if the orbit has just entered the domain, -1 if it has just left it, and zero otherwise. This technique will be powerful for more complicated domains like the boundary of a $n-th$ order generation. The sticking time distribution is then given by the distribution of sizes of $b(t)$ sequences of the type $(1\ 0\ 0\ ..\ 0\ -1)$. The statistics can be performed by using a large number of trajectories (10^5) initiated in the stochastic sea and followed for a given time (10^6), or by using one trajectory followed for a large number of iterations (10^{11}).

These two approaches are equivalent and give similar results. For the sticking time distribution, we will present results using the second option. We initiate an orbit near the hyperbolic point $(0,0)$ and follow it 5×10^{11} steps. In Fig. 9 the plot of the sticking distribution is shown in logarithmic scales, using a bin size of 10 steps. The sticking time distribution exhibits a power law for $t < 10^3$:

$$\Psi_{stick}(\Delta\Gamma;\tau) \sim \frac{1}{\tau^{\gamma_{stick}}} \tag{12}$$

with $\gamma_{stick} = 2.4$ and presents a cross-over within time range $[10^3, 10^4]$. In order to improve bin statistics at larger times, we use bin sizes which increase geometrically with time. The result displayed in Fig. 10 shows that the distribution follows another power law given by $\gamma_{stick} \sim 3.2$. This slope is determined up to a precision of 10%, depending on the time data ranges used for its computation. The given value is obtained for $10^{3.8} < \tau < 10^{5.43}$.

This example shows that different time scales are involved in the sticking process around singular zones (accelerator modes). One may ask the question: Does

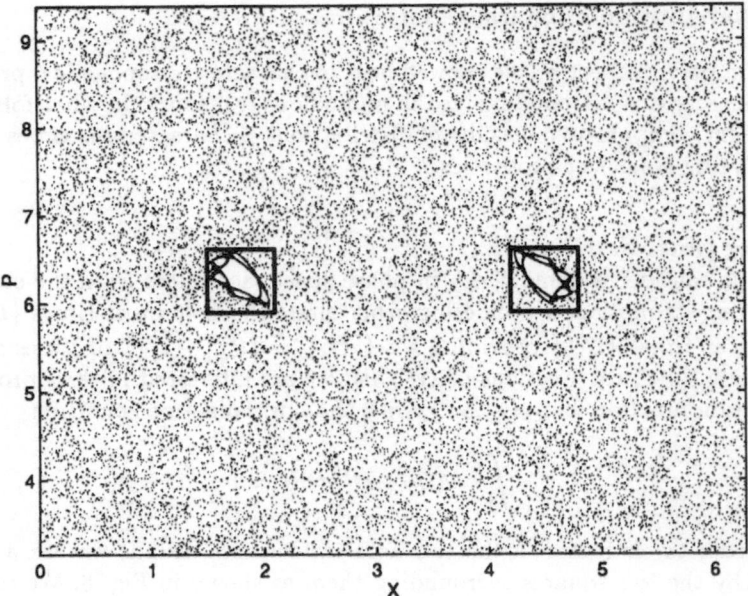

Fig. 8. Domain used for computing the sticking time distribution for $K = 6.476939$.

$\Psi_{stick}(\Delta\Gamma; \tau)$ exhibit an asymptotic behaviour, and if yes what is the transient time ? One can emphasize (Zaslavsky *et al.* 1997) that $\Psi_{stick}(\Delta\Gamma; \tau)$ depends on the topological structure of the singular zone, and that different time scales correspond to characteristic times associated with different accessible fine topological structure. This could be checked by using the boundary of each island chain previously defined as a domain $\Delta\Gamma$, but the computation time required for sufficient statistics is for the moment prohibitive.

Poincaré recurrence time distribution

Let us consider a domain $\Delta\Gamma$ in the stochastic sea. As in the previous section we can consider the set of times at which an orbit enters $\Delta\Gamma$, $\{t_i^{in}\}$ and the set of times at which it leaves it, $\{t_i^{out}\}$. Times defined by the set $\{\tau_i\} = \{t_{i+1}^{in} - t_i^{out}\}$ are called Poincaré recurrence cycles. They represent times which orbits need to perform a first return to the domain $\Delta\Gamma$ after it has left it. We denote their density distribution function $\Psi_{rec}(\Delta\Gamma; \tau)$, and the function

$$\Psi_{rec}(\tau) = \lim_{\Delta\Gamma \to 0} \frac{1}{\Delta\Gamma} \Psi_{rec}(\Delta\Gamma; \tau) \tag{13}$$

can be considered as a probability density of the first return to an infinitesimal phase volume $\Delta\Gamma$ after a time interval τ with the normalization condition

$$\int_0^\infty \Psi_{rec}(\tau)d\tau = 1 \tag{14}$$

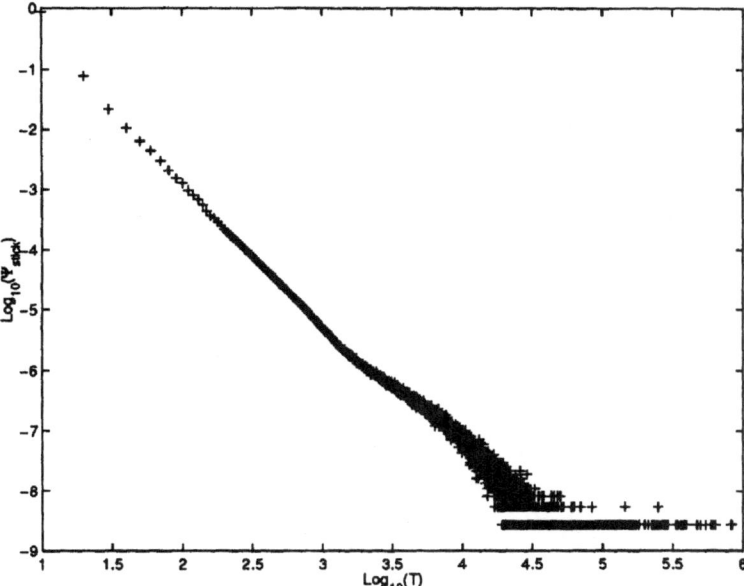

Fig. 9. Sticking time distribution for $K = 6.476939$. The slope of the straight line for $\tau < 10^3$ is 2.4.

The recurrence time distribution characterizes the properties of the phase space outside $\Delta\Gamma$. It has been shown in (Zaslavsky *et al.* 1993, Zaslavsky *et al.* 1997, Afraimovich and Zaslavsky 1997) that it can be related to transport properties.

In the case of chaotic systems with good mixing properties, the recurrence time has a Poissonian distribution :

$$\Psi_{rec}(\tau) = \frac{1}{<\tau>} \exp\left(-\tau/<\tau>\right) \tag{15}$$

Nevertheless, due to the presence of singular zones, one can expect that for large times the recurrence distribution exhibits a power law behaviour, which is confirmed by the following results. Applying the method used above for sticking times, the recurrence time distribution is given by the distribution of sizes of $b(t)$ sequences of type $(-1\ 0\ 0\ ..\ 0\ 1)$. The domain we considered to compute the recurrence time distribution is presented in Fig. 11.

Other domains have been used and have given similar results. The computation consists in initiating 10^5 particles in the stochastic sea and following them up to 10^6 steps. In Fig. 12, $log_{10}(\Psi_{rec}(\tau))$ is shown as a function of τ. The straight line shows that $\Psi_{rec}(\tau)$ follows a Poissonian law for $\tau < 10^3$ and deviates from it at larger times. Let us notice that this time is comparable to the time at which $\Psi_{stick}(\Delta\Gamma; \tau)$ presents a crossover.

In Fig. 13, $\Psi_{rec}(\tau)$ is shown, using as for the sticking time distribution, bin sizes which increase geometrically with time, in order to improve bin statistics

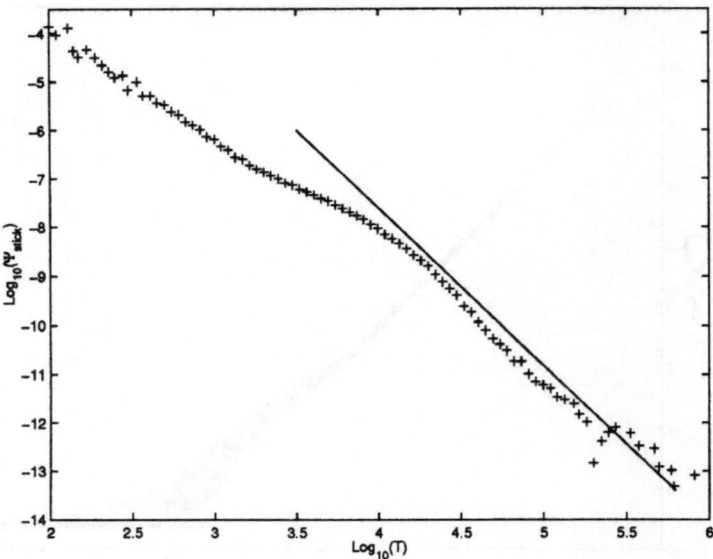

Fig. 10. Sticking time distribution for $K = 6.476939$. The slope of the straight line is 3.2.

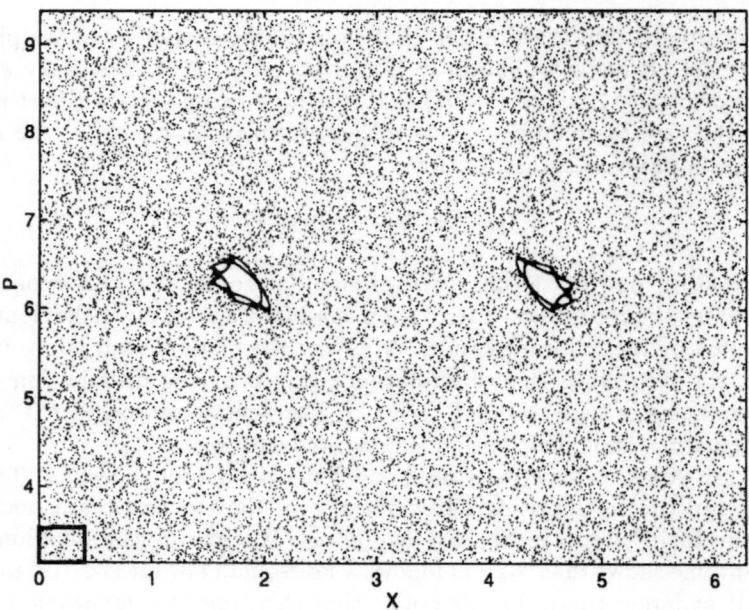

Fig. 11. Domain used for computing the Poincaré recurrence time distribution for $K = 6.476939$.

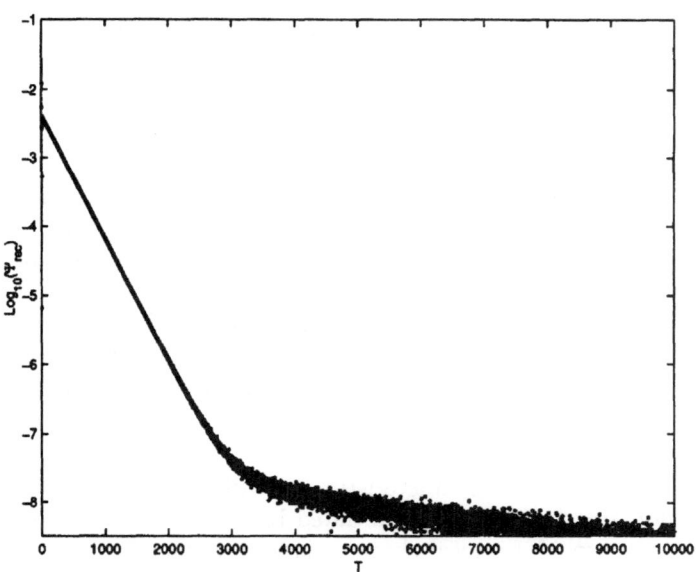

Fig. 12. Poincaré recurrence time distribution for $K = 6.476939$. Poissonian behaviour is observed for $\tau < 2.5 \times 10^3$.

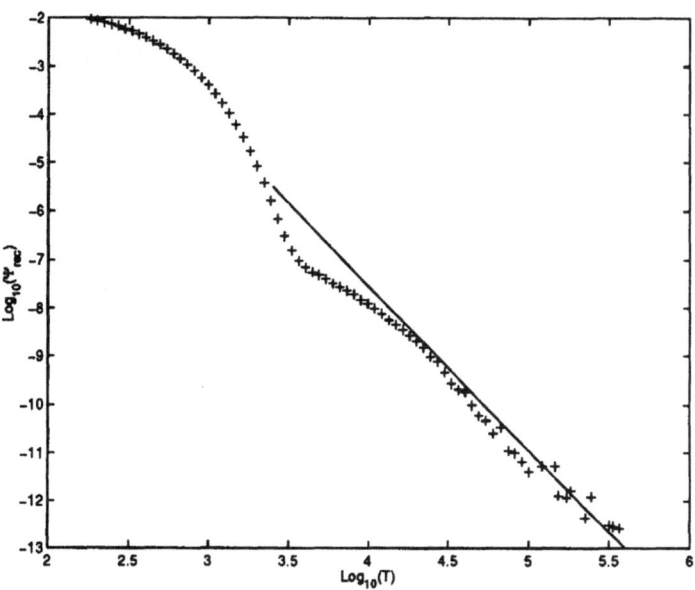

Fig. 13. Poincaré recurrence time distribution for $K = 6.476939$. The slope of the staight line is 3.4.

at large times. A crossover is observed within the time interval $[10^3, 10^4]$ and for times larger than 10^4, $\Psi_{rec}(\tau)$ exhibits a power law :

$$\Psi_{rec}(\tau) \sim \frac{1}{\tau^{\gamma_{rec}}} \tag{16}$$

with $\gamma_{rec} \sim 3.4$. This slope is computed for $T > 10^{4.3}$ with an accuracy comparable to the one obtained for the sticking time distribution, and is in agreement with the Kac theorem which states that mean recurrence time must be finite for compact phase space with ergodic motion (Kac 1958).

The fact that the slope found for sticking time distribution is comparable to the slope found for recurrence time distribution confirms the idea that long recurrence times are due to orbit stickiness around the singular zone.

Escape time distribution

One can define another characteristic time distribution. Considering the phase space domain $\Delta\Gamma$ and orbits initiated in this domain, the set $\{\tau_i^{out}\}$ of first escape times provides a local characteristic of transport. In this approach, one has to specify the domain $\Delta\Gamma$ and where orbits are initiated in the domain. The distribution of escape time $\Psi_{esc}(\Delta\Gamma; \tau)$ is then given by the distribution of sizes of $b(t)$ sequences of type $(0 \ 0 \ .. \ 0 \ -1)$. In what follows we present results obtained for different domains and different initial condition locations. The first domain used to characterize the whole accelerator mode, is a square as seen in Fig. 14.

We uniformly initiate 10^9 orbits in phase space domains represented in Fig. 15.

It can occur that an orbit is initiated in a small 'parasite' island and be stuck forever in the domain. Hence we introduce an observation time after which orbits still in the domain are not included in the statistics. For the present case, the observation time is 10^6.

In Fig. 16 it is shown the normalized escape time density distribution function, using again bins which increase geometrically with time. It exhibits a power law for time larger than 10^4 :

$$\Psi_{esc}(\tau) \sim \frac{1}{\tau^{\gamma_{esc}}} \tag{17}$$

with $\gamma_{esc} \sim 3.2$. This value obtained with 10% accuracy is computed within the time interval $10^{4.2} < \tau < 10^{5.35}$. In order to characterize more precisely the accelerator island and its boundary island chain (BIC), we build domains $\Delta\Gamma$ following the method described in the previous section devoted to the computation of areas. This gives a good representation of the boundary of the island chain. The domains thus defined are used to uniformly initiate orbits. Because each domain only represents one part of the whole chain, we introduce a period of observation as mentioned above to check if an orbit is still in the domain.

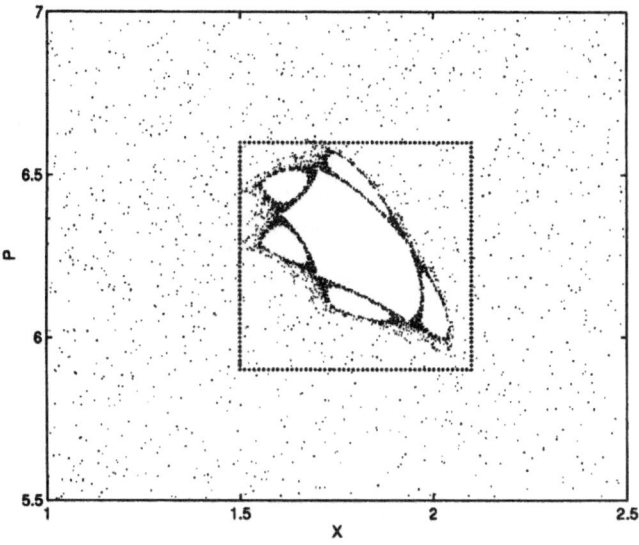

Fig. 14. Domain used for computing the escape time distribution for $K = 6.476939$.

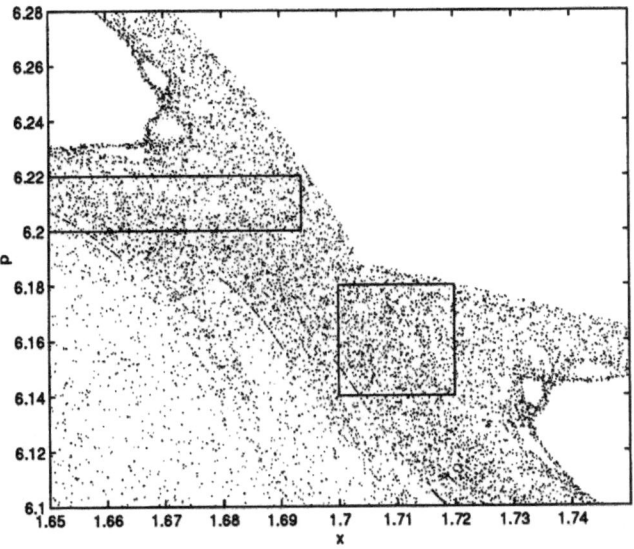

Fig. 15. Domain used to initiate particules for escape time distribution computation.

Let us remark that because at each generation the observation time is multiplied by 11 (period of the islands chain), the total observation time is multiplied by this factor at each generation. We apply this procedure to BIC's 1,2,3. Fig. 17 displays the escape time distributions function associated with each BIC. The observation time for BIC 1 is 10^5, $10^5 \times 11$ for BIC 2 and $10^5 \times 11 \times 11$ for BIC 3. We also note that the escape time distributions of the three generations exhibit the same power law behaviour with an exponent $\gamma_{esc} \sim -1.6$. In this logarithmic scale there is a shift between the distributions of two successive BIC's. This shift is of the order of 10.5 which corresponds to the typical value of the temporal scaling parameter λ_t.

When we use larger observation times, 10^7 for BIC 1 and 10^8 for BIC 2, we observe as shown in Fig. 18, a crossover and a power law behaviour with an exponent $\gamma_{esc} \sim 3.5$ for BIC 1 and BIC 2. Let us emphasize that this agrees with the results obtained for the sticking time and the Poincaré recurrence time distributions studied above. We also note that the γ_{esc} value agrees with the one obtained with the square domain (Fig. 14). Finally, the crossover times of BIC 1 and BIC 2 are also in the ratio λ_t. This confirms in a way the idea that each fine topological structure is associated with a characteristic time scale.

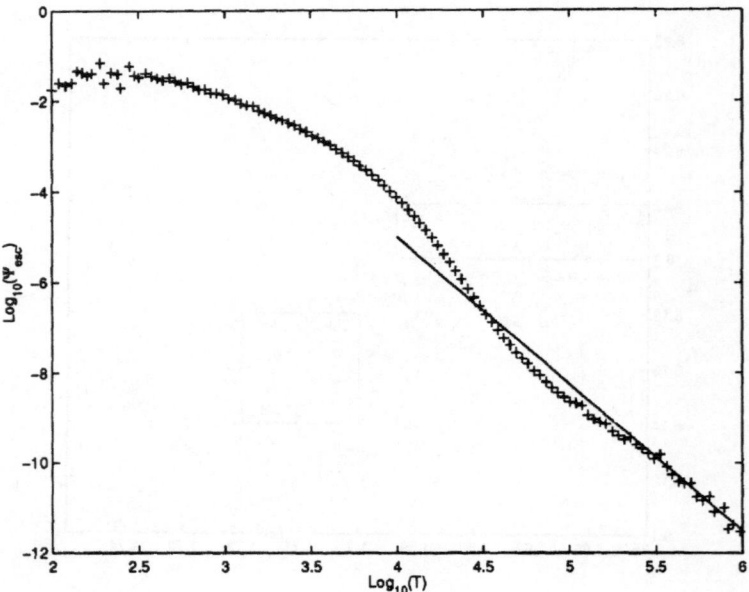

Fig. 16. Escape time distribution for $K = 6.476939$. The slope of the straight line is 3.2.

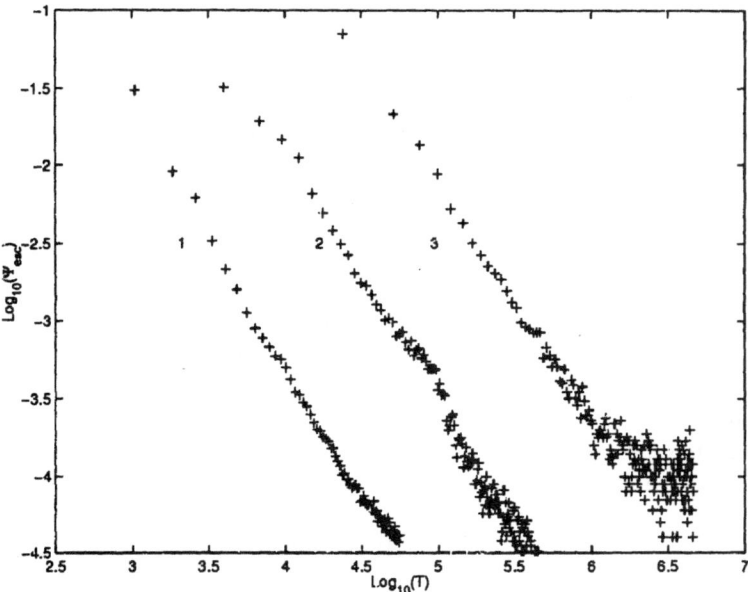

Fig. 17. Escape time distributions at short times for generation 1, 2 and 3, for $K = 6.476939$.

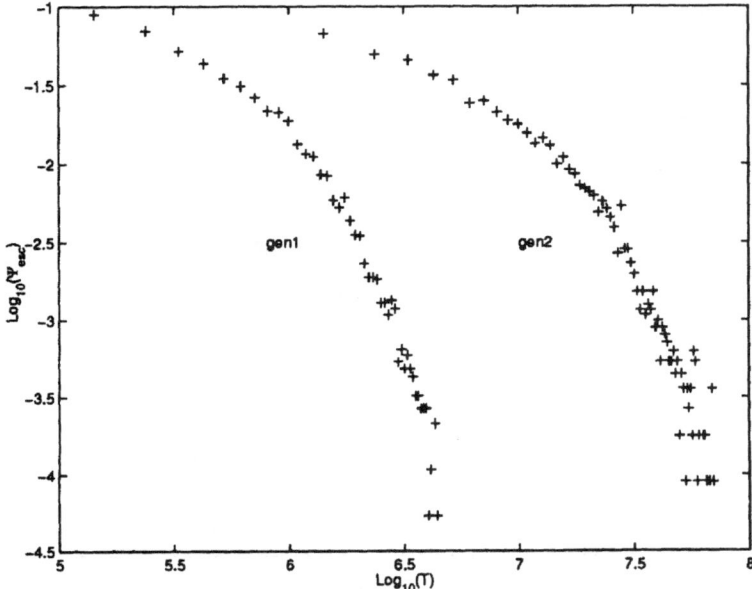

Fig. 18. Escape time distributions at large time for generation 1 and 2, for $K = 6.476939$.

2.3 Transport

In this section, we present numerical computations of time evolution for two quantities which define the transport : $< |p(t)| >$ and $< p^2(t) >$, where brackets mean ensemble average. Non-diffusive transport is characterized by :

$$< p^2(t) > \sim t^\mu \quad , \quad \mu \neq 1 \tag{18}$$

The μ determination involves two kinds of averaging. One is linked to the number of initial conditions, and the other one is associated with time. The previous study of time distributions has shown that the notion of an absolute time asymptotic behaviour should be replaced by an asymptotic limit with respect to a given time scale.
In Fig. 19 is reported the time behaviour of $< |p(t)| >$ for 10^5 orbits initiated near the hyperbolic fixed point $(0,0)$, followed 10^6 steps, showing that :

$$< |p(t)| > \sim t^{\mu/2} \tag{19}$$

In order to check the convergence of $< |p(t)| >$ with respect to the number of initial conditions, we show in Fig. 20 the evolution of $\mu/2$ as a function of this number.

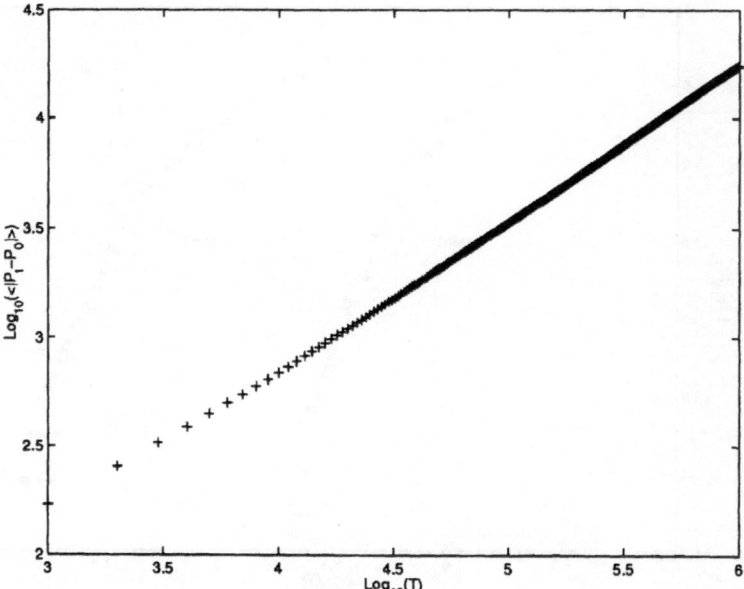

Fig. 19. First order moment $< |p_t - p_0| >$ as a function of time. 10^5 initial conditions used, $K = 6.476939$

The slope evaluation is made using the standard mean square method, for two different time intervals: $5 \times 10^5 < t < 10^6$ and $10^5 < t < 5 \times 10^5$. We see that the convergence is fairly well reached for 10^5 orbits, and that the two slope estimations agree, confirming that within the time interval $10^5 < t < 10^6$ an asymptotic behaviour is also reached. The μ value obtained from this method is $\mu = 1.42 \pm 0.05$. We now consider a large number N (10^5) of initial conditions and study the behaviour of the parameter $\mu/2$ as a function of the time interval $[10^3, t]$. We observe in Fig. 21, that convergence towards $\mu/2 = 0.707$ is reached for $t > 8 \times 10^5$. This value is consistent with the one previously obtained.

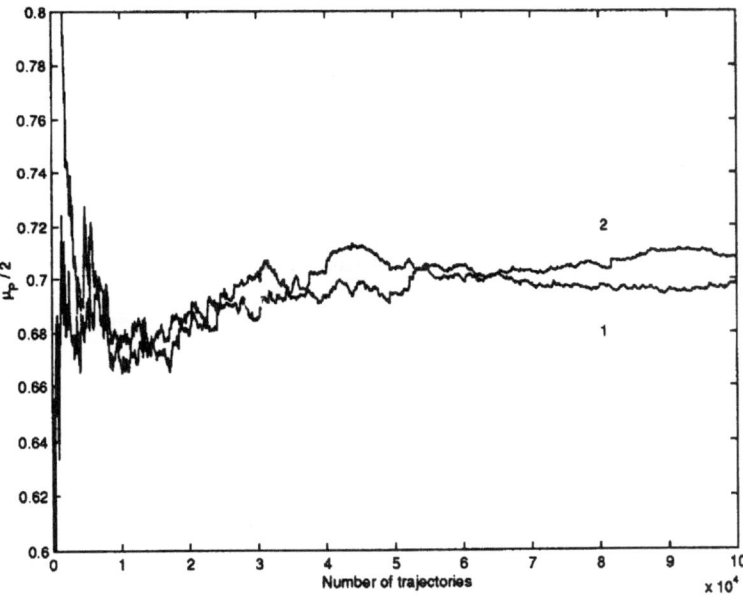

Fig. 20. Convergence of the slope as a function of initial conditions used for its computation, for four different time intervals. Label 1 corresponds to $5 \times 10^5 < T < 10^6$ and label 2 corresponds to $10^5 < T < 5 \times 10^5$.

It has been derived (Zaslavsky *et al.* 1997), using the fractional kinetic equation, and the renormalization group, that in the case of anomalous transport involving self similar island chains,

$$\mu = \frac{ln(\lambda_s)}{ln(\lambda_t)}. \tag{20}$$

This is confirmed in the present case, and has already been confirmed for the web map (Zaslavsky *et al.* 1997, Zaslavsky and Niyazov 1997) and recently for the Cassini billard (Zaslavsky and Edelman 1997).

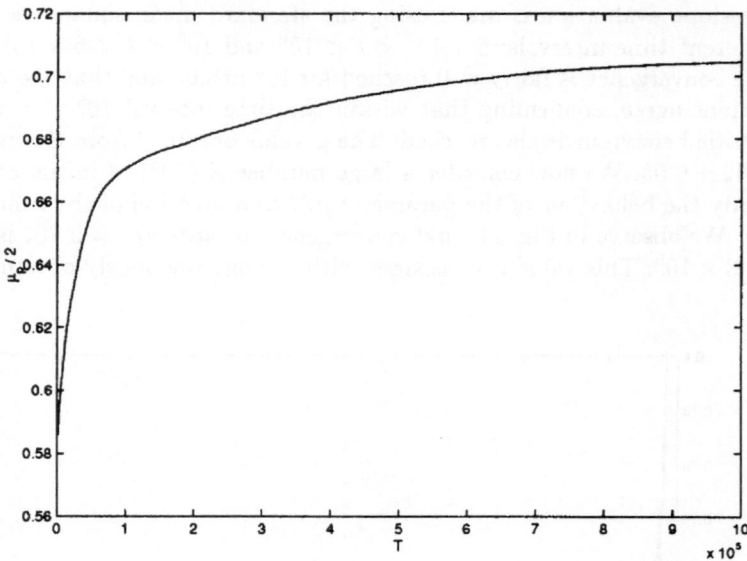

Fig. 21. μ_p computation as a function of time range used, for the curve displayed in Fig. 19.

3 Phase space structures and stickiness

The link between anomalous transport and phase space topology is an important issue. In the case of self similar island chains, when a singular zone exhibits anomalous stickiness properties, the resulting transport can be expressed in terms of topological properties of this chain Eq. (20). However, the self similarity of island chains is not a sufficient condition for anomalous transport. There are also self similar structures which give neither trapping nor anomalous transport. Two very clear examples are given by $K = 1.6615554$, which is a satellite structure of type 6-6-6-6-6-6, and $K = 2.8027452$, of type 4-4-4-4-4-4. In both cases there are large stochastic gaps between islands, so that orbits are not constrained in passing.

In this section we will focus on alternate forms of topological structures responsible for anomalous transport in the standard map. The examination of the case near threshold allows us to find trapping due to complex multi island layers. These topological structures present another kind of trap. Near threshold it is convenient to study motion in angle (x) rather than in action (p) because of prevailing of a ballistic propagation along x. An ensemble average of initial conditions gives the long time behavior of the motion in x as

$$<x^2> \simeq t^\mu, \tag{21}$$

where the brackets indicate an ensemble average with initial conditions taken in the stochastic sea. For K below threshold the motion in x is simply given by flow along Kolmogorov- Arnold-Moser (KAM) surfaces, so $\mu = 2$ (ballistic motion). For very large K the motion is stochastic and diffusive, with $\mu = 1$. In the transition region just above threshold particular values of K give large values of μ due to strong trapping near islands. In the present work we numerically determine the nature of the trapping and hence the cause of the anomalous behavior.

To observe long time trapping, it is simplest to look for the associated flight. We scan values of K and measure average square distance travelled for an ensemble of initial conditions. A Poincaré plot of individual orbits with flights of length greater than 10^7 then identifies the structure responsible for the flight. In Fig. 22 is shown a result of scanning K near threshold and measuring $<x^2>$, versus time. For this purpose the variables x,p are kept on the cylinder, $-\infty < x < \infty$, $-\pi < p < \pi$. Otherwise orbits with large p can contribute significantly to $<x^2>$, but have nothing to do with trapping. Each point is the result of 10^5 steps with 2×10^3 initial conditions, all chosen in the stochastic sea near $x = 0$, $p = 0$. It is seen that all near-threshold transport in x is anomalous, with $\mu > 1$.

There is an overall decrease in μ towards 1 as K increases and the islands vanish, but for some K values there are definite peaks. Figure 22 must be considered as the result of a simulation of a definite time length. Simulations of different lengths produce somewhat different plots, for reasons we will explain, but the peaks persist.

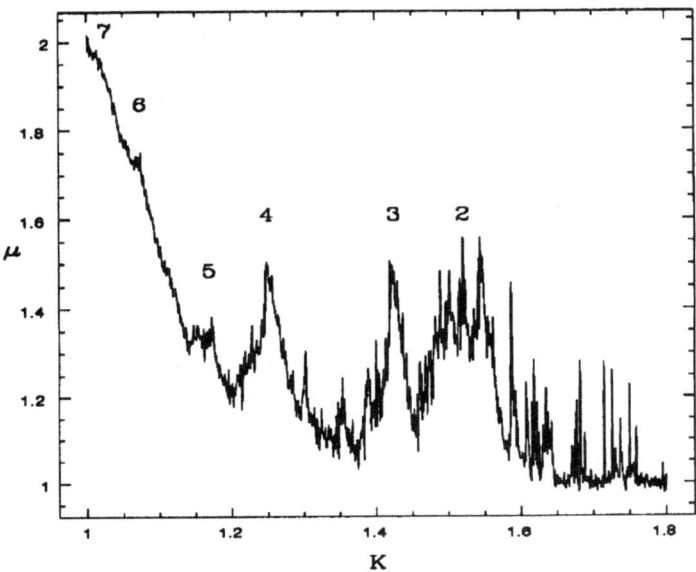

Fig. 22. Exponent μ for anomalous transport in x versus K near threshold.

We find that the major peaks shown in Fig. 22 are all due to satellite island chains born from the same period three island chains located at $p \simeq \pm 2$. Each major peak of Fig. 22 is labeled with the periodicity of the island chain born. The newly formed chains tend to have associated with them a large number of smaller islands with small passages between islands chains, indicating the destruction of a KAM surface at slightly smaller values of K, and the presence of cantori (Hanson *et al.* 1986, Meiss and Ott 1986).

Although most trapping structures are very complex, some of them are simple enough to allow us to extract the essential property which governs their behavior. This appears to be the existence of a heirarchy of phase space domains with geometrically related time scales. One example with both geometrically decreasing size and frequency scales, which we will refer to as a satellite type, has been given in a previous publication (Benkadda *et al.* 1997). In this work we find that the dominant cause of trapping is due to multilayered structures, consisting of many well ordered islands with similar spatial scales but geometrically related time scales. Such a structure is demonstrated in Figs. 23- 25 which will be discussed more later. In the Table 2 are listed some of the structures identified with the peaks shown in Fig. 22, for simulations of 10^6 steps with 10^5 initial conditions. The values shown in Fig. 22 used fewer initial points for shorter times, for computational reasons, and therefore differ somewhat from those of the table.

Table 2 - Major trapping structures and properties

K	islands	A	μ	γ	α
1.02	period 3-7	0.3361	1.8	2.1	2.1
1.06	period 3-6	0.2934	1.6	2.1	3.8
1.16	period 3-5	0.2479	1.4	2.4	3.2
1.2415	period 3-4	0.2179	1.6	2.6	3.1
1.245	period 3-4	0.2239	1.7	2.5	2.6
1.42	period 3-3	0.1068	1.7	2.4	2.7
1.54	period 3-2	0.0267	1.5	2.6	2.9

The peaks in Fig. 22 at the first two values $K = 1.02, 1.06$ are very weak, and the dominance of the identified trap at these values is not as clear as the remaining peaks. Two entries are given for the period 3-4 island, one for K just above the birth value, and one near the maximum. The value A is the area of the island structure. Also given, with only about 10% accuracy, are the anomalous transport exponent μ, and α and γ; the powers of the long time tails for the distributions of time spent within the trap, $t^{-\gamma}$ and outside the trap $t^{-\alpha}$, with $\alpha, \gamma > 2$, in agreement with the Kac theorem (Kac 1958, Meiss 1997). Parameter α describes the next strongest trap in phase space, so $\alpha \geq \gamma$. If no traps existed, the distributions would be Poissonian. Domains including island structures responsible for trapping all exhibit long time power law tails whereas a domain selected at random does not. We also calculated the Poincaré recurrence time for initial conditions chosen in the stochastic sea and found that its probability distribution follows a power law with the same slope γ as the distribution of time spent within the trap.

In general these structures are very complex. A more extensive publication will examine them in detail. But the essential property appears to be the existence of layers of stochastic bands separated by island chains, with geometrically increasing time scales as one penetrates inward from the stochastic sea towards the last bounding KAM surface. An example of a fairly regular multilayered structure is that appearing for K between 1.24 and 1.26 (peak 4 in Fig. 22). The structure arises at the birth of four islands from the period 3 chain and it is surrounded by many layers of island chains. Poincaré plots of a long flight are shown in Figs. 23 and 24.

All strong traps consist of islands born of the period 3 island chain, a fact easily understood. There are two larger island structures, seen in Fig. 24 as a large central island (period one) and the blank spaces bordering $p = \pm\pi$ (period 2), but an orbit trapped around either structure has p alternatively positive and negative, giving $x^2/t \to 0$. Thus the period three traps are the largest structures giving nonzero flight.

The four island structure for $1.24 < K < 1.26$ is regular enough to allow additional quantitative analysis of the trapping dynamics. A Poincaré plot with random initial points, showing a small part of the island chains surrounding the trap, is shown in Fig. 25.

Four different numerical experiments provide sufficient information to make a model. The first is a determination of escape time as a function of initial position, the second and third are determinations of the long time tails for time spent

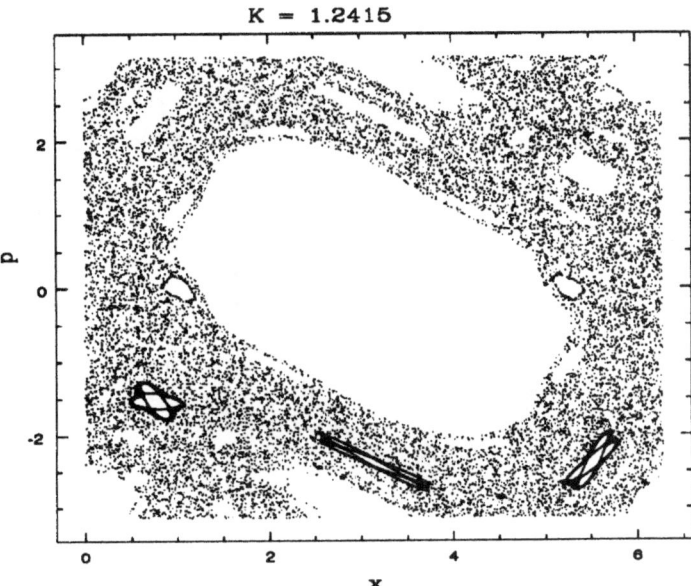

Fig. 23. Poincaré plot with one orbit exhibiting a strong stickiness (flight), $K = 1.2415$.

inside and outside the trap. The fourth is a measure of the relative probability, in each layer, for transition to the layers above or below.

Fig. 26 shows a determination of mean escape times for $K = 1.2415$. Initially 2500 orbits were started near the band labels, $m = 1, 2, ...7$ shown in Fig. 27. We see that the escape time is approximately exponential.

Another numerical experiment reveals a remarkable property of the island chain layers.

By constructing a domain consisting of an annulus bounded by two of the island chains, and starting orbits within this annulus, we find that orbits move at most one layer per step and that the probability of leaving the annulus through the outside edge is much larger than that of leaving through the inside edge. That is, the layers are more difficult to penetrate, by a large ratio, moving inward than moving outward.

From the above information a simple model of the island layer trapping structure can be constructed. Consider a trap consisting of M levels, with $m = 0,1,2...,M$ representing the stochastic layers between adjacent island chains, as shown in Fig. 27. The stochastic sea is level 0 and level M is the last layer before encountering a bounding KAM surface. Characterize each level by probabilities for stepping up $P_+(m)$ and down $P_-(m)$ in m each time step. We know empirically that $P_+(m) << P_-(m)$. To match the exponential nature of the escape

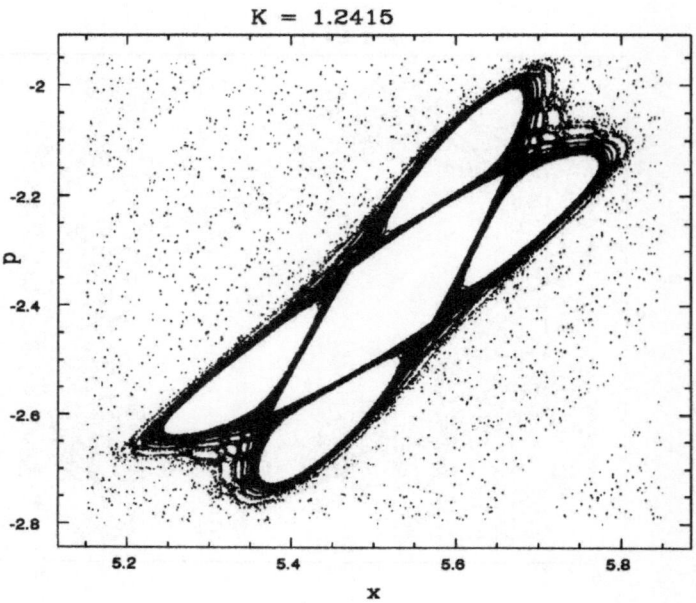

Fig. 24. Poincaré plot with one orbit exhibiting a strong stickiness (flight), $K = 1.2415$.

time we must have $P_-(m) = P_-(m-1)f_m$ with $f_m \ll 1$. The fractions f_m determine the escape time curve. The trapping time distribution is determined by the ratio $r_m = P_-(m)/P_+(m)$, and the slope of the distribution becomes steeper as r_m increases. As a simple model choose f_m and r_m independent of m,

$$P_-(m) = P_0 f^m \, , \qquad P_+(m) = \frac{P_0 f^m}{r} \tag{22}$$

The values $P_0 = .25$, $f = .18$, $r = 15$ give a reasonable fit to the data obtained for $K = 1.2415$.

We do not have a means of estimating these parameters from spatio-temporal scaling laws, as has been done for the self-similar island chains (Benkadda *et al.* 1997, Zaslavsky 1994), or by assuming that the long time trapping is dominated by particular cantori (Meiss and Ott 1986). Of course in an actual layered structure both f and r will be m dependent. As is seen in Fig. 26 the escape time is not exactly exponential, but this simple model is sufficient to capture the main properties of the trap. Note that within the accuracy of the determinations, except for the first two peaks, $\mu \simeq 3/2$ and $\gamma \simeq 5/2$, values found for a system with topology similar to our case (Venkataramani *et al.* 1997).

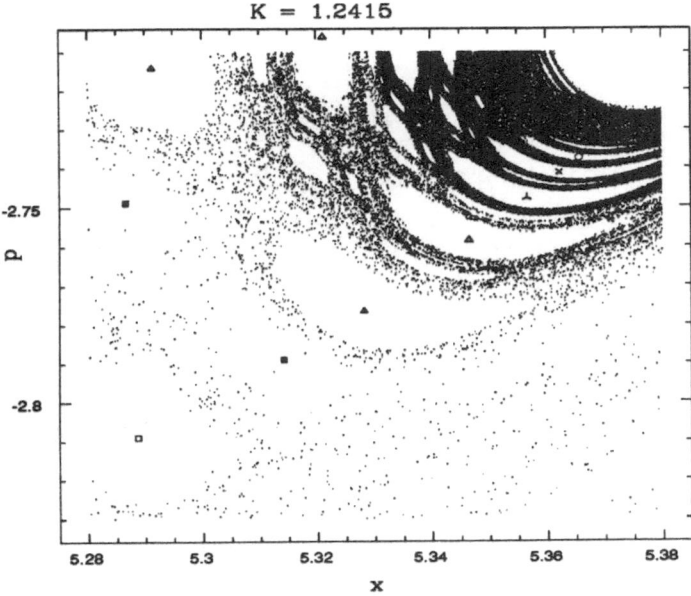

Fig. 25. Poincaré plot with one orbit exhibiting a strong stickiness (flight), $K = 1.2415$.

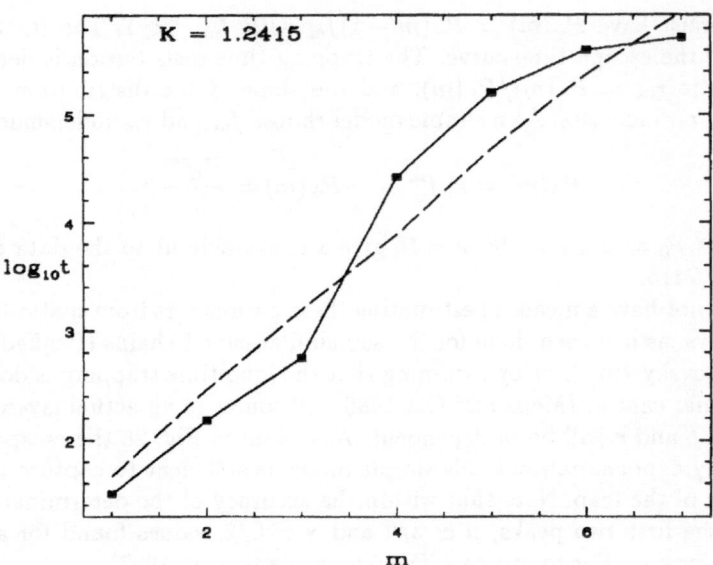

Fig. 26. Escape time versus position, $K = 1.2415$, showing also the model result (dashed line)

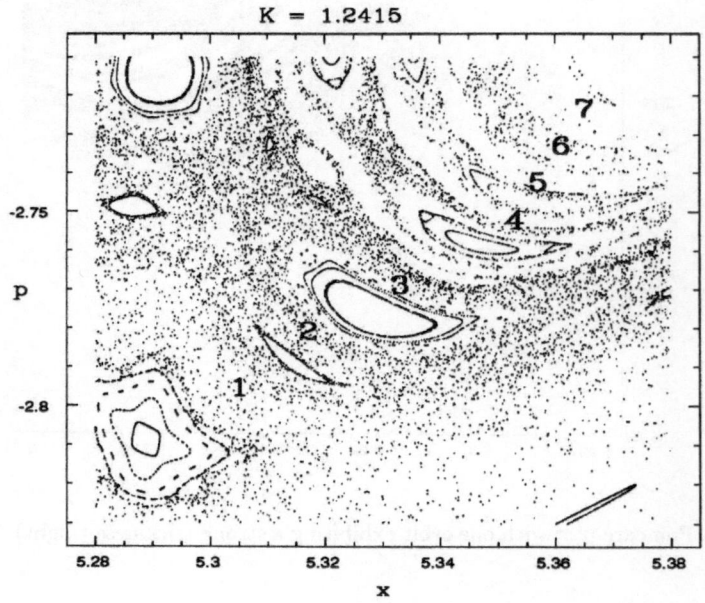

Fig. 27. Initial points for escape time calculation, $K = 1.2415$.

4 Transport and probability density function $P(x,t)$

In this section we present a numerical study of the one dimensional probability density function associated to variable x : $P(x,t)$ and investigate its link to the transport exponent μ. $P(x,t)dx$ gives the probability for a particle to be located between x and $x + dx$, with the following initial condition :

$$P(x,0) = \delta(x) \tag{23}$$

In the case of diffusion processes the solution of the diffusion equation leads to the well known result :

$$P(x,t) = (4\pi Dt)^{-1/2} exp\left[\frac{-x^2}{4Dt}\right] \tag{24}$$

where D is the diffusion coefficient, and for which all moments defined by :

$$< x^m > = \lim_{t\to\infty} \int_{-\infty}^{+\infty} x^m P(x,t)dx \ , \tag{25}$$

are finite. In case of the fractional Brownian motion, $P(x,t)$ leads also to finite moments for all m.

Conversely, processes involving Lévy flights are characterized by the existence of m such that moments of order larger than m are infinite. Hence, the knowledge of $P(x,t)$ gives an important qualitative information about the nature of the processes. We will show that it can also give a quantitative estimate of the anomalous transport exponent μ.

The numerical study of $P(x,t)$ consists in following 10^5 trajectories initially located near the hyperbolic fixed point $(0,0)$. At different times : $t_1 = 10^4$, $t_2 = 5.10^4$, $t_3 = 10^5$, $t_4 = 5.10^5$, $t_5 = 10^6$, the orbit locations are recorded and histograms are built. The number of bins is kept constant and equal to 100 for each time, which implies that bin size depends on time according to $\Delta x = (x_{max}(t) - x_{min}(t))/100$.

In Fig. 29, are shown $log_{10}(P(x,t))$ as a function of x for different times $t_1 \ldots t_5$. Each curve is symmetric as expected showing that orbits can be stuck with equal probability, to each of the island chains of period 3 responsible for anomalous transport. It is clear also that each curve deviates from Gaussian, confirming that the transport is not diffusive. In Fig. 30 are shown $log_{10}(P(x,t))$ as a function of $log_{10}(x)$ for positive x.

$P(x,t)$ exhibits a power law behaviour with different powers. Nevertheless, for $t > 10^4$ the power is fairly stable and $P(x,t)$ has the form :

$$P(x,t) \sim |x|^{-\delta} \ , \quad with \ \ \delta = 2.6 \pm 0.1 \tag{26}$$

This behaviour shows that Lévy statistics must be involved to describe the dynamics and that moments diverge for $m > 2$. In Fig. 30 , all previous curves are

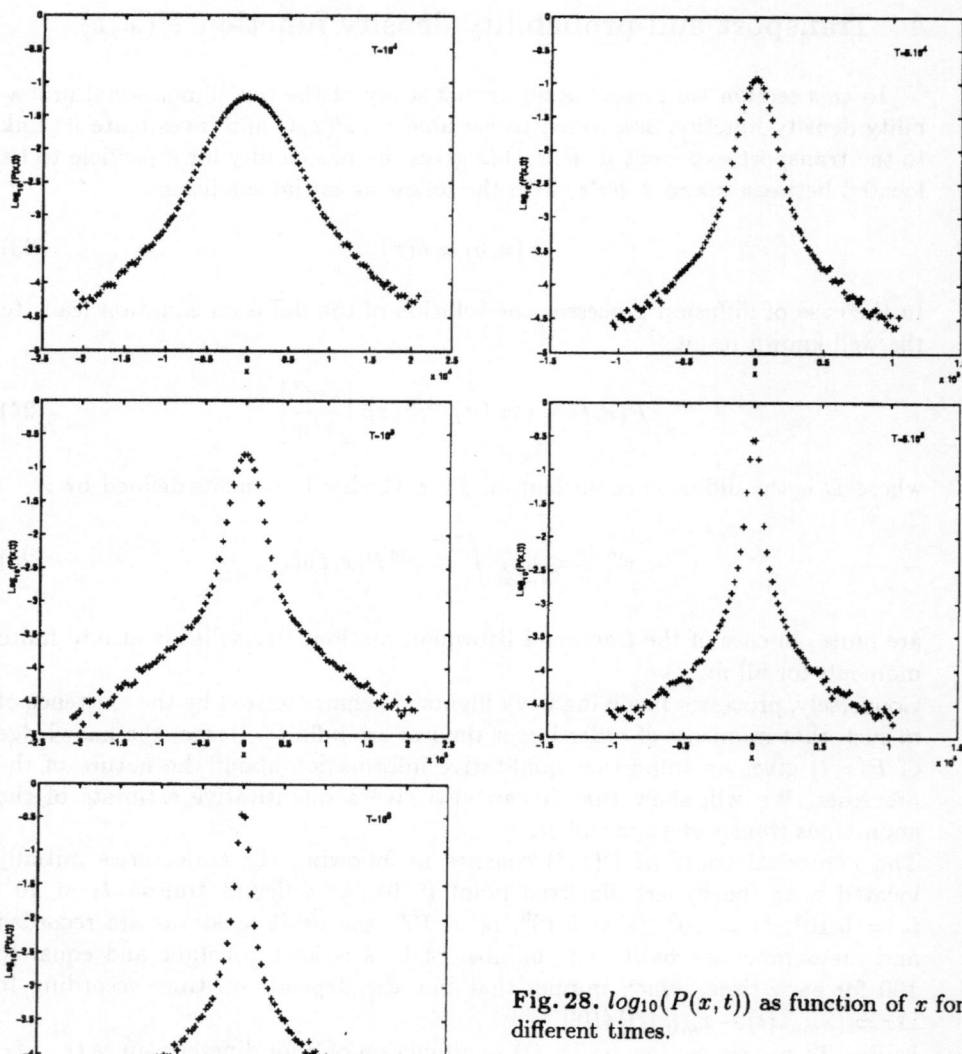

Fig. 28. $log_{10}(P(x,t))$ as function of x for different times.

presented on the same graph showing the same shape shifted in x. Let us assume that $P(x,t)$ can be written in the general form :

$$P(x,t) = \left(\frac{t^\nu}{|x|}\right)^\delta \tag{27}$$

Let us consider then two pairs (x_1,t_1), (x_2,t_2) such that $P(x_1,t_1) = P(x_2,t_2)$, which can be obtained from Fig. 30 by the intersection of straight line $log(P(x,t)) =$

435

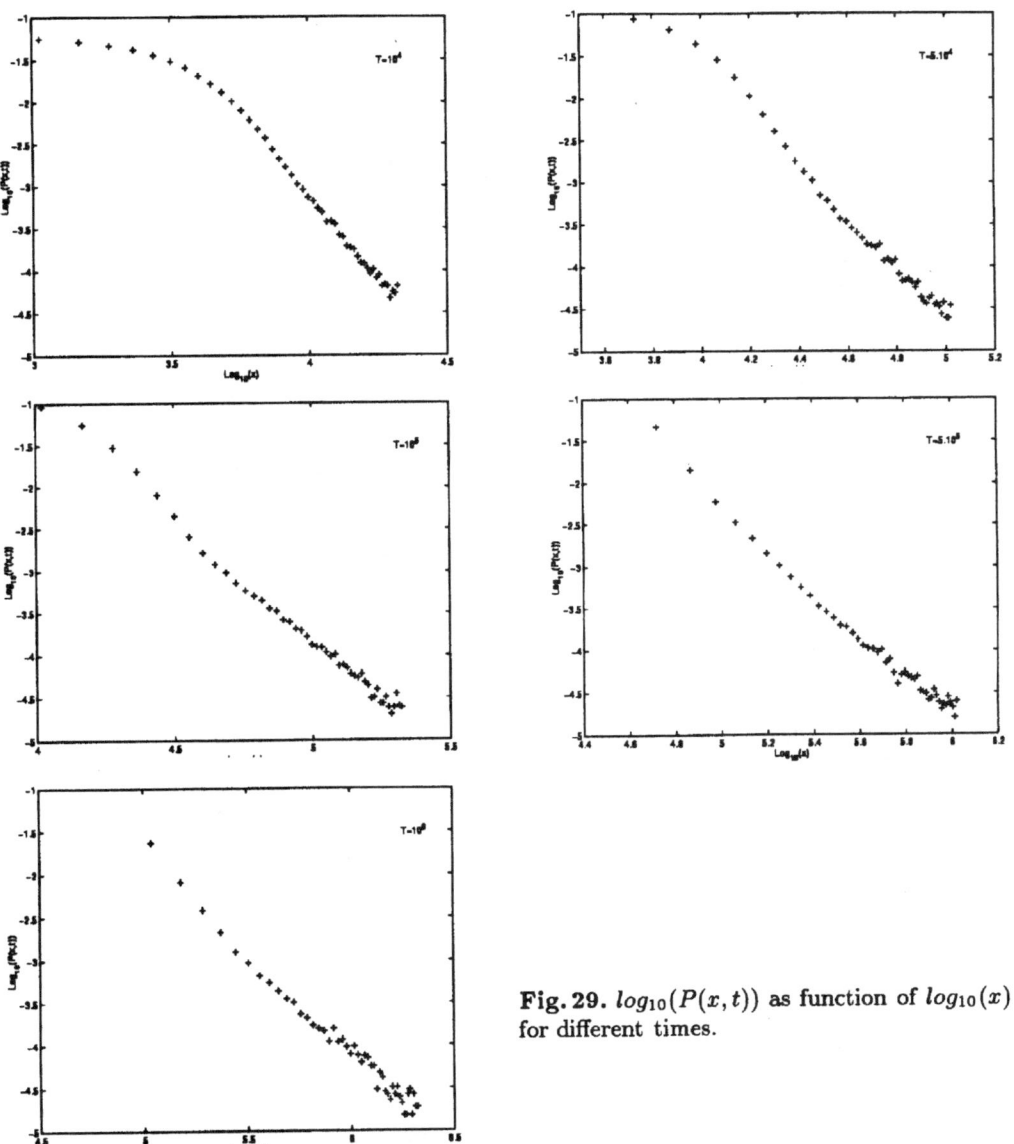

Fig. 29. $log_{10}(P(x,t))$ as function of $log_{10}(x)$ for different times.

const and the two curves $log(P(x,t_1))$, $log(P(x,t_2))$. Using the general expression of $P(x,t)$, we find :

$$\nu = \frac{log(x_1/x_2)}{log(t_1/t_2)} \tag{28}$$

The estimation of ν can be performed for different pairs (x_1,t_1), (x_2,t_2) in order to check its stability and the pertinence of the general expression assumed for $P(x,t)$. This has been done for $P(x_1,t_1)$ and $P(x_2,t_2)$ with $t_1 = 10^5$ and

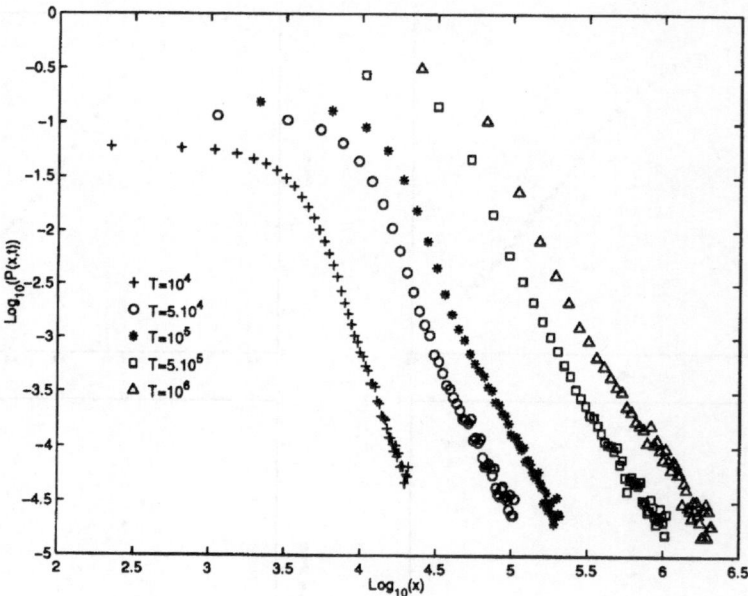

Fig. 30. $log_{10}(P(x,t))$ as function of $log_{10}(x)$ for different times.

$t_2 = 10^6$. The result is shown in Fig. 31 which displays ν as a function of $log(P(x,t)) = const$.

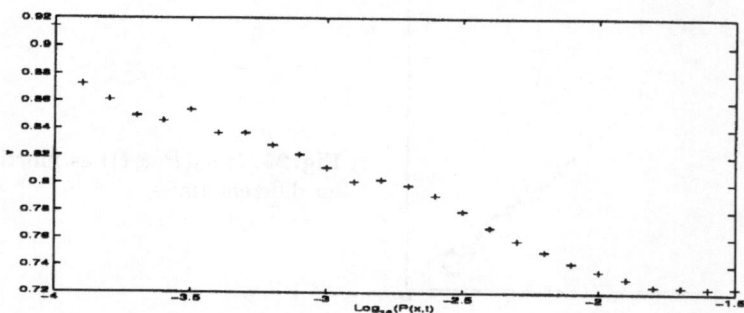

Fig. 31. ν as a function of $log_{10}(P(x,t)) = const$ between $t_1 = 10^5$ and $t_2 = 10^6$.

The shape of this curve shows that the slope that exhibits $P(x,t_2)$ is lower than the slope exhibited by $P(x,t_1)$. Nevertheless, the variation is weak within the range of interest and ν can be estimated :

$$\nu = 0.8 \pm 0.06 \tag{29}$$

As the transport exponent μ (cf Eq. (18)) is related to ν by $\mu = 2\nu$, from Eq. (29) we derive: $\mu = 1.6 \pm .12$ in agreement with the value reported in the Table 2.

5 Conclusion and discussion

The fine structure of the phase space of Hamiltonian systems is the primary origin of anomalous transport. Long-range correlation effects occur from visits of orbits to boundary layers in the vicinity of islands. This phenomenon is generally refered to as the stickiness of the islands. We have shown that for small changes of the stochasticity parameter strong topological changes can occur in the phase portrait of the system, with a corresponding strong change in the transport characteristics. For particular values of map parameter island structures can exist which are very sticky, giving very anomalous transport. As a concrete example we considered in this paper kinetic properties of the standard map near the threshold of the accelerator mode of period 5 and found a value of K producing an exact self-similar chain of islands associated with this accelerator mode. Spatio-temporal properties of this island chain have been numerically determined. We have shown that the kinetics is anomalous near the accelerator mode and corresponds to a superdiffusion process with a characteristic exponent related to the spatio-temporal scaling parameters of the island chain. We also found that the asymptotic behavior of the probability distribution of escape times from boundary domains of the self-similar chain of islands follows a power law, and that these escape times are renormalized according to the generation of the island considered. Moreover it was shown that the exponent of the power law which rules the escape time distribution is related to the transport exponent μ.

Near stochastic threshold of the map, $K \simeq 1$, it is found that the dominant contributions to anomalous transport are from multi-layered island structures of great complexity. The anomalous transport is related to the trapping time distribution which is determined by the scaling of transition rates between stochastic bands deep within the traps. For this reason the transport exponent can depend in a complicated way on the length of the simulation, and may not converge to a well defined value. In cases where the island chains form a fairly regular, ordered set of layers the dynamics of the trap can be successfully modeled.

Note that there is no reason to expect traps to have simple structure. In general, simulations of different length will explore different levels of a trap and lead to somewhat different long time tails. In addition most traps do not possess only one path leading from the stochastic sea to the innermost bounding KAM surface, but many paths with different time scales present along different parts of each one. Thus it is not reasonable for a given continuous time random walk analysis to give accurate results for every time interval, nor for the values of exponents related to characteristic times probability distributions and μ to converge to definite values for experiments of arbitrary length. The real situation in the standard map seems to be infinitely complex, different for each value of K, and unlikely to be described by a simple universal model constructed for one dimensional systems. One can say that real situation is rather multifractal than fractal, nevertheless, different properties of the anomalous transport (Montroll and Shlesinger 1984, Afanas'ev *et al.* 1991, Shlesinger *et al.* 1993) can be efficiently used to make models in corresponding intervals ("windows") of the parameter K.

Acknowledgments
R. White was supported by the U.S. Department of Energy under contract number DE-AC02-76-CHO3073. G. Zaslavsky was supported by the U. S. Navy Grant N00014-96-1-0055 and by the U.S. Department of Energy grant DE-FG02-92ER54184. S. Benkadda and S. Kassibrakis are grateful to the Princeton Plasma Physics Laboratory for hospitality during a visit when part of this work was completed.

References

Afanas'ev V. V., Sagdeev R. Z., and Zaslavsky G. M. (1991): Chaos 1,143.
Afraimovich V. and Zaslavsky G. M. to appear.
Afraimovich V. and Zaslavsky G . M. (1997): Phys. Rev. E 5418.
Beloshapkin V. V. , Zaslavsky G. M. (1983): Phys. Rev. A **97** ,121.
Benettin G., Galgani L. and Strelcyn J. M. (1976): Phys. Lett. A **14** 2338.
Benkadda S., Elskens Y., Ragot B. and Mendonça J. T. (1994): Phys. Rev. Lett. **7** 2859.
Benkadda S., Kassibrakis S., White R. B. and Zaslavsky G. M. (1997): Phys. Rev. E **55**, 4909.
Benkadda S., Gabbai P. and Zaslavsky G. M . (1997) :Phys. of Plasma **4** 2864.
Chirikov, B. V. (1979): Phys. Rep. **52**, 263.
Chirikov B. V. and Shepeliansky D. L. (1984) : Physica D**13** 394.
Easton W. Meiss J. D. and Carver S. (1993): Chaos **3** 153.
Green J. M. (1979): J. Math. Phys. **20**, 1183.
Hanson J. D., Cary J. R. and Meiss J. D. (1 986): Phys. Rev. A **34** 2375.
Ishizaki R., Hata H. and Horita T. (1990) Prog. Theor. Phys. **84** 179.
Kac M. (1958): *Probability and Related Topics in Physical Sciences* (Wiley, New York).
Karney C. C. F. (1983): Physica D **8** 360.
Klafter J. and Zumofen G. (1994): Phys. Rev. E **49** 4873.
Klafter J., Zumofen G. and Shlesinger M. F . (1995) *Lévy flights and Related Topics in Physics* (Springer) 196.
Meiss J. D. (1997): Chaos **7** 139.
Meiss J. D. and Ott E. (1986): Physica D **20** 387.
Montroll E. W. and Shlesinger M. (1984): *Studies in Statistical Mechanics* II (North-Holland, Amsterdam) 1.
Rechester A. B., Rosenbluth M. N. and White R. B. (1981): Phys. Rev. A **23** 2644.
Rechester A. B. and White R. B. (1979): Phys. Let. Rev **42** 1247.
Shlesinger M. F. Zaslavsky G. M. and Klafter J. (1993): Nature **363** 31.
Sinai Ya. (1963): Soviet Math. Dokl. **4** 1818.

Lecture Notes in Physics

For information about Vols. 1–479
please contact your bookseller or Springer-Verlag

Monographs